COMPUTATIONAL PHARMACEUTICAL SOLID STATE CHEMISTRY

COMPUTATIONAL PHARMACEUTICAL SOLID STATE CHEMISTRY

Edited by

YURIY A. ABRAMOV

WILEY

Published by John Wiley & Sons, Inc., Hoboken, New Jersey
Published simultaneously in Canada

Library of Congress Cataloging-in-Publication Data:

Names: Abramov, Yuriy A., editor.
Title: Computational pharmaceutical solid state chemistry / edited by Yuriy A. Abramov.
Description: Hoboken, New Jersey : John Wiley & Sons, 2016. |
Includes bibliographical references and index.
Identifiers: LCCN 2015039015 | ISBN 9781118700747 (cloth)
Subjects: LCSH: Pharmaceutical chemistry. | Solid state chemistry.
Classification: LCC RS404.5 .C66 2016 | DDC 615.1/9–dc23
LC record available at http://lccn.loc.gov/2015039015

Set in 10/12pt Times by SPi Global, Pondicherry, India

10 9 8 7 6 5 4 3 2 1

CONTENTS

LIST OF CONTRIBUTORS

Yuriy A. Abramov Pfizer Worldwide Research & Development, Groton, CT, USA

Frank H. Allen Cambridge Crystallographic Data Centre (CCDC), Cambridge, UK

Bradley D. Anderson Department of Pharmaceutical Sciences, College of Pharmacy, University of Kentucky, Lexington, KY, USA

Anders Broo Pharmaceutical Development, AstraZeneca, Göteborg, Sweden

Viktor Broo Pharmaceutical Development, AstraZeneca, Göteborg, Sweden

Roberta Ceriani State University of Campinas, Cidade Universitária Zeferino Vaz, Campinas-SP, Brazil

Bodhisattwa Chaudhuri Department of Pharmaceutical Sciences; Institute of Material Sciences, University of Connecticut, Storrs, CT, USA

Aurora Cruz-Cabeza The School of Chemical Engineering and Analytical Science, The University of Manchester, Manchester, UK

Larissa P. Cunico SPEED, Department of Chemical & Biochemical Engineering, Technical University of Denmark, Kongens Lyngby, Denmark

Graeme M. Day School of Chemistry, University of Southampton, Southampton, UK

Robert Docherty Pfizer Worldwide Research & Development, Sandwich, Kent, UK

Cheryl L. Doherty Pfizer Worldwide Research & Development, Sandwich, Kent, UK

Maxim V. Fedorov Scottish Universities Physics Alliance (SUPA), Department of Physics, University of Strathclyde, Glasgow, UK

Neil Feeder Cambridge Crystallographic Data Centre (CCDC), Cambridge, UK

Peter T. A. Galek RealVNC Ltd, Cambridge, UK

Rafiqul Gani SPEED, Department of Chemical & Biochemical Engineering, Technical University of Denmark, Kongens Lyngby, Denmark

Robert B. Hammond Institute of Particle Science and Engineering, and Institute of Process Research & Development, School of Chemical and Process Engineering, University of Leeds, Leeds, UK

Andreas Klamt COSMO*logic* GmbH & Co. KG, Leverkusen, Germany; Faculty of Chemistry and Pharmacy, Institute of Physical and Theoretical Chemistry, University of Regensburg, Regensburg, Germany

Joseph F. Krzyzaniak Pfizer Worldwide Research & Development, Groton, CT, USA

Sten O. Nilsson Lill Pharmaceutical Development, AstraZeneca, Göteborg, Sweden

Christoph Loschen COSMO*logic* GmbH & Co. KG, Leverkusen, Germany

Dmitry Lupyan Schrödinger Inc., New York, USA

Suman Luthra Pfizer Worldwide Research & Development, Andover, MA, USA

Luís Mafra CICECO - Aveiro Institute of Materials, Department of Chemistry, University of Aveiro, Aveiro, Portugal

Paul A. Meenan Pfizer Worldwide Research & Development, Groton, CT, USA

Shivangi Naik Research and Development, Kamat Pharmatech, NJ, USA

David S. Palmer Department of Pure and Applied Chemistry, University of Strathclyde, Glasgow, UK

Klimentina Pencheva Pfizer Worldwide Research & Development, Sandwich, Kent, UK

Elna Pidcock Cambridge Crystallographic Data Centre (CCDC), Cambridge, UK

Vasuki Ramachandran Institute of Particle Science and Engineering, and Institute of Process Research & Development, School of Chemical and Process Engineering, University of Leeds, Leeds, UK

Kevin J. Roberts Institute of Particle Science and Engineering, and Institute of Process Research & Development, School of Chemical and Process Engineering, University of Leeds, Leeds, UK

Ekneet Kaur Sahni Pfizer Manufacturing Science and Technology, McPherson, KS, USA

Sérgio Santos CICECO - Aveiro Institute of Materials, Department of Chemistry, University of Aveiro, Aveiro, Portugal

Mariana Sardo CICECO - Aveiro Institute of Materials, Department of Chemistry, University of Aveiro, Aveiro, Portugal

Staffan Schantz Pharmaceutical Development, AstraZeneca, Göteborg, Sweden

Heather Frericks Schmidt Pfizer Worldwide Research & Development, Groton, CT, USA

Woody Sherman Schrödinger Inc., New York, USA

Anjan K. Tula SPEED, Department of Chemical & Biochemical Engineering, Technical University of Denmark, Kongens Lyngby, Denmark

Peter A. Wood Cambridge Crystallographic Data Centre (CCDC), Cambridge, UK

Tian-Xiang Xiang Department of Pharmaceutical Sciences, College of Pharmacy, University of Kentucky, Lexington, KY, USA

PREFACE

It is impossible to imagine that any of the modern industries at the beginning of twenty-first century may survive and progress without application of computational modeling. This observation is reflected in industries such as aviation and space, chemical, oil/gas, materials, and pharmaceuticals. In pharmaceutical industry, molecular modeling applications in Drug Discovery have a 30-year history of great accomplishments; many of society's greatest medicines were made possible, in part, due to computational modeling. However, computational modeling support of Drug Development has emerged only recently in full strength. An explanation for this hindrance is the high complexity of the modeling systems (solid and liquid phases of relatively large and flexible molecules) coupled with high expectations for accurate predictions that accompany late-stage drug development. It is fair to state that in order to overcome these challenges, typical modeling applications in Drug Development are pushing the boundaries of fundamental theories and methods initially tested on the simpler "classical" systems. Therefore, many of the modeling applications in Drug Development became more feasible only with the recent advances in computational power of high-performance computing systems.

There are a number of chemistry books available related to computational materials science and to modeling of molecular solid state, but none of the books cover current pharmaceutical industry applications. The intention of this book is to highlight the importance of the computational pharmaceutical solid-state chemistry and to fill the gap in the current literature. The book examines the state-of-the-art computational approaches to guide and analysis of solid form experiments and to optimize the physical and chemical properties of active pharmaceutical ingredient (API) related to its stability, bioavailability, and formulatability. While all methods and approaches described in the book appear to be state of the art, the book is addressed to a wide audience, including experts within the field and those without much experience in molecular modeling. I anticipate the book will be useful not only to the professional modelers in Drug Development but also to computational chemistry community in Drug Discovery as well as to experimental researchers and students working in the fields of pharmaceutical sciences, solid-state chemistry, materials science, and medicinal chemistry.

The outline of the book is as follows. The book starts with a high-level introduction into the field of computational pharmaceutical solid-state chemistry. Chapters 2–5 consider different computational approaches allowing physical stability analysis (risk assessment) of pharmaceutical solid forms. Chapters 6 and 7 present industrial examples of the application of computational pharmaceutical solid-state chemistry at Pfizer and AstraZeneca, respectively. Chapter 8 considers synthonic engineering of pharmaceutical solid dosage form with a special focus on a surface energy and morphology prediction. Chapters 9–11 consider current state-of-the-art computational approaches of solubility prediction of solid drug-like compounds. Chapter 12 reviews the use of solid-state NMR (SSNMR) for studying small pharmaceutical molecules in synergy with theoretical calculations of NMR parameters. A review of molecular dynamic simulation of the various properties of amorphous pharmaceutical compounds is presented in Chapter 13. Finally, Chapter 14 presents numerical simulations of unit operations in pharmaceutical solid dose manufacturing with a focus on contact drying, film coating, and milling.

I would like to thank all the contributors of this book for writing excellent chapters. I am also grateful to my colleague Brian Samas for valuable discussions and suggestions, which I believe made this book better. Finally, special thanks to my wife and sister (Marina Skornyakova and Elena Abramova, respectively) for their support during the preparation of this book.

Yuriy A. Abramov,
August 2015

EDITOR'S BIOGRAPHY

Dr. Yuriy A. Abramov is a Senior Principal Scientist at Pfizer Inc., with a scientific and technical expertise in the broad area of Computational Science in Drug Discovery and Development. He has a proven track record for developing new ideas and approaches from the concept stage to successful implementation. He has also authored more than 50 publications in peer-reviewed journals and six book chapters.

Yuriy A. Abramov holds a Ph.D. in physical chemistry from the D. Mendeleev University of Chemical Technology and Karpov Institute of Physical Chemistry, Moscow, Russia. He started his career as an Assistant Professor in the Department of Physics at D. Mendeleev University of Chemical Technology followed by a Research Assistant Professor position in the Department of Chemistry at the State University of New York (SUNY), Buffalo, New York.

Yuriy A. Abramov started his career at Pfizer in 2001 as a computational chemist in Drug Discovery, supporting multiple therapeutic areas including inflammation, immunology, and oncology. This led to a computational chemistry position in Pharmaceutical Sciences in 2007. In this position, he developed multiple novel computational approaches to support risk assessment of likelihood of a more stable polymorph to facilitate commercial solid form nomination. In addition to computational pharmaceutical solid-state chemistry, his current interests include, but not limited to, prediction of chemical reactivity and chemical reactions and *in silico* bioengineering.

1

COMPUTATIONAL PHARMACEUTICAL SOLID-STATE CHEMISTRY: AN INTRODUCTION

Yuriy A. Abramov

Pfizer Worldwide Research & Development, Groton, CT, USA

1.1 INTRODUCTION

Traditionally, pharmaceutical industry is focusing on discovery and manufacturing of small-molecule drug compounds. Pharmaceutical industry workflow is characterized by two somewhat overlapping stages—Drug Discovery and Drug Development. At the first stage, a new chemical entity (drug candidate molecule for clinical development) is being discovered and tested on animals. At the end of this stage it is important to make sure that the selected molecule passes preclinical testing such as *in vivo* biological activity in animal models, *in vitro* metabolism, pharmacokinetic profiling in animals, and animal toxicology studies. The drug candidate progresses into an early development stage to pass proof of concept (POC), which refers to early clinical studies on human divided into Phase I and Phase IIa. At this step the candidate molecule becomes an active pharmaceutical ingredient (API) of drug product and is typically formulated in a solid form. The subsequent Drug Development process is focused on drug product and process development to ensure reliable performance, manufacturing, and storage.

Along the pharmaceutical industry workflow path, a drug substance undergoes a significant physical transformation (Fig. 1.1). It starts in early Drug Discovery as a single molecule (ligand) binding to a receptor in order to activate or inhibit

Computational Pharmaceutical Solid State Chemistry, First Edition. Edited by Yuriy A. Abramov.

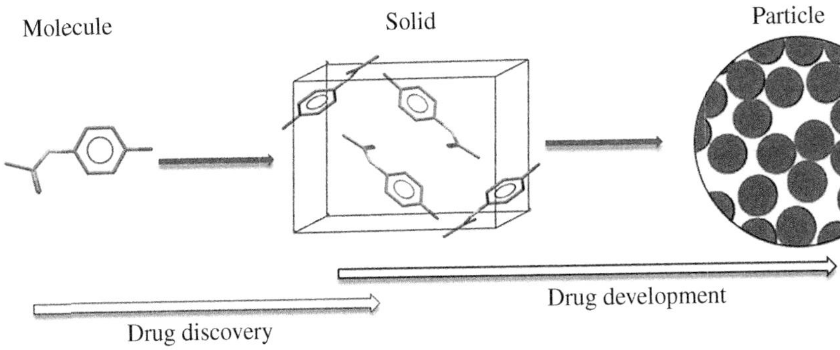

FIGURE 1.1 Physical transformation of a drug substance along the pharmaceutical industry workflow.

the receptor's associated biochemical pathway. Then the drug molecule becomes a biologically active component of a typically solid-state (e.g., crystalline or amorphous) formulation in early Drug Development. Finally, the drug molecule acts as an API of the solid particles of the drug product at the later stages of Drug Development. This transformational pathway reflects the complex nature of the drug design workflow and dictates a diversity of experimental and especially computational methods, which are applied to support Drug Discovery and Drug Development.

The pharmaceutical industry as a whole has faced many challenges in recent years in addition to patent expirations of blockbuster drugs. In particular, the Drug Development branch faces challenges of accelerated development under a high regulatory pressure. An ability to rationalize and guide Drug Development process has become crucial [1]. Computational chemistry methods have become deeply integrated into Drug Discovery over the past 30 years [2, 3]. However, the computational support of Drug Development has emerged only in recent years and is now tasked with the whole spectrum of Drug Development fields including drug formulation and product design, process chemistry, chemical engineering and analytical research and development. This chapter provides a high-level overview of pharmaceutical solid-state landscape and introduces a field of computational modeling in Drug Development, hereinafter called computational pharmaceutical solid-state chemistry (CPSSC).

1.2 PHARMACEUTICAL SOLID-STATE LANDSCAPE

1.2.1 Some Definitions

Approximately 70% of the drug products marketed worldwide are formulated in oral solid dosage forms [4]. The pharmaceutical solid state may be characterized by molecular arrangement displaying long-range order in all directions (crystalline), long-range order in one or two directions (liquid crystals), or only close-range order (amorphous). An overall pharmaceutical solid-state landscape is presented in Figure 1.2. The advantage of formulation of drug substances in crystalline form is

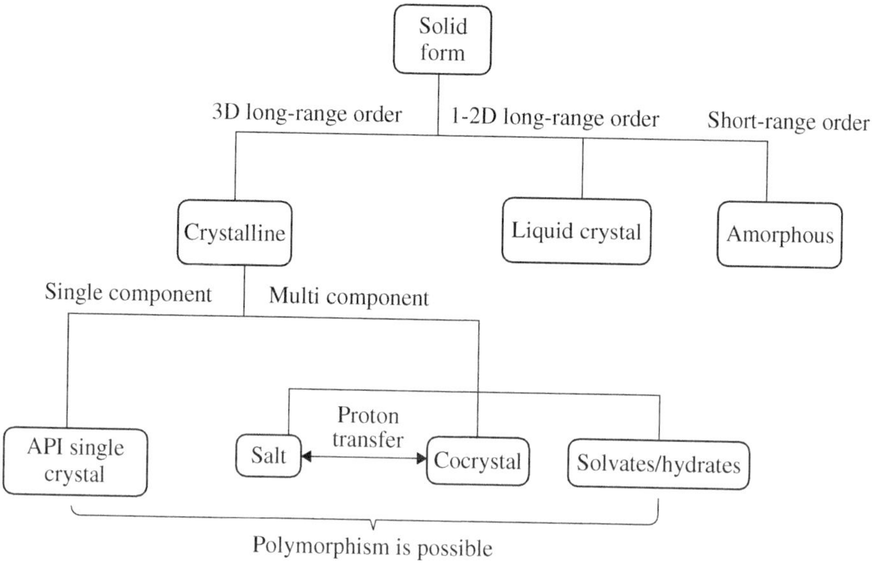

FIGURE 1.2 A typical pharmaceutical solid-state landscape.

dictated by more desirable manufacturing properties: superior stability, purity, and manufacturability relative to amorphous and liquid form formulations. All solid drugs can be subclassified as single- (anhydrous) and multicomponent compounds. Multicomponent substances can be crystalline solvates (including solid hydrates) [5, 6], cocrystals (or co-crystals) [7], and salts [8]. *Solid solvates* (also named *pseudopolymorphs* or *solvatomorphs*) represent crystal structures in which solvent molecules are integrated into the crystal lattice. *Solid hydrates* are the most common pharmaceutical pseudopolymorphs. Pharmaceutical *cocrystals* are defined as stoichiometric multicomponent crystals formed by an API (or an intermediate compound) with at least one cocrystal former (coformer), which is solid at ambient temperature. Within the family of solvates, hydrates, and cocrystals, the components are neutral. Pharmaceutical *salts* are multicomponent materials in which components are ionized *via* proton transfer and are involved in ionic interactions with each other.

Different crystalline structures of one substance (single- or multicomponent) are named *polymorphs* [9, 10]. Polymorphism, which exists as a result of different crystal packing of rigid molecules, is called a *packing polymorphism*. *Conformational polymorphism* is a more common phenomenon for typically flexible drug-like molecules and results from crystallization of different conformers of the same molecule. At a given environmental conditions (temperature, humidity, pressure, etc.) only one solid form is thermodynamically stable (lowest free energy), while all other forms are considered metastable.

The solid-state complexity of a typical distribution of pharmaceutical solid forms was reflected in a recent analysis of 245 polymorph screens performed at Solid State Chemical Information (SSCI) (http://www.ssci-inc.com) [11]. It was demonstrated that about 90% of the compounds screened exhibited multiple crystalline and noncrystalline forms. About half of the compounds screened were polymorphic, and about a third of the compounds exist in hydrated and solvated forms. In cases where cocrystals were attempted for a particular API, 61% of these were able to form cocrystals.

1.2.2 Impact of Solid-State Form on API and Product Properties

Variations of pharmaceutical solid form can result in alternations of physicochemical properties of drug product, which may affect drug performance, safety, and processing [12]. Therefore, solid form selection is strongly regulated by the Food and Drug Administration according to guidelines outlined in an International Conference on Harmonisation (ICH; http://www.ich.org) [13] as well as by other regulatory agencies around the world. Table 1.1 summarizes major properties that may be affected by crystal form change, a selection of these properties are discussed in more detail later.

Solubility and dissolution rate are the key properties of drug product, which are directly related to bioavailability and are often vital for the drug performance. These two properties display a strong dependence on the solid form selected. The largest difference in solubility is observed between crystalline and amorphous pharmaceutical materials and may be as high as several hundred times [14, 15]. Solid crystalline hydrates are known to drop the solubility of the drug relative to its anhydrous form up to 10 times [16]. On the contrary, solid solvates formed from water-miscible solvents are typically more soluble in water than the corresponding nonsolvated form. Generally, dissolution rate is increased significantly in salt and cocrystal solid formulations predominantly due to favorable hydration free energies of counter ion and cocrystal former, respectively [17, 18]. Therefore, salt or cocrystal formulations are the most popular strategies for improving the solubility (dissolution) of poorly soluble drugs [19].

Thermodynamic solubility of a crystalline compound decreases with increased stability (lower free energy) of its polymorphic form. It has been reported that there is a 95% probability that a thermodynamic solubility ratio between a pair of polymorphs is less than twofold [20], although in certain cases it may reach much higher values. At first glance an impact of change of polymorphic form on the solubility and dissolution rate may seem to be less problematic in comparison with amorphous to crystalline or anhydrous to a solid hydrate form transformation. However, in cases where drug absorption is not limited by permeability (BCS classes I and II [21]), depending on the drug dose even 1.5- or 2-fold decrease of solubility due to a switch to a more stable form may have a profound effect on bioavailability of the API (see Section 1.2.3 for discussion of polymorph impact on drug performance). In order to avoid an unexpected interconversion into a less soluble form (with generally different solid-state properties) during manufacturing or shelf life of the drug product, it is a common practice in pharmaceutical industry to perform a stable form screening prior to the selection of a commercial solid form.

Another key property of the drug product, which can be impacted by the solid form, is chemical stability [22]. Drug degradation in solid dosage forms is mostly

TABLE 1.1 Properties Which May be Impacted by Solid Form Variation

Chemical	Physical	Mechanical	Surface	Thermodynamic	Kinetic
Chemical stability/ reactivity	Stability	Compactability	Surface free energy	Solubility	Dissolution rate
Photochemical reactivity	Hygroscopicity	Hardness	Stickiness	Free energy of fusion	Nucleation and crystal growth rates
	Morphology	Powder flow	Interfacial tensions	Melting point	Solid-state reaction rate
	Density	Tableting	Surface area	Heat of fusion	
	Color	Tensile strength		Vapor pressure	
	Refractive index	Cleavage			

determined by the surface characteristics of both the API and the excipient particles. Most pharmaceutical reaction rates are typically the greatest in the amorphous rather than crystalline states due to a higher surface area and molecular mobility. Additionally, amorphous substances show a higher surface energy and may be more hygroscopic, which may be coupled with chemical stability problems [23]. Therefore, an amorphous formulation is generally less preferable than the crystalline one. Chemical reactivity in the solid state may also correlate with the nature of the crystalline form (polymorphic or pseudopolymorphic) and related crystalline morphology [24]. Generally, a stable solid form is more chemically stable than metastable forms.

A change in the solid form may lead to a different crystal morphology, which may have an impact on processibility of the drug product due to the different mechanical and flow properties [25]. For example, needle-shaped crystals are generally undesirable for pharmaceutical applications since they are difficult to process [26].

1.2.3 Challenges of Pharmaceutical Industry Related to Solid Form Selection

A likely dependence of drug performance, processing, and safety on the solid form selection imposes a series of challenges on the pharmaceutical industry. Here three challenges are outlined—solubility improvement, physical stability, and unfavorable solvates and hydrates.

An increasing trend toward low solubility is a major issue for Drug Development as the formulation of poorly soluble compounds can be quite problematic [27]. Aqueous thermodynamic solubility of solid pharmaceutical compound may be defined by two contributions—molecular hydration free energy and lattice (or sublimation) free energy [28]. Consequently, strategies to enhance solubility and drug delivery include molecular modification (lowering hydration free energy) or solid form optimization (crystal packing destabilization or/and lowering hydration free energy). It is only the latter strategy that is applicable at the Drug Development stage. Solid form optimization would typically include counter ion or coformer screening for salt or cocrystal formulation, respectively, of the API with improved dissolution properties. Additionally, amorphous API formulation could be possible *via*, for example, spray-dried dispersion (SDD) technique [29].

Physical instability of pharmaceutical solids is related to interconversion into a new form in the course of handling, manufacturing, processing, or storage, which may have a profound effect on the drug performance and process development. The conversion from one form to another is thermodynamically driven and may take place when a solid form is metastable relative to a more stable form within specific environmental conditions. The most common cases of physical instability are transformation into a stable polymorphic form, desolvation, hydration/dehydration, crystallization of amorphous form or amorphization of a crystalline one. A timeline of events involving physical stability over the past 30 years is presented in Figure 1.3 [9]. In most of the cases, the products were recalled as a result of poor performance. Perhaps, the most famous example of polymorph-induced impact is related to the marketed drug Norvir® (ritonavir). Abbott Laboratories had to stop sales of Norvir in

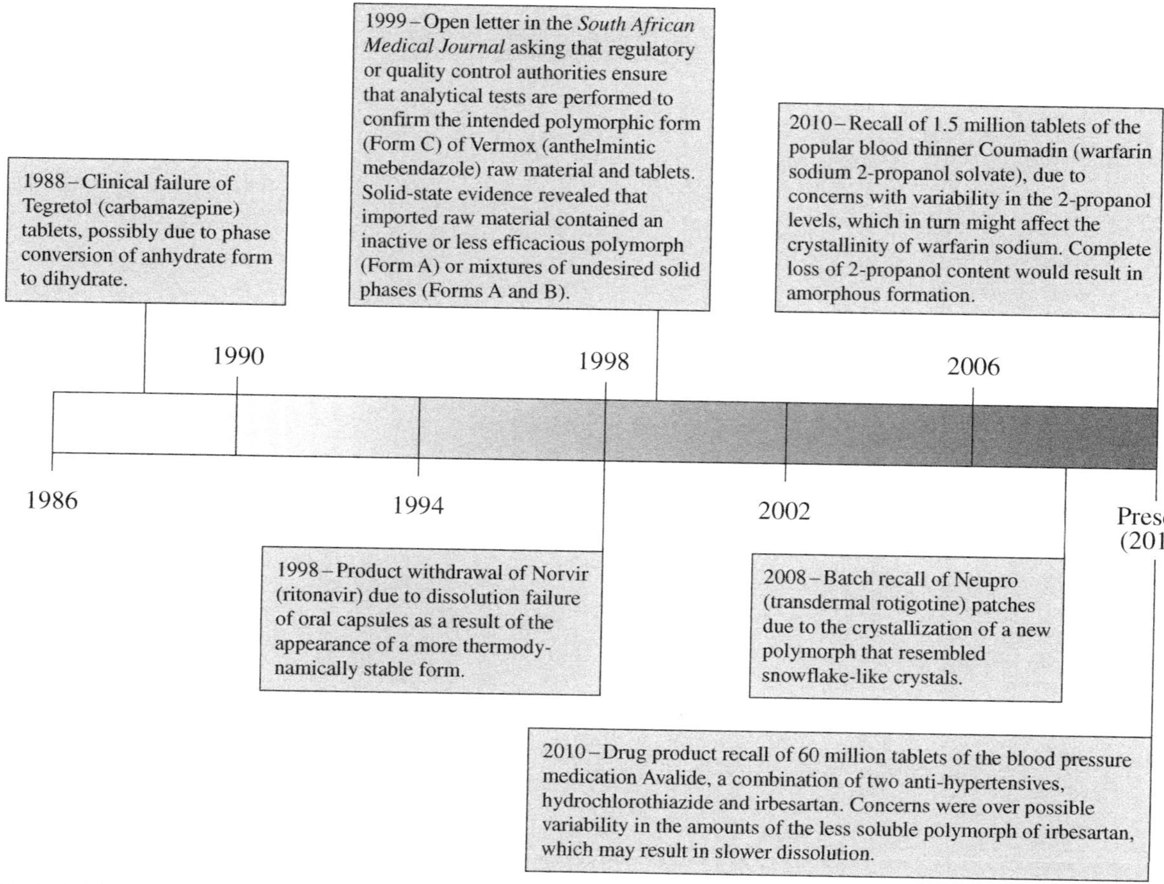

FIGURE 1.3 A timeline of events concerning solid-state issue with polymorphism of pharmaceutical drugs over past 30 years. Adapted from Lee et al. [9]. Reproduced with permission of *Annual Reviews*.

1998 due to a failure in a dissolution test, which was caused by the precipitation of a more stable and less soluble form II of the compound [30].

Some APIs may display a high propensity for forming stable solvates [31]. Though there are marketed drug products that contain solvates such as Prezista®, Crixivan®, and Coumadin®, formulation of a drug product in solvated form is typically undesirable. Solvates (including hydrates) might be subsequently desolvated in a final drying step of the formulation process. In such a situation, the final form could be metastable and may undergo a solid–solid transition during its shelf life. In addition, residual solvent levels in the API must be compatible with ICH guidelines. As a result hydrates and solvates are generally avoided for the reasons mentioned earlier. Therefore, selection of the solvent system for crystallization, which has the lowest probability of forming solvates/hydrates with the API, is a good practice.

1.3 PHARMACEUTICAL COMPUTATIONAL SOLID-STATE CHEMISTRY

Given the complexity of the pharmaceutical solid-state landscape and challenges facing the pharmaceutical industry, an accelerated Drug Development greatly benefits from guidance provided by computational methods. The emerging field of the CPSSC covers the whole spectrum of state-of-the-art computational approaches, which are used to support all steps related to the development of solid-state pharmaceuticals. An outline of these steps in Drug Discovery and Drug Development is presented in Figure 1.4. According to the provided broad definition of the field, the CPSSC covers more than just solid-state calculations. In fact, the CPSSC represents a true multiscale modeling from quantum mechanical studies of molecules (subnanometer scale) to discrete or finite element modeling of solid particles (micron scale) (Fig. 1.5).

Typical CPSSC approaches may be broadly classified into two major categories—those that are used to guide properties and process optimization (engineering) and those that are used for analysis and interpretation of the experimental results. The former category includes all kind of virtual screening approaches—solvent selection for crystallization and desolvation [34, 35], solvent selection for polymorph screening

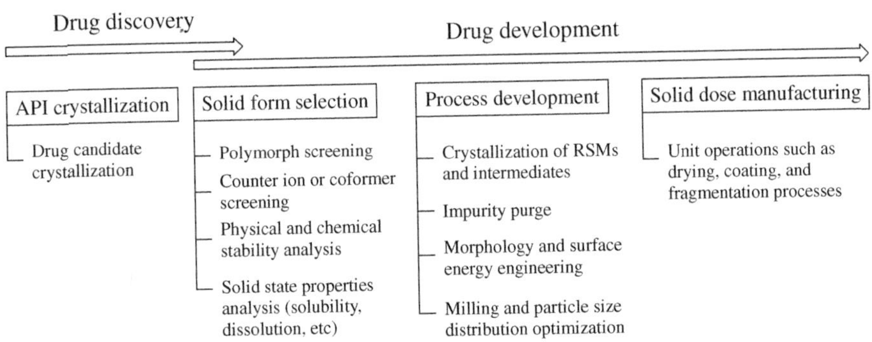

FIGURE 1.4 An outline of stages of solid form development in pharmaceutical industry. RSM is a regulatory starting material.

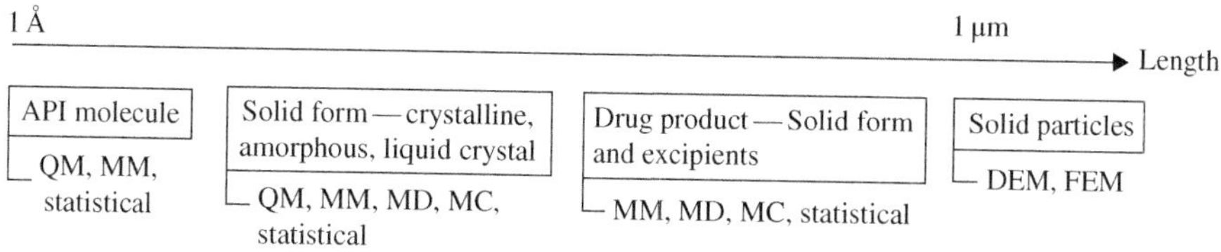

FIGURE 1.5 Multiscale modeling in computational pharmaceutical solid-state chemistry. Here DEM and FEM are discrete and finite element methods; MC, Monte Carlo simulation; MD, molecular dynamics; MM, molecular mechanics; QM, quantum mechanics, respectively; statistical approaches include knowledge-based models based on database analysis (e.g., Cambridge Structure Database [32]) and quantitative structure property relationships (e.g., group contributions models [33a]).

[31b, 36], solvent selection for impurity purge *via* recrystallization [37], cocrystal former and counterion selection for crystallization and solubility improvement [35, 38, 39] as well as for improved relative humidity stability [40], virtual polymorph screening *via* crystal structure prediction (CSP) to explore lattice energy landscape [41], solvent selection for optimization of size and shape distribution of the crystalline product [25, 42], etc. In addition, physical (solubility [43], T_g [44], T_m [33], surface energy [45], etc.) and mechanical [46, 47] properties prediction of solid materials; prediction of excipient effect on API chemical degradation [48]; *in silico* modeling of drug–polymer interaction for amorphous pharmaceutical formulations [49]; and simulations of unit operations in solid dose manufacturing [50] can be also assigned to this category. The second category includes all methods used to support solid form selection *via* risk analysis of physical stability of a commercial solid form [51], *in silico* prediction of pharmaceutical stress (forced) degradation pathways [52], prediction of structure and dynamics in pharmaceutical solids based on analytical methods alternative to single crystal diffraction (SSNMR [53] and PXRD [54]), analysis of source of poor solubility of the drug substance [28], etc. Approaches from the second category are typically used to provide recommendations for a potential experimental follow-up.

As could be expected, challenges facing the pharmaceutical industry contribute to the advancement of the computational solid-state chemistry. For example, some of the virtual screening and other CPSSC methods were developed specifically to help address issues of the pharmaceutical industry. Significant progress has been made recently in many traditional applications (e.g., solubility prediction [55], CSP [56], and morphology prediction [25, 57, 58]) in order to accommodate predictions for complex pharmaceutical systems (solid and liquid multicomponent phases of relatively large and flexible molecules).

1.4 CONCLUSIONS

A complex nature of the pharmaceutical solid-state landscape imposes a series of challenges on the pharmaceutical industry. Computational modeling enables better understanding of the fundamentals of solid-state chemistry and allows an enriched selection of solid form with desired physicochemical and processing properties.

Though the CPSSC is an emerging field, many of the approaches have proved their importance for the industry and are already embedded in the workflows of various pharmaceutical companies. Moreover, though it is currently impossible to build a reliable statistics regarding the use of CPSSC over the whole industry, it is known that some of the methods (like computational support of solid form selection) have already been successfully used to support New Drug Applications (NDAs) of some of the recently approved drugs.

A future outlook of the CPSSC field envisions a wide acceptance of CPSSC support of NDA submissions by all regulatory agencies. Moreover, it is feasible that in addition to guidance of the experimental work, future improvements of the CPSSC field once validated may lead to replacement of some of the experimental studies by accurate predictions.

ACKNOWLEDGMENT

The author wishes to acknowledge helpful discussions with his colleague Mr. Brian Samas.

REFERENCES

[1] Ouyang, D.; Smith, S. C., Eds. *Computational Pharmaceutics: Application of Molecular Modeling in Drug Delivery*. John Wiley & Sons Ltd.: Chichester/Hoboken, NJ, 2015.

[2] Marshall, G. R., *Annu. Rev. Pharmacol. Toxicol.* 1987, *27*, 193–213.

[3] Kenny, P. W., *J. Comput. Aided Mol. Des.* 2012, *26*, 69–72.

[4] Béchard, S.; Mouget, Y. LIBS for the Analysis of Pharmaceutical Materials. In *Laser Induced Breakdown Spectroscopy (LIBS): Fundamentals and Applications*; Miziolek, A. W., Palleschi, V., Schechter, I., Eds.; Cambridge University Press: New York, 2006, pp 314–331.

[5] Griesser, U. J. The Importance of Solvates. In *Polymorphism in the Pharmaceutical Industry*; Hilfiker, R, Ed.; Wiley-VCH: Weinheim, 2006, pp 211–233.

[6] Khankari, R. K.; Grant, D. J., *Thermochim. Acta* 1995, *248*, 61–79.

[7] Schultheiss, N.; Newman, A., *Cryst. Growth Des.* 2009, *9*, 2950–2967.

[8] Stahl, P. H.; Wermuth, C. G., Eds. *Handbook of Pharmaceutical Salts: Properties, Selection, and Use*. Wiley-VCH: Weinheim, 2002.

[9] Lee, A. Y.; Erdemir, D.; Myerson, A. S., *Annu. Rev. Chem. Biomol. Eng.* 2011, *2*, 259–280.

[10] Bernstein, J., *Polymorphism in Molecular Crystals*. Oxford University Press: New York, 2007.

[11] Stahly, G. P., *Cryst. Growth Des.* 2007, *7*, 1007–1026.

[12] Huang, L.-F.; Tong, W.-Q. T., *Adv. Drug Deliv. Rev.* 2004, *56*, 321–334.

[13] Byrn, S.; Pfeiffer, R.; Ganey, M.; Hoiberg, C.; Poochikian, G., *Pharm. Res.* 1995, *12*, 945–954.

[14] Hancock, B. C.; Parks, M., *Pharm. Res.* 2000, *17*, 397–404.

[15] Murdande, S. B.; Pikal, M. J.; Shanker, R. M.; Bogner, R. H., *Pharm. Res.* 2010, *27*, 2704–2714.

[16] Pudipeddi, M.; Serajuddin, A., *J. Pharm. Sci.* 2005, *94*, 929–939.

[17] Maheshwari, C.; André, V.; Reddy, S.; Roy, L.; Duarte, T.; Rodríguez-Hornedo, N., *CrystEngComm* 2012, *14*, 4801–4811.

[18] David, S.; Timmins, P.; Conway, B. R., *Drug Dev. Ind. Pharm.* 2012, *38*, 93–103.

[19] Elder, D. P.; Holm, R.; de Diego, H. L., *Int. J. Pharm.* 2013, *453*, 88–100.

[20] Abramov, Y. A.; Pencheva K. Thermodynamics and Relative Solubility Prediction of Polymorphic Systems. In *Chemical Engineering in the Pharmaceutical Industry: from R&D to Manufacturing*; am Ende, D. J., Ed.; Wiley: New York, 2011, 477–490.

[21] Amidon, G. L.; Lennernäs, H.; Shah, V. P.; Crison, J. R., *Pharm. Res.* 1995, *12*, 413–420.

[22] (a) Byrn, S. R.; Xu, W.; Newman, A. W., *Adv. Drug Deliv. Rev.* 2001, *48*, 115–136; (b) Attwood, D.; Poust, R. I., Chemical Kinetics and Drug Stability. In *Modern Pharmaceutics*; Florence, A. T., Siepmann, J., Eds.; Informa Healthcare: New York, 2009, pp 203–251.

[23] Ohtake, S.; Shalaev, E., *J. Pharm. Sci.* 2013, *102*, 1139–1154.

[24] (a) Okamura M.; Hanano, M., Awazu S., *Chem. Pharm. Bull.* 1980, *28*, 578–584; (b) Walkling, W.; Sisco, W.; Newton, M.; Fegely, B.; Chrzanowski, F., *Acta. Pharma. Technol.* 1986, *32*, 10–12; (c) Krahn, F. U.; Mielck, J. B., *Int. J. Pharm.* 1989, *53*, 25–34; (d) De Villiers, M.; Van der Watt, J.; Lötter, A., *Int. J. Pharm.* 1992, *88*, 275–283; (e) Matsuda, Y.; Akazawa, R.; Teraoka, R.; Otsuka, M., *J. Pharm. Pharmacol.* 1994, *46*, 162–167; (f) Rocco, W. L.; Morphet, C.; Laughlin, S. M., *Int. J. Pharm.* 1995, *122*, 17–25.

[25] Lovette, M. A.; Browning, A. R.; Griffin, D. W.; Sizemore, J. P.; Snyder, R. C.; Doherty, M. F., *Ind. Eng. Chem. Res.* 2008, *47*, 9812–9833.

[26] Variankaval, N.; Cote, A. S.; Doherty, M. F., *AIChE J.* 2008, *54*, 1682–1688.

[27] Williams, H. D.; Trevaskis, N. L.; Charman, S. A.; Shanker, R. M.; Charman, W. N.; Pouton, C. W.; Porter, C. J., *Pharmacol. Rev.* 2013, *65*, 315–499.

[28] Docherty, R.; Pencheva, K.; Abramov, Y. A., *J. Pharm. Pharmacol.* 2015, *67*, 847–856.

[29] Friesen, D. T.; Shanker, R.; Crew, M.; Smithey, D. T.; Curatolo, W.; Nightingale, J., *Mol. Pharm.* 2008, *5*, 1003–1019.

[30] Bauer, J.; Spanton, S.; Henry, R.; Quick, J.; Dziki, W.; Porter, W.; Morris, J., *Pharm. Res.* 2001, *18*, 859–866.

[31] (a) Chekal, B. P.; Campeta, A. M.; Abramov, Y. A.; Feeder, N.; Glynn, P. P.; McLaughlin, R. W.; Meenan, P. A.; Singer, R. A., *Org. Process Res. Dev.* 2009, *13*, 1327–1337; (b) Campeta, A. M.; Chekal, B. P.; Abramov, Y. A.; Meenan, P. A.; Henson, M. J.; Shi, B.; Singer, R. A.; Horspool, K. R., *J. Pharm. Sci.* 2010, *99*, 3874–3886; (c) Samas, B.; Seadeek, C.; Campeta, A. M.; Chekal, B. P., *J. Pharm. Sci.* 2011, *100*, 186–194.

[32] Groom, C. R.; Allen, F. H., *Angew. Chem. Int. Ed.* 2014, *53*, 662–671.

[33] (a) Marrero, J.; Gani, R., *Fluid Phase Equilib.* 2001, *183*, 183–208. (b) Tetko, I. V.; Sushko, Y.; Novotarskyi, S.; Patiny, L.; Kondratov, I.; Petrenko, A. E.; Charochkina, L.; Asiri, A. M., *J. Chem. Inf. Model.* 2014, *54*, 3320–3329.

[34] (a) Kolář, P.; Shen, J.-W.; Tsuboi, A.; Ishikawa, T., *Fluid Phase Equilib.* 2002, *194*, 771–782; (b) Karunanithi, A. T.; Achenie, L. E.; Gani, R., *Chem. Eng. Sci.* 2006, *61*,

1247–1260; (c) Kokitkar, P. B.; Plocharczyk, E.; Chen, C.-C., *Org. Process Res. Dev.* 2008, *12*, 249–256; (d) Modarresi, H.; Conte, E.; Abildskov, J.; Gani, R.; Crafts, P., *Ind. Eng. Chem. Res.* 2008, *47*, 5234–5242.

[35] (a) Abramov, Y. A.; Loschen, C.; Klamt, A., *J. Pharm. Sci.* 2012, *101*, 3687–3697; (b) Loschen, C.; Klamt, A., *J. Pharm. Pharmacol.* 2015, *67*, 803–811.

[36] Abramov, Y. A.; Zell, M.; Krzyzaniak, J. F. Toward a Rational Solvent Selection for Conformational Polymorph Screening. In *Chemical Engineering in the Pharmaceutical Industry: R&D to Manufacturing*; am Ende, D. J., Ed.; John Wiley & Sons, Inc.: New York, 2011, pp 491–504.

[37] Nass, K. K., *Ind. Eng. Chem. Res.* 1994, *33*, 1580–1584.

[38] (a) Musumeci, D.; Hunter, C. A.; Prohens, R.; Scuderi, S.; McCabe, J. F., *Chem. Sci.* 2011, *2*, 883–890; (b) Wood, P. A.; Feeder, N.; Furlow, M.; Galek, P. T.; Groom, C. R.; Pidcock, E., *CrystEngComm* 2014, *16*, 5839–5848.

[39] (a) Parshad, H.; Frydenvang, K.; Liljefors, T.; Larsen, C. S., *Int. J. Pharm.* 2002, *237*, 193–207; (b) Tantishaiyakul, V., *Int. J. Pharm.* 2004, *275*, 133–139; (c) Tantishaiyakul, V., *J. Pharm. Biomed. Anal.* 2005, *37*, 411–415.

[40] Abramov, Y. A., *CrystEngComm* 2015, *17*, 5216–5224.

[41] Ismail, S. Z.; Anderton, C. L.; Copley, R. C.; Price, L. S.; Price, S. L., *Cryst. Growth Des.* 2013, *13*, 2396–2406.

[42] Winn, D.; Doherty, M. F., *AIChE J.* 2000, *46*, 1348–1367.

[43] (a) Klamt, A., *COSMO-RS: From Quantum Chemistry to Fluid Phase Thermodynamics and Drug Design.* Elsevier: Amsterdam, 2005; (b) Ikeda, H.; Chiba, K.; Kanou, A.; Hirayama, N., *Chem. Pharm. Bull.* 2005, *53*, 253–255; (c) Tung, H. H.; Tabora, J.; Variankaval, N.; Bakken, D.; Chen, C. C., *J. Pharm. Sci.* 2008, *97*, 1813–1820.

[44] (a) Barrat, J.-L.; Baschnagel, J.; Lyulin, A., *Soft Matter* 2010, *6*, 3430–3446; (b) Alzghoul, A.; Alhalaweh, A.; Mahlin, D.; Bergström, C. A., *J. Chem. Inf. Model.* 2014, *54*, 3396–3403.

[45] (a) Todorova, T.; Delley, B., *Mol. Simul.* 2008, *34*, 1013–1017. (b)Luner, P. E.; Zhang, Y.; Abramov, Y. A.; Carvajal, M. T., *Cryst. Growth Des. 2012*, 12, 5271–5282.

[46] Beyer, T.; Day, G. M.; Price, S. L., *J. Am. Chem. Soc.* 2001, *123*, 5086–5094.

[47] Shariare, M. H.; Leusen, F. J.; de Matas, M.; York, P.; Anwar, J., *Pharm. Res.* 2012, *29*, 319–331.

[48] Simperler, A.; Kornherr, A.; Chopra, R.; Jones, W.; Motherwell, W. S.; Zifferer, G., *Phys. Chem. Chem. Phys.* 2007, *9*, 3999–4006.

[49] (a) Ahmad, S.; Johnston, B. F.; Mackay, S. P.; Schatzlein, A. G.; Gellert, P.; Sengupta, D.; Uchegbu, I. F., *J. R. Soc. Interface* 2010, *7*, S423–S433; (b) Xiang, T.-X.; Anderson, B. D., *Mol. Pharm.* 2012, *10*, 102–114; (c) Xiang, T. X.; Anderson, B. D., *J. Pharm. Sci.* 2013, *102*, 876–891.

[50] Kremer, D.; Hancock, B., *J. Pharm. Sci.* 2006, *95*, 517–529.

[51] Abramov, Y. A., *Org. Process Res. Dev.* 2012, *17*, 472–485.

[52] (a) Boyd, D. B.; Sharp, T. R. The Power of Computational Chemistry to Leverage Stress Testing of Pharmaceuticals. In *Pharmaceutical Stress Testing: Predicting Drug Degradation*, 2nd ed.; Baertschi, S. W., Alsante, K. M., Reed, R. A., Eds.; Informa Healthcare: New York, 2011, pp 499–539; (b) Kleinman, M. H.; Baertschi, S. W.; Alsante, K. M.; Reid, D. L.; Mowery, M. D.; Shimanovich, R.; Foti, C.; Smith, W. K.; Reynolds, D. W.; Nefliu, M., *Mol. Pharm.* 2014, *11*, 4179–4188.

[53] Baias, M.; Dumez, J.-N.; Svensson, P. H.; Schantz, S.; Day, G. M.; Emsley, L., *J. Am. Chem. Soc.* 2013, *135*, 17501–17507.

[54] Datta, S.; Grant, D. J., *Nat. Rev. Drug Discov.* 2004, *3*, 42–57.

[55] Palmer, D. S.; McDonagh, J. L.; Mitchell, J. B.; van Mourik, T.; Fedorov, M. V., *J. Chem. Theory Comput.* 2012, *8*, 3322–3337.

[56] Kazantsev, A. V.; Karamertzanis, P. G.; Adjiman, C. S.; Pantelides, C. C.; Price, S. L.; Galek, P. T.; Day, G. M.; Cruz-Cabeza, A. J., *Int. J. Pharm.* 2011, *418*, 168–178.

[57] (a) Deij, M.; van Eupen, J.; Meekes, H.; Verwer, P.; Bennema, P.; Vlieg, E., *Int. J. Pharm.* 2008, *353*, 113–123; (b) Schmidt, C.; Ulrich, J., *J. Cryst. Growth* 2012, *353*, 168–173.

[58] Hammond, R.; Pencheva, K.; Ramachandran, V.; Roberts, K., *Cryst. Growth Des.* 2007, *7*, 1571–1574.

2

NAVIGATING THE SOLID FORM LANDSCAPE WITH STRUCTURAL INFORMATICS

PETER T. A. GALEK[1], ELNA PIDCOCK[2], PETER A. WOOD[2], NEIL FEEDER[2], AND FRANK H. ALLEN[2]

[1] *RealVNC Ltd, Cambridge, UK*
[2] *Cambridge Crystallographic Data Centre (CCDC), Cambridge, UK*

2.1 INTRODUCTION

The physical properties of a material depend on the nature and mutual arrangement of its constituents. In crystalline materials these constituents, usually molecules or ions, are arranged in essentially infinite, repeating three-dimensional (3D) patterns determined by space group symmetry. However, these same constituents can often adopt multiple 3D patterns to form different crystal structures – the phenomenon of polymorphism [1]. Different polymorphic arrangements, despite being built from the same constituents, can lead to materials with very different physical properties. Polymorphs can therefore have different stabilities, solubilities, bioavailabilities and storage characteristics, and any change in the crystalline form of an active pharmaceutical ingredient (API) can seriously affect its efficacy as a drug. Hence, polymorphism is a crucial factor in drug delivery to patients, a process that relies substantially on the delivery of APIs in crystalline forms.

There are several well-documented cases of the conversion of existing marketed drugs to previously unknown polymorphs, for example ritonavir [2] and rotigotine [3], with serious medical, social and financial consequences. It is therefore crucial for drug development scientists to understand, as far as possible, the solid form landscape,

Computational Pharmaceutical Solid State Chemistry, First Edition. Edited by Yuriy A. Abramov.

that is the inherent form diversity, of each API to ensure robust and reliable delivery of the medicine. The widely used experimental techniques for ameliorating the risk of late-stage polymorphism are collectively termed 'screening', which comprise crystallisation experiments using a wide variety of solvents, physical conditions and crystallisation methods. However, it is impossible to guarantee that all polymorphs have been found and the scope of a screen is often limited by time constraints and budgetary considerations. This is a major issue in risk management for new pharmaceuticals as identified in Guideline Q9 of the International Conference on Harmonisation of Technical Requirements for Registration of Pharmaceuticals for Human Use (ICH) [4].

Computational assessment of the likelihoods of occurrence and the relative stabilities of polymorphs is not necessarily more effective than the experimental approach. Whilst great advances have been made in the field of *ab initio* crystal structure prediction (CSP), as documented in five international blind tests spanning the years 1999–2010 [5], it is still not routinely possible to predict whether a molecule is likely to be polymorphic or to confirm whether the most thermodynamically stable structure has been found experimentally, especially for molecules of the complexity of a typical drug. It is possible to compute the polymorph landscape for a specific flexible molecule, but the calculations require considerable expertise, and the timescales and computing resources can render CSP impractical for application to even a limited portfolio of candidate APIs.

Another route for the inference and examination of polymorph landscapes involves analysis of existing crystal structures of compounds that are similar in some way to the API(s) under study. The structures of nearly 800,000 carbon-containing small molecules have been reported in the literature and numerical, chemical and bibliographic data for these structures have been collected, curated and organised by the Cambridge Crystallographic Data Centre (CCDC) to form the Cambridge Structural Database (CSD) [6]. Thus the CSD contains millions of discrete pieces of information about intramolecular geometry and conformation as well as similarly extensive information on the intermolecular interactions of atoms and chemical functional groups. Software tools included in the CSD System (CSDS) (see Section **2.2**) allow easy access to all data, particularly to distributions of geometrical parameters, both bonded and non-bonded, and to the frequencies of occurrence of a wide variety of functional group interactions. Research applications of the CSD have generated some 3000 publications since the late 1970s, and many of these applications are reviewed elsewhere [7]. The knowledge contained in the CSD has been used extensively in the pharmaceutical industry, most notably in the drug discovery arena. A series of papers by Stahl and co-workers [8–10] elegantly highlight the utility of structural data in the design of drug molecules. More recently the CCDC has been investigating the use of structural knowledge in the later stages of the drug discovery and development process: when the solid form, not just the active ingredient, is under scrutiny.

By evaluating a structure in the context of existing knowledge in the CSD, it is relatively straightforward to identify both common and unusual structural features, for example an unusual conformation of a molecule, ring or functional group, a geometrically unusual hydrogen-bonded interaction, or an unusual donor–acceptor combination, which can be regarded as suggesting that alternative crystal forms where molecules aggregate without these compromises might possibly exist [11–13]. Comparative CSD analysis can give answers easily and quickly and can influence and advance the decision-making process with respect to risk mitigation.

Another area of significant interest to the pharmaceutical and agrochemical industries [14] is that of cocrystallisation of the active ingredient with an acceptable coformer. Cocrystals offer a route to access new solid forms and therefore new physical properties. This contrasts with polymorphism which, when observed, can be difficult to control and rarely represents an opportunity to significantly enhance physical properties. A study of drug solubility showed that the ratio of polymorph stabilities was typically less than two [15]. Cocrystallisation has shown promise in the tuning of a range of physical properties including dissolution rate, compressibility and physical stability.

From an academic point of view, a coformer could be any neutral organic molecule, so the number of potential coformers is vast. In the pharmaceutical industry, however, the list of potential coformers is likely to be restricted to those that are regarded as safe for human consumption (i.e. the Generally Regarded as Safe, or GRAS, list [16]) which, nevertheless, still encompasses hundreds of compounds. This number of potential coformers means that design is crucial as there will always be a limit of the number of cocrystallisations that can be attempted, whether that limit is based on material availability, time or cost. A method for reducing the list of coformers *in silico* to those most likely to cocrystallise is therefore valuable.

The CSD-related scientific and software tools developed for polymorph risk mitigation, and cocrystal design, are the central focus of this chapter. We begin with a brief summary of the CSDS, and then discuss: (i) the development and application of H-bond propensity analysis, (ii) the study of H-bond landscapes and (iii) informatics-based cocrystal screening. In each case we provide case studies to exemplify the methodology. Ongoing development areas and new opportunities are noted in the section 'Conclusions and Outlook'.

2.2 THE CSD SYSTEM

The CSD [6], at the time of writing, contains information on nearly 800,000 crystal structures and increases by around 50,000 structures annually. The database covers all published single-crystal studies of organic and metal-organic small molecules determined by X-ray (single crystal and powder) and neutron diffraction. Several thousand otherwise unpublished structures are also included. All bibliographic, 2D chemical connectivity and 3D crystallographic data are checked and evaluated before inclusion. In the database (and in this chapter), structures are assigned a *reference code* of the form ABCDEFnn to identify the chemical compound and its publication history. The CSD data itself forms part of the complete CSDS that additionally comprises a range of standard software tools: (i) *ConQuest* [17], for searching all CSD information fields, performing 2D substructure and 3D geometry-constrained searches; (ii) *Mercury* [17, 18], a comprehensive structure visualiser, with facilities for visual and numerical analyses [19] of structural information at both the molecular and intermolecular levels; (iii) *Mogul* [20], a knowledge base of intramolecular geometry that contains greater than 20 million bond lengths, valence angles and torsions organised into chemically searchable distributions, each relating to a specific chemical environment or ring; and (iv) *IsoStar* [21], a library of graphical and numerical information about non-bonded interactions derived from the CSD and from protein-ligand complexes in the Protein Data Bank (PDB) [22]. *IsoStar* provides more than 25,000 interactive 3D scatterplots showing the distribution of one of 48 contact groups, for example an H-bond donor, around a central group, with the 300 central groups covering a very wide range of chemical functionality. Data made available *via* the tools of the CSDS are fundamental to the scientific approaches referred to in the remainder of this chapter.

2.3 HYDROGEN-BOND PROPENSITY: THEORY AND APPLICATIONS TO POLYMORPHISM

2.3.1 Methodology

The aim of this development is to take a simple 2D chemical formula as a target and then use CSD information relating to similar compounds as a knowledge base to predict which, if any, of its potential donors and acceptors might form H-bonds in putative crystal structures. The importance of such an answer is clear: H-bonds are strong, reliable interactions which pervade organic structures, and studies have shown that H-bonds between the best donor and acceptor pairs will normally be observed [23]. Crystals that display other interactions at the expense of the best donor–acceptor pair are unusual and should be distinguishable from expected, stable forms as, for example metastable polymorphs, cocrystals or solvates. Recent experience has shown that unforeseen new polymorphs, involving changes in H-bonding with respect to existing formulations, can occur: ritonavir is a well-known example [2], which proved to be hugely problematic and costly. This methodology uses a set of purely 1D and 2D QSAR-like chemical descriptors and allows both novel and existing compounds to be assessed in a quantifiable manner based on their H-bonding possibilities. By using closely similar compounds to the target, each analysis is bespoke for that target, providing a flexible and accurate depiction of H-bonding which is simpler and easier to use and assimilate than computational tools or experimental screening.

The method [24] identifies donor (D) and acceptor (A) pairs, which are assigned a *propensity* to form an H-bond in a structure, based on their respective chemical environments. This is modelled by a probability function and trained using knowledge of H-bonds in related crystal structures. The model parameters enable the method to be truly predictive from the starting point of a chemical diagram. The procedure involves data set curation, model derivation and predictive assessment. Whilst it can be practical to curate sets of CSD structures for model training manually using substructure searches *via ConQuest* (Section **2.2**), an automated chemical similarity search was implemented as part of a wizard in *Mercury* [18] using bit-screens based on chemical fingerprints, giving the user ease of control over the size of data set and definition of chemical groups. Full details of the probability model function are

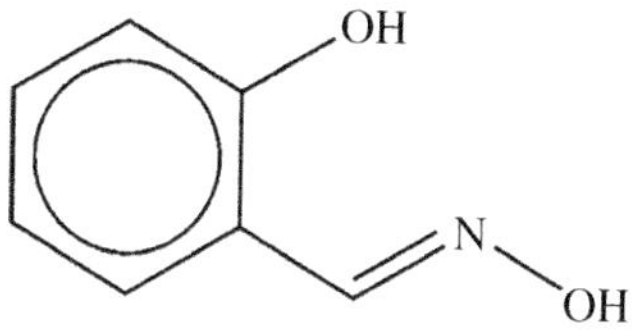

FIGURE 2.1 Chemical diagram of salicylaldoxime. The phenolic OH and oxime-N are in a favourable geometric arrangement for intramolecular H-bond formation.

available elsewhere [24]. Four kinds of attributes have been identified to capture influences on intermolecular H-bonding: (1) a competition function, accounting for the presence or absence of other donors and acceptors; (2) a steric density function for both D and A groups, assessing steric hindrance and related to accessible surface; (3) functional group categories for both D and A groups, capturing chemical and electronic influences; and (4) an aromaticity function, quantifying potential disruptions of H-bonding by donor to π-density interactions [25].

Intramolecular H-bonds are treated differently [26], due to their more universal nature across the entire set of compounds in the CSD, thus transcending the bespoke data sets used in the intermolecular case. Hence, for intramolecular H-bonds, we have been able to create excellent predictive models for large proportions of the entire database, eliminating the need for user intervention in their application. The form of the model function involves (i) the size of the potential intramolecular ring; (ii) a text description of π-electron delocalisation, for example *S–U–S* denotes a potential motif consisting of saturated–unsaturated–saturated bonds; (iii) a binary flag representing the presence of any flexible σ bonds; (iv) *D* and *A* chemical function types defined using SYBYL atom types [27]; and (v) the number of H-bond donors. In the case of the commonly used organic ligand salicylaldoxime (Fig. 2.1), the propensity for an intramolecular H-bond between the —OH donor and =N— acceptor has the high value of $\pi = 0.92$. The model indicates that a potential six-membered ring mediated by an aromatic bond is a major factor. However, there is a small (0.08) probability of the interaction failing to form, an unlikely outcome that only occurs in the metastable (high pressure) crystal structure of salicylaldoxime (SALOXM09 [28]).

The H-bond propensity methodology has been extensively tested and validated [24, 26]. First, the statistical nature of the method was used to analyse thoroughly the predictive ability and transferability of the approach to ‘real’ target compounds outside of the original training set. Second, we prepared test sets by dividing the CSD according to the age of structures. We were then able to compare predictions using a model based on ‘historical’ data and the complete current database. This process resulted in a loss in predictive ability compared with the training set of just 2–3%.

2.3.2 Case Study 1: Ritonavir

The appearance of a second polymorph of ritonavir (Norvir®) during production of a formulation based on the existing form I was a significant blow for Abbott Laboratories [2]. A hydrogen-bond propensity analysis was carried out [29] based on

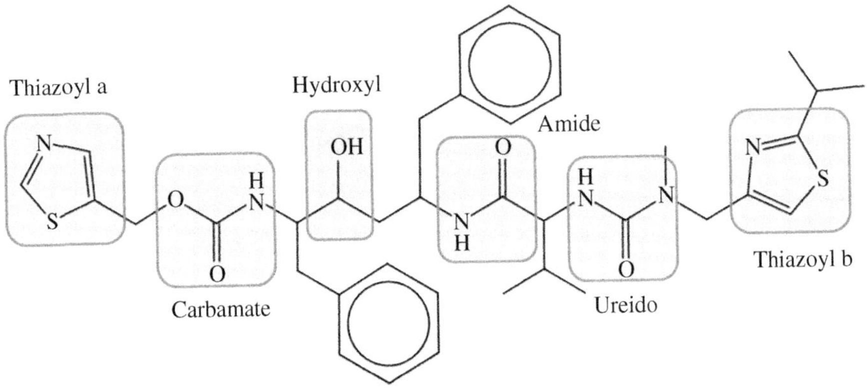

FIGURE 2.2 Designation of chemical group types in ritonavir [29].

the CSD as of January 2008 to see if it could have helped to avert this crisis. Crystal structures were retrieved that contained at least one of the six unique chemical environments that have the ability to form H-bonds (Fig. 2.2). We applied filters to obtain reliable CSD training data: all 3D coordinates determined, R-factor<0.10, no ionic structures, no X-ray powder studies, no metal atoms present, no unresolved errors from CSD evaluation, no crystallographic disorder and no polymeric bonds. Redeterminations were also excluded to avoid structural bias. The treatment of intramolecular H-bonds was not available for this study, so structures with these bonds were also removed. The final training data set consisted of 836 CSD structures. The sulphur atom on the thiazoyl moiety and the alkoxy oxygen of the carbamate (Fig. 2.2) were excluded as potential acceptors since very few such instances occur in the CSD. In the training data, an 'other' variable was assigned to any potential donor or acceptor not identified as a specific modelled group. 'Aromatic nitrogen' and 'ether' acceptors proved to be common in the structures surveyed (639 and 2854 instances, respectively) and were also included as model variables.

Table 2.1 shows the set of potential donor–acceptor pairs and their corresponding H-bond propensities, with the known H-bonds in forms I and II compared with our predictions as an assessment of relative stability. The assertion is that the structure with the most viable H-bonds is likely to be the more stable. The striking feature concerning form I is that, of its four H-bonds, two are most unusual: the hydroxyl–thiazoyl interaction is very unlikely with $\pi=0.114$, and the ureido–ureido interaction also has a low propensity of $\pi=0.224$. Nevertheless, form I *is* an observed crystal structure and the remaining two H-bonds with $\pi>0.5$ seem to provide the structural stability that is needed. Our contention given this analysis is that prior knowledge of the structure of form I, for example during experimental solid form screening, would have directed effort to further searches for other polymorphic forms.

Such a post-rationalisation was also carried out by scientists at Abbott Laboratories who were able to find a stable formulation of the new form II (after it proved almost impossible to reproduce form I). They were also able to understand the physical differences between the polymorphs in a manual assessment of the relative nature of

TABLE 2.1 Predicted Hydrogen-Bond Propensities (π) for the Potential Donor–Acceptor Pairs of Ritonavir

Donor Group	Acceptor Group	π	+/–[a]	Form I	Form II
Amide	Carbamate	0.618	0.094	✗	✗
Amide	Hydroxyl	0.551	0.052	✗	✓
Carbamate	Carbamate	0.538	0.090	✓	✗
Hydroxyl	Carbamate	0.537	0.090	✗	✗
Amide	Amide	0.501	0.055	✓	✗
Amide	Ureido	0.499	0.072	✗	✗
Carbamate	Hydroxyl	0.470	0.078	✗	✗
Hydroxyl	Hydroxyl	0.469	0.037	✗	✗
Carbamate	Amide	0.420	0.083	✗	✓
Hydroxyl	Amide	0.419	0.045	✗	✗
Carbamate	Ureido	0.418	0.088	✗	✗
Hydroxyl	Ureido	0.417	0.058	✗	✓
Ureido	Carbamate	0.319	0.086	✗	✓
Ureido	Hydroxyl	0.263	0.041	✗	✗
Ureido	Amide	0.225	0.040	✗	✗
Ureido	Ureido	0.224	0.044	✓	✗
Amide	Thiazoyl a	0.152	0.054	✗	✗
Amide	Thiazoyl b	0.142	0.050	✗	✗
Carbamate	Thiazoyl a	0.115	0.044	✗	✗
Hydroxyl	Thiazoyl a	0.114	0.039	✓	✗
Carbamate	Thiazoyl b	0.107	0.041	✗	✗
Hydroxyl	Thiazoyl b	0.106	0.036	✗	✗
Ureido	Thiazoyl a	0.049	0.020	✗	✗
Ureido	Thiazoyl b	0.046	0.018	✗	✗

[a] The error bars represent a 95% confidence interval following a χ^2 distribution.

the H-bonding and overall packing between the two forms [2, 30]. The ritonavir saga raised warning flags in the pharmaceutical industry about the crucial nature of the solid form landscape, and in the ensuing 15 years many other compounds have shown similar trends and thus carry the same potential impact.

2.4 HYDROGEN-BOND LANDSCAPES: DEVELOPING THE PROPENSITY APPROACH

2.4.1 Methodology

Thus far we have demonstrated the relationship of H-bond propensity, optimal intermolecular bonding and the likelihood of polymorphism. Combining propensities with models of how many H-bonds may be formed by a given atom in a functional group (i.e. the H-bond coordination environment) allows the *in silico* generation of all chemically reasonable structures. Comparing a given solid form to other possible structures in the resultant *H-bond landscape* provides a powerful analysis of whether a more stable form is possible or whether other competitive polymorphs might exist. The tool can also provide knowledge-based prediction of cocrystals, hydrates and solvates. H-bonds that are possible in a pure form can easily be compared with those involving other components (homo-molecular vs. hetero-molecular interactions) to reveal whether or not there is a preference for recognition between an active ingredient and a solvent, an excipient or a coformer.

Here, we determine an *H-bond coordination number*, *n* [31], as the expected number of discrete H-bonds that a given donor or acceptor atom might form. We then model the likelihood of occurrence of different *n*-values for a range of different D or A atom types. H-bond coordination depends on physical, chemical and topological influences affecting both the donor and acceptor capacity of an atom, and we have developed around 120 statistical models to describe H-bond coordination based on CSD data for common chemical environments. Our approach classifies D and A atoms according to unique SYBYL atom types [27], using a subset of organic structures in the CSD that have at least one H-bond donor (24,502 structures containing O, N or S connected to 1 or more H atoms, R factor$<$0.05, no errors, no disorder, all atomic 3D coordinates determined, no powder structures). As a result, we can now answer what is the likelihood, p_c, of a given H-bond D or A atom (e.g. the C=O acceptor of a carboxylic acid) forming two H-bonds [$p_c(n=2)=0.87$], or no H-bonds [$p_c(n=0)=0.02$]. A different atom, for example O-methoxy, has $p_c(n=2)=0.08$, but $p_c(n=0)=0.67$. Information at this level gives further powerful insight into structural stability by comparing the co-ordination achieved in known structures with the co-ordination predicted for its donors and acceptors. This type of analysis adds to the H-bond propensity predictions since we are now able to assess which H-bond pairings may or may not be mutually exclusive, that is how often can each donor or acceptor be 'used'?

For a given compound, H-bond propensity and coordination number predictions can be combined by creating a set of 'structural hypotheses' – a shorthand for the number of unique ways we allow the D and A atoms to be paired. Hypotheses are enumerated by iterating over donors and acceptors and pairing potential H-bonds when $p_c(n)>p_{min}$, a threshold below which participation of that atom is no longer feasible. Once the unique pairings are generated, we are free to score them based on the overall propensity values and coordination likelihoods. In practice this is done by taking the mean propensity for the observed bonds on the *x*-axis, and taking a mean p_c for the donors and acceptors according to the observed number of H-bonds formed at each atom site. This value is then inverted such that we have a maximum at 0 and minimum at −1 for the *y*-axis. This was an aesthetic decision such that our most likely pairings feature in the bottom right of the landscapes, in a bid to resemble the energy density plots of CSP calculations.

2.4.2 Case Study 2: Metastable versus Stable Form of Piroxicam

This study examines the compound piroxicam (Fig. 2.3), a non-steroidal anti-inflammatory drug. Although there are relatively few donors and acceptors (2 and 5, respectively), a rather surprising number of H-bond hypotheses can be generated.

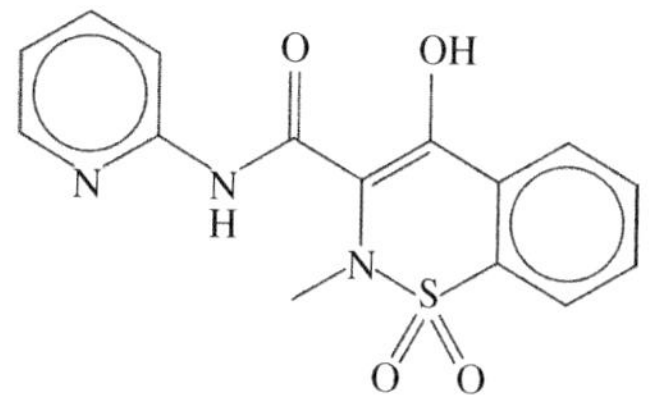

FIGURE 2.3 Chemical diagram of piroxicam.

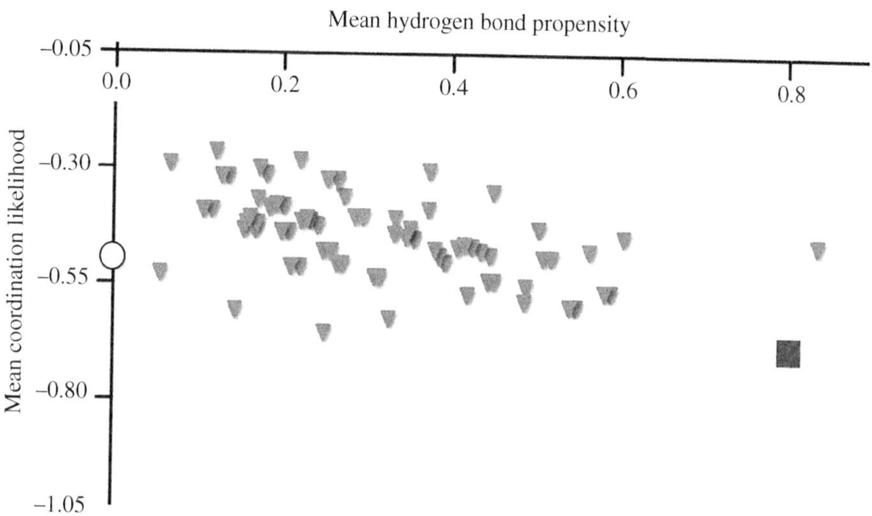

FIGURE 2.4 The H-bond landscape for piroxicam showing a spread of hypothetical structures (grey triangles). Optimal H-bond outcomes appear at the bottom right of the chart, whereas unlikely outcomes appear at the upper left. The thermodynamically stable form I corresponds to the dark grey square to the bottom right, metastable form II corresponds to the white circle on the left-hand side.

Polymorphic form II (BIYSEH06 [32]) exhibits only an intramolecular H-bond, which places the structure to the left of the distribution of possible structures (Fig. 2.4, white circle). The abundance of data points towards the lower right-hand side of Figure 2.4 indicates more likely proposed usage of the donors and acceptors, any one of which could represent an undiscovered form. An analogous result for a newly discovered crystalline form would indicate a metastable phase, and further screening should be employed. The structure of form I (BIYSEH03 [33]) is observed to have a different packing and H-bond arrangement. Placing form I in the same landscape gives a contrasting picture (Fig. 2.4, dark grey square): the H-bonding observed in this structure represents the best outcome in both propensity and coordination scores. Experimentally it is found that form I is the most stable [34].

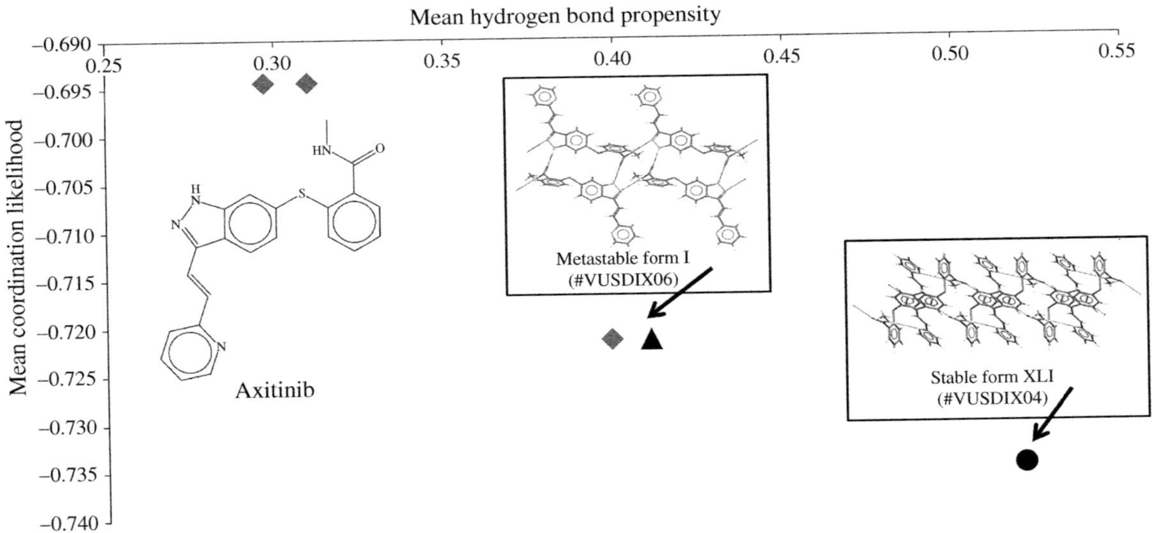

FIGURE 2.5 Chart of mean hydrogen-bond propensity versus mean coordination likelihood for highly likely putative donor/acceptor pairings for axitinib. The arrangement for the metastable Forms I and VI is represented by the triangle, whilst the *most* likely arrangement represented by the circle corresponds to Forms XXV, IV and the thermodynamically stable Form XLI.

2.4.3 Case Study 3: Exploring the Likely Hydrogen-Bond Landscape of Axitinib (Inlyta®)

The extensive polymorphism of axitinib (a tyrosine kinase inhibitor for the treatment of various types of cancer) has been thoroughly investigated experimentally [35, 36] and understood through structural analysis and state-of-the-art solid-state computational modelling approaches [37]. It was found that differing modes of H-bonding were a key feature of the polymorphs and their thermodynamic stability. A CSD-based structural informatics approach also has been used to build a polymorphic risk assessment for axitinib [37]. In Figure 2.5, we show the polymorph landscape calculated for axitinib using the methods described in Section 2.4.1. Here the arrangements have been filtered to include only those that use individual coordination likelihoods greater than 0.1. This simplifies the chart considerably leaving only the highly likely arrangements. Inspection of the chart reveals that one arrangement represents the actual D/A pairings observed for the metastable form I (VUSDIX06 [36]). This point at 0.41/–0.72 (black triangle in Fig. 2.5) is found in the central part of the chart and a further arrangement represented by the point 0.52/–0.73 (black circle in Fig. 2.5) would appear to be more likely. This point does represent the H-bonding arrangement observed in three further polymorphs of axitinib, including the thermodynamically stable form, denoted as form XLI (VUSDIX04 [36]), further validating this approach to polymorph risk assessment.

2.5 INFORMATICS-BASED COCRYSTAL SCREENING

2.5.1 Methodology

In 1990, Etter [23a] published a seminal paper on H-bonding which included the proposal that all good proton donors and acceptors are used in H-bonding. A few years later, in 1995, Desiraju [38] introduced another crucial concept – that of the supramolecular synthon, which is a structural unit (involving intermolecular interactions) shown to be robust enough to be exchanged from one structure to another. Using these two key ideas, subsequent researchers such as Aakeröy [39] and Zaworotko [14a, b, 40] have established successful knowledge-based cocrystal design strategies.

Building upon these design concepts, we have now developed a transferable two-step method for cocrystal screening based on molecular complementarity descriptors and H-bond propensities, which takes into account additional factors that influence cocrystal formation. In the first step, molecular descriptors for the active ingredient and coformers are calculated. The complementarity descriptors capture the shape of the molecule through ratios of its extent – long (L), medium (M) and short (S) along the axes of its principal components, the dipole moment and the number of nitrogen (N) or oxygen (O) atoms as a fraction of the non-hydrogen atoms in the molecule. For novel targets, the molecular conformation can be determined by analysing the range of torsion angles for similar molecules in the CSD Mogul library [20] (see Section **2.2**). Mogul knowledge has been successfully employed in this manner in a recent blind test of computational CSP methods [41]. Coformers with a descriptor value significantly different from those of the active ingredient are judged to fail and are removed from the list of potential cocrystals. The cut-off values and details of the descriptors can be found elsewhere [42].

The second step of the *in silico* screen is to calculate H-bond propensities. For a given active ingredient (A) and a coformer (B), three sets of H-bond propensity calculations are performed: for A on its own, for B on its own and for the two-component A:B system. A multi-component (MC) score can be calculated by subtracting the propensity value of the most likely pure form interaction (AA or BB propensity) from the equivalent value for cocrystal interactions (AB or BA propensity). A H-bonding-based drive towards cocrystal formation is indicated if the MC score is positive.

$$\text{MC score} = \max\left(\text{AB propensity}\right) - \max\left(\text{AA propensity}, \text{BB propensity}\right)$$

In both steps of the cocrystal screening method described earlier, coformers are treated individually allowing the effects of differences in, for example, substitution patterns to be considered in the analysis.

2.5.2 Case Study 4: Paracetamol

Here we apply the earlier two-step screening approach to paracetamol and a list of 35 screened coformers obtained from the literature [43], including 14 of which are reported experimentally to interact in the solid state. Table 2.2 shows the results of the application of the molecular descriptor methodology. The differences between the descriptors calculated for the active ingredient and the coformers are given, along with the judgement as to whether the coformer passed or failed the screening. Of the 35 coformers, 10 fail the molecular descriptor screen and are removed. No coformer fails the screen on the basis of the dipole moment. Of the 25 coformers which pass, 9 are found to form cocrystals experimentally.

The 25 molecules that passed step one then proceed to the second step of the computational screen, the multi-component H-bond propensity analysis. The methodology for generating H-bond propensity values is as outlined in Section **2.3**: in the same way, relevant CSD structures are gathered, only here; data sets are prepared for model training for each unique AA, BB and AB system = 1 + 25 + 25 iterations; and H-bond propensities are calculated in each case. The MC score for each AB pair is then determined (Table 2.3). Of the 25 coformers that passed the molecular descriptor test 21 are found to have an MC score ≥ 0. We observe the range of positive scores is fairly evenly spread between 0.22 (caffeine) and 0.00 (resorcinol, theophylline). The most negative MC score is found for 3-isochromanone (−0.4) where the most likely H-bond is A—A, indicating the two oxygen acceptors of the coformer are much less likely to form H-bonds with paracetamol.

If the results were used to design a screening experiment, a range of MC score cut-off values could be considered depending on the amount of time and resources for screening. A starting point would be to choose the coformers that returned an MC score ≥ 0, which would identify 20 experiments. From the relevant literature [43], 9 of those 20 experiments would result in a cocrystal. By performing the informatics-based screening outlined here we have reduced the number of experiments

TABLE 2.2 Calculated Differences between Molecular Descriptors for 35 Experimentally Tested Coformers and Paracetamol

Compound	Decision to Proceed with Coformer	ΔM/L Ratio	ΔS	ΔS/L Ratio	ΔDipole Moment (Debye)	ΔFNO
1,2-Bis-4-pyridyl-ethane	Yes	0.05	0.11	0.02	3.73	0.13
1,4-Dioxane	No	0.30	0.85	***0.42***	3.74	0.06
1-Naphthol	Yes	0.24	0.75	0.00	1.60	0.18
2,5-Dihydroxy-benzoic acid	Yes	0.21	0.76	0.00	2.71	0.09
3-Isochromanone	Yes	0.18	0.80	0.15	0.15	0.09
3S-*Cis*-3,6-dimethyl-1, 4-dioane-2,5-dione	No	***0.33***	1.46	***0.33***	3.25	0.13
4,4-Bipyridine	Yes	0.05	0.76	0.03	3.74	0.11
4,4-Trimethylene-dipyridine	Yes	0.10	0.02	0.07	2.32	0.14
Adipic acid	Yes	0.13	0.00	0.00	3.72	0.13
Ascorbic acid	Yes	0.14	1.21	0.21	0.10	0.23
Benzoic acid	Yes	0.12	0.76	0.01	2.74	0.05
Caffeine	Yes	0.28	0.16	0.08	2.73	0.16
Citric acid	Yes	0.07	2.11	0.26	1.95	0.27
Fumaric acid	Yes	0.01	0.76	0.01	1.10	0.23
Imidazole	No	***0.32***	0.76	0.17	2.42	0.13
Isonicotinamide	Yes	0.18	0.71	0.04	2.46	0.06
Maleic acid	Yes	0.08	0.73	0.03	0.47	0.23
Malic acid	Yes	0.05	0.71	0.16	0.45	0.28
Malonic acid	No	0.09	0.01	0.17	2.04	***0.30***
Melamine	No	***0.31***	0.76	0.02	3.73	***0.39***
Morpholine	No	***0.38***	0.82	0.40	1.88	0.06
N,N-dimethyl-piperazine	Yes	0.12	0.79	0.19	3.74	0.02
Nicotinamide	Yes	0.12	0.75	0.01	1.30	0.06
N-methyl-morpholine	No	0.28	0.80	***0.31***	1.59	0.01
Oxalic acid	No	0.14	0.76	0.09	3.74	***0.40***
Phenazine	Yes	0.04	0.74	0.06	3.70	0.13
Piperazine	No	***0.32***	0.82	***0.34***	3.74	0.06
Pyrazine	Yes	0.30	0.75	0.17	3.74	0.06
Resorcinol	Yes	0.24	0.76	0.03	0.23	0.02
Saccharin	No	***0.32***	1.45	0.27	1.60	0.06
Succinic acid	Yes	0.02	0.03	0.08	2.66	0.23
Theobromine	Yes	0.20	0.02	0.07	2.39	0.19
Theophylline	Yes	0.26	0.11	0.06	3.02	0.12
1,4-Di-4-pyridyl-ethylene	Yes	0.05	0.72	0.09	3.73	0.13
1,4-Diaminocyclohexane	Yes	0.14	1.09	0.22	2.70	0.02

A calculated difference in molecular descriptor which indicates a cocrystal is unlikely (falls outside of the allowed pass mark) is highlighted in bold and italics.

performed by 43% (from 35 to 20) and retained 64% of the cocrystal hits (9 out of 14). Moreover, a set of screening experiments prioritised in line with the MC scores would return a cocrystal by the second experiment and six cocrystals in the first eight experiments.

TABLE 2.3 Multi-component HBP Screening Results Summary for Paracetamol with Potential Coformer Molecules Ranked in Order of Their MC Score

Rank	Coformer	MC Score	Statistical Uncertainty	Most Likely Interaction	Cocrystal Observed Experimentally	CSD Refcode
1	Caffeine	0.22	0.10	A:B	No	—
2	4,4-Trimethylene-dipyridine	0.21	0.07	A:B	Unclear[a]	N/A
3	Pyrazine	0.21	0.09	A:B	No	—
4	1,2-Bis-4-pyridyl-ethane	0.20	0.07	A:B	Yes	KETZAM
5	4,4-Bipyridine	0.19	0.07	A:B	Yes	MUPQAP
6	1,4-Di-4-pyridyl-ethylene	0.19	0.07	A:B	Yes	KETYUF
7	Phenazine	0.19	0.15	A:B	Yes	LUJSOZ
8	Citric acid	0.15	0.05	B:A	Yes	AMUBAM
9	Malic acid	0.12	0.04	B:A	No	—
10	Adipic acid	0.11	0.06	B:A	No	—
11	2,5-Dihydroxy-benzoic acid	0.11	0.07	B:A	No	—
12	Benzoic acid	0.11	0.07	B:A	No	—
13	1,4-Diamino-cyclohexane	0.10	0.08	B:A	Yes	WIGCEW
14	Theobromine	0.10	0.12	A:B	No	—
15	Succinic acid	0.09	0.06	B:A	No	—
16	Fumaric acid	0.09	0.07	B:A	No	—
17	Maleic acid	0.09	0.07	B:A	No	—
18	*N,N*-dimethyl-piperazine	0.05	0.07	A:B	Yes	MUPPIW
19	Resorcinol	0.00	0.09	A:A	No	—
20	Theophylline	0.00	0.11	B:B	Yes	KIGLUI
21	Ascorbic acid	−0.03	0.06	B:B	No	—
22	1-Naphthol	−0.07	0.09	A:A	No	—
23	Nicotinamide	−0.11	0.04	B:B	No	—
24	Isonicotinamide	−0.12	0.04	B:B	No	—
25	3-Isochromanone	−0.40	0.08	A:A	No	—

Each computed MC score also has an associated estimated uncertainty based on χ^2 statistics resulting from model fitting. The size of the uncertainty is related to the amount of contributory data from the CSD.
[a]The cocrystallisation experiment produced a sticky substance unsuitable for PXRD analysis, rather than starting materials or a verifiable cocrystal [43c].

By using only the difference between the maximum propensities for AA, AB or BB H-bonds we are ignoring the fact that cocrystal networks can be formed of both homo (AA or BB) *and* AB interactions. This situation is found in the observed paracetamol: theophylline cocrystal, for example. By considering more than one H-bonded interaction and looking at possible combinations of donors with acceptors the predictability of this informatics-based screening tool may be improved.

2.5.3 Case Study 5: AMG 517 – Sorbic Acid Cocrystal

AMG 517 is a potent and selective VR1 antagonist developed for the treatment of acute and chronic pain by Amgen, Inc. [44] The AMG 517 free base was found to be insoluble in water and in buffers at physiological pH, significantly limiting the exposure in animal studies. In contrast, however, a sorbic acid cocrystal of AMG 517 was found to provide increased solubility which dramatically improved exposure. Evaluation of the cocrystal form in rats revealed that a 30 mg/kg dose had a comparable AMG 517 exposure to a 500 mg/kg dose of the free base. Whilst this is an important example of how a cocrystal can provide enhanced physiochemical properties for a new chemical entity, the discovery of the AMG 517: sorbic acid cocrystal owed a great deal to serendipity – it was first crystallised from a formulated suspension of the free base in which sorbic acid was present simply as a preservative. Here we demonstrate how the methodologies described in Section 2.5.1 could have been used to rationally select sorbic acid as a highly suitable coformer for AMG 517, providing further validation of our general approach.

The molecular structures of AMG 517 and sorbic acid are shown in Figure 2.6, whilst the crystal packing arrangement in the 1:1 cocrystal (TEGFOC [44]) is represented in Figure 2.7. Close packing is achieved through a degree of shape complementarity between the AMG 517 and sorbic acid molecules, whilst the arrangement is further stabilised through a heterocyclic H-bonded motif. Using CSD knowledge, these key aspects of the crystal structure are predictable and therefore provide an indicator that sorbic acid is a suitable coformer for AMG 517.

To generate the molecular complementarity descriptors, the representative conformation of AMG 517 was selected by using the CSD Mogul library [20]. A sufficiently feasible conformation was obtained as can be seen in an overlay of the Mogul-derived and actual AMG 517 conformations from the cocrystal structure (Figure 2.8). Sorbic acid is a relatively rigid molecule and so its conformation was selected from the crystal structure of free sorbic acid (LEZHUT [45]).

The results of the molecular descriptor analysis listed in Table 2.4 indicate that AMG517 and sorbic acid do indeed have a molecular complementarity that would promote cocrystal formation. H-bond propensity values calculated for the AMG 517

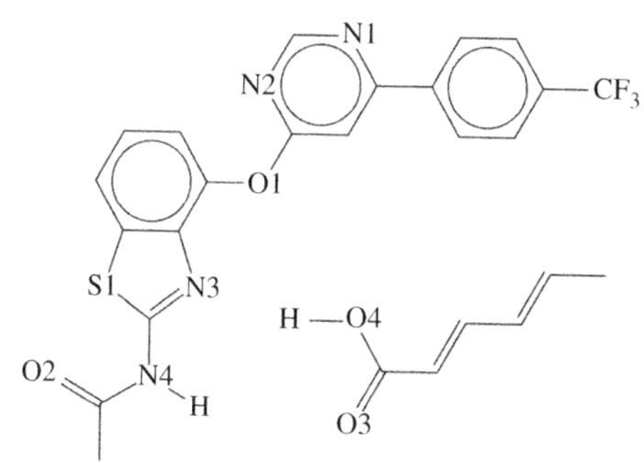

FIGURE 2.6 Schematic representation of AMG 517 and sorbic acid.

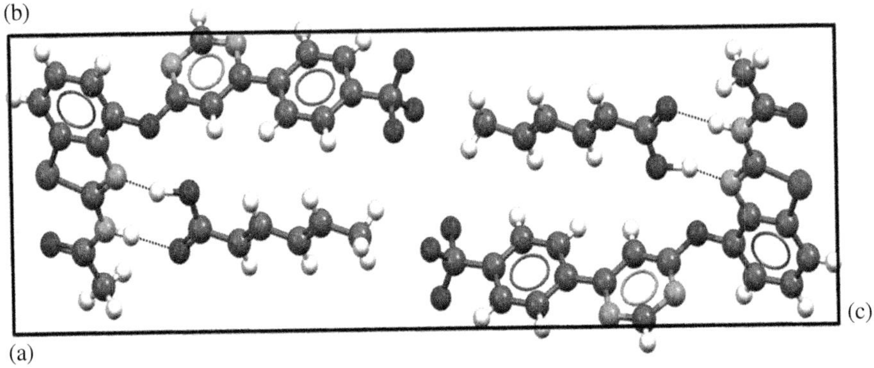

FIGURE 2.7 The packing arrangement for the 1 : 1 AMG 517: sorbic acid cocrystal (TEGFOC [44]) viewed down the crystallographic *a*-axis (figure prepared using *Mercury* [18]).

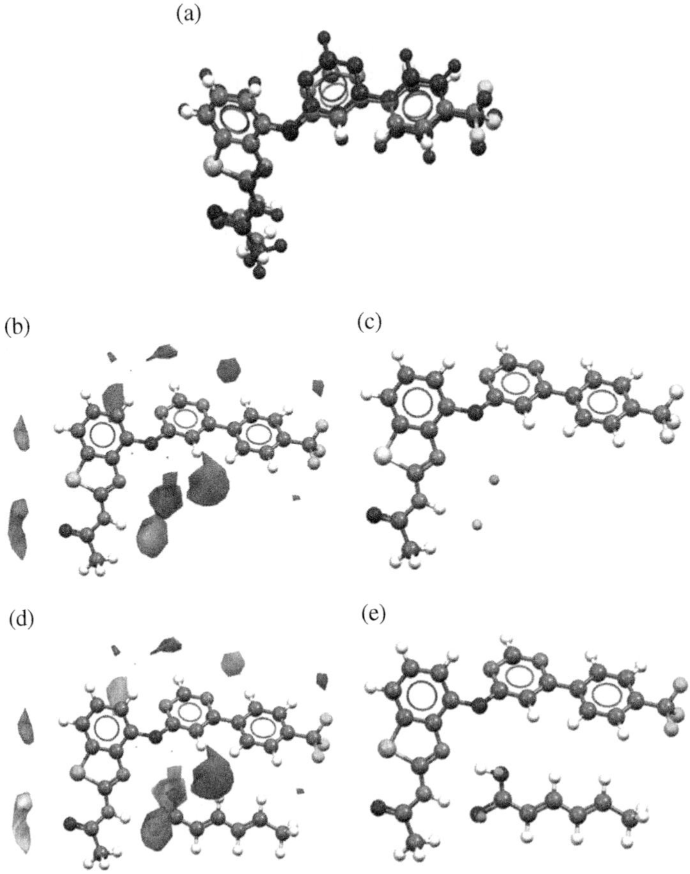

FIGURE 2.8 Conformation and *IsoStar* Full Interaction Maps of AMG 517. (a) Overlay of the AMG 517 molecule from the sorbic acid cocrystal with a likely conformation suggested by Mogul analysis (dark). Overlay RMSD = 0.392 Å. (b) Contour maps for hydrogen-bond donor (cyan) and acceptor (orange) probes, (c) maximum hot spots in the donor (cyan) and acceptor (orange) maps, (d) overlay of a sorbic acid molecule onto the contour maps and (e) overlay of a sorbic acid molecule onto the hot spots (figures prepared using *Mercury* [18]). (*see insert for color representation of the figure.*)

TABLE 2.4 Molecular Complementarity Descriptors for AMG 517 and Sorbic Acid

	ML Axis Ratio	S Axis (/Å)	SL Axis Ratio	Dipole Moment (Debye)	Fraction of Nitrogen and Oxygen
Sorbic acid	0.50	4.17	0.38	1.03	0.25
AMG 517	0.66	6.46	0.35	1.65	0.20
Abs(AMG 517-sorbic acid)	0.16	2.29	0.03	0.62	0.05
Pass criterion	<0.31	<3.23	<0.275	<5.94	<0.294
Decision to proceed with coformer	Yes	Yes	Yes	Yes	Yes

TABLE 2.5 Hydrogen-Bond Propensities for the AMG 517 Sorbic Acid (SA) System

Donor	Acceptor	Propensity
N4 (AMG)	**O3 (SA)**	**0.69**
N4 (AMG)	N2 (AMG)	0.52
N4 (AMG)	N1 (AMG)	0.50
O4 (SA)	O3 (SA)	0.47
N4 (AMG)	O2 (AMG)	0.41
N4 (AMG)	N3 (AMG)	0.31
O4 (SA)	N2 (AMG)	0.29
O4 (SA)	N1 (AMG)	0.28
O4 (SA)	O2 (AMG)	0.21
N4 (AMG)	O4 (SA)	0.20
O4 (SA)	**N3 (AMG)**	**0.15**
O4 (SA)	O4 (SA)	0.09
N4 (AMG)	O1 (AMG)	0.01
N4 (AMG)	S1 (AMG)	0.01
O4 (SA)	O1 (AMG)	0.00
O4 (SA)	S1 (AMG)	0.00

The observed donor–acceptor pairings for hydrogen bonding are highlighted in bold (see Fig. 2.6 for atomic numbering scheme).

sorbic acid system are presented in Table 2.5. They indicate that the interaction involving the AMG517 hetero-amide and the acid moiety of sorbic acid is the most likely donor–acceptor pairing, providing a strong indication that cocrystal formation would be likely. This donor–acceptor pairing is indeed observed in the AMG 517: sorbic acid crystal structure, as shown in Figure 2.7.

The likely geometry of intermolecular interactions and their combination into supramolecular synthons [37] can also be studied using a method built on *IsoStar* data called Full Interaction Map (FIM) analysis [46]. The FIM tool breaks down the molecule into a set of central groups, assembles a set of *IsoStar* interaction maps for selected donor/acceptor/hydrophobic probes around each central group and then combines these to give the likely full 3D interaction environment around the molecule [46]. From such maps we can start to understand how a coformer may satisfy the requirements of the interaction map of an API and thus identify the supramolecular synthons that might drive cocrystal formation. Such maps are shown in Figure 2.8 for AMG 517. Here we identify the 'hot spots' in the H-bond donor and acceptor maps as the most likely positions for such functional groups and notice that a carboxylic acid group, as found in sorbic acid, is able to satisfy this geometry particularly convincingly and is thus a good coformer candidate.

2.6 CONCLUSIONS AND OUTLOOK

Solid form development in the pharmaceutical industry, where candidates are taken from discovery and turned into drug products, is a process which takes many years [47a] and costs significant sums of money [47b]. Traditionally, the solid form that is chosen for scale-up and commercialisation is established through experimental methods such as crystallisations at different conditions and competitive slurrying. Introducing new technologies to aid decision making around risk assessment and form selection is one approach to expedite the progress of drugs through the discovery and development pipelines.

Informatics-based software tools offer a unique strategy to complement the currently available solid form development technologies, benefitting from the foundations of high-quality, relevant structural data gathered by the community over more than 40 years. It has been illustrated here how solid form informatics can be applied to aid both polymorph risk assessment and multi-component form design. The tools are designed to be straightforward to use and are based on chemico-structural knowledge rather than high-level computational/theoretical techniques. As such the tools are potentially much more amenable for routine use by 'bench-top' solid form scientists on the short timescales most relevant to their working practices. Naturally this allows the feedback of complementary information from the informatics into decision-making processes.

Continued scientific and technical focus in the areas of solid form control, risk assessment and design is the remit of CCDC's Crystal Form Consortium (CFC), established in 2008, complementing CCDC's successful membership in the Pfizer Institute for Pharmaceutical Materials Science (PIPMS), which began operations in 2002. There is much scope for further development. Salts are ubiquitous in drug products but choosing the correct counter-ion for an API is generally done through experimental screening with little design. Charged species have proven to be a greater challenge for computational methods, such as CSP, so building on knowledge of the CSD's thousands of salt structures [48] is being investigated. Similarly, assessment of the stability of hydrates [49] and prediction of the likelihood of a hydrated form [49] are two areas well-suited to knowledge-based methods, and current work is to look into how the tools might be generalised for the assessment of common organic solvates [50]. An informatics-led prediction of the geometric placement of solvents, coformers or counter-ions upon inclusion into the crystal lattice [51] is of relevance for salts, hydrates, solvates and cocrystals.

We have shown here that great progress has been made in the past decade towards harnessing crystallographic knowledge for solid form risk assessment and design. There is still, however, much that can further be done to form a holistic picture of solid form assessment. We anticipate, as has been proven, that relevant, accurate and accessible data will be a vital resource.

REFERENCES

[1] Bernstein, J. *Polymorphism in Molecular Crystals*. Oxford University Press, Oxford, UK, 2002.

[2] Bauer, J.; Spanton, S.; Henry, R.; Quick, J.; Dziki, W.; Porter, W.; Morris, J. *Pharm. Res.*, 2001, *18*, 859–866.

[3] Rascol, O.; Perez-Lloret, S. *Expert Opin. Pharmacother.*, 2009, *10*, 677–691. See also http://www.fda.gov/Safety/MedWatch/SafetyInformation/SafetyAlertsforHumanMedicalProducts/ucm094861.htm (accessed 28 January 2014).

[4] ICH. 2015. International Conference on Harmonisation of Technical Requirements for Registration of Pharmaceuticals for Human Use (ICH). http://www.ich.org/ (accessed 28 January 2014).

[5] Bardwell, D.A.; Adjiman, C.S.; Arnautova, Y.; Bartashevich, E.; Boerrigter, S.X.M.; Braun, D.E.; Cruz-Cabeza, A.J.; Day, G.M.; Della Valle, R.G.; Desiraju, G.R.; van Eijck, B.P.; Facelli, J.C.; Ferraro, M.B.; Grillo, D.; Habgood, M.; Hofmann, D.W.M.; Hofmann, F.; Jose, K.V.J.; Karamertzanis, P.G.; Kazantsev, A.V.; Kendrick, J.; Kuleshova, L.N.; Leusen, F.J.J.; Maleev, A.V.; Misquitta, A.J.; Mohamed, S.; Needs, R.J.; Neumann, M.A.; Nikylov, D.; Orendt, A.M.; Pal, R.; Pantelides, C.C.; Pickard, C.J.; Price, L.S.; Price, S.L.; Scheraga, H.A.; van de Streek, J.; Thakur, T.S.; Tiwari, S.; Venuti E.; Zhitkov, I.K. *Acta Crystallogr., Sect. B: Struct. Sci.*, 2011, *67*, 535–551, and references therein.

[6] (a) Allen, F.H. *Acta Crystallogr.*, 2002, *B58*, 380–388. (b) Groom, C.R.; Allen, F.H. *Angew. Chem. Int. Ed. Engl.*, 2014, *53*, 662–671.

[7] (a) Allen, F.H.; Motherwell, W.D.S. *Acta Crystallogr., Sect. B: Struct. Sci.*, 2002, *58*, 407–422. (b) Allen, F.H.; Taylor, R. *Chem. Soc. Rev.*, 2004, *33*, 463–475. (c) Wong, R.; Allen, F.H.; Willett, P. *J. Appl. Crystallogr.*, 2010, *43*, 811–824.

[8] Brameld, K.A.; Kuhn, B.; Reuter, D.C.; Stahl, M. *J. Chem. Inf. Model.*, 2008, *48*, 1–24.

[9] Bissantz, C.; Kuhn, B.; Stahl, M. *J. Med. Chem.*, 2010, *53*, 5061–5084.

[10] Kuhn, B.; Mohr, P.; Stahl, M. *J. Med. Chem.*, 2010, *53*, 2601–2611.

[11] Chisholm, J.A.; Pidcock, E.; van de Streek, J.; Infantes, L.; Motherwell, W.D.S.; Allen, F.H., *CrystEngComm*, 2006, *8*, 11–28.

[12] Pidcock, E.; Chisholm, J.A.; Wood, P.A.; Galek, P.T.A.; Fábián, L.; Korb, O.; Cruz-Cabeza, A.J.; Liebeschuetz, J.W.; Groom, C.R.; Allen, F.H. The Cambridge Structural Database System and its applications in supramolecular chemistry and materials design. In *Supramolecular Chemistry, from Molecules to Nanomaterials*. Steed, J.W.; Gale, P.A., Eds; Wiley: New York, 2012, 2927–2946.

[13] Galek, P.T.A.; Pidcock, E.; Wood, P.A.; Bruno, I.J.; Groom, C.R. *CrystEngComm*, 2012, *14*, 2391–2403.

[14] (a) Almarsson, Ö.; Zaworotko, M.J. *Chem. Commun.*, 2004, 1889–1896. (b) Vishweshwar, P.; McMahon, J.A.; Bis, J.A.; Zaworotko, M.J. *J. Pharm. Sci.*, 2006, *95*, 499–516. (c) Stanton, M.K.; Bak, A. *Cryst. Growth Des.*, 2008, *8*, 3856–3862.

[15] Pudipeddi, M.; Serjuddin, A.T.M. *J. Pharm. Sci.*, 2005, *94*, 929–939.

[16] GRAS. The GRAS list. http://www.fda.gov/Food/IngredientsPackagingLabeling/GRAS/ (accessed 28 January 2014).

[17] Bruno, I.J.; Cole, J.C.; Edgington, P.R.; Kessler, M.; Macrae, C.F.; McCabe, P.; Pearson, J.; Taylor, R. *Acta Crystallogr., Sect. B: Struct. Sci.*, 2002, *58*, 389–397.

[18] (a) Macrae, C.F.; Edgington, P.R.; McCabe, P.; Pidcock, E.; Shields, G.P.; Taylor, R.; Towler, M.; van de Streek, J. *J. Appl. Crystallogr.*, 2006, *39*, 453–457. (b) Macrae, C.F.; Bruno, I.J.; Chisholm, J.A.; Edgington, P.R.; McCabe, P.; Pidcock, E.; Rodriguez-Monge, L.; Taylor, R.; van de Streek, J.; Wood, P.A. *J. Appl. Crystallogr.*, 2008, *41*, 466–470.

[19] Sykes, R.A.; McCabe, P.; Allen, F.H.; Battle, G.M.; Bruno, I.J.; Wood, P.A. *J. Appl. Crystallogr.*, 2011, *44*, 882–886.

[20] (a) Bruno, I.J.; Cole, J.C.; Kessler, M.; Luo, J.; Motherwell, W.D.S.; Purkis, L.H.; Smith, B.R.; Taylor, R.; Cooper, R.I.; Harris, S.E.; Orpen, A.G. *J. Chem. Inf. Comput. Sci.*, 2004, *44*, 2133–2144. (b) Cottrell, S.J.; Olsson, T.S.G.; Taylor, R.; Cole, J.C.; Liebeschuetz, J.W. *J. Chem. Inf. Model.*, 2012, *52*, 956–962.

[21] Bruno, I.J.; Cole, J.C.; Lommerse, J.P.M.; Rowland, R.S.; Taylor, R.; Verdonk, M.L. *J. Comput. Aided Mol. Des.*, 1997, *11*, 525–537.

[22] Berman, H.M.; Westbrook, J.; Feng, Z.; Gilliland, G.; Bhat, T.N.; Weissig, H.; Shindyalov, I.N.; Bourne, P.E. *Nucleic Acids Res.*, 2000, *28*, 235–242.

[23] (a) Etter, M.C. *Acc. Chem. Res.*, 1990, *23*, 120–126. (b) Etter, M.C. *J. Phys. Chem.*, 1991, *95*, 4601–4610. (c) Etter, M.C.; MacDonald, J.C.; Bernstein, J. *Acta Crystallogr., Sect. B: Struct. Sci.*, 1990, *46*, 256–262. (d) Etter, M.C.; Reutzel, S.M. *J. Am. Chem. Soc.*, 1991, *113*, 2586–2598.

[24] (a) Galek, P.T.A; Fábián, L.; Allen, F.H.; Motherwell, W.D.S.; Feeder, N. *Acta Crystallogr., Sect. B: Struct. Sci.*, 2007, *63*, 768–782. (b) Galek, P.T.A.; Fábián L.; Allen, F.H. *Acta Crystallogr., Sect. B: Struct. Sci.*, 2009, *65*, 68–85. (c) Galek, P.T.A.; Fábián L.; Allen, F.H. *CrystEngComm*, 2010, *12*, 2091–2099.

[25] Desiraju, G.R., Steiner, T. Other Weak and Non-Conventional Hydrogen Bonds. In *The Weak Hydrogen Bond in Structural Chemistry and Biology*. Oxford University Press: New York, 1999, p. 122.

[26] Galek, P.T.A.; Fábián L.; Allen, F.H. *CrystEngComm*, 2010, *12*, 2091–2099.

[27] Clark, M.; Cramer, R.D. III; Van Opdenbosch, N. *J. Comp. Chem.*,1989, *10*, 982–1012.

[28] Wood, P.A.; Forgan, R.S.; Henderson, D.; Parsons, S.; Pidcock, E.; Tasker, P.A.; Warren, J.E. *Acta Crystallogr., Sect. B: Struct. Sci.*, 2006, *62*, 1099–1111.

[29] Galek, P.T.A.; Allen, F.H.; Fábián, L.; Feeder, N. *CrystEngComm*, 2009, *11*, 2634–2639.

[30] (a) Chemburkar, S.R.; Bauer, J.; Deming, K.; Spiwek, H.; Patel, K.; Morris, J.; Henry, R.; Spanton, S.; Dziki, W.; Porter, W.; Quick, J.; Bauer, P.; Donaubauer, J.; Narayanan, B.A.; Soldani, M.; Riley, D.; McFarland, K. *Org. Proc. Res. Dev.*, 2000, *4*, 413–417. (b) Morissette, S.L.; Soukasene, S.; Levinson, D.; Cima, M.J.; Almarson, O. *Proc. Natl. Acad. Sci. U. S. A.*, 2003, *100*, 2180–2184.

[31] Galek, P.T.A.; Chisholm, J.A.; Pidcock, E.; Wood, P.A. *Acta Crystallogr., Sect. B: Struct. Sci.*, 2014, *70*, 91–105.

[32] Vrecer, F.; Vrbinc, M.; Meden, A. *Int. J. Pharm.*, 2003, *256*, 3–15.

[33] Sheth, A.R.; Bates, S.; Muller, F.X.; Grant, D.J.W. *Cryst. Growth Des.*, 2005, *5*, 571–578.

[34] Vrecer, F.; Srcic, S.; Smid-Korbar, J. *Int. J. Pharm.*, 1991, *68*, 35–41.

[35] Chekal, B.P.; Campeta, A.M.; Abramov, Y.A.; Feeder, N.; Glynn, P.P.; McLaughlin, R.W.; Meenan, P.A.; Singer R.A. *Org. Process Res. Dev.*, 2009, *13*, 1327–1337.

[36] Campeta, A.M.; Chekal, B.P.; Abramov, Y.A.; Meenan, P.A.; Henson, M.J.; Shi, B.; Singer, R.A.; Horspool, K.R. *J. Pharm. Sci.*, 2010, *99*, 3874–3886.

[37] Abramov, Y.A. *Org. Process Res. Dev.*, 2013, *17*, 472–485.

[38] Desiraju, G.R. *Angew. Chem. Int. Ed. Engl.*, 1995, *34*, 2311–2327.

[39] Aakeröy C.B.; Salmon, D.J. *CrystEngComm*, 2005, *7*, 439–448; Aakeröy, C.B. *Acta Crystallogr., Sect. B: Struct. Sci.*, 1997, *53*, 569–586; Aakeröy, C.B.; Beatty, A.M.; Leinen, D.S. *Cryst. Growth Des.*, 2001, *1*, 47–52.

[40] Cheney, M.L.; Weyna, D.R.; Shan, N.; Hanna, M.; Wojtas, L.; Zaworotko, M.J. *J. Pharm. Sci.*, 2011, *100*, 2172–2181.

[41] Kazantsev, A.; Karamertzanis, P.G.; Adjiman, C.S.; Pantelides, C.C.; Price, S.L.; Galek, P.T.A.; Day G.M.; Cruz-Cabeza, A.J. *Int. J. Pharm.*, 2011, *418*, 168–178.

[42] Fábián, L. *Cryst. Growth Des.*, 2009, *9*, 1436–1443.

[43] (a) Karki, S.; Friscic, T.; Fabian, L.; Laity, P.R.; Day, G.M.; Jones, W. *Adv. Mater.*, 2009, *21*, 3905–3909. (b) Oswald, I.D.H.; Allan, D.R.; McGregor, P.A.; Motherwell, W.D.S.; Parsons, S.; Pulham, C.R. *Acta Crystallogr., Sect. B Struct. Sci.*, 2002, *58*, 1057–1066. (c) Srirambhatla, V.K.; Kraft, A.; Watt, S.; Powell, A.V. *Cryst. Growth Des.*, 2012, *12*, 4870–4879. (d) Childs, S.L.; Stahly, G.P.; Park, A. *Mol. Pharm.*, 2007, *4*, 323–328. (e) Elbagerma, M.A.; Edwards, H.G.M.; Munshi, T.; Scowen, I.J. *CrystEngComm*, 2011, *13*, 1877–1884.

[44] Bak, A.; Gore, A.; Yanez, E.; Stanton, M.; Tufekcic, S.; Syed, R.; Akrami, A.; Rose, M.; Surapaneni, S.; Bostick, T.; King, A.; Neervannan, S.; Ostovic, D.; Koparkar, A. *J. Pharm. Sci.*, 2008, *97*, 3942–3956.

[45] Cox, P.J. *Acta Crystallogr., Sect. C Cryst. Struct. Commun.*, 1994, *50*, 1620–1622.

[46] Wood, P.A.; Olsson, T.S.G.; Cole, J.C.; Cottrell, S.J.; Feeder, N.; Galek, P.T.A.; Groom, C.R.; Pidcock, E. *CrystEngComm*, 2013, *15*, 65–72.

[47] (a) Pammolli, F.; Magazzini, L.; Riccaboni, M. *Nat. Rev. Drug Discov.*, 2011, *10*, 428–438. (b) Paul, S.M.; Mytelka, D.S.; Dunwiddie, C.T.; Persinger, C.C.; Munos, B.H.; Lindborg, S.R.; Schacht, A.L. *Nat. Rev. Drug Discov.*, 2010, *9*, 203–214.

[48] Haynes, D.A.; Jones, W.; Motherwell, W.D.S. *CrystEngComm*, 2005, *55*, 342–345.

[49] (a) Infantes, L.; Chisholm, J.; Motherwell, W.D.S. *CrystEngComm*, 2003, *5*, 480–486. (b) Mascal, M.; Infantes, L.; Chisholm, J. *Angew. Chem. Int. Ed.*, 2006, *45*, 32–36. (c) Infantes, L.; Fabian L.; Motherwell, W.D.S. *CrystEngComm*, 2007, *9*, 65–71.

[50] Cruz Cabeza, A.J.; Day, G.M.; Motherwell, W.D.S.; Jones, W. *J. Am. Chem. Soc.*, 2006, *128*, 14466–14467.

[51] Korb, O.; Wood, P.A. *Chem. Commun.*, 2010, *46*, 3318–3320.

3

THEORETICAL HYDROGEN-BONDING ANALYSIS FOR ASSESSMENT OF PHYSICAL STABILITY OF PHARMACEUTICAL SOLID FORMS

YURIY A. ABRAMOV
Pfizer Worldwide Research & Development, Groton, CT, USA

3.1 INTRODUCTION

According to recent IUPAC recommendations [1] hydrogen bond (H-bond) is defined as "an attractive interaction between a hydrogen atom from a molecule or a molecular fragment X–H in which X is more electronegative than H and an atom or a group of atoms in the same or a different molecule, in which there is evidence of bond formation." A typical H-bond may be depicted as X–H···Y, where X–H designates an H-bond donor (HBD) group, while Y is an H-bond acceptor (HBA) atom. This very general definition covers a wide diversity of H-bonds from a very strong, having a partially covalent character, to a weak, having energies slightly exceeding van der Waals interactions [2, 3]. Overall H-bonding is considered to be one of the strongest and most specific (directional) nonbonding attractive intermolecular interaction and is responsible for most recognition phenomena in biological systems [4]. An importance of the H-bonding was early recognized by Pauling, who proposed that the secondary structure of proteins is determined by H-bonding between backbone amino acid residues [5]. Pharmaceutical small molecules typically contain a certain number of donors and acceptors [6] and therefore readily form H-bonding interactions between themselves and with surrounding media.

Computational Pharmaceutical Solid State Chemistry, First Edition. Edited by Yuriy A. Abramov.

Consequently, H-bonding impacts a wide range of properties critical to drug design and development, including potency [7, 8], selectivity [7, 9, 10], solubility [6, 11], partitioning [12], solid form crystallization [13], and physical stability [14, 15]. Therefore, the proper quantification of H-bond energies is of crucial importance for the understanding and prediction of properties of pharmaceutical systems.

H-bonding analysis for assessment of physical stability of pharmaceutical solid forms is the central focus of this chapter. Among other reported cases [16], the importance of thermodynamically stable form selection in the pharmaceutical industry can be illustrated by well-known examples of polymorph-induced impacts on marketed drugs Norvir® (ritonavir) and Neupro® (rotigotine patches). In the former case, Abbott Laboratories had to stop sales of Norvir in 1998 due to a failure in a dissolution test, which was caused by the precipitation of a more stable and less soluble form II of ritonavir [17, 18]. In the latter example, undesirable crystallization of a previously unknown stable form of rotigotine was found in the patches that were used to administer the drug. This newly discovered more stable form of rotigotine caused UCB to suspend the marketing of this drug in the United States [19].

A simplified classification of polymorph stability based on selected dominant interaction analyses has become a common practice. As an example, a good polymorph stability ranking of the systems, which are dominated by nonspecific van der Waals interactions, can be provided by the crystal density rule [20]. According to this rule a polymorph with a higher density is assumed to be more stable at 0 K than the other polymorph(s) with a lower density. However, pharmaceutical small molecules typically display H-bonding interactions in the solid forms. Since these are one of the strongest and the most specific intermolecular interactions, it is H-bonding, rather than van der Waals interactions, that plays a dominant role in both the crystallization kinetics [13] and thermodynamic stability of pharmaceutical solids [21]. As a result of the H-bonding contribution, pharmaceutical systems are known to violate the density rule. This inconsistency was recognized by Cambridge Crystallographic Data Centre (CCDC) scientists who developed a statistical knowledge-based approach for analyses of organic crystal stability [14, 15] based on statistical analysis of H-bonds in the Cambridge Structural Database (CSD) [22, 23]. Based on this analysis, a structure that adopts the strongest (highest propensity) H-bond(s) displays a low likelihood of the existence of a more stable form. In contrast, the presence of only weak(er) (low(er) propensity) H-bonds indicates a high risk of existence of a more stable form and that an additional study is likely to be required.

The results of knowledge-based calculations may depend on the availability of crystallographic observations for certain H-bonding functional groups in the CSD database. That is why, in order to obtain better statistics for model fitting, some functional groups are often presented in a generalized form, resulting in a loss of a specific nature of the H-bonding centers in the molecule of interest. This issue may be resolved by an independent theoretical approach, which would treat all H-bonding features explicitly to allow a good ranking of H-bonding propensities and would complement risk assessment of a likelihood of a missed stable form [24].

An outline of the chapter is as follows. A brief overview of available experimental scales of the H-bonding strength ranking will be discussed in Section 3.2. This will be followed by discussion of different theoretical approaches, which were developed to accurately reproduce experimental observations, as well as of results of high-level theoretical calculations (Section 3.3). Finally, applications of theoretical H-bonding analysis to pharmaceutical solid form selection will be presented in Section 3.4.

3.2 EXPERIMENTAL SCALES OF H-BONDING BASICITY AND ACIDITY

3.2.1 In Solution Phase

From experimental and theoretical points of view, the quantification of H-bonding energy strength is not straightforward since individual interaction between HBD and HBA groups defining H-bond cannot be uniquely separated from all other contributions to the total intermolecular interaction energy. Instead, it is a common practice to describe the H-bonding strength by the free energy ΔG^0_{HB} and equilibrium constant K_{HB} of the following complexation reaction, which involves only nonbonding interactions:

$$XH + Y \leftrightarrow XH^{\delta+} \cdots Y^{\delta-} \tag{3.1}$$

$$\Delta G^0_{HB} = -RT \ln K_{HB} \tag{3.2}$$

This approximation is more valid for simpler 1 : 1 donor–acceptor complexes with single H-bonding centers per molecule, especially for the cases where H-bonding interaction is more pronounced relative to contribution of all non-H-bonding groups of the interacting molecules.

Historically, Taft [25–27] and Arnett [28] pioneered application of thermodynamic data for building hydrogen-bond basicity and affinity scales. The work toward the same objective was later continued by groups of Abraham [29, 30], Laurence [31, 32], and Raevsky [33]. In order to determine donor (acidity) and acceptor (basicity) strength scales, Abraham et al. used linear free-energy relationships (LFERs) between experimentally available equilibrium constants for 1 : 1 donor–acceptor complexes in an inert solvent system CCl_4 [30, 34, 35]. The equilibrium constants for a set of 89 acids against 45 given reference bases allowed Abraham et al. to construct H-bond acidity scales, $\log K^H_A$ and α^H_2 [34, 36]. In a similar way, log$K$ constants for a set of 215 bases against 34 reference acid probes led to the construction of H-bond basicity scales, $\log K^H_B$ and β^H_2 [30, 35]. The α^H_2 and β^H_2 scales conveniently range from 0 to 1 and are related to $\log K^H_A$ and $\log K^H_B$ scales, respectively, by a simple linear transformation:

$$\alpha^H_2\left(\beta^H_2\right) = \frac{\log K^H_A\left(\log K^H_B\right) + 1.1}{4.636} \tag{3.3}$$

The overall solute H-bond acidity and basicity scales, as properties of the whole molecule with multiple H-bonding centers, were denoted as $A = \Sigma\alpha^H_2$ and $B = \Sigma\beta^H_2$, respectively [37, 38].

More recently, Laurence et al. reported the development of the pK_{BHX} H-bond basicity scale based on a set of 1338 experimental values related to 1164 HBAs [32]. This approach uses the H-bonding free energies determined in CCl_4 for a large number of chemically diverse H-bond acceptor molecules using 4-fluorophenol as the reference donor. The pK_{BHX} scale has a meaning similar to the $\log K_B^H$ H-bond basicity scale.

The aforementioned acidity (HBD strength) and basicity (HBA strength) experimental scales were derived from the solution state, and their transferability to a scale of the H-bonding interactions in organic crystals is typically assumed though should be proven. These experimental solution–based scales were useful for testing theoretical approaches for an accurate description and ranking of the H-bonding interaction energies. Some examples of such predictions will be discussed in Section 3.3.

3.2.2 In Solid-State Phase

A pharmaceutical molecule in a crystalline phase is necessarily involved in multiple simultaneous interactions with the surrounding molecules. As a result, the energy of H-bonds in the solid state cannot be directly measured [39]. The most relevant information about the geometry and propensity of H-bonding in pharmaceutical solids relies on experimentally observed (X-ray or neutron diffraction) crystal structures, which involve organic molecules with both HBA and HBD functional groups. From a large number of structures in the CSD database (nearly 700,000), it is possible to perform statistical surveys of H-bonded geometries and build H-bonding propensity (ranking) models [40].

A probabilistic approach to analyses of organic crystal stability was recently developed [14] based on statistical analysis of H-bonds in the CSD crystallographic database. This knowledge-based method analyzes potential H-bonding functional groups of a molecule to make predictions on statistically likely pairwise H-bonding interactions in the stable crystalline form. Model construction is carried out via logistic regression with a linear description of variable parameters to best reproduce all true and false H-bonding observations in the selected training set. The fitted H-bonding propensity (LHP) model is applied to compute a probability measure (propensity) for the formation of H-bond between a specified donor and acceptor atom of a molecule. The knowledge-based H-bond propensity models were used for testing theoretical approaches discussed in Sections 3.3 and 3.4.

3.3 THEORETICAL STUDY OF H-BONDING STRENGTH IN SOLUTION AND IN SOLID STATE

The availability of the experimental H-bonding scales enabled the optimization and selection of different theoretical approaches for an optimal H-bond strength description. Here we briefly cover the most important approaches in the application to H-bonding in gas phase or solution, as well as in solid state.

3.3.1 Supermolecular Approach

For quantitative description of H-bond interactions in solution or in gas phase, it is a common practice to define the H-bond energy by intermolecular interaction energy. In the supermolecular approach, H-bonding energies are described by the free energy ΔG_{HB} of the complexation reaction (3.1), which is estimated according to the following equation:

$$\Delta G_{HB} = \Delta G_{AD} - (\Delta G_A + \Delta G_D) + ZPE + BSSE \tag{3.4}$$

Here ΔG_{HB} is the total interaction free energy of two molecules in solution, ΔG_{AD} is the free energy of the geometry optimized H-bonded dimer, and ΔG_A and ΔG_D are the free energies of the geometry optimized individual H-bond acceptor and donor molecules, respectively. Zero-point energy (ZPE) correction summarizes the electronic ground-state energies of nucleus' vibration at an absolute zero temperature. Basis set superposition error (BSSE) correction tends to remove an error introduced by an effectively larger basis set of each molecule in the complex relative to that in the unbound state [41]. An entropic contribution to the ΔG_{HB} in Equation 3.4 is typically introduced by the ZPE correction only, as well as by solvation energy terms in case the calculations are performed in a solution state.

Though in quantum chemical calculations unconstrained geometry relaxation of the individual HBD and HBA molecules, as well as ZPE and BSEE corrections, is thought to be needed, it was demonstrated that all or some of these contributions may be neglected providing a similar or even better correlation with the experimental results [42–45]. Overall, there are many reports of supermolecular calculations that were able to predict the experimental H-bond strength in solution or gas phase with a reasonable accuracy. For example, in a recent study Rahaman et al. [44] demonstrated that the regression analysis suggested by Abraham et al. for constructing the HBD (α_2^H) and HBA (β_2^H) strength scales from experimental data can be applied to derive analogous H-bond donor and acceptor strength scales using gas phase energies from supermolecular calculations at MP2/6-31++G(d,p) level of theory. The accuracy of the method in predicting Abraham's α_2^H and β_2^H parameters was quite good, displaying root mean squared deviations (RMSDs) of only 0.0693 and 0.0677, respectively.

In summary, the supermolecular approach provides a reasonable good accuracy of H-bond strength prediction and could be used together with the experimental scales as a benchmark for descriptor-based predictions discussed in the following section.

3.3.2 Descriptor-Based Approaches

Despite of successes of the supermolecular approach in the H-bond strength prediction, it has the following limitations. The approach is quite expensive and for realistic pharmaceutical systems it would require sampling of different binding conformations of the supermolecular complex. Therefore, simplified cheaper approaches, which are based on a correlation of certain molecular descriptor or descriptors combination with the experimental (or supermolecular derived) H-bonding strength, are attractive.

Historically, multiple theoretical descriptor–based approaches to H-bond strength ranking were proposed. That includes approaches based on group contribution method [46], electrostatic potentials [47], electrophilic superdelocalizability and self-atom polarizability [48], Quantum Theory of Atoms In Molecules (QTAIM) descriptors [49–51], the two-center shared electron number σ and the product of ionization potential [45, 52], and local quantum mechanical molecular parameters, which quantify electrostatic, polarizability, and charge transfer contributions to H-bonding [53, 54].

An alternative approach [24, 55, 56] of H-bonding ranking is based on COSMO-RS fluid thermodynamics theory [57]. Abramov [24, 58] performed a study of the correlation of two independent schemes of atomic charge calculations (polarization σ H-bonding (σ_{HB}) charges and electrostatic potential (ESP) fitted charges, Q_{esp}) with Abraham's experimental H-bonding scales α_2^H and β_2^H, and with some available results from theoretical supermolecular calculations [59]. The COSMO-RS σ_{HB} charges are based on the acceptor or donor atomic polarization charge density in the infinite dielectric above or below, respectively, a threshold value. These are surface charges that directly reflect both polarization and surface accessibility effects. Q_{esp} are partial atomic point charges, which are fitted to better reproduce the molecular electrostatic potential at the molecular surface [60]. The ESP-type charges are popular in force field development to better describe electrostatic intermolecular interactions [61].

It was demonstrated that a nice ranking of H-bonding interaction energies may be performed by a product of the σ_{HB} charges of the interacting donor and acceptor atoms [24]. As an example, correlations of theoretical H-bond interaction energies [59] between substituted phenols and neutral or ionic probes with phenolic hydroxyl donor and acceptor charges are presented in Figure 3.1. The best correlations were found between the σ_{HB} charges of the acceptor and donor and the interaction energies of substituted phenols with the ionic probes (Fig. 3.1c and d), for which H-bonding contributions are expected to be the highest. At the same time the Q_{esp} atomic charges provided a relatively poor ranking of all considered intermolecular interactions (Fig. 3.1a and b). Klamt et al. reported a systematic study, which demonstrates a very good correlation of 2465 DFT/COSMO H-bond energies derived from the supermolecular calculations with a product of polarization σ charge densities of donor and acceptor atoms [55] (Fig. 3.2). The fitted relationship yielded a root mean square deviation (RMSD) of only 0.36 kcal/mol. These results provided an excellent justification of the applicability of COSMO-RS fluid thermodynamics theory for H-bonding propensity ranking.

3.3.3 Solid-State H-bonding Strength

In general, both supermolecular and descriptor-based approaches, discussed earlier, could be applied to the study of H-bonding strength in a crystalline solid state. An advantage of the latter approaches is that they typically do not require actual crystal structure information to rank propensities of the H-bonding interactions in the solid state. This consideration makes descriptor-based approaches the most appropriate for

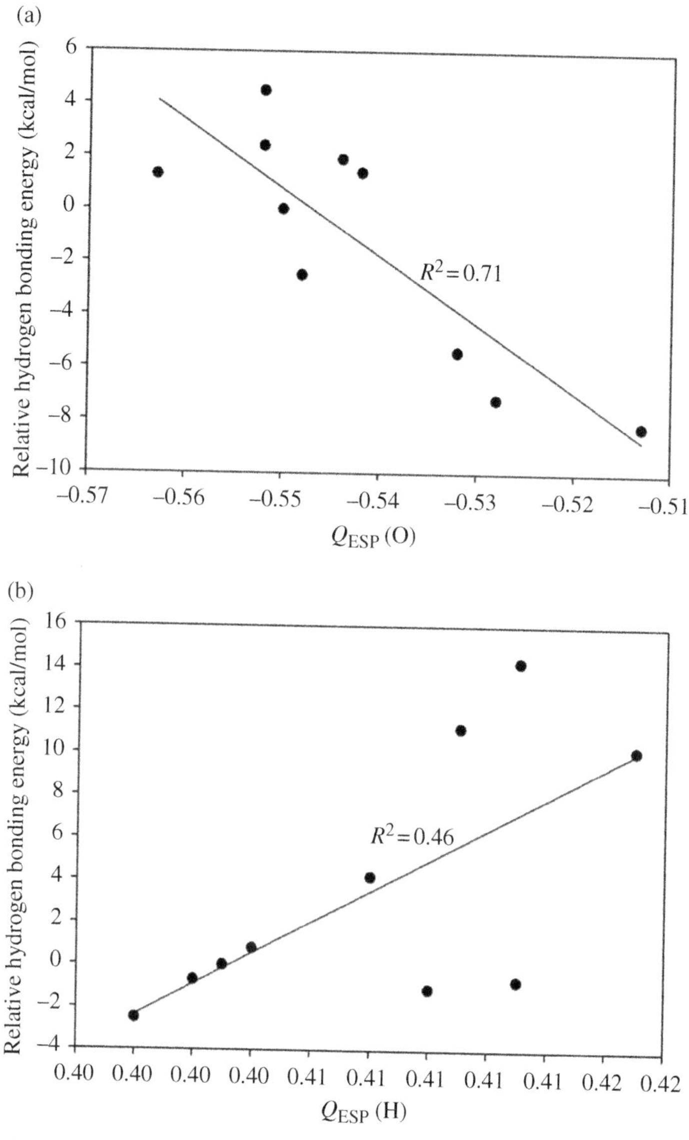

FIGURE 3.1 Correlations of theoretical H-bonding interaction energies [59] between substituted phenols and neutral or ionic probes with the charges of the H-bonding centers on phenolic hydroxyl (O acceptor and (O)H donor). Energies are presented relative to the unsubstituted system. ESP charges are used to describe interactions of substituted phenols with protonated methylamine and acetate anion probes in (a) and (b), respectively. H-bonding interactions of substituted phenols with methanol (▲) and protonated methylamine (•) donor probes are ranked by σ_{HB} acceptor charges in (c). H-bonding interactions of substituted phenols with formaldehyde (▲) and acetate anion (•) acceptor probes are described by σ_{HB} donor charges in (d). ESP and σ_{HB} charges were calculated at PBE/DNP and PBE/DNP/COSMO levels of theory, respectively. Source: Adapted from Abramov [24]. Reproduced with permission of American Chemical Society.

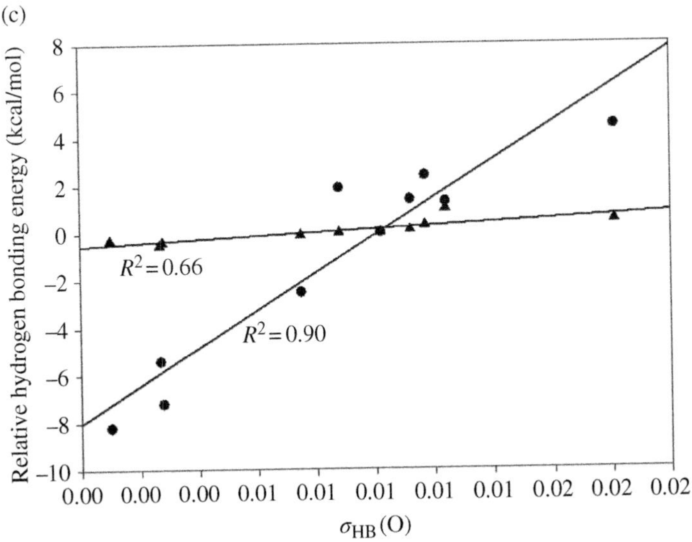

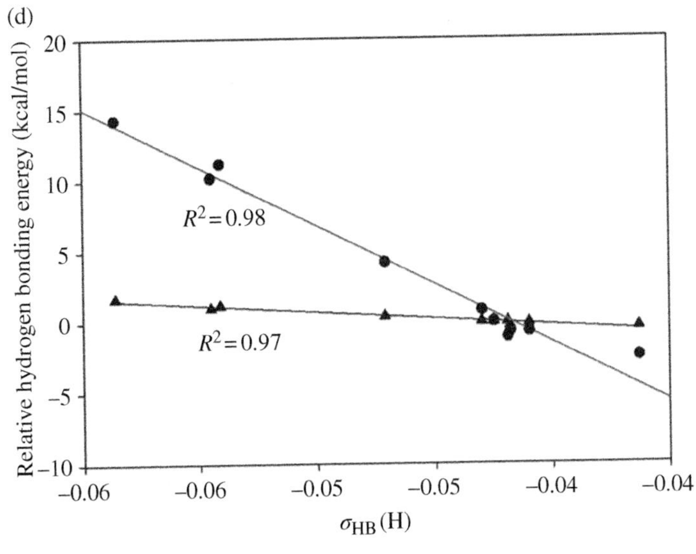

FIGURE 3.1 (Continued)

risk assessment of the existence of an unknown more stable form of a pharmaceutical compound. Two of the descriptor-based approaches (COSMO-RS σ_{HB} and QTAIM) will be considered in more detail in this section.

In order to validate the applicability of the COSMO-RS approach to H-bond strength ranking in molecular crystals, Abramov calculated σ_{HB} charges for 86 organic molecules from the CSD database. It is expected from general consideration as well as from the knowledge-based approach findings [14, 15] that one should observe a high probability

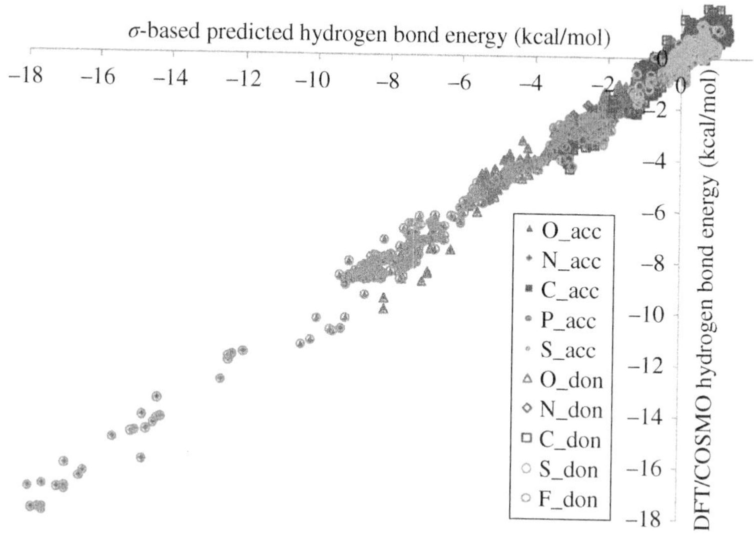

FIGURE 3.2 Scatter plot of the DFT/COSMO H-bonding energy versus the predicted HB energy based on a product of donor and acceptor polarization charge densities. Source: Adapted from Klamt et al. [55]. Reproduced with permission of Royal Society of Chemistry. (*see insert for color representation of the figure.*)

of pairing of the strongest HBDs and HBAs, assuming that majority of these crystals correspond to the stable forms. H-bonding energy between the strongest acceptor (the most positive σ_{HB} charge) and strongest donor (the most negative σ_{HB} charge of donating H) centers of interacting molecules in each crystal was represented by a product of the corresponding σ_{HB} charges. The lower the value of the product of these charges, the higher the H-bonding energy. It was found that the overall probability of paring of the strongest donor and acceptor is quite high exceeding 70% (Fig. 3.3). Moreover, the probability of the strongest HBD and HBA paring dramatically increases with the increase of the strength of the H-bonding interaction (Fig. 3.3). Therefore, it was demonstrated that σ_{HB} charge analysis is a valid tool for H-bonding propensity evaluation in molecular crystals, and it could be used for analysis of a likelihood of a missing stable form as a complementing approach to knowledge-based analysis, crystal structure prediction (CSP), and other computational techniques (Section 3.4) [24].

Perhaps, the most popular descriptor-based approaches to estimate H-bonding energies in crystals are based on QTAIM theory [62, 63]. Within this theory, a bonding interaction is defined by a maximum electron density path (bond path) connecting two interacting atoms. A saddle point in the charge density along the bond path is called a bond critical point (BCP). The existence of BCP between the acceptor atom and the donating hydrogen, as well as the charge density properties at this point, defines the major criteria of the H-bond existence [64]. It was demonstrated [49, 65] that H-bonding energy is proportional to multiple properties at the BCP, including electronic potential energy density, V_b [65]; electronic kinetic energy

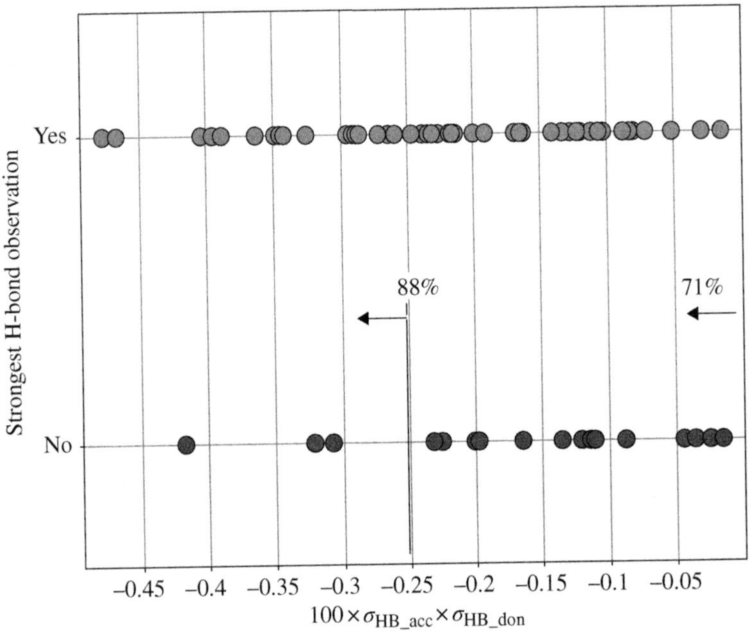

FIGURE 3.3 A scatter plot of true (gray) and false (black) observations of H-bonding between the strongest acceptor and donor centers in 86 molecular crystals versus the strength of H-bonding interaction, as measured by a product of the corresponding σ_{HB} charges.

density, G_b [49]; and electron density, ρ_b [49, 51]. Espinosa et al. [65] suggested to estimate the energy of weak and moderate H-bonding interactions in molecular crystals displaying A=H···O bonds (A=O,N,C) as follows (in a.u.):

$$E_{HB} \sim -0.5V_b \tag{3.5}$$

Here E_{HB} is a dissociation energy of the H-bonded complex, defined as the enthalpy difference between the monomers and the complex.

Mata et al. [49] have recently proposed a relationship between H-bonding energy and electronic kinetic energy density at the BCP based on gas phase studies of of FH…FR (R = H, Li, Al,Cl, CCH) complexes (in a.u.):

$$E_{HB} \sim 0.429G_b \tag{3.6}$$

In the same study, a relationship between E_{HB} and charge density at the BCP was defined as (in a.u.) [49]:

$$E_{HB} \sim 0.163\rho_b \tag{3.7}$$

Verner et al. [66] performed periodic boundary conditions predictions at B3LYP/6-311G** level of theory for 18 molecular crystals containing 28 intermolecular O–H···O H-bonds to test accuracy and applicability of the proposed linear relationships between E_{HB} and V_b and G_b. The authors concluded that relationship (3.5) is applicable only for weak and moderate H-bonds (E_{HB} < 45 kJ/mol) overestimating the energies by approximately 20%. At the same time an alternative relationship (3.6) is applicable even for strong H-bonds (E_{HB} < 85 kJ/mol) overestimating the E_{HB} values only by approximately 10%.

These two linear relationships between E_{HB} and V_b and G_b appear to be quite simple (BSSE-free) and useful approximations, which enable the evaluation of the H-bonding energy in solid state using theoretical [66] or experimental (derived from X-ray or synchrotron diffraction experiment [67, 68]) energy densities. In the latter case the electronic kinetic energy density distribution, G_b, is derived from the following accurate approximation for closed-shell interactions (like H-bond is) in terms of experimental electron density, ρ, and its Laplacian $\nabla^2\rho$ at the BCP (in a.u.) [69]:

$$G_b = \left(\frac{3}{10}\right)\left(3\pi^2\right)^{2/3}\rho^{5/3} + \left(\frac{1}{6}\right)\nabla^2\rho \tag{3.8}$$

Using Equation 3.7 and the local form of the virial theorem [62] (in a.u.),

$$2G + V = \left(\frac{1}{4}\right)\nabla^2\rho, \tag{3.9}$$

the electronic potential energy density (V_b) can be also estimated at intermolecular BCPs as follows [62, 65] (in a.u.):

$$V_b = \left(\frac{1}{4}\right)\nabla^2\rho - 2G_b \tag{3.10}$$

However, while QTAIM-based approaches (3.5) and (3.6) allow H-bonding strength ranking in molecular crystals, they require actual H-bonding geometry information to perform the predictions. Therefore, the drawback of these descriptor-based methods is that they cannot be directly used for the assessment of the likelihood of an unknown missed stable form.

3.4 APPLICATION TO SOLID FORM SELECTION

The H-bonding propensity analysis is one of the major computational approaches that are used to complement experimental polymorph screening by facilitating the assessment of the likelihood of a missed stable form [24]. The polymorphic stability classification by the model is based on the rationale of best donor/best acceptor pairing and the corresponding hierarchy of observed and absent H-bonds in a particular case. Based on this approach, a structure that adopts the highest propensity H-bond(s) reflects a low likelihood of a missed stable form. In contrast, the lack of strongest donor and acceptor pairing indicates a high likelihood of a different stable form and an additional polymorph screening is likely to be required.

While many of the earlier discussed computational methods could be potentially used for the H-bonding propensity prediction, currently only the σ_{HB} charge analysis was systematically applied to support pharmaceutical solid form selection [24, 70]. Several examples of application of H-bonding propensity analysis to support solid form selection are considered later, including three pharmaceutical cases—axitinib, crizotinib, and ritonavir. All σ_{HB} charges were calculated by COSMOtherm software (Eckert, F.; Klamt, A. COSMOtherm, version C3.0, release 01.12; COSMOlogic GmbH & Co. KG: Leverkusen, Germany, 2012) using polarization surface charge densities predicted at the PBE/DNP/COSMO level of theory [71, 72].

3.4.1 Examples of Theoretical H-Bonding Analysis to Support Solid Form Selection

Axitinib. Axitinib (Inlyta®) is an oncology drug that inhibits vascular endothelial growth factor (VEGF). Axitinib has been observed to form five anhydrous conformational polymorphs and numerous solvates [70]. An application of H-bonding propensity analysis based on σ_{HB} charges to support the solid form selection of axitinib was reported by Chekal et al. [70a] and Campeta et al. [70b]. The resulting σ_{HB} charges are presented in Figure 3.4a. According to these results, the strongest donor–acceptor pair for this molecule should be the pyrazole amine–amide oxygen H-bond. This result is in excellent agreement with the predictions based on the knowledge-based model described earlier [24]. In fact, pyrazole amine–amide oxygen H-bond was observed in all known axitinib crystal structures, including stable form XLI. Therefore, in case of axitinib the σ_{HB} charges analysis by itself does not allow the stability ranking of the five known anhydrous forms. This example outlines a limitation of H-bonding propensity analysis, which is addressed in more detail in Section 3.4.2.

Ritonavir. Ritonavir (Norvir®) is an antiretroviral drug from the protease inhibitor class used to treat HIV infection and AIDS. σ_{HB} charges of ritonavir, presented in Figure 3.4b, suggest that the strongest H-bonding pairing should take place between the ureido acceptor and hydroxyl donor, followed by the amide acceptor and carbamate donor. These favorable H-bonds are absent in the metastable form I, while they do take place in the stable form II. Therefore, a retrospective σ_{HB} analysis properly indicates the high risk of form I selection as a commercial solid form of ritonavir. It should be noted that H-bonding propensity ranking based on this analysis disagrees with the results of the reported knowledge-based model [15] though both approaches indicate a high likelihood of form I as a metastable form.

Crizotinib. Crizotinib (Xalcory®) is an ALK/C-MET receptor mediator developed for the treatment of ALK-positive non-small-cell lung cancer. The anhydrous free base form A is the only crystalline form found after extensive polymorph and hydrate screening. H-bonding propensity analyses described later was performed to complement the experimental polymorph screening to ensure that no alternative stable anhydrous form was missed [24]. The σ_{HB} charges for crizotinib are presented in Figure 3.4c. The strongest donor is the primary amine and the strongest acceptors are piperidine and pyridine nitrogens. Strong H-bonding interactions between these

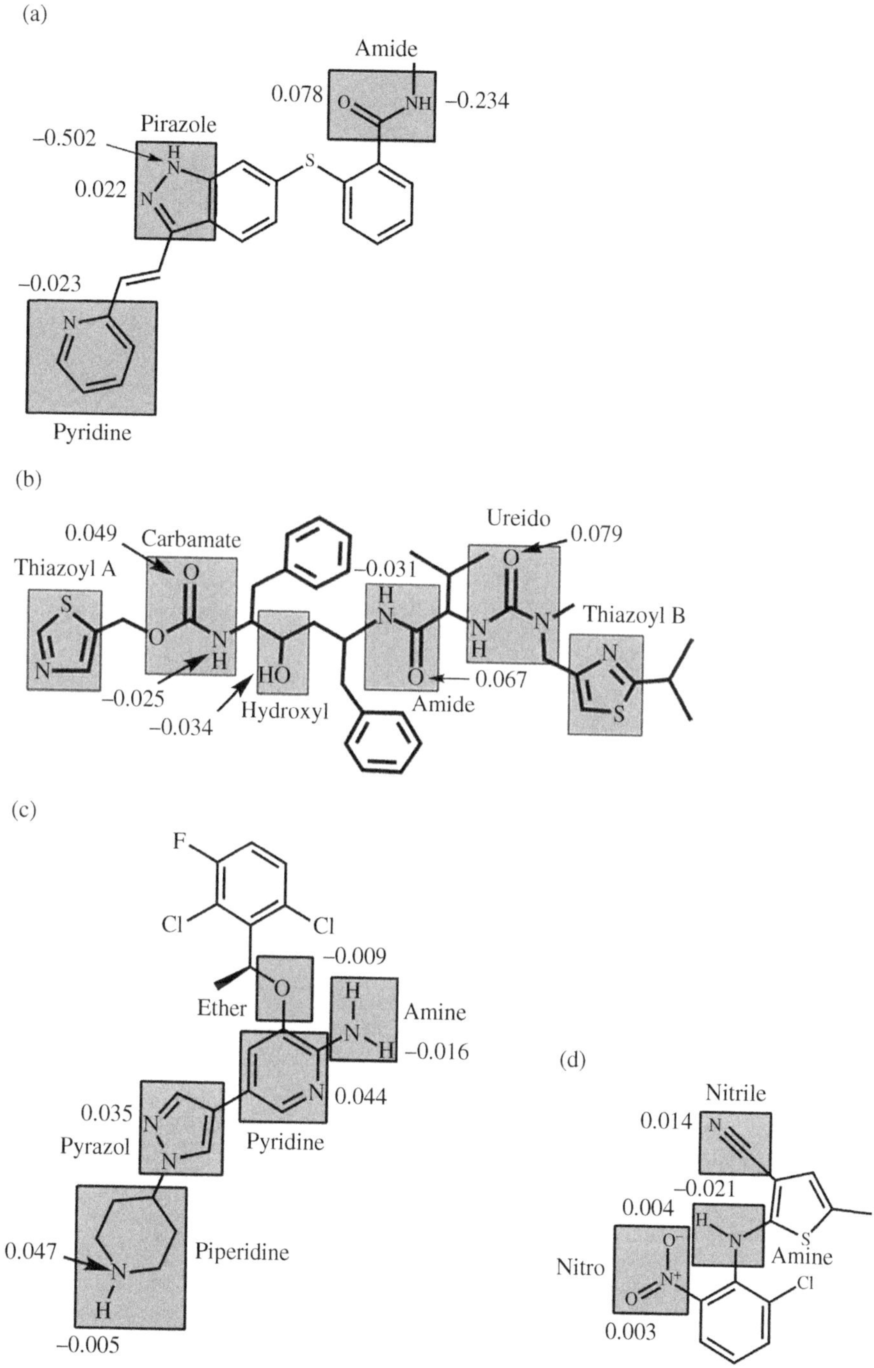

FIGURE 3.4 Examples of theoretical H-bonding propensity analysis for axitinib (a), ritonavir (b), crizotinib (c), and ROY (d). The σ_{HB} charges were calculated at the BP/TZVP/COSMO level of theory. Charges of only up to three pairs of the strongest donors and acceptors are presented for clarity. Source: Adapted from Abramov [24]. Reproduced with permission of American Chemical Society.

acceptors and the primary amine were indeed observed in crystal structure of form A. In addition, weak H-bonds are formed between the ether acceptor and primary and secondary amine donors. The pyrazole ring was not predicted to be a strong acceptor and is not involved in H-bonding in form A. The top-ranked H-bonding interactions predicted by the σ_{HB} analysis are in a good overall agreement with the knowledge-based predictions [24]. In result, the H-bonding propensity analysis demonstrates that the strongest H-bonding interactions are satisfied in form A of crizotinib, supporting a low likelihood of finding a more stable polymorph.

ROY. 5-Methyl-2-[(2-nitrophenyl)amino]-3-thiophenecarbonitrile (ROY) is perhaps the most famous nonpharmaceutical example of multiple polymorphs, existing at ambient temperature [73, 74]. The σ_{HB} charge analysis (Fig. 3.4d) demonstrates that cyano nitrogen is the strongest acceptor in the molecule, which has a preference of intermolecular H-bonding interaction with the amino donor in the stable form of ROY. This analysis is supported by experimental observation that the most stable yellow polymorph (Y) is indeed stabilized by such a H-bond in contrast to metastable YN, ON, OP, and R forms [74].

3.4.2 Consideration of Limitations of Hydrogen-Bonding Propensity Approach

There are several important limitations in stability ranking based on H-bonding propensity analysis, in addition to the fact that H-bonding is not the only contribution to the stability of drug-like crystals. Generally, the models based on the propensity of the H-bonding interaction cannot account for the enantiotropic relationship between the polymorphic forms at ambient temperature [24]. In addition, an inherent limitation of the H-bonding propensity models is the inability to estimate the stability ranking of the polymorphs displaying similar H-bonding patterns in the solid state. There are several known examples of the drug-like molecules, which preserve at least the strongest H-bonds for two or more polymorphic forms. Among them are five anhydrous forms of axitinib [70], rotigotine forms I and II [75], paracetamol forms I and II [76, 77], nimodipine forms I and II [78, 79], and four anhydrous forms of carbamazepine [80].

One of the possible ways to address this issue is to combine a CSP study and H-bonding propensity analysis. For example, the final selection of three predicted structures of compound XX in the fifth blind test was performed based on both the lowest lattice energy and knowledge-based approach considerations [81]. Alternatively, the results of the initial crystal packing generation, which do not allow the preferred strongest H-bonding interactions, could be filtered out from a following higher level CSP analysis. In that case, the final polymorph ranking will be based on lattice energy minimization results while fully satisfying H-bonding propensity expectations.

In an attempt to overcome the earlier limitation, it was also suggested to take an extra step beyond H-bonding propensity analysis and focus on the estimation of the relative strength of crystallographically observed intermolecular H-bonds in different polymorphic forms of drug-like molecules [82]. For that, a prediction of the most stable form may be performed based on the relative strength of the strongest H-bonding interaction, as measured by QTAIM descriptors at the corresponding

BCP (Eqs. 3.5–3.7). For example, linear relationships between ΔH_{fus} data and V_b, G_b, and ρ_b properties of the H-bond between the pyrazole amine and amide oxygen were found for four axitinib polymorphs (Fig. 3.5). The found relationships describe reasonably well the actual rank order of the experimental stabilities of the anhydrous

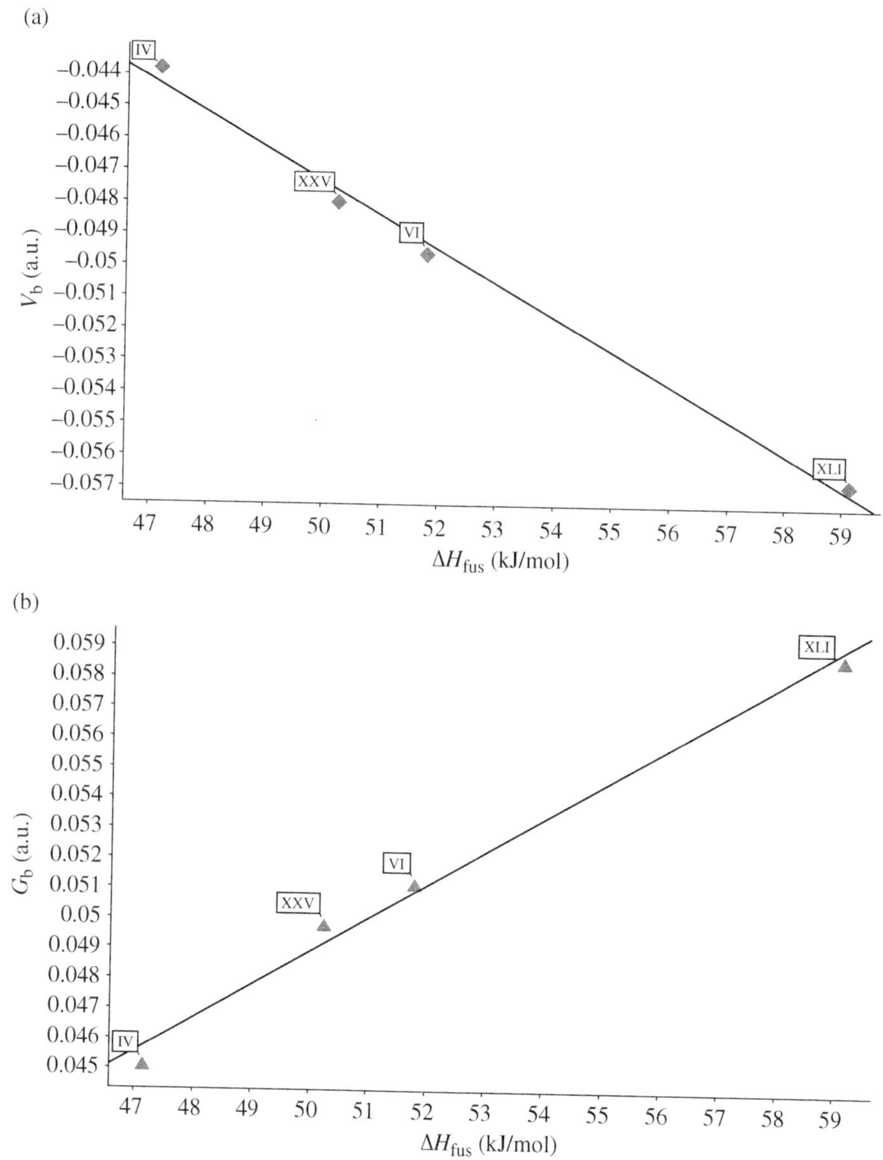

FIGURE 3.5 Correlation between heats of fusion of axitinib polymorphs and relative strength of the corresponding H-bonding interactions between pyrazole amine and amide oxygen, as measured by the QTAIM descriptors V_b (a), G_b (b), and ρ_b (c). Source: Adapted from Abramov [82]. Reproduced with permission of American Chemical Society.

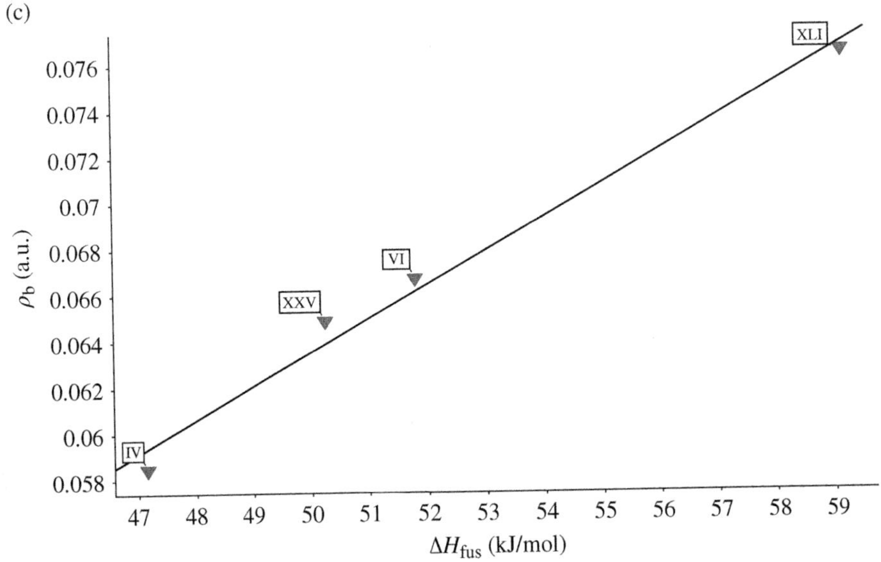

FIGURE 3.5 (Continued)

forms at ambient temperature, XLI>XXV≥VI>IV>I [70b]. This observation reflects the dominant contribution of the strongest H-bonds to relative stabilities of the related polymorphs. The proposed QTAIM descriptor-based approach may help predicting the relative stabilities of drug polymorphs from experimentally observed or computationally generated crystal structures.

3.5 CONCLUSION

The selection of the commercial solid form is a key deliverable in the pharmaceutical industry. Changes in polymorphic form during pharmaceutical development can have a negative impact on a drug's performance: solubility and bioavailability, chemical and physical stability, and mechanical properties. Therefore, it is generally desirable to identify the thermodynamically stable crystal form under normal manufacturing and storage conditions to ensure that the form does not change during the life cycle of the drug product. In practice, an experimental polymorph screening is performed to identify thermodynamically stable crystal form. In addition, different computational approaches are applied to complement and guide stable form screening in order to assess the likelihood of missing a stable form at early stages of drug development. These approaches cover H-bonding propensity analysis, CSP, risk of conformational polymorphism, and rational solvent selection for stable form screening [24]. In order to mitigate limitations of each of these approaches, it is recommended to perform a general computational support of solid form selection based on a combination of all of these tools [24].

This chapter reviewed only H-bond propensity analysis approaches to complement polymorph screening by facilitating the risk assessment of a missed more stable form. A focus was given to different theoretical approaches of H-bonding ranking, which were developed to accurately reproduce experimental observations as well as high-level theoretical calculations. Successful applications of COSMO-RS and QTAIM-based methods to support the pharmaceutical solid form selection were demonstrated.

Despite the great progress that has been made in the H-bonding propensity analysis field, there is still further development that can be undertaken to locate the best approaches and to better understand limitations of their applicability.

ACKNOWLEDGMENT

The author is grateful to Mr. Brian Samas for valuable discussions.

REFERENCES

[1] Arunan, E.; Desiraju, G. R.; Klein, R. A.; Sadlej, J.; Scheiner, S.; Alkorta, I.; Clary, D. C.; Crabtree, R. H.; Dannenberg, J. J.; Hobza, P., *Pure Appl. Chem.* 2011, *83*, 1637–1641.

[2] Jeffrey, G. A. *An Introduction to Hydrogen Bonding*. Oxford University Press: New York, 1997.

[3] Desiraju, G. R.; Steiner, T., *The Weak Hydrogen Bond in Structural Chemistry and Biology*. Oxford University Press: New York, 2001.

[4] Saenger, W.; Jeffrey, G., *Hydrogen Bonding in Biological Structures*. Springer-Verlag: Berlin, 1991.

[5] Pauling, L.; Corey, R. B.; Branson, H. R., *Proc. Natl. Acad. Sci. U. S. A.* 1951, *37*, 205–211.

[6] Lipinski, C. A.; Lombardo, F.; Dominy, B. W.; Feeney, P. J., *Adv. Drug Deliv. Rev.* 2012, *64*, 4–17.

[7] Fersht, A. R.; Shi, J.; Knill-Jones, J.; Lowe, D.; Wilkinson, A.; Blow, D.; Brick, P.; Carter, P.; Waye, M.; Winter, G., *Nature* 1985, *314*, 235–238.

[8] Williams, M.; Ladbury, J. Hydrogen Bonds in Protein–Ligand Complexes, In *Protein–Ligand Interactions: From Molecular Recognition to Drug Design*; Böhm, H.-J., Schneider, G. Eds; Wiley-VCH: Weinheim, pp 137–161.

[9] Brzozowski, A. M.; Pike, A. C.; Dauter, Z.; Hubbard, R. E.; Bonn, T.; Engström, O.; Öhman, L.; Greene, G. L.; Gustafsson, J.-Å.; Carlquist, M., *Nature* 1997, *389*, 753–758.

[10] Williams, S. P.; Sigler, P. B., *Nature* 1998, *393*, 392–396.

[11] Abraham, M. H.; Ibrahim, A.; Zissimos, A. M.; Zhao, Y. H.; Comer, J.; Reynolds, D. P., *Drug Discov. Today* 2002, *7*, 1056–1063.

[12] Shalaeva, M.; Caron, G.; Abramov, Y. A.; O'Connell, T. N.; Plummer, M. S.; Yalamanchi, G.; Farley, K. A.; Goetz, G. H.; Philippe, L.; Shapiro, M. J., *J. Med. Chem.* 2013, *56*, 4870–4879.

[13] Cui, P.; Zhang, X.; Yin, Q.; Gong, J., *Ind. Eng. Chem. Res.* 2012, *51*, 13663–13669.

[14] Galek, P. T.; Fabian, L.; Motherwell, W. S.; Allen, F. H.; Feeder, N., *Acta Crystallogr. Sect. B: Struct. Sci.* 2007, *63*, 768–782.

[15] Galek, P. T.; Allen, F. H.; Fábián, L.; Feeder, N., *CrystEngComm* 2009, *11*, 2634–2639.

[16] Lee, A. Y.; Erdemir, D.; Myerson, A. S., *Annu Rev. Chem. Biomol. Eng.* 2011, *2*, 259–280.

[17] Bauer, J.; Spanton, S.; Henry, R.; Quick, J.; Dziki, W.; Porter, W.; Morris, J., *Pharm. Res.* 2001, *18*, 859–866.

[18] Kempf, D. J.; Marsh, K. C.; Denissen, J. F.; McDonald, E.; Vasavanonda, S.; Flentge, C. A.; Green, B. E.; Fino, L.; Park, C. H.; Kong, X.-P., *Proc. Natl. Acad. Sci. U. S. A.* 1995, *92*, 2484–2488.

[19] Rascol, O.; Perez-Lloret, S., *Expert Opin. Pharmacother.* 2009, *10*, 677–691.

[20] Burger, A.; Ramberger, R., *Microchim. Acta* 1979, *72*, 259–271.

[21] Bernstein, J., *Polymorphism in Molecular Crystals*. Oxford University Press: Oxford, 2007.

[22] Allen, F. H., *Acta Crystallogr. Sect. B: Struct. Sci.* 2002, *58*, 380–388.

[23] Groom, C. R.; Allen, F. H., *Angew. Chem. Int. Ed.* 2014, *53*, 662–671.

[24] Abramov, Y. A., *Org. Process Res. Dev.* 2012, *17*, 472–485.

[25] Gurka, D.; Taft, R. W., *J. Am. Chem. Soc.* 1969, *91*, 4794–4801.

[26] Taft, R. W.; Gurka, D.; Joris, L.; Von Schleyer, P. R.; Rakshys, J., *J. Am. Chem. Soc.* 1969, *91*, 4801–4808.

[27] Joris, L.; Mitsky, J.; Taft, R., *J. Am. Chem. Soc.* 1972, *94*, 3438–3442.

[28] (a) Arnett, E. M.; Joris, L.; Mitchell, E.; Murty, T.; Gorrie, T.; Von Schleyer, P. R., *J. Am. Chem. Soc.* 1970, *92*, 2365–2377; (b) Arnett, E. M.; Mitchell, E. J.; Murty, T., *J. Am. Chem. Soc.* 1974, *96*, 3875–3891.

[29] Abraham, M. H.; Duce, P. P.; Prior, D. V.; Barratt, D. G.; Morris, J. J.; Taylor, P. J., *J. Chem. Soc., Perkin Trans. 2* 1989 (10), 1355–1375.

[30] Abraham, M. H.; Grellier, P. L.; Prior, D. V.; Morris, J. J.; Taylor, P. J., *J. Chem. Soc., Perkin Trans. 2* 1990 (4), 521–529.

[31] Laurence, C.; Berthelot, M., *Perspect. Drug Discovery Des.* 2000, *18*, 39–60.

[32] Laurence, C.; Brameld, K. A.; Graton, J.; Le Questel, J.-Y.; Renault, E., *J. Med. Chem.* 2009, *52*, 4073–4086.

[33] Raevsky, O., *J. Phys. Org. Chem.* 1997, *10*, 405–413.

[34] Abraham, M. H.; Grellier, P. L.; Prior, D. V.; Duce, P. P.; Morris, J. J.; Taylor, P. J., *J. Chem. Soc., Perkin Trans. 2* 1989 (6), 699–711.

[35] Abraham, M. H.; Greillier, P. L.; Prior, D. V.; Morris, J. J.; Taylor, P. J.; Laurence, C.; Berthelot, M., *Tetrahedron Lett.* 1989, *30*, 2571–2574.

[36] Abraham, M. R.; Duce, P. P.; Grellier, P. L.; Prior, D. V.; Morris, J. J.; Taylor, P. J., *Tetrahedron Lett.* 1988, *29*, 1587–1590.

[37] Abraham, M. H., *Chem. Soc. Rev.* 1993, *22*, 73–83.

[38] Abraham, M. H.; Du, C. M.; Platts, J. A., *J. Org. Chem.* 2000, *65*, 7114–7118.

[39] Steiner, T., *Angew. Chem. Int. Ed.* 2002, *41*, 48–76.

[40] Macrae, C. F.; Bruno, I. J.; Chisholm, J. A.; Edgington, P. R.; McCabe, P.; Pidcock, E.; Rodriguez-Monge, L.; Taylor, R.; van de Streek, J.; Wood, P. A., *J. Appl. Crystallogr.* 2008, *41*, 466–470.

[41] Boys, S. F.; Bernardi, F., *Mol. Phys.* 1970, *19*, 553–566.

[42] Hao, M.-H., *J. Chem. Theory Comput.* 2006, *2*, 863–872.

[43] Nocker, M.; Handschuh, S.; Tautermann, C.; Liedl, K. R., *J. Chem. Inf. Model.* 2009, *49*, 2067–2076.

[44] Rahaman, O.; Doren, D. J.; Di Toro, D. M., *J. Phys. Org. Chem.* 2014, *27*, 783–793.

[45] Wendler, K.; Thar, J.; Zahn, S.; Kirchner, B., *J. Phys. Chem. A* 2010, *114*, 9529–9536.

[46] Platts, J. A.; Butina, D.; Abraham, M. H.; Hersey, A., *J. Chem. Inf. Comput. Sci.* 1999, *39*, 835–845.

[47] Kenny, P. W., *J. Chem. Inf. Model.* 2009, *49*, 1234–1244.

[48] Gancia, E.; Montana, J. G.; Manallack, D. T., *J. Mol. Graph. Model.* 2001, *19*, 349–362.

[49] Mata, I.; Alkorta, I.; Espinosa, E.; Molins, E., *Chem. Phys. Lett.* 2011, *507*, 185–189.

[50] Green, A. J.; Popelier, P. L., *J. Chem. Inf. Model.* 2014, *54*, 553–561.

[51] Ayoub, A. T.; Tuszynski, J.; Klobukowski, M., *Theor. Chem. Acc.* 2014, *133*, 1–7.

[52] Thar, J.; Kirchner, B., *J. Phys. Chem. A* 2006, *110*, 4229–4237.

[53] Schwöbel, J.; Ebert, R.-U.; Kühne, R.; Schüürmann, G., *J. Chem. Inf. Model.* 2009, *49*, 956–962.

[54] Schwöbel, J. A.; Ebert, R. U.; Kühne, R.; Schüürmann, G., *J. Phys. Org. Chem.* 2011, *24*, 1072–1080.

[55] Klamt, A.; Reinisch, J.; Eckert, F.; Hellweg, A.; Diedenhofen, M., *Phys. Chem. Chem. Phys.* 2012, *14*, 955–963.

[56] Klamt, A.; Reinisch, J.; Eckert, F.; Graton, J.; Le Questel, J.-Y., *Phys. Chem. Chem. Phys.* 2013, *15*, 7147–7154.

[57] Klamt, A., *COSMO-RS: From Quantum Chemistry to Fluid Phase Thermodynamics and Drug Design*. Elsevier: Amsterdam, 2005.

[58] Abramov, Y. A.; Mustakis, J.; am Ende, D. J., *COSMO-RS Symposium*, Wermelskirchen, Germany, March 31–April 1, 2009.

[59] Reynisson, J.; Mcdonald, E., *J. Comput. Aided Mol. Des.* 2004, *18*, 421–431.

[60] Singh, U. C.; Kollman, P. A., *J. Comput. Chem.* 1984, *5*, 129–145.

[61] Cornell, W. D.; Cieplak, P.; Bayly, C. I.; Gould, I. R.; Merz, K. M.; Ferguson, D. M.; Spellmeyer, D. C.; Fox, T.; Caldwell, J. W.; Kollman, P. A., *J. Am. Chem. Soc.* 1995, *117*, 5179–5197.

[62] Bader, R. F. W. *Atoms in Molecules: A Quantum Theory*. Oxford University Press: New York, 1994.

[63] Matta, C. F.; Boyd, R. J., Eds. *The Quantum Theory of Atoms in Molecules: From Solid State to DNA and Drug Design*. Wiley-VCH: Weinheim, 2007.

[64] Koch, U.; Popelier, P., *J. Phys. Chem.* 1995, *99*, 9747–9754.

[65] Espinosa, E.; Molins, E.; Lecomte, C., *Chem. Phys. Lett.* 1998, *285*, 170–173.

[66] Vener, M.; Egorova, A.; Churakov, A.; Tsirelson, V., *J. Comput. Chem.* 2012, *33*, 2303–2309.

[67] Gatti, C.; Macchi, P., Eds. *Modern Charge-Density Analysis*. Springer: New York, 2012.

[68] Coppens, P.; Abramov, Y.; Carducci, M.; Korjov, B.; Novozhilova, I.; Alhambra, C.; Pressprich, M. R., *J. Am. Chem. Soc.* 1999, *121*, 2585–2593.

[69] Abramov, Y. A., *Acta Crystallogr. Sect. A: Found. Crystallogr.* 1997, *53*, 264–272.

[70] (a) Chekal, B. P.; Campeta, A. M.; Abramov, Y. A.; Feeder, N.; Glynn, P. P.; McLaughlin, R. W.; Meenan, P. A.; Singer, R. A., *Org. Process Res. Dev.* 2009, *13*, 1327–1337; (b) Campeta, A. M.; Chekal, B. P.; Abramov, Y. A.; Meenan, P. A.; Henson, M. J.; Shi, B.; Singer, R. A.; Horspool, K. R., *J. Pharm. Sci.* 2010, *99*, 3874–3886.

[71] Andzelm, J.; Kölmel, C.; Klamt, A., *J. Chem. Phys.* 1995, *103*, 9312–9320.

[72] Perdew, J. P.; Burke, K.; Ernzerhof, M., *Phys. Rev. Lett.* 1996, *77*, 3865.

[73] Chen, S.; Guzei, I. A.; Yu, L., *J. Am. Chem. Soc.* 2005, *127*, 9881–9885.

[74] Yu, L., *Acc. Chem. Res.* 2010, *43*, 1257–1266.

[75] (a)Avdagić, A.; Grebenar, I.; Marković, M.; Pavličić, D.; Šiljković, Z.; Tuksar, M. Preparation of crystalline rotigotine base. WO2009063170 A1, 2009; (b) Wolff, H. M.; Quere, L.; Riedner, J. Polymorphic form of rotigotine and process for production. US Patent 8,592,477 B2, 2012.

[76] Drebushchak, T. N.; Boldyreva, E. V., *Z. Kristallogr.* 2004, *219*, 506–512.

[77] Nichols, G.; Frampton, C. S., *J. Pharm. Sci.* 1998, *87*, 684–693.

[78] Langs, D. A.; Strong, P. D.; Triggle, D. J., *J. Comput. Aided Mol. Des.* 1990, *4*, 215–230.

[79] Wang, S.; Herbette, L.; Rhodes, D., *Acta Crystallogr. Sect. C: Cryst. Struct. Commun.* 1989, *45*, 1748–1751.

[80] Grzesiak, A. L.; Lang, M.; Kim, K.; Matzger, A. J., *J. Pharm. Sci.* 2003, *92*, 2260–2271.

[81] Kazantsev, A. V.; Karamertzanis, P. G.; Adjiman, C. S.; Pantelides, C. C.; Price, S. L.; Galek, P. T.; Day, G. M.; Cruz-Cabeza, A. J., *Int. J. Pharm.* 2011, *418*, 168–178.

[82] Abramov, Y. A., *J. Phys. Chem. A* 2011, *115*, 12809–12817.

4

IMPROVING FORCE FIELD PARAMETERS FOR SMALL-MOLECULE CONFORMATION GENERATION

DMITRY LUPYAN[1], YURIY A. ABRAMOV[2], AND WOODY SHERMAN[1]
[1] *Schrödinger Inc., New York, USA*
[2] *Pfizer Worldwide Research & Development, Groton, CT, USA*

4.1 INTRODUCTION

An important part of pharmaceutical development is predicting possible polymorphs that can arise when a compound or mixture of compounds are exposed to different conditions (time, temperature, humidity, solvents, etc.). The ability to formulate drugs in a solid crystal form is highly desirable in the pharmaceutical industry due to the ease of administering pills, the knowledgebase around small-molecule crystal forms, and the associated infrastructure in the pharmaceutical industry. The crystal forms that can be adopted by a compound are intimately related to their energetically accessible three-dimensional (3D) conformations. The crystal packing, even for a single conformation, can take on a single form or multiple forms based on the regular geometric arrangements of the molecules. When a single conformation of molecule exists in multiple crystal forms, it is referred to as packing polymorphism. When different conformations of a molecule take on different crystal forms, which is a commonly observed phenomenon for molecules with conformational flexibility, it is referred to as conformational polymorphism. As such, the ability to accurately predict 3D conformations of small molecules is critical to polymorph prediction and computational crystal structure prediction (CSP) in general.

Computational Pharmaceutical Solid State Chemistry, First Edition. Edited by Yuriy A. Abramov.

Polymorphism is ubiquitous across different sizes and atomic compositions of the chemical entities involved—compounds that exist in multiple crystal forms have been observed in all crystalline materials, including small molecules, proteins, polymers, minerals, and metals. Polymorphism is related to allotropy, which refers to different crystal forms of the same element, such as carbon having the ability to exist in multiple forms (diamond, graphite, graphene, and fullerenes). As such, allotropy is polymorphism of elements (atoms), whereas polymorphism more generally refers to any solid material existing in different crystalline forms. In this chapter, we primarily focus on conformations of small molecules of the size of typical drug-like molecules (molecular weight < 500 Da) with the aim of accurately predicting conformations, which could then be used as inputs for polymorph prediction programs.

The study of polymorphism in some areas of materials science using computational techniques is commonplace [1]; however, the computational prediction of polymorphs is less common in the pharmaceutical industry, even though polymorphism of pharmaceutical compounds is a common phenomenon [2]. In fact, polymorphism can impact the performance of a drug in terms of solubility, bioavailability [3–5], stability [6, 7], and mechanical properties [8]. The limited application of CSP in drug discovery and development perhaps is a consequence of the difficulty of the problem. Indeed, CSP for a rigid molecule can be challenging; drug-like molecules typically have six to eight rotatable bonds and many energetically accessible conformations, which adds significant complexity to the CSP search problem. While advances have been made, the accuracy of CSP programs is not at the level where broad-scale application in the pharmaceutical industry would be expected. Nonetheless, significant progress in CSP has been made in recent years, especially when computational and experimental approaches are combined [9–11].

Experimental polymorph screening in the pharmaceutical field is typically performed to identify the most thermodynamically stable crystal form under normal manufacturing and storage conditions to ensure that this form does not change during the life cycle of the drug product, as changes can affect the properties of the administered drug. However, in many cases polymorphs can be kinetically trapped, leading to metastable polymorphs that can transform over long time periods to more stable polymorphs. Computational approaches to CSP offer promise in this respect since an accurate method with sufficient sampling of barriers should be able to explore all thermodynamically stable states. While the need for accurate CSP methods is clearly present, the application of computational CSP methods in the pharmaceutical industry is still fairly uncommon. Perhaps one of the reasons that computational CSP is less common in pharmaceutical research is because drug molecules tend to be flexible, having many intramolecular degrees of freedom (six to eight rotatable bonds in typical drug-like small molecules), as compared with the relatively more rigid molecules in some other materials science fields (dyes, semiconductors, light-emitting materials, etc.). Indeed, there is a need for more accurate generation of small-molecule conformations that can be used for CSP in pharmaceutical development.

The accurate generation of small-molecule conformations is an important part of computational CSP. In order to predict a correct polymorph structure, one must be able to generate an accurate conformation of the small molecule, either prior to the sampling of the crystalline packing environment or concurrently. Accurate CSP requires the correct conformation of the small molecule and the associated lattice packing of the molecules within the crystal. In addition to generating an accurate conformation, which could theoretically be done with exhaustive enumeration of all states, the energy must be assessed to determine the energetically accessible states at standard temperatures and pressures. While one could assess the energetics of a large number of conformational states using quantum mechanics (QM), treating all reasonable conformations (typically in the thousands, or more, for drug-like molecules) with QM is currently impractical with the computational resources available to most pharmaceutical labs.

Given the requirement for both speed and accuracy in conformation generation, physics-based molecular mechanics force fields for small molecules [12–16], which are parameterized to reproduce the underlying QM at a fraction of the computational costs, tend to be employed. In addition, experimental data from the Cambridge Structural Database (CSD) [17, 18] offers an excellent data source to study small-molecule conformations and molecular interactions, both for force field parameterization and method validation. In this chapter, we use physics-based molecular mechanics force fields to generate low-energy conformations, independent of the crystal packing environment, and compare the geometries to the conformation found in the CSD. We also use QM as a reference to assess the accuracy that one could hope to attain by studying molecules in the absence of the crystal packing. In this chapter we specifically study molecular interactions that are notoriously challenging for force fields (namely, close S···O and X···O interactions, where X is a halogen) and show that more accurate conformations can often be attained with improved force field parameterization.

There are many areas in the pharmaceutical industry where conformation generation is used, and in general, the required accuracy for a small-molecule conformation depends on the area of application. For example, in molecular docking, a root mean squared deviation (RMSD) of 1.0–2.0Å to the crystallographic conformation is typically considered to be "accurate" since such a conformation usually reproduces the important characteristics of the protein–ligand crystal structure (hydrogen bonds, nonpolar contacts, etc.). However, for small-molecule CSP more accurate conformations are usually needed, due to the tight packing in the solid state and the need to predict the entire environment of the system (in protein–ligand complexes the protein generally provides a semirigid constraint to limit the possible conformations and orientations of the small molecule). A conformation with an RMSD to the crystal structure of 0.50Å might be good enough to generate an accurate polymorph in some cases, although a lower RMSD (perhaps even below 0.25Å) is usually preferred. Figure 4.1 shows examples of RMSDs between crystal structures and computationally generated conformations. As can be seen, with conformations above 0.50Å there can be significant differences in key rotameric states and atomic positions, which would make the accurate reproduction of a crystal structure unlikely.

There are several reasons that the high level of accuracy for CSP is not needed in docking. First, in docking the receptor provides a relatively rigid scaffold into which

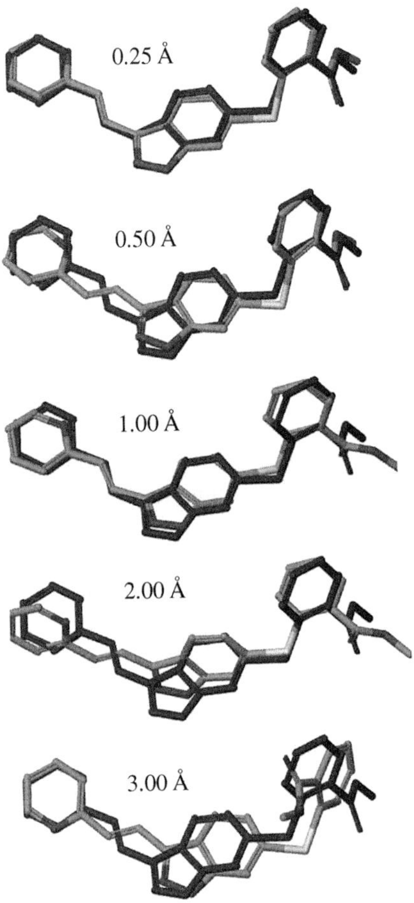

FIGURE 4.1 Example of different conformation accuracy levels, as measured by root mean squared deviation (RMSD). This example shows computationally generated conformations of axitinib (light gray) at various levels of dissimilarity to one particular crystallographic structure (dark gray).

the compound is docked. While some flexibility of the receptor may take place upon ligand binding [19–22], which could accommodate conformational variations in the small-molecule binder, in many cases the receptor can in fact be treaded rigidly [23–25], thereby greatly reducing the conformational search space. Even when there are flexible residues in the receptor, many of the key binding motifs are more rigid, thereby constraining the system and simplifying the problem. In addition, many binding sites are at least partially solvent exposed, giving more space for a rigid conformation of the small molecule to adjust to the binding site even if the conformation is not perfect. The need to simultaneously search the molecule conformations and intermolecular packing of the molecule within the crystal cell makes CSP a challenging problem with significant room for further improvement.

The importance of small-molecule polymorphism in the pharmaceutical industry can be illustrated by two known examples of polymorph-induced impacts on marketed drugs—Norvir (ritonavir) and Neuro (rotigotine patches). In the first case, Abbott Laboratories had to stop sales of Norvir in 1998 due to a failure in a dissolution test, which was caused by the precipitation of a more stable form II of the compound [26, 27]. In the second example, undesirable crystalline material of rotigotine was found in the manufactured patches that were used to administer the drug. These crystals formed as a result of a previously unknown stable polymorph of the drug substance, thereby leading UCB to suspend the marketing of this drug in the United States [28]. The financial implications of compounds having problems with polymorphism late in the development phase can be enormous, further emphasizing the importance of accurately being able to predict polymorphs.

Computational CSP is beginning to show promise as a useful approach to complement experimental polymorph screening as well as being able to assess the risk of discovering a more stable form [29–33]. However, the currently available CSP methods rely heavily on accurate input conformations for the molecules of interest, which are typically held rigid during the initial crystal packing generation. As mentioned earlier, in drug discovery applications such as molecular docking or pharmacophore modeling, a conformation with heavy-atom RMSD of 1.0–2.0 Å to the bioactive conformation is generally considered to be acceptable; for polymorph prediction the accuracy requirements tend to be much more sensitive to the input conformation, with good predictions typically requiring input conformations within 0.25–0.50 Å of the crystallographic conformation. This highlights why proper selection of the starting molecular conformations for CSP is of great importance [30, 34–36].

Here, we study two specific interactions that are typically treated poorly with force fields and describe how the force field representation of these molecules can be improved, thereby leading to better prediction of the low-energy conformations. First, we study close sulfur–oxygen (S···O) nonbonded contacts (distance less than the sum of their van der Waals (vdW) radii), and we develop modified parameters for the OPLS_2005 force field to treat this specific interaction better in order to generate conformations close to those found in CSD structures [37]. Next, we briefly study intramolecular halogen–oxygen (X···O) interactions, which are generally not treated well by force fields and show that improvements can be realized with a similar approach to that taken with the close S···O contacts, although the improvements are not as pronounced. The general approach outlined in this chapter, which involves characterization of the interaction of interest and fitting to QM calculations, can be readily extended to other cases. We conclude with a brief description of the protocols employed in this work and highlight a new OPLS force field, OPLS2, which shows significant advantage over other small-molecule force fields without the need for refitting. In summary, using physics-based force fields rather than QM to assess the energetics of small-molecule conformations can be sufficient to capture important conformational energetics, but customization may be required, especially in the case of more exotic interactions. Accurate treatment of small-molecule conformations and energetics is an important part of solid-state chemistry in the pharmaceutical sciences, and the approaches presented in this chapter should provide a basis for improvements in this area.

4.2 METHODS

In this chapter, we study small-molecule conformations and energetics using different levels of theory (molecular mechanics force field, semiempirical, and QM) to reproduce conformations found in the CSD. While our eventual goal is to predict the actual crystal form, here we focus on the conformation of the small molecule, with only one example to show the use of an accurate conformation to be used in the prediction of the solid form. We study two particular interaction types that tend to be poorly reproduced with standard force fields: close intramolecular S···O interactions and halogen X···O interactions; however, the methods presented here are generalizable to other interactions. We study two heavily used small-molecule force fields, the MMFFs variant of MMFF [38, 39] and the OPLS_2005 version of OPLS [16, 40], and demonstrate how modifications to the latter can improve the generation of structures with low RMSD relative to small-molecule crystal structures.

The OPLS force field has a standard functional form, with bond and angle geometries treated using a harmonic potential and torsions treated with a sinusoidal potential. Nonbonded interactions are treated with a 6–12 Lennard-Jones potential for the van der Waals (vdW) energy and Coulomb's law for the electrostatics with fixed atomic point charges. There are no special terms for hydrogen bonds, as they can generally be modeled by the electrostatics. The functional form of the OPLS force field is similar to that in other small-molecule force fields, such as CHARMM [41] and Amber [42]. The MMFFs [43] force field was developed by Halgren at Merck and is based on the earlier MM3 [44] force field developed by Allinger. Bonds, angles, torsions, and electrostatics in MMFFs are treated similarly to OPLS. However, the vdW potential takes the form of a 6–9 potential, which has the effect of softening the repulsive interaction at short interatomic distances. We also study the semiempirical method PM3 [45, 46] and QM with the DFT [47, 48] level of theory and B3LYP-MM hybrid functional, which has been shown to accurately account for dispersion interactions with minimal computational overhead compared with standard DFT functionals [49].

For the first part of the study (close S···O interactions), a total of 309 neutral organic molecules with the SCxC=O motif (where "x" is either nitrogen or carbon, and sulfur is divalent), and R-factor below 0.1 were selected from the CSD database version 5.30 [17, 18]. From this set, 202 molecules displayed close S···O intramolecular nonbonding contacts, defined as the S···O distance being shorter than the sum of S and O vdW radii (3.32 Å). Further filtering for a planar arrangement of S—C and C=O bonds (an absolute value of S—C—C—O pseudo-torsion angle did not exceed 20°), which was the motif of greatest interest to us in this study, resulted in a subset of 132 molecules. Finally, molecules with more than 30 heavy atoms (one case) or those containing two or more SCxC=O motifs (one case) were filtered out, resulting in a final count of 130 molecules. The Supporting Information of the original publication describing this work [37] contains a table with the 130 molecules and their associated properties, such as S···O distance, SCxC=O angle, molecular weight, and formal charge.

To characterize the S···O interaction, we first performed gas phase minimizations of the crystal structure conformations using a single copy of each of the 130 CSD molecules described earlier. For the halogen X···O data set, we identified molecules from the CSD database that contain a [F,Br,I,Cl]CxC=O motif with close 1–5 interaction between a halogen and oxygen atoms. A cutoff of 4 Å was used to define the contact, as it is slightly larger than the sum of the vdW for halogen–oxygen atom types. This resulted in 223 molecules for the halogen X···O data set (58 F···O, 108 Cl···O, 27 Br···O, and 30 I···O).

For each data set, force field minimizations were performed with MacroModel [50, 51] using a maximum of 500 steps and the Polak-Ribiere Conjugate Gradient (PRCG) method, which tends to perform well on small molecules and is the default in MacroModel for molecules with less than 500 atoms. Semiempirical minimizations were performed with the PM3 method [45, 46] available within Schrödinger Suite 2010. QM minimizations were performed using Jaguar (Schrödinger, Inc., Portland, OR) with the B3LYP-MM/6-31G** method and level of theory [49]. The B3LYP-MM method has been parameterized on a large data set of CCSD(T) nonbonding interaction energies and has been shown to be an efficient and effective way to account for dispersion interactions.

A two-step minimization scheme was used for both MM and QM geometry optimizations. The first step allowed for minimization of hydrogen atoms with the heavy atoms restrained using a harmonic potential (50 kcal/(mol Å^2)), which helps correct systematically shortened covalent bonds to hydrogens that have been observed in X-ray diffraction analysis [52]. After the hydrogens are fully minimized to convergence, a full all-atom minimization was performed with all restraints removed. The initial step was found to be necessary in order to get reasonable initial hydrogen positions before the full molecule was allowed to relax and resulted in lower RMSD of the atomic coordinates to the initial crystal structure as compared to a full minimization without this step. The minimization process described earlier was performed to analyze the force field accuracy if a correct local minimum could be found.

Next, a conformational search was performed on each molecule using the force fields to determine the ability to predict accurate conformations. Conformational searches were performed with MacroModel using the Low-Mode Conformational Search (LMCS) method in global search mode, and all conformations within 5.0 kcal/mol from lowest-energy conformation were retained [53]. The LMCS method follows the low-mode vector, but every 2 Å it applies a few steps of steepest descent minimization to relieve any strain due to distorted bond lengths and bond angles introduced by the move. LMCS is an efficient search method that has the advantage that ring structures and variable torsion angles do not have to be specified. LMCS works by exploring the low-frequency eigenvectors of the system, which are expected to follow "soft" degrees of freedom, such as torsions.

To fit the torsional potential for the modified OPLS_2005 force field, we scanned the torsion of interest at 30° increments and used QM to compute the reference potential. QM geometry optimizations were run with the B3LYP/6-31G** method and basis set to obtain minimized geometries using the program Jaguar. Then, single-point energies were computed for each geometry at the localized MP2 (LMP2) level of theory with the cc-pVTZ(-f) basis set. Parameters for the torsion of interest were generated using internally developed nonlinear curve fitting code to minimize the energetic deviation from the quantum mechanical torsional potential. For the work here, a substructure of axitinib (without the ylethenyl-indazol fragment) was used to parameterize the CA—CA—S—CA torsional potential, where "S" is a sulfur atom and "CA" is an SP2 aromatic carbon atom. The torsion parameter was fitted with a single Fourier coefficient (V2), which does not capture the local minima in each of the two peaks that appear on the QM potential energy surface. However, for this work our focus was on reproducing the location of the minima and the relative barrier heights as opposed to the subtleties around the peaks, since the relevant structures for conformation generation in the context of CSPs will be relatively low in energy. The torsional parameters added in this work are shown in Table 4.1.

To allow for closer S···O contact distances, we created a special pair potential for the vdW energy between a carbonyl oxygen and a divalent sulfur. Figure 4.2 shows the shape of multiple vdW curves with different sigma values, where sigma defines the point that the vdW potential becomes positive. Smaller sigma values that allow closer contacts are shifted to the left. We selected a value of $r=2.50$Å for the S···O vdW interaction because it is associated with a minimum interaction energy at an atomic separation of 2.72Å, consistent with the median value of 2.72 ± 0.11Å for the S···O distance of the molecules in this study. In order to not deteriorate the interaction energy surface of sulfur or oxygen with other atoms (e.g., water molecules), the sigma value described earlier was only applied to S···O interactions that involved a carbonyl oxygen and a divalent sulfur. This ensures that parameters previously tuned to reproduce solvation free energies, condensed phase properties, and quantum mechanical potentials would be unaffected. The specific atom pair vdW potential was implemented within the MacroModel program using an approach similar to the NBFIX option available in CHARMM. This allows for explicit control over the vdW parameters for specific pairwise interactions without altering the form of the potential for these atoms interacting with other atoms. This is particularly important for the case of S···O interactions presented here, since these atoms are common in medicinally relevant molecules and we do not want to change the force field parameters for their interactions with anything else.

TABLE 4.1 CA—CA—S—CA Torsion Fitting Parameters for Axitinib

Force Field	Torsion Definition	V2 Term	V4 Term
OPLS_2005	XX—CA—S—CA	0.5	0.5
$OPLS_{S\cdots O}$	CA—CA—S—CA	1.84	0.0

The OPLS_2005 force field describes this torsion using a wildcard, denoted by "XX". "CA" is an atom type for an SP2 aromatic carbon atom, originally described by Jorgensen and Severance [54]. "S" is for the sulfur. $OPLS_{S\cdots O}$ denotes the OPLS_2005 parameter set containing the new torsion type.

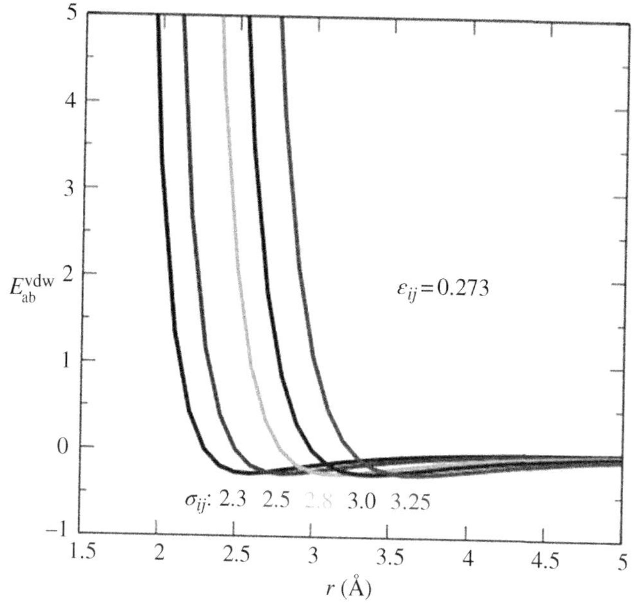

$$E_{ab}^{vdw} = \sum_{i}^{ona} \sum_{j}^{onb} \left(\frac{A}{r_{ij}^{12}} \right) - \left(\frac{B}{r_{ij}^{6}} \right) \qquad A = 4\varepsilon_{ij}\sigma_{ij}^{12} \qquad B = 4\varepsilon_{ij}\sigma_{ij}^{6}$$

FIGURE 4.2 Van der Waals (vdW) interaction energy profile between divalent sulfur and carbonyl oxygen atoms in kcal/mol. Using a Lennard-Jones 6–12 potential, the vdW energy between atoms i and j in OPLS_2005 can be expressed in terms of the parameters σ_{ij} and ε_{ij}. The total energy (E_{ab}^{vdW}) is the sum over all pairwise nonbonded interactions. Shown here are curves for values of σ ranging from 2.3 to 3.25 Å, with the latter value ($\sigma_{ij}=3.25$) being the default for OPLS_2005. To reproduce the most probable distance observed in the CSD between the sulfur and oxygen, a σ value of 2.5 Å was chosen for this work. The ε parameter was not altered from the default OPLS_2005 value of 0.273 kcal/mol. Source: Adapted from Lupyan et al. [37]. Reproduced with permission of Springer.

To improve the Coulomb interactions, we attenuated the electrostatic energies using a distance-dependent dielectric constant. Because sulfur and oxygen both have negative partial atomic charges in the fixed charge force field representations in MMFFs and OPLS_2005, there is a natural repulsion between them. While reversal of the charge sign on one atom would create an attractive force, it would be inconsistent with the quantum mechanical electrostatic potential profile for sulfur and oxygen. Therefore, attenuating the polar interactions using a distance-dependent dielectric allowed for the sulfur and oxygen repulsion to be less unfavorable while maintaining their assigned partial atomic charges. Combining each of the aforementioned enhancements we produced a modified OPLS force field, termed $OPLS_{S\cdots O}$, which allows for more accurate conformations to be generated for small molecules with close S⋯O interactions without a priori knowledge of the crystal packing.

Then, we used the Polymorph Predictor module in Materials Studio 6.5 (Accelrys Software Inc.) to predict crystal forms of axitinib in the corresponding space groups using the MacroModel conformations described earlier as input. "Fine" quality settings were selected for the Polymorph Predictor sampling and the COMPASS force field was used [55]. In the first step of predictions, a Monte Carlo simulated annealing packing algorithm generates starting structures, treating the molecule as a rigid unit. Next, a geometry minimization of each structure is performed, optimizing unit cell parameters and relaxing molecular geometries, while imposing the space group symmetry constraints. Finally, clustering is applied to all minimized structures based on interatomic distances to reduce the set to 500 diverse structures that are scored with the COMPASS force field and ranked by total energy. Since the COMPASS force field does not include the new S···O interaction parameters developed in this work, the results rely strongly on the starting molecular conformations. In addition, since the internal energies of the axitinib crystallographic conformations are not well described by the COMPASS force field, the calculated total energies reflect only a preliminary ranking of the polymorphic forms. A postprocessing of the generated crystal structures at a higher level of theory would be required for a reliable ranking of the axitinib polymorphs and will be the focus of future work.

4.3 RESULTS AND DISCUSSION

In this chapter, we examine close intramolecular contacts from the CSD to understand how well they are treated with molecular mechanics force fields. We then attempt to improve the force field, first by introducing specific modifications to more accurately reproduce the CSD and quantum mechanical geometries for close S···O and halogen X···O interactions. In both cases that force field modifications can lead to improved reproduction of the CSD geometries. We then present the protocol taken for these two classes of interactions and explain how it could be applied to other interaction types. Finally, we present a new force field, OPLS2, which has a greater coverage of small-molecule parameters and shows improved results for small-molecule solvation free energies and reproduction of quantum mechanical energy profiles.

4.3.1 Close S···O Interactions

In this section, we focus on small molecules in the CSD that contain a close sulfur–oxygen (S···O) contact. Specifically, each molecule in the data set of this study contains the chemical motif SCxC=O (i.e., a sulfur and oxygen connected by four bonds) and has a close contact between the S and O (the distance being less than the sum of vdW radii for these two atoms, 3.32 Å). In total, 130 CSD molecules satisfied these criteria and were further analyzed to study the relationship between the geometries and energetics. First, we minimized each molecule using the MMFFs and OPLS_2005 force fields. In addition, minimizations were performed with the semiempirical method PM3 and using quantum mechanics with the B3LYP-MM/6-31G** method and basis set. Figure 4.3 shows the distribution of S···O distances for the 130 molecules

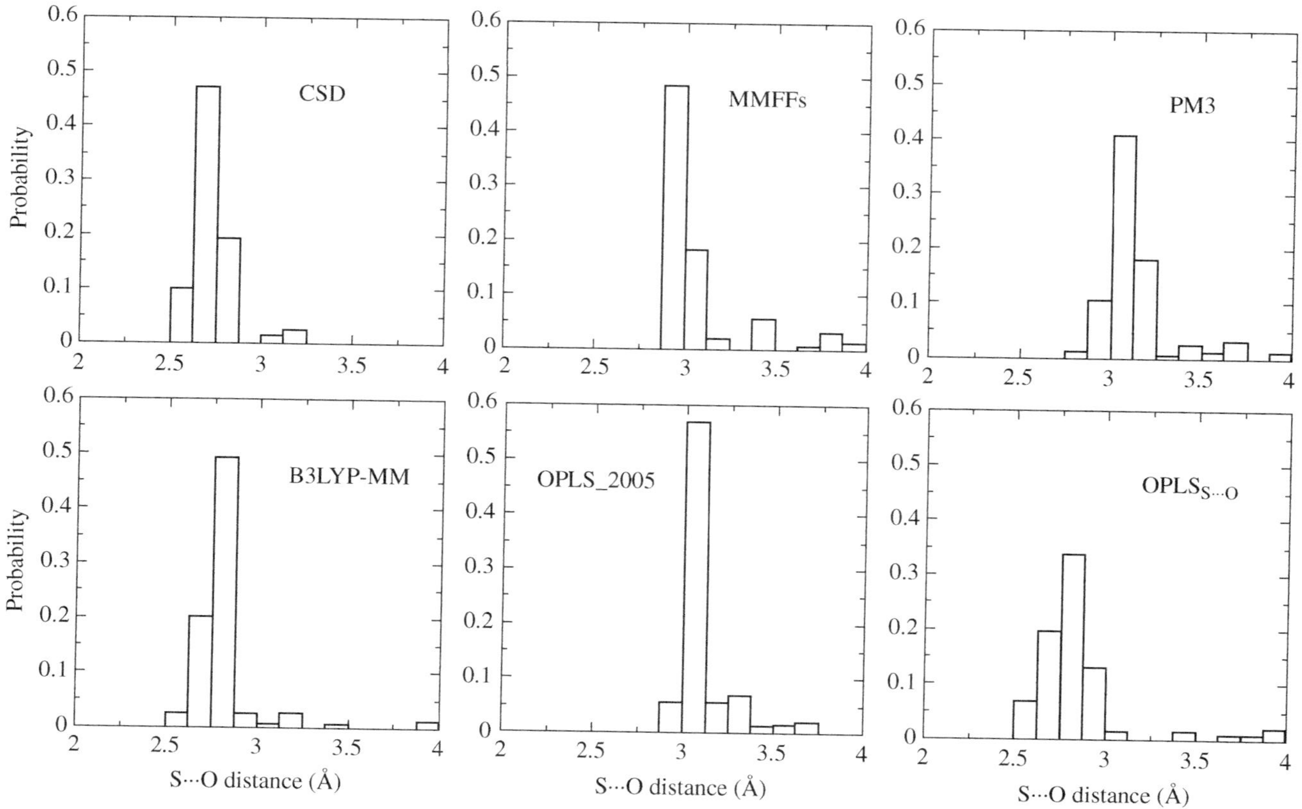

FIGURE 4.3 Distribution of S···O distances in 0.125 Å bins for the 130 molecules in the S···O study. Distributions are shown for the original Cambridge Structural Database (CSD) conformations as well as the geometries obtained through minimizing with various levels of theories, including molecular mechanics (MMFFs, OPLS_2005, and OPLS$_{S\cdots O}$), semiempirical (PM3), and quantum mechanics (B3LYP-MM). OPLS$_{S\cdots O}$ denotes the OPLS_2005 force field with additional parameters developed in this work. The B3LYP-MM and OPLS$_{S\cdots O}$ distributions are in close agreement and also similar to the distribution from the CSD. Source: Adapted from Lupyan et al. [37]. Reproduced with permission of Springer.

based on their crystallographic conformations in the CSD. The peak in the distribution is at 2.75 Å, which is considerably shorter than the optimal distance based on the vdW parameters for these two atom types in the OPLS force field (3.32 Å). This close distance would create a large energetic penalty, which would be reflected in altered geometries upon energy minimization.

Indeed, minimizing each of the 130 molecules produces the distribution shown in Figure 4.3, which shows a significant bias toward longer distances than observed in the CSD. While there are some molecules with an S—O distance less than 3.0 Å using the force fields, none of the molecules approach the 2.75-Å maximum probability distance observed in the CSD conformations. Interestingly, the semiempirical method (PM3) performs equally poorly as the molecular mechanics force fields. On the other hand, the quantum mechanics method (B3LYP-MM) produces S—O distances much more consistent with the observed geometries from the CSD. The agreement between the quantum mechanics and CSD geometries suggests that the close S···O distance is not a consequence of crystal packing, which could hold the two atoms closer together than they would prefer to be in the gas phase, but that does not appear to be the case here.

To assess the nature of the chemical motif of interest in greater detail, we also studied the planarity of the angle between the C—S and C═O bond vectors. The distribution from the CSD is shown in Figure 4.4, along with statistics for the geometries of the molecules after minimization using MMFFs, OPLS_2005, PM3, and B3LYP-MM. As was the case for the distances, a difference is observed in the distributions between the CSD conformations and the minimized force field conformations. The center in the distribution for the CSD conformations is close to planar (i.e., 0°) with a relatively narrow peak, while the distributions after minimization with OPLS_2005 and MMFFs distributions are much broader. In fact, many molecules have angles greater than 20° after the force field minimization. Specifically, 38.6% of the molecules have an angle greater than 20° for the OPLS_2005 minimized structures and 19.1% for MMFFs, whereas none of the CSD structures in this work have an angle greater than 20°. The PM3 semiempirical method shows 14.2% of the molecules greater than 20°. The quantum mechanics structures are closer to the crystal, with only 4.4% of the molecules adopting planarity greater than 20° and most molecules being close to planar.

To improve the force field treatment of these molecules we modified the OPLS_2005 force field by altering the torsional parameters, vdW treatment, and electrostatics (see Section 4.2) with an aim to more accurately reproduce the experimental CSD geometries. The parameter modifications included fitting of the torsional profile to a quantum mechanically derived energy surface, scaling the vdW potential to match the S···O distance profile observed in the CSD, and attenuating the electrostatic potential using a distance-dependent dielectric. Figure 4.5 shows the substructure used for the CA—CA—S—CA torsion angle parameterization and the associated potential energy surface. Figure 4.5 shows the improved agreement between the modified OPLS force field (called $OPLS_{S \cdots O}$ here) and quantum mechanics relative to either OPLS_2005 or MMFFs. As seen in Figures 3.3 and 3.4, the distribution with the modified force field is closer to the CSD distribution than the OPLS_2005, MMFFs, or PM3 distributions, with an average S···O distance of 2.84 ± 0.37 Å and

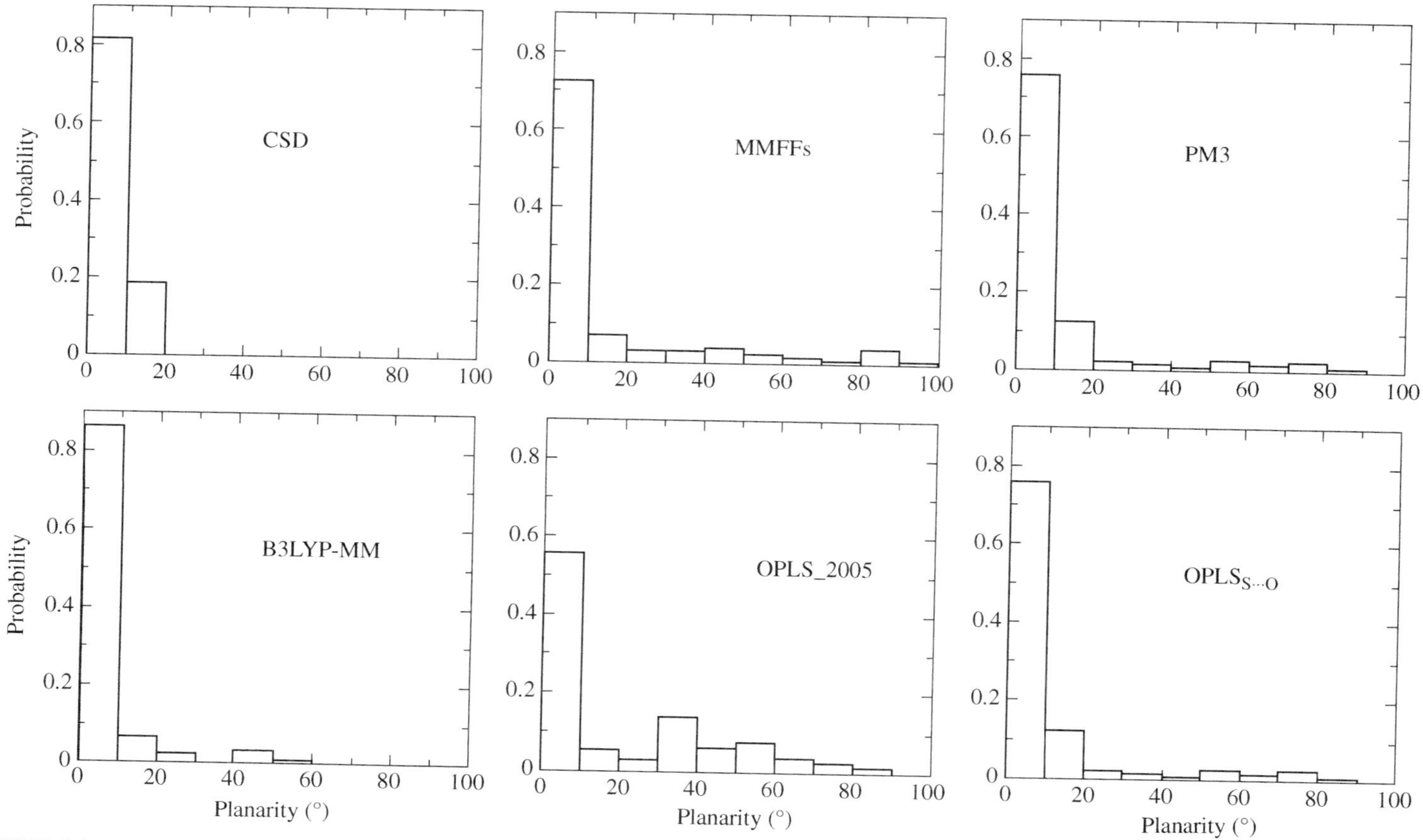

FIGURE 4.4 Distribution of SCxC=O motif angle (deviation from planarity between the S—C bond and C=O bond) for the 130 CSD molecules. The geometry of each molecule was energy minimized *in vacuo* using different levels of theory and force field parameters. See Figure 4.3 for a description of the different plots. Source: Adapted from Lupyan et al. [37]. Reproduced with permission of Springer.

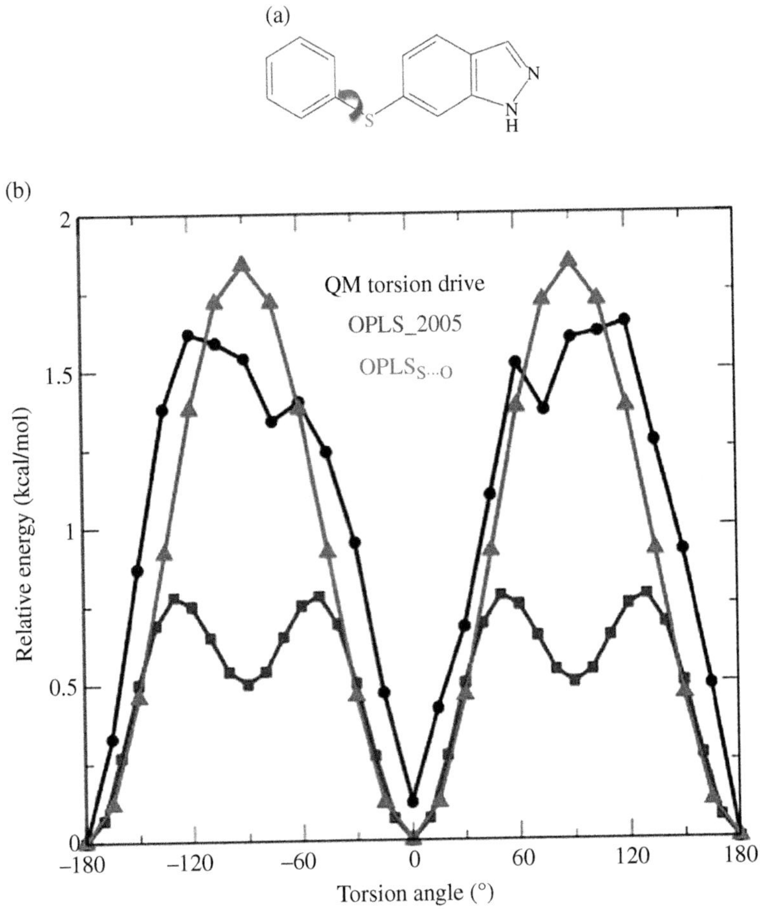

FIGURE 4.5 Axitinib structure and torsional energy profile. (a) Substructure of axitinib used to fit the new torsion parameter (see Section 4.2 for details). (b) Potential energy surface for the CA—CA—S—CA torsion angle in the SCCC=O motif. The QM energy (dark gray circles), OPLS_2005 (medium gray squares), and refitted OPLS$_{S\cdots O}$ (light gray triangles). Source: Adapted from Lupyan et al. [37]. Reproduced with permission of Springer.

only 11.5% of the angles being greater than 20°. While these statistics are slightly worse than the quantum mechanics results, as compared with the geometries from the CSD, they are much better than OPLS_2005 or MMFFs. Figure 4.6 shows box plots of the RMSD distribution for the 130 molecules with OPLS$_{S\cdots O}$ and the other methods studied here, which also shows an improvement using the OPLS$_{S\cdots O}$ parameters compared with OPLS_2005 and MMFFs.

Figure 4.7 shows an example structure from the CSD (refcode: FOKWEI; compound name: ethyl 3-(benzoylamino)-3-(isopropylthio)acrylate) that illustrates the problem observed with OPLS_2005 and MMFFs for most of the molecules in this study. The experimental CSD structure exhibits an S···O distance of 2.77 Å and

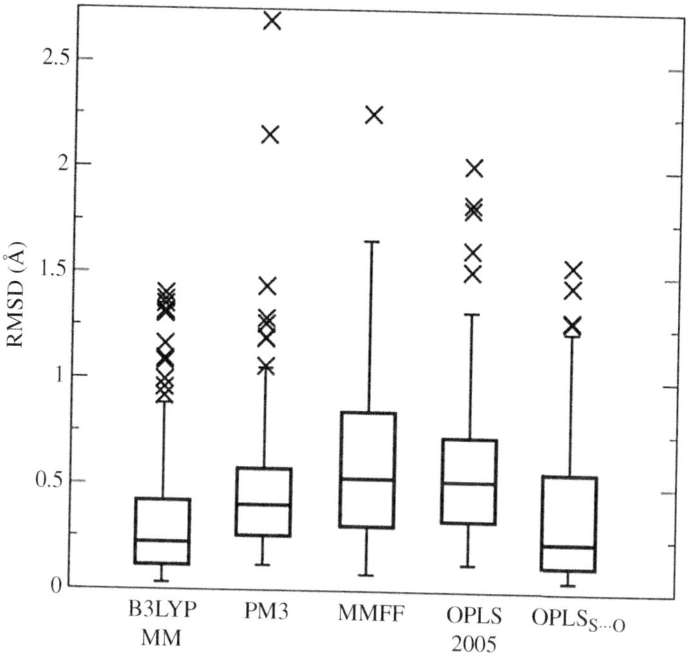

FIGURE 4.6 RMSD box plots of the 130 test set structures using different levels of theory for geometry minimization. The median RMSD is denoted as a horizontal line through each box. The bottom and the top of the box mark the lower and upper quartile, respectively. The whiskers extend to 1.5*IQR (interquartile range) and the molecules with RMSD larger than that are the outliers, shown with "x" marks. Source: Adapted from Lupyan et al. [37]. Reproduced with permission of Springer.

planarity of 7.2°. The structures minimized with MMFFs and OPLS_2005 have S···O distances at 3.09 Å and 3.05 Å with the planarity of the motif at 12.9 and 32.2°, respectively. The heavy-atom RMSD with respect to CSD structure is 0.75 Å for the MMFFs-minimized structure and 0.61 Å for OPLS_2005-minimized structure. The PM3 method performs better than the force fields in this case, with S···O distance of 2.93 Å, planarity of 8.8°, and RMSD of 0.37 Å from the CSD structure. The B3LYP-MM/6-31G** optimized geometry is much closer to the crystal, with S···O distance of 2.75 Å, planarity of 3.2°, and RMSD of 0.11 Å. Finally, the modified OPLS$_{S\cdots O}$ force field parameters developed in this work produces an S···O distance of 2.83 Å, a planarity of 6.1°, and an RMSD of 0.24 Å. These results are better than the force fields (MMFFs and OPLS_2005) and between the PM3 and quantum mechanics calculations.

To test the applicability of the modified force field to generate accurate conformations for CSP on a pharmaceutically relevant molecule, we studied axitinib, developed at Pfizer to treat cancer by targeting the vascular endothelial growth factor (VEGF). Axitinib is a particularly interesting compound from the perspective of studying small-molecule crystal structures because it forms five anhydrous polymorphs [56, 57]. Figure 4.8 shows the conformations of axitinib in the five

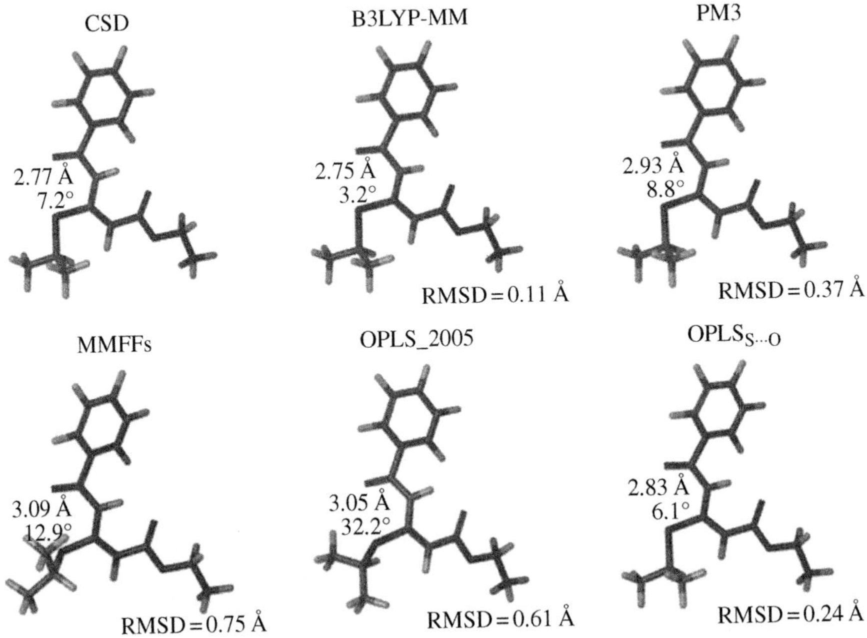

FIGURE 4.7 Lowest-energy structures of a CSD molecule (CSD refcode: FOKWEI) containing a close S···O interaction, minimized with different levels of computational theory. Numerical values to the left of each molecule show the S···O distance and SCNC=O planarity. The RMSD with respect to the CSD conformation is shown to the bottom right of each structure. Source: Adapted from Lupyan et al. [37]. Reproduced with permission of Springer.

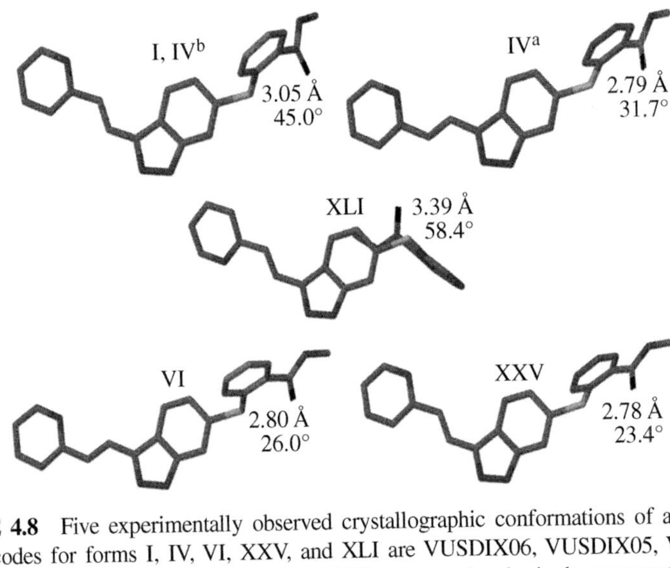

FIGURE 4.8 Five experimentally observed crystallographic conformations of axitinib. The CSD refcodes for forms I, IV, VI, XXV, and XLI are VUSDIX06, VUSDIX05, VUSDIX03, VUSDIX, and VUSDIX04, respectively. Form IV has two molecules in the asymmetric unit cell. Form IV conformation "b" (IV^b) is nearly identical to the conformation of form I. S···O distance and SCCC=O motif planarity are shown for each conformation. Source: Adapted from Lupyan et al. [37]. Reproduced with permission of Springer. (*see insert for color representation of the figure.*)

polymorph structures, and Table 4.2 shows the crystallographic (CSD) S···O distance and SCxC=O planarity. In addition, Table 4.3 shows the RMSD matrix between all five structures, highlighting that they are all different conformations. Each of these has a close S···O distance, ranging from 2.8 to 3.4 Å and the planarity ranges from 23 to 58°, with the largest value of distance and angle corresponding to the lowest-energy crystal form. This is a good test for the modified force field because the angle is outside the range we observed in the 130-molecule CSD training set and, therefore, if the S···O distance parameter was overfit, it would become apparent here.

Table 4.2 also shows the ability of different force fields to generate low RMSD axitinib conformations compared with the crystal structures. We find that each of the five observed crystal conformations is reproduced using the new $OPLS_{S \cdots O}$ force field. The average heavy-atom RMSD for the closest structure to each of the five crystal conformations is 0.33 Å, and the maximum RMSD is 0.54 Å with $OPLS_{S \cdots O}$. This is in contrast to the results obtained using MMFFs and OPLS_2005, where the average RMSD is 0.94 and 0.80 Å, and the maximum is 1.38 and 1.18 Å, respectively. We find that the $OPLS_{S \cdots O}$ conformational search geometries are almost as accurate as the quantum mechanics geometry minimizations of the CSD crystal structures, which have an average RMSD of 0.21 Å and a maximum of 0.37 Å.

Finally, the ultimate objective of this work is to generate conformations that can be repacked in the correct crystal form. To do this, a limited CSP study was performed on axitinib, as follows. First, we took the structures from the conformational search using each force field with the lowest RMSD to the crystal forms I, VI, XXV, and XLI. These conformations were used as input for the Polymorph Predictor in an attempt to reproduce the crystal forms in the corresponding space groups. Form IV was not considered in this limited study because it has two molecules in the asymmetric unit cell (termed IV^a and IV^b), which makes the computational CSPs substantially more challenging and time consuming.

The $OPLS_{S \cdots O}$ conformations were correctly packed to reproduce three out of four forms: forms I, VI, and XLI. In fact, each of the predicted forms was ranked among the top few solutions from the 500 generated polymorphs as ranked by COMPASS force field. The predicted crystal forms displayed a good overlay with the experimental crystal structures (see Fig. 4.9), with RMSDs of 0.66, 0.54, and 0.47 Å for the overlay of 10 molecules in the crystal. In contrast, the CSPs based on conformations generated by the MMFFs and OPLS_2005 force fields failed to reproduce the crystal structures of forms I and XLI within any of the top 500 solutions. Only form VI was reproduced starting from the MMFFs conformation, resulting in an RMSD of 0.54 Å for the overlay of 10 molecules.

Due to the high level of computational time required, a reranking of the generated crystal structures at a higher level of theory (e.g., DFT with the dispersion energy corrections [58, 59]) was not considered in the current limited study. For the same reason, a full CSP using the Polymorph Predictor program considering multiple space groups for each starting conformation was not preformed.

TABLE 4.2 Tabulated Properties for Axitinib Conformations[a]

	CSD		B3LYP-MM			MMFFs			OPLS_2005			$OPLS_{S\cdots O}$		
Form	Dist	Angle	Dist	Angle	RMSD	Dist	Angle	RMSD	Dist	Angle	RMSD	Dist	Angle	RMSD
XXV	2.75	23.4	2.87	28.4	0.30	3.04	37.4	0.94	3.36	31.7	1.18	2.77	21.9	0.31
VI	2.76	26.0	2.82	31.1	0.29	3.21	41.5	1.38	3.40	38.0	1.11	2.77	27.8	0.21
XLI	3.34	58.4	3.36	59.2	0.24	3.27	66.8	0.73	3.21	45.9	0.56	3.15	57.4	0.35
IV[b]	2.79	31.7	2.85	32.8	0.42	3.12	39.4	0.47	3.27	44.2	0.60	2.86	30.1	0.54
I,IV[c]	3.06	45.8	3.01	39.2	0.26	3.37	42.5	1.21	3.25	53.3	0.54	3.02	46.8	0.25

Form IV has two molecules in the asymmetric unit cell.

[a] Source: Adapted from Lupyan et al. [37]. Reproduced with permission of Springer.

[b] Unique conformation.

[c] Conformation is very close to the conformation of form I and has similar values for the distance and angle reported.

"Dist" is the S···O distances (in Å) and "Angle" is the planarity of the SCxC=O motif (in degrees). Values are shown for the CSD crystal structure conformation and the minimized geometries for each of the methods. RMSD values (in Å) are shown with respect to CSD structures.

TABLE 4.3 Pairwise RMSD of Five Axitinib Crystal Conformations, in Angstrom (Å)[a]

	XXV	VI	XLI	IV[b]	I, IV[c]
XXV	0				
VI	2.90	0			
XLI	3.06	4.12	0		
IV[b]	2.75	0.86	3.99	0	
I, IV[c]	2.63	1.13	3.71	0.62	0

Form IV has two molecules in the asymmetric unit cell.
[a] Source: Adapted from Lupyan et al. [37]. Reproduced with permission of Springer.
[b] Unique conformation.
[c] Conformation is very close to the conformation of form I and has similar values for the distance and angle reported.

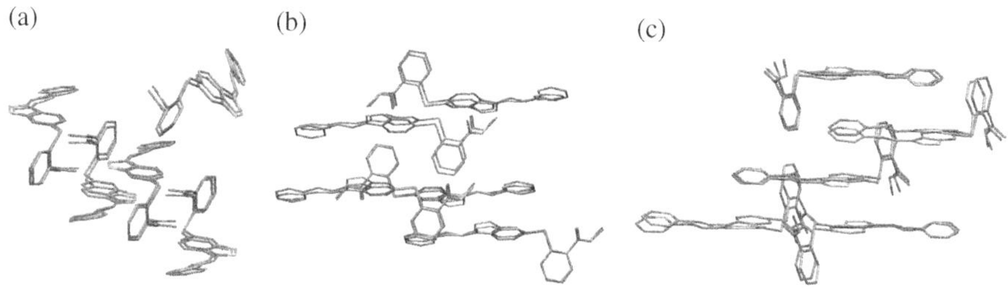

FIGURE 4.9 Structural overlay of CSD and predicted crystal packing geometries of axitinib using conformations generated using the $OPLS_{S\cdots O}$ force field parameters developed in this work. Structures of axitinib forms XLI (a), VI (b), and I (c) are shown. Source: Adapted from Lupyan et al. [37]. Reproduced with permission of Springer.

4.3.2 Halogen X···O Interactions

Here, we extend our work to explore the nonbonded parameters of halogen atoms commonly used in medicinal biology. Halogen atoms play a particularly important role in medicinal chemistry, and a number of natural products, drugs on the market, and compounds in clinical studies contain halogen atoms [60–63]. Halogen substituents are often used in lead optimization stage to optimize lipophilicity of the molecule [64], introduce specific steric contributions [65], or displace water molecules [65–68]. In addition, halogen variants are used in biomolecular systems to improve interactions, such as halogen bonding and multipolar interactions. Such properties have been exploited in several drugs classes, specifically for anesthetic halothane, inhibitors of HIV retrovirus, nonsteroidal anti-inflammatory drugs, and others [60–62, 69]. However, halogen interactions are challenging to model accurately within the framework of traditional molecular mechanics force fields [70, 71].

We first identified molecules from the CSD database that contain a [F,Br,I,Cl] CxC=O signature motif with close 1–5 interaction between a halogen and oxygen atoms. A cutoff of 4 Å was used to define the contact, as it is slightly larger than the sum of the vdW for halogen–oxygen atom types. This resulted in 223 molecules for the halogen X···O data set (58 F···O, 108 Cl···O, 27 Br···O, and 30 I···O). Some molecules contained multiple halogen X···O interactions, in which case the shortest interaction pair was used for the analysis. The default OPLS_2005 parameters (σ for the distance in Å and ε for the well depth in kcal/mol) for each of the halogens interacting with a carbonyl oxygen are as follows: F $\sigma=2.85$ $\varepsilon=0.061$, Cl $\sigma=3.40$ $\varepsilon=0.30$, Br $\sigma=3.47$ $\varepsilon=0.47$, and I $\sigma=3.75$ $\varepsilon=0.60$. The other force field terms (bond stretch, angle, and torsion) were not modified in this work.

For each of the halogens (F, Cl, Br, and I), we performed a similar analysis to the S···O work earlier. Tables 4.4–4.7 show the statistics for each halogen with the different force fields. While the deviation of the distances from the CSD values appears

TABLE 4.4 Fluorine Distances (Å) and Correlation (R^2) with CSD Structures

	F···O Distances (Å)	R^2 w.r.t. CSD
CSD	3.03±0.19	N/A
MMFFs	3.10±0.23	0.72
OPLS_2005 ($\sigma=2.90$)	3.14±0.27	0.70
+NBFIX ($\sigma=2.73$)	3.09±0.29	0.72
+NBFIX ($\sigma=2.63$)	3.06±0.21	0.73
OPLS2.0	3.11±0.24	0.80

TABLE 4.5 Chlorine Distances (Å) and Correlation (R^2) with CSD Structures

	Cl···O Distances (Å)	R^2 w.r.t. CSD
CSD	3.12±0.15	N/A
MMFFs	3.21±0.31	0.31
OPLS_2005	3.14±0.31	0.59
OPLS2.0	3.17±0.20	0.64

TABLE 4.6 Bromine Distances (Å) and Correlation (R^2) with CSD Structures

	Br···O Distances (Å)	R^2 w.r.t. CSD
CSD	3.08±0.12	N/A
MMFFs	3.21±0.12	0.13
OPLS_2005	3.05±0.44	0.44
OPLS2.0	3.07±0.10	0.79

TABLE 4.7 Iodine Distances (Å) and Correlation (R^2) with CSD Structures

	I···O Distances (Å)	R^2 w.r.t. CSD
CSD	3.24±0.41	N/A
MMFFs	3.52±0.43	0.80
OPLS_2005	3.29±0.41	0.97
OPLS2.0	3.29±0.42	0.96

to be relatively small for all halogens, for all cases we see an improvement in the correlation (R^2) between the CSD and calculated geometries. Most of the distances for the calculated geometries are within the variance of the CSD distances, even with the default MMFFS and OPLS_2005 force field parameters; however, a more detailed analysis shows that there are differences between the force fields. To more clearly emphasize the differences between the force fields for the halogen work here, we look at the correlation with experiment rather than the absolute deviation of the distances. By doing this, we see that the correlation for MMFFs is relatively poor for Cl and Br. In all cases, OPLS2.0 (described in more detail later) produces the highest correlation (with the exception of iodine, where OPLS_2005 produces a comparable correlation of $R^2=0.97$, compared with 0.96 for OPLS2.0).

It is possible that these small differences are not significant, but for tight crystal packing (as in polymorphs) even tenths of an Angstrom can make a difference. In addition, errors can accumulate, so it is important to get an accurate representation for each atom and interaction type. We see that MMFFs consistently overpredicts the X···O distances by 2–9% compared with the CSD distances, which is significant, especially if it alters torsional angles connected to other parts of the molecule that have an amplifying impact due to the lever effect. It should be noted that in this chapter we did not take special consideration of the electronic properties of halogens, which are not well represented by the standard force field atom-centered charges; this topic has been studied by others via the introduction of off-site charges that better models the sigma hole [71].

Similar to the previous work on S···O interactions, we see that the default parameters for traditional force fields (MMFFs and OPLS_2005) do not adequately reproduce the distributions observed in the CSD in some cases (see Fig. 4.10). While the effect is less dramatic than for the close S···O contacts, there is still a clear lack of agreement between the CSD geometries and those obtained from field minimizations, and even small errors could result in an inaccurate CSP.

4.3.3 Generalization of the Approach to Other Interactions

While the earlier work on close S···O contacts and halogen X···O interactions provides specific parameters that can be used to improve force field treatment of certain interactions in small molecules, there are potentially many other interactions that could be amended with a special force field treatment. The approach presented earlier offers a straightforward way to improve specific aspects of a force field for the generation of accurate small-molecule conformations. The approach can be broken down into the following three steps:

1. Identification of the problematic interaction
2. Improvement of the force field using QM and experimental data
3. Validation of the improved parameters

Identification of the problematic interaction: In many cases the deficiency in the force field will be clear. For example, in the case of the close S···O interactions it was

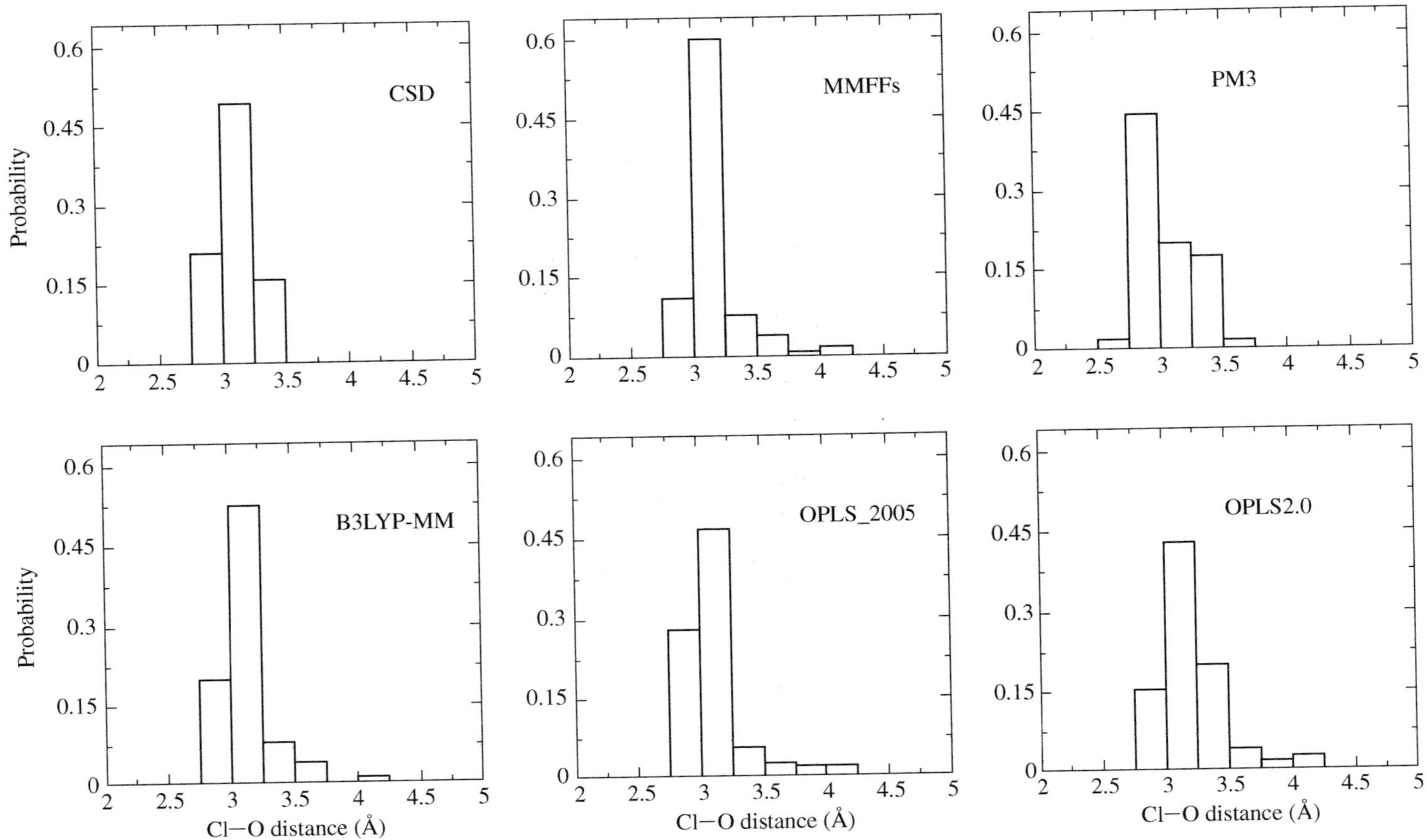

FIGURE 4.10 Distribution of Cl···O distances from the CSD and different minimization methods.

not possible to generate the crystallographic conformation of the small molecules with this motif, and minimization of the crystal structure conformation led to a significant change in the coordinates. In addition, the energy of the crystallographic conformation, as assessed by the force field, was significantly higher in energy than the minimized conformation to such an extent that the crystallographic conformation would not be energetically accessible. Finally, quantum mechanics corroborated the crystallographic data, thereby building confidence that there was indeed a force field problem that required attention.

In other cases the deficiency can be less obvious. For example, determining whether a specific hydrogen bond is stabilizing or destabilizing, and to what degree, involves a subtle balance between interaction energy, internal strain, desolvation penalty, and other energetic contributions. In addition, the experimental structures that we aim to reproduce are typically attained under conditions that are not identical to those used to generate the parameters (i.e., the fit is being done based on low-temperature data and therefore includes mostly potential energy, while the experiment may be done at room temperature with solvent, buffer, etc.), making it hard to determine if the force field treatment is inaccurate, and to what degree. Furthermore, it is possible to get the geometry of an interaction correct while still not predicting the energetics accurately, which could have implications in comparing energies between different states and computing kinetic properties. In such cases, it may be necessary to compare the results from more rigorous calculations, like free energy calculations [72–76], with experimental data.

Improvement of the force field: Once the problem has been identified, the force field reparameterization can begin. In general, the objective is to modify as few parameters as possible, thereby reducing the risk of adversely affecting the treatment of other molecules. For example, in the close S···O work we only modified the pair potential between sulfur and a carbonyl oxygen, thus preventing changes in the force field that would affect interactions of sulfur with water molecules, which would alter the previously optimized solvation properties of sulfur-containing molecules. Force fields are generally fit to specific types of data, such as condensed phase properties or quantum mechanical data, which are important to maintain as new parameters are added to improve properties that were not included in the original force field parameterization.

Validation of the improved parameters: There are several ways in which one can test force field improvements. Direct comparison to quantum mechanical data offers one possibility, although the same data are often used to fit the force field parameters, so the validation work must be done carefully to ensure some level of independence in the testing. This can be done, for example, by fitting to one molecule with the missing parameter and testing on other molecules that have the same interaction type but with other differences in the molecules. Another approach is comparing with experimental data, although as with quantum mechanical data, it is important to compare with an independent test set. Independence does not only mean that the molecules are different, but that the actual parameter change is assessed as directly as possible without confounding variables. For example, for the close S···O interactions, we optimized the force field to get the correct geometry of the S···O motif in close interactions, but to validate the approach we also compared the predicted geometries of small-molecule conformations containing the S···O motif that were not making close interactions. As such, if we overfit the force field to get the precise geometry of the close S···O interactions right, then we would likely deteriorate the geometries of the molecules in which the S and O were not making close interactions.

4.3.4 An Improved OPLS Force Field (OPLS2)

The OPLS family of force fields adopts a functional form that represents the potential energy of the system as the sum of bond, angle, torsion, and nonbonded terms. Parameters for the bond and angle terms aim to reproduce molecular geometries and vibrational frequencies, while the torsion term aims to reproduce the energetics of conformational rearrangement. The nonbonded terms include electrostatic interactions using atom-centered partial charges and a Lennard-Jones potential representing dispersion and electronic repulsion. Nonbonded parameters are developed against a combination of *ab initio* gas phase properties and experimental condensed phase properties that sensitively probe these intermolecular interactions.

The objective in developing the present iteration of OPLS (OPLS2.0) [77] is to improve the accuracy for drug-like molecules, which should improve the general quality of physics-based approaches including the prediction of binding free energies, docking poses, conformational preferences, and energy landscapes. To do so, the OPLS2.0 force field has substantially expanded the data sets used in its parameterization. The present set includes *ab initio* data on more than 11,000 molecules including their optimized geometries, rotational profiles, and electrostatic potentials with the aim of more thoroughly covering the diversity of chemical functionalities that comprise drug-like molecules. In comparison, the data sets used to parameterize the torsional terms in MMFF and OPLS_2005 are based on 140 and 631 rotational profiles, respectively.

In addition to the expanded parametrical coverage, OPLS2.0 uses semiempirical charges with bond charge corrections (CM1A-BCC) to help account for changes in charge distributions that result from variations in functional group substitutions. The charges are obtained from a combination of the Cramer-Truhlar CM1A charge model [78] and specifically fit bond charge correction terms (BCC) [79] that are parameterized against the OPLS-AA charges for a core set of 112 molecules and the electrostatic potential at the HF/6-31G* level of theory for the OPLS2.0 training set. In a subsequent refinement step, the BCC terms are adapted to minimize the errors with regard to the absolute solvation free energy using a training set of 153 molecules. All structures used to perform the CM1A calculations are prepared via a conformational search and subsequent minimization without the electrostatic term and only repulsive vdW parameters for interactions between atoms to prevent collapse of the molecules from intramolecular interactions.

Validation of the OPLS2.0 force field includes comparison to quantum mechanical energy profiles and experimental solvation free energies. For the latter, we performed explicit solvent molecular dynamics free energy perturbation simulations on a set of 239 diverse small molecules [80]. Compared with other popular force fields

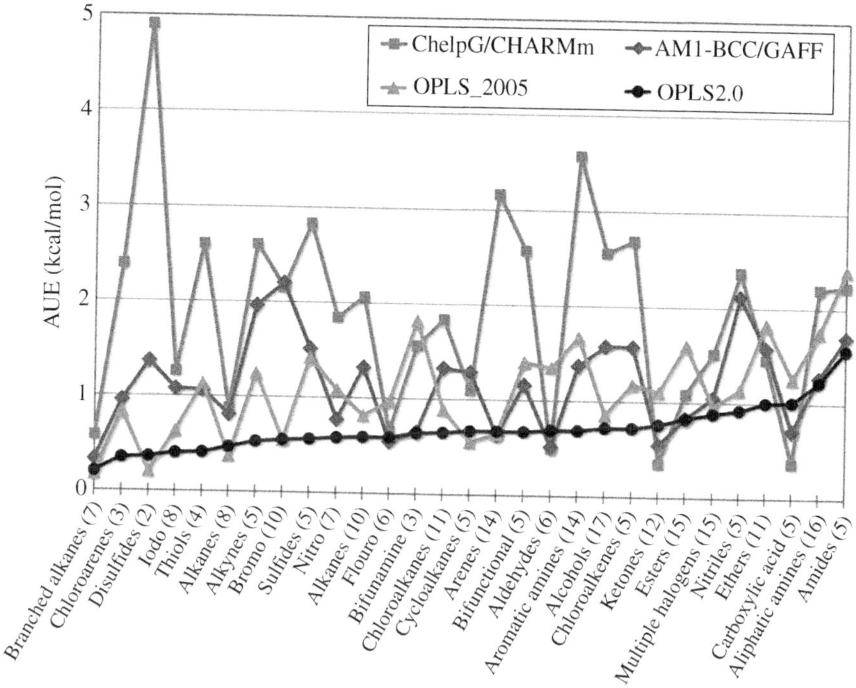

FIGURE 4.11 Comparison of absolute solvation free energy errors for different force fields. The average unsigned error (AUE, kcal/mol) for the different functional classes of compounds compared with experiment is shown for CHARMm/ChelpG (squares; medium light gray), GAFF/AM1-BCC (diamonds; medium dark gray), OPLS_2005 (triangles; lightest gray), and OPLS2.0 (circles; black). The number of compounds for each chemical class is shown in parentheses. Compound classes are sorted in order of increasing OPLS2.0 AUE.

(OPLS_2005, AFF, and CHARMm-MSI), OPLS2.0 produces the best correlation with experimental data (R^2=0.95, slope=0.96) and the lowest average unsigned errors (0.7 kcal/mol; see Fig. 4.11) [16]. Important classes of compounds that performed suboptimally with OPLS_2005 show significant improvements with OPLS2.0. These improvements should result in general improvements for small-molecule conformational search, which should benefit the CSP.

4.4 CONCLUSION

The need for high-quality force field parameters is critical to rapidly generate accurate small-molecule conformations, which are needed for reliable computational CSP. In this chapter, we presented specific force field improvements that lead to better small-molecule conformations for close S···O and halogen X···O interactions. We customized the force field by fitting to quantum mechanical data and comparing with conformations from the CSD, which is a general approach that can be extended to other types of interactions. We also presented a new force field, OPLS2.0, which has a much greater coverage of torsional space than other popular force fields and also has an improved charge assignment method (CM1A-BCC).

The approach presented here is straightforward and can be applied to other classes of molecules and other force fields. In short, the force field should reproduce the underlying quantum mechanical energy landscape as accurately as possible. This will ensure that most molecules and molecular interactions are handled well. However, in the cases where special interactions are not treated well, specific parameters can be developed based on quantum mechanics calculations and/or experimental data (i.e., CSD structures).

Computational CSP is an emerging area of research with great potential to impact drug development. Indeed, the ability to accurately predict crystal structures and polymorphs could accelerate the development process and minimize liabilities that may arise from unforeseen polymorphs. However, significant challenges still exist in the field of CSP. In addition to the need for accurate force fields, there is also a need for advanced search algorithms for small-molecule conformations and crystal packing. Furthermore, many drug molecules are formulated as a mixture, which complicates the problem considerably. We have not attempted to investigate the actual crystal packing problem in this work. However, we presented evidence demonstrating that with highly accurate conformations of a small molecule it is possible to predict the polymorph using existing tools. These encouraging results need to be further validated on a larger data set and will likely lead to the identification of other challenges in the field of CSP, which could trigger efforts to develop new or improved crystal packing algorithms.

REFERENCES

[1] Desiraju, G. R., *Science* 1997, *278*, 404–405.

[2] Bernstein, J. *Polymorphism in Molecular Crystals*. Oxford University Press: New York, 2007.

[3] Abramov, Y. A.; Pencheva, K., Thermodynamics and Relative Solubility Prediction of Polymorphic Systems. In *Chemical Engineering in the Pharmaceutical Industry: R&D to Manufacturing*; am Ende, D. J., Ed. John Wiley & Sons, Inc.: Hoboken, NJ, 2011; pp. 477–490.

[4] Kobayashi, Y.; Ito, S.; Itai, S.; Yamamoto, K., *Int. J. Pharm.* 2000, *193*, 137–146.

[5] Brittain, H. G., Ed. *Polymorphism in Pharmaceutical Solids*. Informa Healthcare: New York, 2009.

[6] Singhal, D.; Curatolo, W., *Adv. Drug Deliv. Rev.* 2004, *56*, 335–347.

[7] Crowley, K. J.; Zografi, G., *J. Pharm. Sci.* 2002, *91*, 492–507.

[8] Beyer, T.; Day, G. M.; Price, S. L., *J. Am. Chem. Soc.* 2001, *123*, 5086–5094.

[9] Baias, M.; Dumez, J.-N.; Svensson, P. H.; Schantz, S.; Day, G. M.; Emsley, L., *J. Am. Chem. Soc.* 2013, *135*, 17501–17507.

[10] Datta, S.; Grant, D. J., *Nat. Rev. Drug Discov.* 2004, *3*, 42–57.

[11] Baias, M.; Widdifield, C. M.; Dumez, J.-N.; Thompson, H. P.; Cooper, T. G.; Salager, E.; Bassil, S.; Stein, R. S.; Lesage, A.; Day, G. M., *Phys. Chem. Chem. Phys.* 2013, *15*, 8069–8080.

[12] Weiner, S. J.; Kollman, P. A.; Case, D. A.; Singh, U. C.; Ghio, C.; Alagona, G.; Profeta, J., S.; Weiner, P., *J. Am. Chem. Soc.* 1984, *106*, 765–784.

[13] Vanommeslaeghe, K.; Hatcher, E.; Acharya, C.; Kundu, S.; Zhong, S.; Shim, J.; Darian, E.; Guvench, O.; Lopes, P.; Vorobyov, I., *J. Comput. Chem.* 2010, *31*, 671–690.

[14] Jorgensen, W. L.; Chandrasekhar, J.; Madura, J. D.; Impey, R. W.; Klein, M. L., *J. Chem. Phys.* 1983, *79*, 926–935.

[15] Jorgensen, W. L.; Tirado-Rives, J., *J. Am. Chem. Soc.* 1988, *110*, 1657–1666.

[16] Shivakumar, D.; Williams, J.; Wu, Y.; Damm, W.; Shelley, J.; Sherman, W., *J. Chem. Theory Comput.* 2010, *6*, 1509–1519.

[17] Allen, F. H., *Acta Crystallogr., Sect. B Struct. Sci.* 2002, *58*, 380–388.

[18] Groom, C. R.; Allen, F. H., *Angew. Chem. Int. Ed.* 2014, *53*, 662–671.

[19] Sherman, W.; Beard, H. S.; Farid, R., *Chem. Biol. Drug Des.* 2006, *67*, 83–84.

[20] Sherman, W.; Day, T.; Jacobson, M. P.; Friesner, R. A.; Farid, R., *J. Med. Chem.* 2006, *49*, 534–553.

[21] Csermely, P.; Palotai, R.; Nussinov, R., *Trends Biochem. Sci.* 2010, *35*, 539–546.

[22] Thoma, J. A.; Koshland Jr, D., *J. Am. Chem. Soc.* 1960, *82*, 3329–3333.

[23] Friesner, R. A.; Banks, J. L.; Murphy, R. B.; Halgren, T. A.; Klicic, J. J.; Mainz, D. T.; Repasky, M. P.; Knoll, E. H.; Shelley, M.; Perry, J. K.; Shaw, D. E.; Francis, P.; Shenkin, P. S., *J. Med. Chem.* 2004, *47*, 1739–1749.

[24] Halgren, T. A.; Murphy, R. B.; Friesner, R. A.; Beard, H. S.; Frye, L. L.; Pollard, W. T.; Banks, J. L., *J. Med. Chem.* 2004, *47*, 1750–1759.

[25] Meng, E. C.; Gschwend, D. A.; Blaney, J. M.; Kuntz, I. D., *Proteins: Struct., Funct., Bioinf.* 1993, *17*, 266–278.

[26] Bauer, J.; Spanton, S.; Henry, R.; Quick, J.; Dziki, W.; Porter, W.; Morris, J., *Pharm. Res.* 2001, *18*, 859–866.

[27] Kempf, D. J.; Marsh, K. C.; Denissen, J. F.; McDonald, E.; Vasavanonda, S.; Flentge, C. A.; Green, B. E.; Fino, L.; Park, C. H.; Kong, X. P., *Proc. Natl. Acad. Sci. U. S. A.* 1995, *92*, 2484.

[28] Rascol, O.; Perez-Lloret, S., *Expert Opin. Pharmacother.* 2009, *10*, 677–691.

[29] Blagden, N.; Davey, R. J., *Cryst. Growth Des.* 2003, *3*, 873–885.

[30] Ouvrard, C.; Price, S. L., *Cryst. Growth Des.* 2004, *4*, 1119–1127.

[31] Price, S. L., *Adv. Drug Deliv. Rev.* 2004, *56*, 301–319.

[32] Price, S. L., *Phys. Chem. Chem. Phys.* 2008, *10*, 1996–2009.

[33] Cross, W. I.; Blagden, N.; Davey, R. J.; Pritchard, R. G.; Neumann, M. A.; Roberts, R. J.; Rowe, R. C., *Cryst. Growth Des.* 2002, *3*, 151–158.

[34] Abramov, Y. A.; Zell, M.; Krzyzaniak, J. F. Toward a Rational Solvent Selection for Conformational Polymorph Screening. In *Chemical Engineering in the Pharmaceutical Industry: R&D to Manufacturing*; am Ende, D. J., Ed. John Wiley & Sons, Inc.: Hoboken, NJ, 2011; pp. 491–504.

[35] Cooper, T. G.; Hejczyk, K. E.; Jones, W.; Day, G. M., *J. Chem. Theory Comput.* 2008, *4*, 1795–1805.

[36] Day, G. M.; Motherwell, W. S.; Jones, W., *Phys. Chem. Chem. Phys.* 2007, *9*, 1693–1704.

[37] Lupyan, D.; Abramov, Y. A.; Sherman, W., *J. Comput. Aided Mol. Des.* 2012, *26*, 1195–1205.

[38] Halgren, T. A., *J. Comput. Chem.* 1996, *17*, 520–552.

[39] Halgren, T. A., *J. Comput. Chem.* 1996, *17*, 490–519.

[40] Jorgensen, W. L.; Maxwell, D. S.; Tirado-Rives, J., *J. Am. Chem. Soc.* 1996, *118*, 11225–11236.

[41] Momany, F. A.; Rone, R., *J. Comput. Chem.* 1992, *13*, 888–900.

[42] Cornell, W. D.; Cieplak, P.; Bayly, C. I.; Gould, I. R.; Merz, K. M.; Ferguson, D. M.; Spellmeyer, D. C.; Fox, T.; Caldwell, J. W.; Kollman, P. A., *J. Am. Chem. Soc.* 1995, *117*, 5179–5197.

[43] Beachy, M. D.; Chasman, D.; Murphy, R. B.; Halgren, T. A.; Friesner, R. A., *J. Am. Chem. Soc.* 1997, *119*, 5908–5920.

[44] Allinger, N. L.; Yuh, Y. H.; Lii, J. H., *J. Am. Chem. Soc.* 1989, *111*, 8551–8566.

[45] Stewart, J. J. P., *J. Comput. Chem.* 1989, *10*, 209–220.

[46] Stewart, J. J. P., *J. Comput. Chem.* 1989, *10*, 221–264.

[47] Dreizler, R. M.; Engel, E., *Density Functional Theory: An Advanced Course*. Springer: Berlin, 2011.

[48] Parr, R. G.; Yang, W. *Density-Functional Theory of Atoms and Molecules*. Oxford University Press: New York, 1989.

[49] Schneebeli, S. T.; Bochevarov, A. D.; Friesner, R. A., *J. Chem. Theory Comput.* 2011, *7*, 658–668.

[50] Mohamadi, F.; Richards, N. G. J.; Guida, W. C.; Liskamp, R.; Lipton, M.; Caufield, C.; Chang, G.; Hendrickson, T.; Still, W. C., *J. Comput. Chem.* 1990, *11*, 440–467.

[51] Chang, G.; Guida, W. C., Still, W. C.; *J. Am. Chem. Soc.* 1989, *111*, 4379–4386.

[52] Speakman, J. C. Neutron Diffraction. In *Molecular Structure by Diffraction Methods*, Volume *6*; Sutton, L. E.; Truter, M. R., Eds. Royal Society of Chemistry: London, 1978; pp. 117–131.

[53] Kolossváry, I.; Guida, W. C., *J. Am. Chem. Soc.* 1996, *118*, 5011–5019.

[54] Jorgensen, W. L.; Severance, D. L., *J. Am. Chem. Soc.* 1990, *112*, 4768–4774.

[55] Sun, H.; Ren, P.; Fried, J., *Comput. Theor. Polym. Sci.* 1998, *8*, 229–246.

[56] Campeta, A. M.; Chekal, B. P.; Abramov, Y. A.; Meenan, P. A.; Henson, M. J.; Shi, B.; Singer, R. A.; Horspool, K. R., *J. Pharm. Sci.* 2010, *99*, 3874–3886.

[57] Chekal, B. P.; Campeta, A. M.; Abramov, Y. A.; Feeder, N.; Glynn, P. P.; McLaughlin, R. W.; Meenan, P. A.; Singer, R. A., *Org. Process Res. Dev.* 2009, *13*, 1327–1337.

[58] Abramov, Y. A., *J. Phys. Chem. A* 2011, *115*, 12809–12817.

[59] Neumann, M. A.; Perrin, M. A., *J. Phys. Chem. B* 2005, *109*, 15531–15541.

[60] Xu, Z.; Yang, Z.; Liu, Y.; Lu, Y.; Chen, K.; Zhu, W., *J. Chem. Inf. Model.* 2014, *54*, 69–78.

[61] Sirimulla, S.; Bailey, J. B.; Vegesna, R.; Narayan, M., *J. Chem. Inf. Model.* 2013, *53*, 2781–2791.

[62] Parisini, E.; Metrangolo, P.; Pilati, T.; Resnati, G.; Terraneo, G., *Chem. Soc. Rev.* 2011, *40*, 2267–2278.

[63] Silverman, R. B.; Holladay, M. W. *The Organic Chemistry of Drug Design and Drug Action*. Academic Press: San Diego, 2014.

[64] Le, T. D.; Arlauskas, R. A.; Weers, J. G., *J. Fluor. Chem.* 1996, *78*, 155–163.

[65] Ren, J.; He, Y.; Chen, W.; Chen, T.; Wang, G.; Wang, Z.; Xu, Z.; Luo, X.; Zhu, W.; Jiang, H.; Shen, J.; Xu, Y., *J. Med. Chem.* 2014, *57*, 3588–3593.

[66] Abel, R.; Salam, N. K.; Shelley, J.; Farid, R.; Friesner, R. A.; Sherman, W., *ChemMedChem* 2011, *6*, 1049–1066.

[67] Barillari, C.; Taylor, J.; Viner, R.; Essex, J. W., *J. Am. Chem. Soc.* 2007, *129*, 2577–2587.

[68] Lockett, M. R.; Lange, H.; Breiten, B.; Heroux, A.; Sherman, W.; Rappoport, D.; Yau, P. O.; Snyder, P. W.; Whitesides, G. M., *Angew. Chem. Int. Ed.* 2013, *52*, 7714–7717.

[69] Lu, Y.; Shi, T.; Wang, Y.; Yang, H.; Yan, X.; Luo, X.; Jiang, H.; Zhu, W., *J. Med. Chem.* 2009, *52*, 2854–2862.

[70] Rendine, S.; Pieraccini, S.; Forni, A.; Sironi, M., *Phys. Chem. Chem. Phys.* 2011, *13*, 19508–19516.

[71] Jorgensen, W. L.; Schyman, P., *J. Chem. Theory Comput.* 2012, *8*, 3895–3901.

[72] Pearlman, D. A.; Case, D. A.; Caldwell, J. W.; Ross, W. S.; Cheatham, T. E., *Comput. Phys. Commun.* 1995, *91*, 1–41.

[73] Chipot, C., Pohorille, A., Eds. *Free Energy Calculations. Theory and Applications in Chemistry and Biology*. Springer Verlag: Berlin, New York, 2007.

[74] Kollman, P., *Chem. Rev.* 1993, *93*, 2395–2417.

[75] Jorgensen, W. L., *Acc. Chem. Res.* 1989, *22*, 184–189.

[76] Wang, L.; Wu, Y.; Deng, Y.; Kim, B.; Pierce, L.; Krilov, G.; Lupyan, D.; Robinson, S.; Dahlgren, M. K.; Greenwood, J.; Romero, D. L.; Masse, C.; Knight, J. L.; Steinbrecher, T.; Beuming, T.; Damm, W.; Harder, E.; Sherman, W.; Brewer, M.; Wester, R.; Murcko, M.; Frye, L.; Farid, R.; Lin, T.; Mobley, D. L.; Jorgensen, W. L.; Berne, B. J.; Friesner, R. A.; Abel, R., *J. Am. Chem. Soc.* 2015, *137*, 2695–2703.

[77] Shivakumar, D.; Harder, E.; Damm, W.; Friesner, R. A.; Sherman, W., *J. Chem. Theory Comput.* 2012, *8*, 2553–2558.

[78] Storer, J. W.; Giesen, D. J.; Cramer, C. J.; Truhlar, D. G., *J. Comput. Aided Mol. Des.* 1995, *9*, 87–110.

[79] Jakalian, A.; Jack, D. B.; Bayly, C. I., *J. Comput. Chem.* 2002, *23*, 1623–1641.

[80] Shivakumar, D.; Deng, Y.; Roux, B., *J. Chem. Theory Comput.* 2009, *5*, 919–930.

5

ADVANCES IN CRYSTAL STRUCTURE PREDICTION AND APPLICATIONS TO PHARMACEUTICAL MATERIALS

GRAEME M. DAY

School of Chemistry, University of Southampton, Southampton, UK

5.1 INTRODUCTION

Crystal structure prediction (CSP) is one of the greatest challenges for the application of computational methods to the solid state. The goal of CSP is the prediction of the likely crystal structures of a molecule, given only chemical information regarding the molecule(s) being crystallized. Typically, the starting information that is required by CSP methods is the bonding between atoms, as expressed in a chemical sketch of the molecule. No three-dimensional information about the molecular conformation, apart from stereochemistry, is generally required so that the methods are as general as possible and can be performed in advance of any structural characterization having been performed, or even before synthesis has been completed successfully. As a goal, the predicted structures should agree with what would be obtained from a diffraction-based structural model to as high an accuracy as possible; typically, errors in lattice parameters of up to 3–4% are judged as acceptable and average errors in atomic positions of up to about 0.3 Å [1]. Such a level of accuracy is required so that predicted crystal structures can serve as a starting point for predicting the physicochemical properties of the crystal.

The prediction of how the chemical properties of a molecule determine its crystal packing is a broader goal in the area of crystal engineering, where most

Computational Pharmaceutical Solid State Chemistry, First Edition. Edited by Yuriy A. Abramov.

research activity relates to the development and verification of empirical rules describing the close packing of molecules and the association of functional groups *via* reliable intermolecular interactions (e.g., hydrogen bonding, halogen bonding, and π–π stacking). Computational methods for CSP can contribute to providing a quantitative foundation to crystal engineering and have developed thanks to the development of realistic physical models of the interactions between atoms, their implementation in efficient software packages, and the availability of computational resources that have allowed this computationally demanding problem to be addressed.

The purpose of this chapter is to describe computational methods for CSP, recent progress that has been made in these methods, and to illustrate the use of CSP in various applications, with a particular focus on applications that are relevant to pharmaceutical materials.

5.1.1 Motivation

A large part of the motivation behind developments in CSP has been the perceived challenge of such a demanding problem. In 1988, Maddox asserted in his *Nature* editorial [2] that it is "One of the continuing scandals in the physical sciences" that "it remains in general impossible to predict the structure of even the simplest crystalline solids from a knowledge of their chemical composition." While he was referring to inorganic solids in his statement, the general consensus in the fields across inorganic and organic materials research is that this situation had not progressed significantly over the next 10–15 years [3, 4]. Such a seemingly unmet challenge naturally attracts continued research. Even without considering the practical applications of CSP, the goal of structure prediction has merits: the pursuit of reliable methods has pushed the development of simulation methods that have impacted on other areas of computational chemistry. As an example, the use of elaborate, atomic multipole-based electrostatic models for intermolecular interactions has been shown to dramatically improve the reliability of predictions [5, 6]; the evidence from CSP studies has helped motivate more widespread use of multipolar electrostatics in molecular simulations [7]. Applications of CSP to organic molecular solids have also recently served as challenging tests of solid-state electronic structure calculations, such as periodic implementations of density functional theory (DFT) [8–11], and of global optimization methods designed for exploring high-dimensional energy landscapes [12–15].

Several applications for CSP within an academic or industrial research setting can be envisioned. These applications range from providing structural models that might aid the determination of crystal structures in combination with experimental data, to the more ambitious aim of performing solid form screening *in silico* and assessing the likely solid forms of a molecule prior to synthesis. These applications are discussed in more detail in a later section, before which an overview of methodologies is presented.

5.2 CRYSTAL STRUCTURE PREDICTION METHODOLOGIES

The most general and successful method that has been developed for CSP is referred to as global lattice energy minimization [16]. This approach involves locating and assessing the relative stabilities of all low-energy local minima on the lattice energy landscape. The global energy minimum refers to the lowest energy structure on the lattice energy landscape, and the usual assumption is that the structure corresponding to the global minimum is the most likely observable crystal structure [16, 17]. The description of the crystal structure prediction problem as global minimization is somewhat misleading; the global minimum is important, but we are always also interested in the other low-energy structures, which represent potential polymorphs. It is the entire landscape of putative crystal structures, sometimes referred to as the crystal energy landscape [18], which is required.

The process of exploring the crystal energy landscape can conceptually be broken down into three steps (Fig. 5.1): (i) prediction of the molecular geometry, (ii) a search for physically reasonable crystal-packing alternatives of the molecule being studied (trial structures), and (iii) an assessment of the relative stabilities of these computer-generated crystal structures. While most methodologies involve some overlap of these three general steps, we discuss them as three distinct steps here for simplicity.

5.2.1 Molecular Geometry

Many applications of CSP involve performing predictions before any crystal structure of the molecule is known, or perhaps even before the compound has been synthesized. In this situation, the starting point for the prediction usually only involves the chemical diagram or connectivity between constituent atoms. Therefore, the construction of

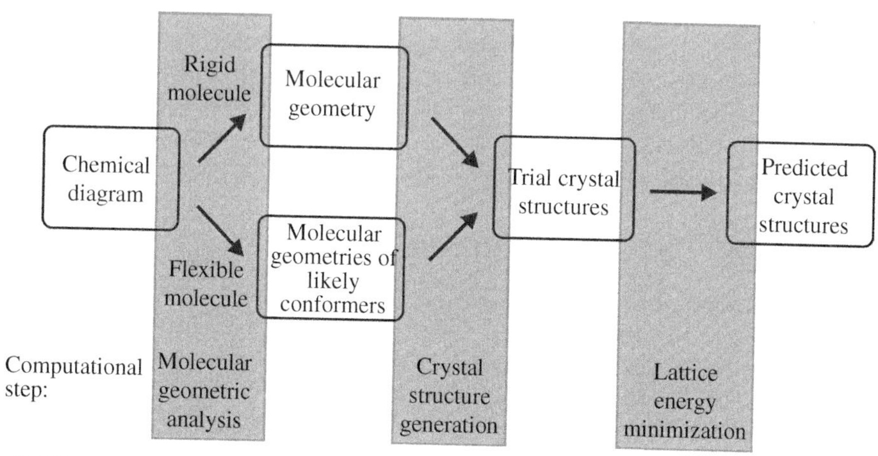

FIGURE 5.1 Schematic representations of the steps of crystal structure prediction (CSP) by global lattice energy minimization.

trial crystal structures requires that this chemical information is converted into a realistic representation of the three-dimensional geometry of the molecule.

The construction of the molecular geometry is now a fairly trivial step for small molecules with only one conformation; a starting model can be built from typical bond lengths and angles, and this starting geometry can be refined by geometry optimization using electronic structure methods.

5.2.1.1 Rigid Molecule Assumption For small molecules, CSP is often simplified by making the *rigid molecule assumption*, at least in some stages of the prediction process. This approximation assumes that the molecular geometry in the crystal remains undistorted from that of the isolated molecule. Molecules for which CSP is realistic are always small enough that the geometry of the isolated molecule can be calculated with good accuracy using quantum chemical methods, based on either wavefunction theory (e.g., Hartree–Fock, perturbation theory methods such as MP2) or DFT. If the rigid molecule approximation is realistic, the molecular structure obtained by geometry optimization of the isolated molecule can be kept fixed through all subsequent steps of generating and optimizing crystal structures.

The importance of this simplification should become clear in the discussions of crystal structure generation and energy minimization that follow. At the stage of generating initial crystal structures, the rigid molecule approximation reduces the number of structural degrees of freedom. Including fewer degrees of freedom during structure searching reduces the computational difficulty and expense significantly. At the final stage of lattice energy minimization and the ranking of structures on calculated stability, the rigid molecule lattice energy approach removes the requirement for an intramolecular energy model when comparing sets of predicted crystal structures. As a consequence, only differences in calculated intermolecular interactions must be considered, simplifying the assessment of relative stabilities and eliminating a source of error in the evaluation of differences in intramolecular energies between crystal structures. It is because of these simplifications that many early applications of CSP focused on small, rigid molecules.

While it is a computationally convenient approximation, the rigid molecule approach is only justified under certain conditions. For discussion, we partition the total lattice energy of a crystal structure into intermolecular and intramolecular components:

$$E_{\text{lattice}} = E_{\text{intramolecular}} + E_{\text{intermolecular}} \tag{5.1}$$

Any structural change that lowers the lattice energy, E_{lattice}, is energetically favorable. Therefore, distortion of the molecular geometry away from the geometry of the isolated molecule can occur as long as this leads to a lower total lattice energy. Since the isolated molecule–optimized molecular structure is, by definition, at a minimum in intramolecular energy, any distortion away from this geometry must increase the intramolecular energy, $E_{\text{intramolecular}}$. Therefore, such a distortion can only be energetically favorable in a crystal structure if $E_{\text{intermolecular}}$ is lowered by more than the increase in $E_{\text{intramolecular}}$; the *intra*molecular energy penalty must be balanced out by a resulting improvement in *inter*molecular interactions. Therefore, for a molecule to be considered rigid in the context of its crystal packing, any intramolecular change must come at a higher energetic cost than the gain in intermolecular interactions that is associated with this distortion in any of its potential crystal structures.

Where molecular flexibility must be included in a calculation, it is necessary to consider what changes in molecular geometry can be caused by the intermolecular interactions in a crystal. The potential flexibility of a molecule can be considered in terms of the three types of intramolecular distortion: bond stretching or compression, angle bending, and changes in torsion angles describing the rotation about bonds (including changes in ring conformations and improper torsions describing the pyramidalization around atoms). The cost of each type of distortion can be estimated through force field or quantum mechanical calculation on simple molecules; for example, a 0.05 Å change in a C—C single bond costs approximately 5 kJ/mol, while a C—C—C bond angle can change by 5–10° at the same energetic cost [19, 20]. By comparison, changes in torsion angles are much softer: a 10° change in the central torsion angle of butane amounts to less than 1 kJ/mol increase in energy [20]. From these values, and the expected improvement in intermolecular interactions that can be achieved, we expect bond lengths to be minimally affected by crystal packing and angles to potentially be changed by a few degrees, while torsion angles can be expected to be significantly distorted by intermolecular interactions.

Two types of studies are useful for verifying these expectations: the comparison of molecular geometric parameters in different crystalline phases and investigations of the differences between gas phase and crystalline phase molecular geometries. The former type of investigation can be based on experimentally determined crystal structures by comparing geometrical parameters (i) of the same molecule in different solid forms (polymorphs, solvates or co-crystals), (ii) between crystallographically independent molecules in the same crystal structure (i.e., crystal structures with more than one molecule in the asymmetric unit of the crystal), and (iii) comparing geometric parameters that would be symmetrically equivalent in the idealized molecular geometry, but where the full point group symmetry of the molecule is not maintained in its crystal structure. From such studies, Kitaigorodskii concluded that covalent bond lengths are unaffected by crystal packing, while bond angles can be altered by up to a few degrees [19]. The comparison of gas phase and crystalline molecular geometries based on either experimental data or high-level computational studies provides a similar picture [21–23]. Therefore, it is usually a reasonable approximation to fix bond lengths in a molecule at gas phase values. The same is usually true for bond angles, apart from where the atoms are involved in strong, directional interactions such as hydrogen bonding, when the intramolecular cost of distorting an angle can be balanced by improving the geometry, and hence the energy of the intermolecular interaction.

In contrast to bond lengths and angles, crystal packing can have an important effect on the torsion angles in a molecule; the rotation about single bonds comes at a small energetic cost, but can lead to large changes in overall molecular shape. A molecule can, therefore, adjust its shape to improve close packing and to optimize the relative arrangement of functional groups to improve intermolecular interactions. A simple example is biphenyl, where the angle between aromatic rings is near 40–45° in its gas phase geometry [24, 25], but flattens to a planar conformation in its room temperature crystal structure to improve its intermolecular interactions in its crystal structure [26, 27]. A similar crystal packing effect has been shown for the ring conformations of cyclobutane rings, whose gas phase geometry is puckered, while flattening of the ring can improve intermolecular interactions to more than compensate for the increase in molecular energy [27]. Such changes in molecular geometry should therefore be considered during the generation of crystal structures and during their lattice energy minimization.

One example of how this information is sometimes used is that flexible molecules can be treated as a set of linked, rigid fragments, where the fragment geometries are fixed at their gas phase geometries, while reorientation of fragments with respect to each other is allowed by rotation about linking bonds [28, 29]. Another example is the CrystalOptimizer [29–31] approach to lattice energy minimization of crystals structures of flexible molecules: the molecule is treated quantum mechanically, with its geometry optimized in isolated molecule calculations, subject to certain intramolecular degrees of freedom being distorted by the intermolecular interactions in the crystal structure.

5.2.1.2 Conformational Analysis Apart from a few early exceptions [32–34], successful CSP studies have been limited to fairly small, rigid molecules until the past decade. This was partly due to the necessity of making the rigid molecule approximation. This limitation to rigid molecules has restricted the application of CSP methods, particularly in the field of pharmaceutical materials, where typical molecules are becoming larger and more flexible year on year.

In practice, completely rigid molecule CSP is becoming less and less common: some degree of molecular flexibility is now almost always included at some stage of the calculations, either in generating trial structures or in their lattice energy minimization. However, the adaptation of methods to large, very flexible molecules has been slow.

Conformationally flexible molecules have been a particularly difficult challenge for CSP methods for two reasons. To understand these, we must consider two aspects of conformational flexibility. One of these is the possibility of low-energy distortions of a molecule from its ideal isolated molecule geometry, the nature of which has been discussed earlier. The second aspect of flexibility that must be considered is the existence of multiple local minima (i.e., conformers) on the intramolecular energy surface. Either of these two aspects of conformational flexibility effectively increases the dimensionality and size of the search space that must be explored in locating all possible low-energy crystal packing possibilities, thus increasing the difficulty of exploring all crystal packing possibilities during the generation of trial crystal structures. Crystal structure generation methods and the dimensionality of search space are discussed in more detail later. The second impact on the CSP of flexible molecules is the added requirement to accurately model the balance between inter- and intramolecular energies in the computer-generated crystal structures. We first address the impact of flexibility on the crystal structure search space. Advances relating to the second of these issues are dealt with in a later section.

The presence of multiple conformers (distinct minima on the intramolecular energy surface) impacts on the crystal structure search space because there is no restriction that constrains a molecule to crystallize using the lowest energy conformer that the molecule would adopt in isolation (the gas phase or an isolated molecule calculation). This fact is evident from the occurrence of conformational polymorphism [35, 36], where a given molecule is observed to crystallize in alternative crystal structures using different conformers. One can imagine that any conformer that is populated under the thermodynamic conditions used in crystallization has the opportunity to form the basis of the molecule's crystal structure. The particular set of conformers that are available to a molecule might depend strongly on crystallization conditions, such as temperature or solvent of crystallization.

Under the usual assumption used in CSP that the most likely crystal structures are those with lowest lattice energy, a conformer other than the global minimum conformer is possible as long as it can make up the increase in *intra*molecular energy by improved *inter*molecular interactions, such that the total lattice energy (Eq. 5.1) is as low as possible. Since the intramolecular energies calculated using electronic structure methods provide the total energy involved in bringing electrons and nuclei together, while intermolecular energies include relatively weak interactions, the absolute magnitudes of $E_{\text{intramolecular}}$ and $E_{\text{intermolecular}}$ are very different. Therefore, it is usually convenient to work with *relative* intramolecular energies:

$$E_{\text{lattice}} = \Delta E_{\text{intra}} + E_{\text{Intermolecular}} \tag{5.2}$$

where ΔE_{intra} is the molecular energy given relative to the lowest energy possible conformer of the molecule being studied. This corresponds to the intramolecular energy change that would be involved in moving the molecule from its crystalline geometry to its most stable gas phase geometry.

To discuss ΔE_{intra} further, and the choice of molecular conformers in CSP, we refer to the schematic intramolecular energy surface in Figure 5.2. The conformers that are possible for the given molecule, corresponding to the local energy minima with respect to any intramolecular geometry change, are numbered in order of increasing $E_{\text{Intramolecular}}$. The general intramolecular coordinate shown on the horizontal axis encompasses all possible geometrical distortions, such as angle bending, torsional changes about rotatable bonds, and changes in ring conformations. ΔE_{intra} is defined with respect to the global minimum energy conformer, labeled 1. The molecular geometry, as found in the crystal structure, is labeled as point A in Figure 5.2. Since the molecule can be distorted by intermolecular packing forces in its crystal structure, the crystalline molecular geometry need not correspond to a local energy minimum on the intramolecular energy surface. Conceptually, we extract the molecular geometry from its crystal structure and perform a local energy minimization of the resulting isolated molecule, that is, adjusting the atomic coordinates to reach the nearest local minimum on the intramolecular energy surface without crossing any energy barriers. The result of this process is the conformer that is related to the molecular conformation found in the crystal (point B, Fig. 5.2). In the example shown in the schematic, the conformer

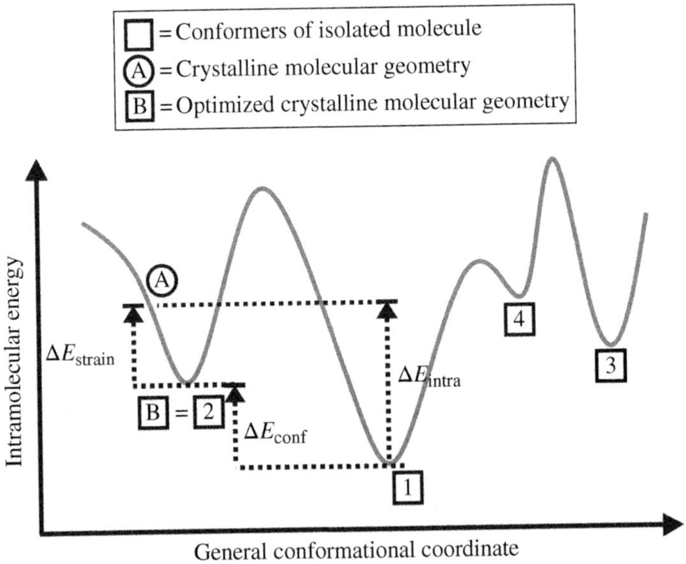

FIGURE 5.2 Schematic representation of a conformational energy surface. Numbered points represent conformers of the isolated molecule. Points A and B represent the crystalline molecular geometry and its associated conformer. ΔE_{strain} and ΔE_{conf} are defined in the text. Source: Thompson and Day [37], http://pubs.rsc.org/en/content/articlehtml/2014/sc/c4sc01132e. Used under CC-BY 3.0 http://creativecommons.org/licenses/by/3.0/.

that is related to the crystalline molecular geometry is not the global minimum conformer, but the second lowest energy conformer available to the molecule.

We can now partition ΔE_{intra} into two contributions, which we refer to as ΔE_{conf} and ΔE_{strain}. These relate to the two types of conformational flexibility introduced earlier. ΔE_{conf} is the energy of the conformer associated with the crystalline molecular geometry, relative to the conformational global energy minimum: the energy difference between two minima on the intramolecular energy surface. ΔE_{strain} is the increase in intramolecular energy (strain) associated with the distortion of the molecular geometry away from its associated conformer due to intermolecular interactions in its crystal structure (Fig. 5.2).

For CSP, it is important to understand the range of these two quantities, ΔE_{conf} and ΔE_{strain}, that are possible. Theoretically possible molecular conformers that lie above some limit in ΔE_{conf} must become irrelevant when it comes to crystallization. The maximum ΔE_{conf} that must be considered therefore dictates how many different molecular conformers must be considered as possibly leading to low-energy crystal structures. Most approaches to CSP require a separate series of calculations (molecular geometry optimization, trial crystal structure generation, and lattice energy minimization) for each distinct conformer of a molecule. Hence, the time required for performing CSP scales approximately linearly with the number of relevant conformers. Worryingly, the number of possible conformers tends to increase rapidly with molecular size. The example data in Table 5.1, taken from Thompson and Day's recent study of molecular conformations in the crystal structures of flexible molecules [37],

TABLE 5.1 Numbers of Computer-Generated Isolated Molecule Conformers for a Selection of Flexible Molecules from Thompson and Day [37], Before and After Limiting Conformer Selection by ΔE_{conf}

Molecule	N_{conf}	N_{conf} (Within 26 kJ/mol in ΔE_{conf})
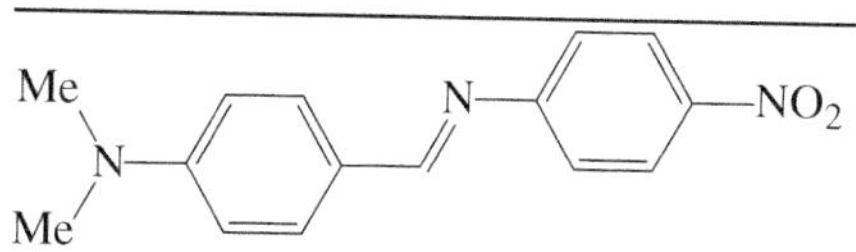	2	2
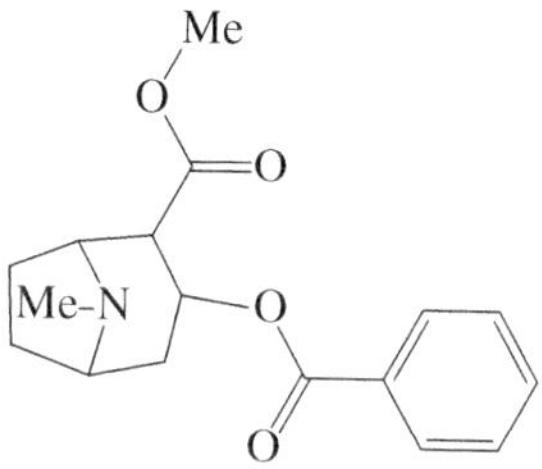	15	7
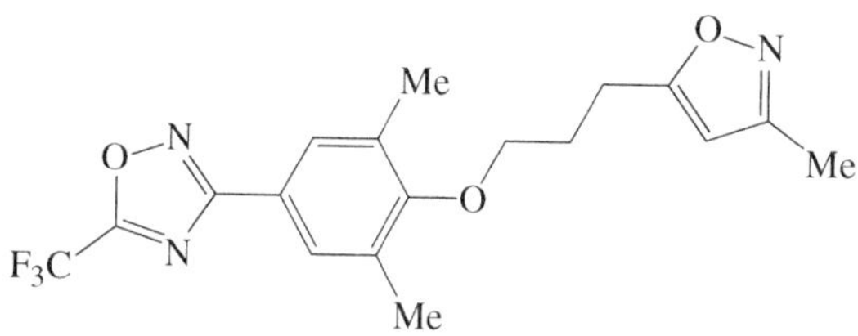	126	124
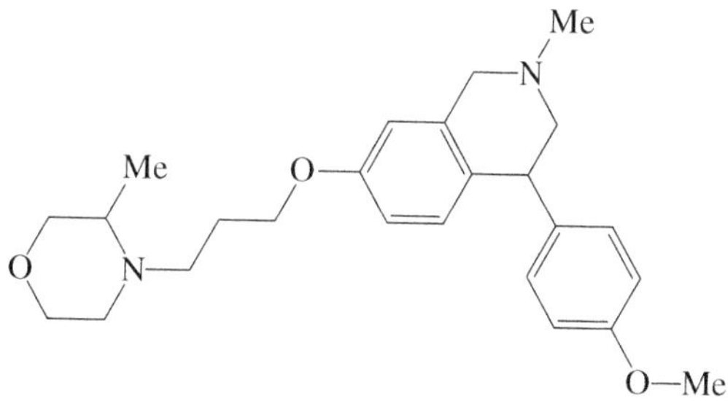	2418	623

demonstrates how the number of possible conformers increases rapidly with increasing flexible degrees of freedom. N_{conf} is the number of conformers for each molecule that are located in a systematic conformational search on the isolated molecule. Increasing the number of rotatable (exocyclic) single bonds and conformationally flexible rings leads to a rapid explosion in the number of possible molecular conformers. These numbers demonstrate how conformational flexibility can quickly become the bottleneck that makes CSP unfeasible for very flexible molecules. The computing time required to generate and lattice energy minimize crystal structures for a single molecular conformer is typically on the order of days using moderate computing resources. Molecules with hundreds or thousands of relevant molecular conformers therefore correspond to years of required computational effort and are therefore beyond the scope of current

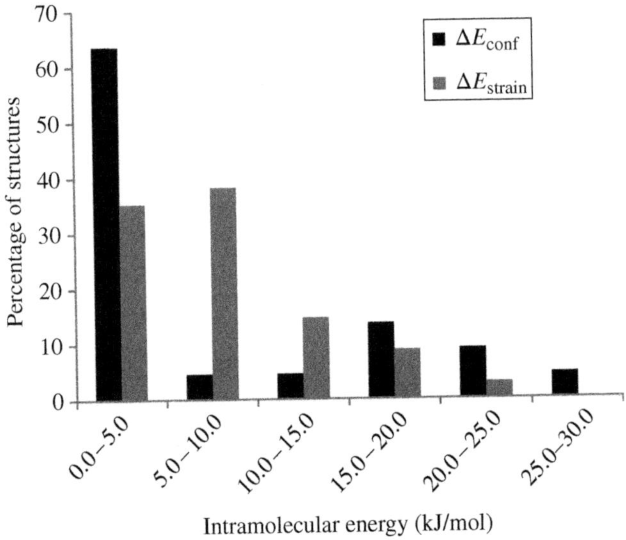

FIGURE 5.3 Distributions of ΔE_{strain} and ΔE_{conf} in crystal structures of flexible molecules. Source: Thompson and Day [37], http://pubs.rsc.org/en/content/articlehtml/2014/sc/c4sc01132e. Used under CC-BY 3.0 http://creativecommons.org/licenses/by/3.0/.

methods. Some of the most challenging molecules studied to date using CSP methods have required 40–80 conformers to be included in the calculations [29, 38].

A good understanding of what ΔE_{conf} is possible in crystal packing can focus computational effort on the most relevant conformers and reduce the computational cost of tackling large, flexible molecules. A recent study, based on single molecule and solid-state DFT calculations, aimed at addressing this question. As a part of this study, the full lists of gas phase conformers of 15 flexible molecules with known crystal structures (many of the molecules being polymorphic) were generated and geometry optimized. These lists of possible conformers were then compared to the conformer obtained by geometry optimizing the molecule starting from the geometry (or geometries, for polymorphic molecules) found in the known crystal structures [37]. The resulting conformer that is associated with the crystalline geometry was then located in the list of all possible conformers, allowing an assessment of ΔE_{conf} for each of the 29 crystal structures of the 15 molecules. The distribution of ΔE_{conf} from this study is shown in Figure 5.3. One important finding was that it is rare that flexible molecules crystallize using their lowest energy (global minimum) gas phase conformer: only 6 of the 15 molecules adopt their lowest energy conformer in one of their crystal structures. ΔE_{conf} is usually small: approximately two thirds of crystal structures adopt a molecular conformer that is within 5 kJ/mol of the gas phase global minimum. However, a number of molecules adopt high-energy conformations, with ΔE_{conf} ranging up to 25.6 kJ/mol.

This study helps define an energetic cutoff for molecular conformers that must be considered in CSP: conformers with $\Delta E_{conf} > 26$ kJ/mol are unlikely to result in observed crystal structures, presumably because this is near the limit of what intramolecular energy penalty can be compensated by improved intermolecular interactions. The effect of restricting conformers to within this limit is shown in Table 5.1; the savings resulting from exclusion of high-energy conformers can be considerable for some molecules, but the largest molecules would still require hundreds of conformers to be exhaustively studied.

One caveat to the earlier study is that the molecules studied had no capacity to form intramolecular hydrogen bonds. Where intramolecular hydrogen bonding is possible, all low-energy conformers of the isolated molecule are expected to form the intramolecular hydrogen bond, while the crystalline conformer might break this interaction in favor of intermolecular hydrogen bonding; in these cases, ΔE_{conf} can be in an excess of 50–60 kJ/mol [39].

Another important observation regarding conformers of flexible molecules in their crystal structures is that molecules tend to adopt extended conformers in the solid state [37]. This can be explained by the increased accessible surface area that an extended conformer has available to form intermolecular interactions. Thompson suggests that the intermolecular stabilization available to a conformer can be estimated prior to predicting its possible crystal structures, by calculating its accessible surface area. Therefore, the selection of relevant conformers for CSP can be adjusted by including this estimate of intermolecular stabilization, biasing the choice of conformers to those where ΔE_{conf} can be compensated by intermolecular interactions. The suggested form of this biased energy ranking was [37]:

$$\Delta E_{conf,biased} = \Delta E_{conf} + m\Delta A_{Connolly} \tag{5.3}$$

where $\Delta A_{Connolly}$ is the difference in Connolly surface area of the conformer relative to the global minimum energy conformer of the same molecule. The constant m has a suggested value of 0.75 kJ/(mol Å^2), based on the variation in sublimation enthalpies with molecular surface area of a set of rigid hydrocarbons [37]. Re-ranking conformers based on $\Delta E_{conf,biased}$ found that all conformers seen in observed crystal structures are now located within 5 kJ/mol of the lowest energy conformer (compared to 26 kJ/mol based on ΔE_{conf}), and that the resulting lists of conformers that must be considered are considerably shorter. This is an important step and, as we continue to increase our understanding of conformational preferences in crystal structures, the range of molecules to which CSP can be applied is expanding.

Thompson's study of conformations in crystal structures also provides information on typical strain energies, ΔE_{strain}, in molecular crystals, by reporting the change in intramolecular energy during optimization of the crystalline molecular geometry to its nearest local minimum. The amount of molecular strain was found to range up to about 20 kJ/mol (Fig. 5.3).

A complementary approach to energy-based conformational search methods is the informatics approach, where likely conformations of a molecule are built up from the distributions of geometric parameters (bond lengths, angles, torsions, and ring conformations) in chemically similar molecules to the molecule under investigation. Databases of crystal structures form the richest source of information on molecular

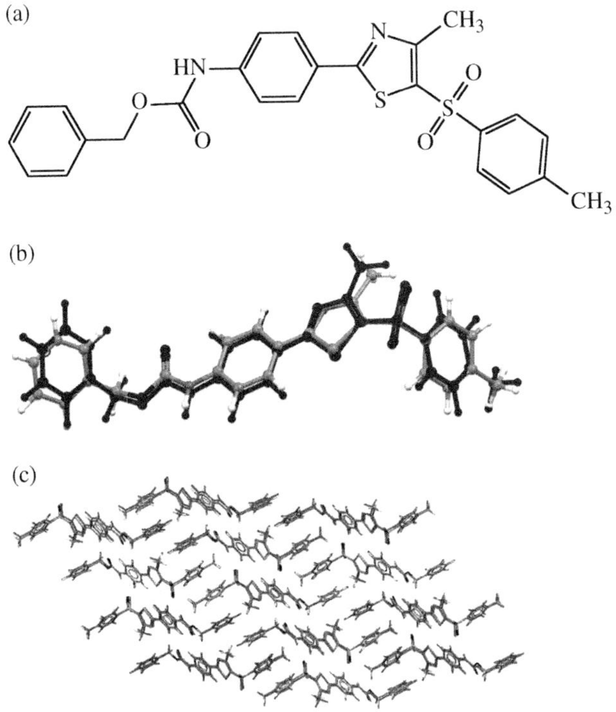

FIGURE 5.4 Molecule XX (benzyl-(4-(4-methyl-5-(p-tolylsulfonyl)-1,3-thiazol-2-yl)phenyl) carbamate) from the fifth blind test of crystal structure prediction. (a) Chemical diagram, (b) overlay of one of the 48 database generated conformations (red) with the conformation in the observed crystal structure, (c) overlay of the CSP global minimum in lattice energy (green) with the observed structure from X-ray diffraction. Source: Kazantsev et al. [29]. Reprinted with permission of Elsevier. (*see insert for color representation of the figure.*)

geometric preferences, and powerful software tools exist to help interrogate the information held in databases, for example, the MOGUL software [40], which gives access to a library of geometric parameters from the Cambridge Structural Database (CSD) [41]. The assumption in using such data is that distributions of geometric parameters seen in observed crystal structures represent the underlying probability distributions of the values for these parameters in as-yet undetermined crystal structures. Therefore, likely conformers of a molecule can be built up from distributions of observed geometric parameters and used to seed the generation of crystal structures.

A recent example of such an approach was in the successful prediction of the crystal structure of benzyl-(4-(4-methyl-5-(*p*-tolylsulfonyl)-1,3-thiazol-2-yl)phenyl) carbamate, a large, flexible pharmaceutical-like molecule in the fifth CSP blind test (Fig. 5.4) [1, 29], often referred to by its blind test reference number, molecule XX. These blind tests are organized every few years to evaluate the current state of CSP methods, by setting a series of molecules as targets, whose crystal structures are withheld until participants have submitted their predictions. Molecule XX was included in the latest, 2010 blind test as a flexible, pharmaceutical-like target. In one of the two successful methodologies, database distributions around all flexible torsion angles were generated and 42 starting conformers of molecule XX were built from all combinations of maxima in these distributions. After crystal structure generation with all 42 conformers and flexible-molecule lattice energy minimization of the resulting crystal structures, the global minimum in lattice energy was found to reproduce the observed structure extremely well (Fig. 5.4c). The success of this method led from the fact that one of the database-derived conformers very closely reproduced the conformer in the true crystal structure (Fig. 5.4b).

In many cases, the energy-based approach and informatics approach should be expected to lead to similar conformers: the maxima in distributions of geometric parameters drawn from crystallographic data typically lie near the minima in a calculated energy profile for that parameter. Indeed, the only other success for molecule XX in the CSP blind test used an energy-based conformational analysis [29]. The advantage of using crystallographic data is in speed, but relies on having the structures of enough sufficiently similar molecules from which a probability distribution can be built. Energy models, at least when based on nonempirical, quantum mechanical energy calculations, are more generally applicable when exploring new chemistry, and there are insufficient chemically similar molecules from which reliable distributions of geometric parameters can be obtained.

The informatics approach to predicting molecular conformation has a clear advantage when the crystalline environment systematically biases molecular geometries away from what would be calculated for an isolated "gas phase" molecule. One example is that of the amino acids, which invariably adopt a zwitterionic form in their crystal structures, whereas the zwitterion is unstable in the gas phase and does not even correspond to high energy, local minimum [42–44]; unconstrained optimization of α-amino acids starting from a zwitterionic geometry revert to a non-zwitterionic geometry with no energy barrier. In this situation, gas phase conformational preferences become irrelevant to predicting their crystal structures. However, sufficient crystal structures of amino acids are known so that their conformational preferences in these crystal structures can be used to guide the initial choice of conformers, leading to successful CSP [45–47], even under blind prediction conditions [48].

5.2.2 Crystal Structure Searching

Once a molecular geometry, or ensemble of conformers, has been produced, the central computational step involved in CSP is the generation of trial crystal structures (Fig. 5.1). The goal of crystal structure generation methods is to produce all possible, physically reasonable arrangements of molecules in a translationally repeating arrangement.

Observable crystal structures must correspond to local minima on the lattice energy surface. A local minimum corresponds to a point from which any change in structure results in an increase in lattice energy. Local lattice energy minima are located by generating trial crystal structures, which are each locally lattice energy minimized, that is, moved downhill on the energy surface to find the nearest energy minimum, without crossing any energy barriers. The generation of trial crystal structures involves exploring all structural parameters, which define the crystal packing of a molecule and to create enough starting points so that all relevant local minima are reached after local lattice energy minimization. In practice, it is rarely realistic to locate all local energy minima; their number can be very large, particularly in the high-energy regions of the energy surface, typically corresponding to poorly packed structures. However, a useful method should aim to find those energy minima that are low enough in energy to correspond to observable crystal structures.

The degrees of freedom that must be sampled during structure generation include the six bulk parameters defining the crystallographic unit cell: three cell lengths (a, b, and c) and three angles (α, β, and γ), and the internal structure of the unit cell—the positions and orientations of molecules within the unit cell (6 degrees of freedom per molecule)—as well as any flexible intramolecular, conformational degrees of freedom. One can immediately see that this is a high-dimensional problem. Excluding intramolecular degrees of freedom, a crystal structure containing G molecules in the unit cell (G is used in preference to Z [49], whose crystallographic meaning is the number of formula units in the unit cell, so differs from Z for co-crystals, solvates, and salts) is defined by $6+6G-3$ degrees of freedom, 6 defining the unit cell parameters and $6G-3$ to position the molecules in the unit cell (3 of the total $6G$ degrees of freedom correspond to combinations of molecular translations that translate the entire unit cell in space, leaving the structure and energy unchanged).

The difficulty of sampling an energy landscape sufficiently well to locate all local minima is known to increase very quickly with the dimensionality of the phase space that must be searched. Generating a set of trial structures that leads to all local lattice energy minima being located in such a high-dimensional space has been one of the big problems for CSP. The scaling of the difficulty of this problem with the dimensionality of the search can be illustrated most easily by considering a simple grid–based sampling: bounds are set for each degree of freedom, and trial structures are built at regular intervals for each parameter between these bounds. A grid taking N values for each structural parameter for a problem with D dimensions involves N^D combinations of the values of each structural parameter, each corresponding to a trial structure that must be lattice energy minimized. Single component crystal structures of organic molecules commonly contain 2, 4, or 8 molecules in the unit cell, corresponding to 15, 27, and 51 degrees of freedom, assuming no intramolecular degrees of freedom. Even for very coarse grids, sampling a few values for each degree of freedom, the number of trial structures required for a grid search is in excess of 10^9 for the simplest systems; energy minimizing this number of trial crystal structures is unmanageable, even using large-scale high-performance computing resources.

One simplification to the problem comes through the use of space group symmetry; the common values mentioned earlier for numbers of molecules in the unit cell are dictated by space group preferences for organic molecules, which are clear from surveys of the CSD [50]. This unequal distribution of observed crystal structures among the 230 possible space groups reflects favorable combinations of symmetry elements that enable close packing of molecules [19], and is often taken advantage of in structure searching strategies. Rather than perform the generation and lattice energy minimization with common values of G, it is usual to generate structures in a set of the most commonly observed space groups. One advantage of using space group symmetry is that the generation of crystal structures can be restricted to the most commonly observed space groups, under the assumption that space group statistics from already known crystal structures reflect the probability that low-energy crystal structures can be found within that space group.

The second advantage of using space group symmetry is in the reduction in the dimensionality of the search space. Within a particular space group, the structure generation problem is greatly simplified. First, space group symmetry usually places constraints on the unit cell parameters, fixing the values of some unit cell angles in all but triclinic crystal systems and sometimes reducing the number of independent unit cell lengths. In the extreme case of a cubic unit cell, the 6 degrees of freedom defining the unit cell are reduced to 1 independent cell length, with all cell angles fixed at 90°. In more commonly observed monoclinic and orthorhombic crystal systems for organic molecules, the number of unit cell degrees of freedom is reduced to 4 and 3, respectively. The more important simplification afforded by space group symmetry comes from only having to consider the position and orientation of molecules in the crystallographic asymmetric unit. Once the molecules in the asymmetric unit are positioned, all other molecules in the unit cell can be generated from the space group symmetry operations. Given that most single-component crystal structures have only one molecule in the asymmetric unit, there are only 6 degrees of freedom (3 describing the molecular position and 3 describing its orientation) required to define the molecular arrangement within the unit cell. The dimensionality of searches within common space groups is now typically reduced to between 9 and 12.

The choice of space group symmetries is clearly important. Early CSP studies typically only considered 4–10 of the very commonly observed space groups [51–54]. It is now more common that 20–60 space groups are included in CSP studies [38, 55–60], and some recent studies have considered all 230 space groups in their search [8, 9, 61]. It is worth noting that, even when all space groups are considered, the reduced dimensionality of generating crystal structures without a space group lowers the complexity of the search in comparison to searching without space group symmetry. A further consideration in space group selection is that approximately 10% of reported molecular crystal structures contain more than one independent molecule in the asymmetric unit ($Z' > 1$) [62]. $Z' > 1$ is naturally difficult to sample, given the extra degrees of freedom that must be considered in positioning two or more molecules independently. However, this situation should be included in predictions, and successful predictions have been reported with Z' up to 4 [28, 61, 63].

Having defined the structural degrees of freedom and dimensionality of the space to be searched, we now come to the methods used to generate trial crystal structures. Many methods have been developed for this step, an example of which is the grid search method described earlier. In practice, grid searches are rarely used. Early pioneering approaches, such as Williams' "Ab initio molecular packing analysis" [64] and Dzyabchenko's prediction of benzene's packing modes [51], used the random generation of structures: trial structures are generated with randomly chosen values for all structural degrees of freedom, chosen within sensible physical bounds (e.g., to ensure a reasonable density). Many structures will be unphysical, perhaps because of overlapping molecules, but can be rapidly rejected; the rest are subjected to lattice energy minimization. An advantage over grid-based searches is that the number of trial structures can be continuously increased, whereas a grid fixes the number of trial structures in advance. Therefore, the structure generation can be monitored, lattice energy minimizing as the search proceeds, with new structures added until no unique, low-energy crystal structures are located.

The method might seem fairly brute force in nature, but the random sampling approach has endured and accounts for a large proportion of modern studies [29, 56, 58, 65–67]. A reason for its success can be rationalized by an observation that low-energy structures are usually easily located, due to having wide basins of attraction (the volume in multidimensional structural space that lead to a particular local minimum) [68]. van Eijck tested the random structure generation method, using between 5,000 and 50,000 trial structures per space group, and found the approach to be fairly effective on problems of up to 20 degrees of freedom [49]. This number of degrees of freedom covers structure generation for a rigid molecule with up to two independent molecules in any space group. Another attractive feature of the random search is that the generation of every structure is independent from the variables describing any other trial structure: this makes structure generation and energy minimization an almost perfectly parallelizable problem that scales very well on large high-performance computing resources or distributed computing. The main modification to the random structure generation method that has occurred over the years has been to replace (pseudo)random sequences by low discrepancy and quasi-random sequences [55, 60, 68, 69]. The principal motivation behind this change is that these sequences are designed to provide the most evenly distributed set of structures at all points in the search. This ensures, as well as possible, that the full range of packing possibilities has been sampled.

Of course, CSP has also been a playground for the range of global optimization algorithms: genetic algorithms [15], particle swarm optimization [70], metadynamics [14], and simulated annealing [71, 72]. Keeping in mind that the goal is to produce the full set of energetically feasible crystal structures during a search, methods should not be judged using the popular criterion of which is fastest at finding the global minimum. Low-energy crystal structures can often be structurally very different, in terms of molecular conformation, intermolecular interactions, unit cell dimensions, and space group symmetry, so a useful method should balance speed with the necessary breadth of sampling the structural degrees of freedom. As yet, there is no clear evidence that these more advanced global optimization methods offer a substantial advantage over random or quasirandom search methods. In fact, some of these might be too aggressive at hunting the global minimum, at the expense of a broad sampling of all low-energy packing possibilities.

5.2.3 Structure Ranking

A principal assumption of CSP by lattice energy minimization is that the observed crystal structure (or structures, where polymorphism occurs) corresponds to the lowest energy possible crystal packings of the molecule in question. Thus, the quality of the model used to evaluate the energies of predicted crystal structures is key to the success of CSP studies. The ranking of crystal structures is particularly challenging because the possible crystal structures of a given molecule are often found to be separated by less than a kilojoule per mole.

The most commonly applied method is the atom–atom potential, or force field, method for calculating lattice energies. Initially considering intermolecular interactions only, the total lattice energy is assessed as a sum over intermolecular interactions, U_{MN}, which in turn are computed as a sum over atom–atom interactions, U_{ik}:

$$U_{\text{intermolecular}} \approx \sum_{M<N}^{Nmol} U_{MN} = \sum_{i\in M,k\in N} \sum_{M<N}^{Nmol} U_{ik} \tag{5.4}$$

The functional form of atom–atom interactions must contain terms describing the exchange-repulsion, the attractive dispersion interaction between atoms, and the electrostatic interaction:

$$U_{ik} = U_{ik}^{\text{rep-disp}} + U_{ik}^{\text{electr}} \tag{5.5}$$

Electrostatic interactions can be most simply modeled as the Coulomb interaction between partial atomic charges, while the repulsion–dispersion part is usually described by a Lennard-Jones or, more accurately, an *exp-6* form, each of which contains parameters that must be fixed. High-quality empirically fitted parameter sets have been developed, where the atom–atom interactions are parameterized to reproduce the structures, sublimation enthalpies and, sometimes, further observable properties of organic molecular crystals [73, 74]. Their use has been very effective in CSP. Nonempirical approaches to fitting intermolecular force fields, where the parameters are derived from quantum mechanical calculations, have occasionally been applied for CSP [75–78], but these are currently limited to small molecules, so currently lack relevance for typical pharmaceutical molecules.

The most important development in force fields that has advanced the field of CSP has been the application of anisotropic atom–atom force fields [79], where the interaction between atoms depends on their mutual orientation as well as their separation. In terms of the physical contributions to intermolecular interactions, anisotropy is most important in the electrostatic component [80], which can be modeled using distributed multipole methods, where each atomic site is assigned a dipole, quadrupole, and sometimes higher angular momentum functions, in addition to its partial charge [81]. These distributed multipole methods are necessary to capture details of the molecular charge distribution such as lone pairs and π-electron density. As with atomic partial charge models, the charge distribution in a molecule is determined from a quantum mechanical calculation on the isolated molecule.

The failings of simpler models that lack directionality in the atom–atom interactions can be very pronounced in CSP of hydrogen bonding molecules. A molecule can find many ways of filling space to give a crystal structure with reasonable density, and the subset of structures in which hydrogen bond acceptors and donors are in close proximity can still be large. A model that lacks directionality in the electrostatic model does not distinguish structures in which hydrogen bond donors approach a lone pair from structures with any other acceptor orientation around the donor. Thus, their energies are not distinguished in such a simple energy model. With an anisotropic electrostatic model, in which atomic dipoles and quadrupoles can adequately represent lone pair density, structures with favorable hydrogen bond geometries are preferentially stabilized [6, 28]. It has been shown that, for hydrogen bonding molecules, the proportion of observed crystal structures that are located at, or within 1 kJ/mol of the lattice energy global minimum increases by about 15% when an atomic point charge model is replaced by atomic multipoles [6]. It is also noticeable in the results of CSP blind tests that the most consistently successful force field methods employ anisotropic electrostatic models [1, 82, 83].

The remaining limitation of the atom–atom method described earlier is that interactions are assumed to be pairwise additive: the interaction between any pair of atoms or molecules does not depend on the positions of the other molecules in the crystal. The pairwise additive approximation is appealing because of the computational savings that result from ignoring many-body interactions. However, the cooperativity of interactions can be important and is ignored in these models. The most rigorous solution is to include a polarizability on each atom so that its multipole moments can adjust in response to the electric field generated by all other atoms in the crystal, giving the induction energy contribution to the lattice energy. This approach is currently rarely applied, due to the expense and technical difficulty of having to reevaluate the induction energy at every step of a lattice energy minimization. A much simpler approach that has started to become popular is to evaluate the charge distribution in a molecule from a calculation where the molecule is embedded in a polarizing continuum, with a dielectric constant that is typical of the organic solid state ($\varepsilon \approx 3$) [46].

The atom–atom force field method can be extended to include intramolecular interactions, through bond stretch, angle bending, and torsional energy terms. However, another result that has emerged from numerous literature studies, as well as the blind tests, is that such intramolecular force fields are unable to model intramolecular energies with sufficient accuracy for CSP. In their place, hybrid approaches have been developed, coupling electronic structure methods (usually DFT) to describe the molecular geometry and intramolecular energy, with atom–atom force fields. The approach was first proposed by van Eijck in his studies of glycol, glycerol, and monosaccharides [33, 84] and has since been successfully applied to a range of sometimes quite flexible molecules [29, 58]. While the expense of the quantum mechanical calculations necessary for evaluating the intramolecular energy in tens of thousands of crystal structures would be prohibitive, the energies can be reused between crystal structures with the same, or very similar molecular geometries, by storing all quantum mechanical results in a database and building up approximate models for estimating the energies of molecular geometries that are close to a geometry, which has been sampled in a previously encountered crystal structure [31].

The main alternative approach to force field methods that has emerged in the past few years is the application of solid-state (periodic) DFT calculations for the lattice energy minimization and energetic evaluation of predicted crystal structures. The main weakness of DFT is its failure to account for the attractive dispersion interactions between molecules. While this is a serious limitation for molecular crystals, where dispersion can account for most of the binding energy, DFT can be supplemented by a force field "dispersion correction" [10]. These dispersion-corrected DFT methods have shown remarkable success, particularly in the fourth CSP blind test, where their use led to the successful prediction of all four target molecules [9, 82]. The cost of lattice energy minimization with solid-state DFT is many orders of magnitude more expensive than force field methods. Therefore, the approach developed by Neumann and co-workers is to generate and energy minimize crystal structures with a force field that has been tailored to the molecule under investigation [85]. As long as the force field is sufficiently accurate, only the lowest energy structures from the force field search must be refined using solid-state DFT calculations.

A final remaining consideration in the thermodynamic ranking of crystal structures is that the stability of a crystal structure at a given temperature and pressure is governed by the *free* energy, not simply the lattice energy. Including pressure in the ranking of crystal structures is straightforward, through including a PV term to the calculated lattice energy, during lattice energy minimization and ranking of predicted structures. In this way, CSP can be used to predict polymorphs that emerge at high pressures [86, 87].

Including the effects of temperature is more challenging, as it requires an assessment of the heat capacity and entropy of each predicted crystal structure. Differences in these properties will arise either through differences in disorder (contributing to configurational entropy) or differences in frequencies of the lattice vibrations between structures. While disorder in crystal structures can sometimes be inferred from families of structurally related structures on the crystal energy landscape [88–90], predicted structures are usually assumed to be perfectly ordered. The influence of lattice vibrational contributions in CSP was assessed by van Eijck [32], who showed that vibrational entropy differences could be significant and lead to important re-ranking of the stability of predicted crystal structures. Recently, Nyman and Day evaluated lattice energy and entropy differences between polymorphs of over 500 organic molecules (Fig. 5.5), demonstrating that lattice vibrational contributions to free energy differences can be as large as 3 kJ/mol at room temperature [91]. While this contribution is usually smaller than static lattice energy difference, which range up to about 10 kJ/mol, it causes a re-ranking of polymorph stability in about 10% of cases. Lattice vibrational energy contributions are currently only rarely included in CSP studies [92, 93], but should see an increased usage as the methods for calculating the vibrational frequencies themselves [94] improve in accuracy and efficiency.

5.3 APPLICATIONS OF CRYSTAL STRUCTURE PREDICTION

Crystal structure prediction methods do not aim to replace experimental studies but are a complementary technique to experimental investigations. The roles that CSP can play in such investigations can fall into three categories: to characterize an existing material, usually as an aid to crystal structure determination from

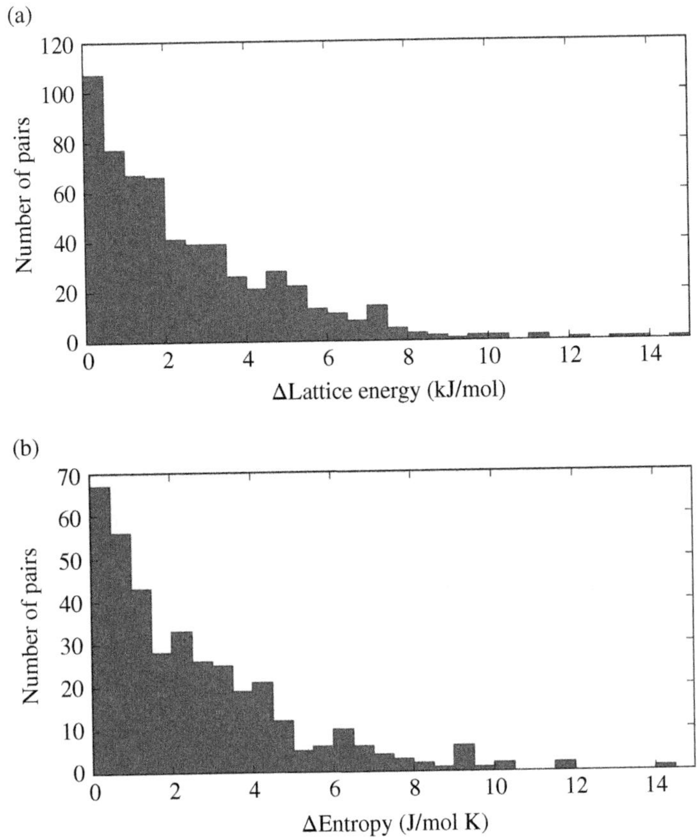

FIGURE 5.5 Distributions of calculated differences in (a) lattice energy and (b) lattice vibrational entropy between pairs of known polymorphs of organic molecules. Source: Nyman and Day [91], http://pubs.rsc.org/en/content/articlehtml/2015/ce/c5ce00045a. Used under CC-BY 3.0 http://creativecommons.org/licenses/by/3.0/.

limited experimental data; to inform solid form (polymorph, salt or co-crystal) screening; and to guide the discovery of new crystal forms with targeted properties.

5.3.1 Crystal Structure Determination

The first, and earliest, application of CSP was to aid in crystal structure determination of an uncharacterized material, combined with experimental observations that, on their own, are insufficient to provide a structure. The idea is that CSP calculations can provide a set of plausible, energetically feasible structures that could be used as starting models to refine against experimental data. Under the assumption that the set of predicted structures is complete, and sufficiently accurate that the experimental observable can be predicted with confidence, the correct structure can be identified by comparing simulated properties of the predicted structures with the observed data.

This method has most commonly been applied where the experimental data consists of a powder X-ray diffraction pattern that is not of high enough quality to allow direct determination of the crystal structure. As just a few of many examples, Schmidt determined the structure of the pigment methylbenzimidazolonodioxazine from an unindexable powder pattern [95], Wu and co-workers determined the crystal structure of the diastereomeric salt L-ephedrine D-tartrate by combining CSP with synchrotron powder X-ray diffraction [96], Perrin, Neumann and co-workers determined the structure of the unstable form III of paracetamol (acetaminophen) [61], and the second polymorph of scyllo-inositol was determined from a mixed-phase powder X-ray diffraction pattern using CSP-generated models [97]. The method has also been applied to solve the structure of an acetic acid:theobromine co-crystal with unknown stoichiometry of its components using CSP calculations at various compositions to generate models from which diffraction patterns were simulated [98]. More recently, CSP has been combined with various other experimental methods for determining crystal structures, such as transmission electron microscopy [99], which can provide diffraction patterns from sub-micron-sized crystallites, leading to the structure of a new polymorph of theophylline from picogram quantities of material [59].

Developments in electronic structure methods have led to accurate methods for predicted NMR chemical shifts from a crystal structure [100]. This has contributed to the development of methods for determining structures from solid-state NMR. The assigned ^{1}H chemical shifts have been shown to provide enough information to select the observed crystal structure of many drug-like molecules from sets of CSP-generated structures [101, 102], and this combination of CSP with solid-state NMR was recently used to determine the structure of a polymorph of 4-[4-(2-adamantylcarbamoyl)-5-tert-butyl-pyrazol-1-yl]benzoic acid, a pharmaceutical compound developed for the treatment of Type 2 diabetes [38].

The combination of experimental data with CSP partly relaxes the requirement for very accurate final energies of the predicted structures. As long as the correct crystal structure is located within a reasonably small subset of low-energy predicted structures, it can be left to the experimental data to identify the structure. This was the case in the CSP-NMR study of 4-[4-(2-adamantylcarbamoyl)-5-tert-butyl-pyrazol-1-yl] benzoic acid, where a lower quality intramolecular model was used to speed up the prediction process. This model overestimated the energy differences between crystal structures containing different conformers, but a good structural model could still be identified by combining energetic assessment with comparison of simulated and measured NMR chemical shifts [38].

Another aspect of structural characterization where CSP can provide insight is the nature of disordered structures. CSP calculations generate an ensemble of perfectly ordered structures. Configurational disorder in molecular crystals can occur where a whole molecule occupies different orientations, or a fragment of a large, flexible molecule can take different orientations with little change in total lattice energy. In these cases, multiple structures can be identified on the crystal energy landscape with very similar overall packing and lattice dimensions, but different molecular orientations [88, 89]. A similar situation can occur in layered crystal structures, where multiple related lattice energy minima differ by the relative arrangement of layers of molecules [57, 103]. The structures and calculated energies resulting from CSP studies can help interpret such disordered structures, the features of crystal packing that lead to disorder and sometimes help refine structural models against diffraction data.

5.3.2 Solid Form Screening

The second type of common application of CSP in a pharmaceutical materials context is to use the calculations to anticipate the likely solid forms of a molecule. This is usually in the context of polymorphism; different crystal packings of a molecule can occur, sometimes concomitantly, but more commonly due to changes in crystallization conditions. Many important properties, such as solubility, dissolution rate, tabletability, and crystal habit can depend strongly on polymorph. Therefore, an unanticipated change in polymorph can be disastrous for a pharmaceutical molecule. The consequences can be particularly difficult to deal with if the new polymorph has a more stable structure, lowering solubility and sometimes leading to the disappearance of previously known polymorphs [104, 105]. In this context, CSP can be applied to assess the likelihood that other polymorphs could exist for a given molecule, and the risk that the thermodynamically most stable structure has yet to be identified.

The energy range of the crystal energy landscape in which relevant potential polymorphs of a molecule can be found can be limited by studies on known polymorph pairs, which show that 95% of polymorphs are separated by less than 7.2 kJ/mol and over 98% by less than 10 kJ/mol [91]. One major problem in the anticipation of polymorphism is that CSP studies almost always lead to many more putative crystal structures in such an energy range than there are observed polymorphs. The reasons why CSP tends to overpredict polymorphism have recently been discussed in detail by Price [106]. Two main considerations are that (i) limitations to the energy models that are currently used, including the fact that many local energy minima on the lattice energy surface might be unstable once thermal motion is considered [107] and that (ii) CSP currently takes no account of the kinetic route to the low-energy crystal structures. Low-energy crystal structures might exist on the energy landscape, but be particularly inaccessible under typical crystallization conditions. Related to this is the possibility that many CSP structures are potentially observable, but the right experimental conditions that lead to these structures have not yet been investigated. Despite these considerations, there have been several reports where polymorphs of pharmaceutical molecules have been predicted in advance of their eventual experimental realization. As an example, Zaworotko and co-workers discovered a second polymorph of aspirin during attempted co-crystallization studies [108], the structure of which had been predicted a year earlier by CSP [109]. In a similar story, Lutker and Matzger reported on the polymorphism of the anticonvulsant oxcarbazepine [110], where the newly discovered form II was in excellent agreement with the second lowest energy predicted polymorph of Cruz-Cabeza and co-workers [111].

In neither of these examples were the experiments that led to the new polymorph prompted or guided by the CSP results. However, as confidence grows in the reliability of CSP methods, experimental polymorph screening might be conducted as a result of interesting predicted polymorphs, as in the case of a predicted [111, 112] and subsequently discovered [112] polymorph of 10,11-dihydrocarbamazepine. These interdisciplinary approaches, where CSP is applied early on in solid form screening to assess a molecule's propensity for polymorphism and likely packing motifs, should become more widely adopted as a part of polymorph screening.

Beyond polymorphism, the solid form diversity of a drug molecule can be deliberately increased by forming different salts or, more recently, co-crystals of the active pharmaceutical ingredient. Considering their widespread use as the solid form for pharmaceutical molecules, CSP studies on organic salts have been fairly limited. This can be explained by the added difficulty of structure generation with independent cation and anion species in the unit cell, as well as challenges in accurately modeling the strong electrostatic and polarization interactions between charged species. Nevertheless, some successful applications of CSP to salt systems have been reported [113–115], demonstrating that salt CSP is possible. Co-crystal prediction presents similar challenges to salts, with the complication that the stoichiometry of a molecule and its co-former is not fixed by charge balance, as in the case of salts. Therefore, the prediction of co-crystal structures requires an assessment of different compositions during CSP. The resulting issue of comparing the stabilities of predicted structures with different compositions is resolved by assessing their stability relative to the same molar ratio of the pure crystal structures of the potential co-crystal formers [98, 116]. Thus, such calculations can also help assess whether a co-crystal is likely to form at all, in preference to separate crystallization of the pure components [98, 116–122]. Again, these calculations are most powerful when the CSP results are able to guide experiments to new solid forms. Recently, CSP was used to suggest a seeding approach to the realization of an elusive co-crystal of caffeine with benzoic acid, using the predicted co-crystal global minimum to select heteronuclear seeds that led to a co-crystal that could not otherwise be produced [122].

The prediction of solvate formation, where solvent of crystallization is incorporated into the crystal lattice, can be addressed using similar methods to co-crystals: two component CSP at a range of stoichiometries of the organic molecule and solvent, followed by comparison to the energies of the pure phases [98, 116]. A complication for solvates is the range of possible stoichiometries, as well as variable stoichiometry solvates, where the solvate molecules fill a pocket or channel in the host crystal structure formed by the main molecule, in which the amount of solvent present could vary, depending on crystallization and environmental conditions. In such cases, an alternative approach to predicting solvate formation is required. It has been shown that the solvent-free host structure of these inclusion solvates corresponds to local energy minima of the crystal energy landscape of the host molecule [123, 124]. Therefore, crystalline solvate prediction could be addressed through analysis of the higher energy predicted structures that result from CSP for structures with pockets or channels that could accommodate solvent molecules [124].

5.4 SUMMARY

The purpose of this chapter has been to overview current methods in CSP by global lattice energy minimization, their current limitations, and the recent developments that are making these computational methods applicable to a range of molecular organic materials, including pharmaceuticals. Our understanding of the crystal packing of organic molecules continues to develop, as do the theoretical and algorithmic ingredients to CSP. With these developments, our confidence in using these methods in an applied context increases. The applications of structure prediction in structure determination and solid form screening have been summarized. These methods are already at a stage where they can provide important insight into the solid forms of pharmaceutical molecules, can be applied to molecules of pharmaceutical relevance, and should continue to find more widespread use in the pharmaceutical industry.

REFERENCES

[1] Bardwell, D. A.; Adjiman, C. S.; Arnautova, Y. A.; Bartashevich, E.; Boerrigter, S. X. M.; Braun, D. E.; Cruz-Cabeza, A. J.; Day, G. M.; Della Valle, R. G.; Desiraju, G. R.; van Eijck, B. P.; Facelli, J. C.; Ferraro, M. B.; Grillo, D.; Habgood, M.; Hofmann, D. W. M.; Hofmann, F.; Jose, K. V. J.; Karamertzanis, P. G.; Kazantsev, A. V.; Kendrick, J.; Kuleshova, L. N.; Leusen, F. J. J.; Maleev, A. V.; Misquitta, A. J.; Mohamed, S.; Needs, R. J.; Neumann, M. A.; Nikylov, D.; Orendt, A. M.; Pal, R.; Pantelides, C. C.; Pickard, C. J.; Price, L. S.; Price, S. L.; Scheraga, H. A.; van de Streek, J.; Thakur, T. S.; Tiwari, S.; Venuti, E.; Zhitkov, I. K., *Acta Crystallogr. Sect. B: Struct. Sci.* 2011, *67*, 535–551.

[2] Maddox, J., *Nature* 1988, *335*, 201.

[3] Dunitz, J. D., *Chem. Commun.* 2003, 545–548.

[4] Gavezzotti, A., *Acc. Chem. Res.* 1994, *27*, 309–314.

[5] Mooij, W. T. M.; Leusen, F. J. J., *Phys. Chem. Chem. Phys.* 2001, *3*, 5063–5066.

[6] Day, G. M.; Motherwell, W. D. S.; Jones, W., *Cryst. Growth Des.* 2005, *5*, 1023–1033.

[7] Cardamone, S.; Hughes, T. J.; Popelier, P. L. A., *Phys. Chem. Chem. Phys.* 2014, *16*, 10367–10387.

[8] Kendrick, J.; Stephenson, G. A.; Neumann, M. A.; Leusen, F. J. J., *Cryst. Growth Des.* 2013, *13*, 581–589.

[9] Neumann, M. A.; Leusen, F. J. J.; Kendrick, J., *Angew. Chem. Int. Ed.* 2008, *47*, 2427–2430.

[10] Neumann, M. A.; Perrin, M.-A., *J. Phys. Chem. B* 2005, *109*, 15531–15541.

[11] Lund, A. M.; Pagola, G. I.; Orendt, A. M.; Ferraro, M. B.; Facelli, J. C., *Chem. Phys. Lett.* 2015, *626*, 20–24.

[12] Pillardy, J.; Arnautova, Y. A.; Czaplewski, C.; Gibson, K. D.; Scheraga, H. A., *Proc. Natl. Acad. Sci. U. S. A.* 2001, *98*, 12351–12356.

[13] Wawak, R. J.; Pillardy, J.; Liwo, A.; Gibson, K. D.; Scheraga, H. A., *J. Phys. Chem. A* 1998, *102*, 2904–2918.

[14] Raiteri, P.; Martoňák, R.; Parrinello, M., *Angew. Chem. Int. Ed.* 2005, *44*, 3769–3773.

[15] Zhu, Q.; Oganov, A. R.; Glass, C. W.; Stokes, H. T., *Acta Crystallogr. Sect. B: Struct. Sci.* 2012, *68*, 215–226.

[16] Day, G. M., *Crystallogr. Rev.* 2011, *17*, 3–52.

[17] Lommerse, J. P. M.; Motherwell, W. D. S.; Ammon, H. L.; Dunitz, J. D.; Gavezzotti, A.; Hofmann, D. W. M.; Leusen, F. J. J.; Mooij, W. T. M.; Price, S. L.; Schweizer, B.; Schmidt, M. U.; van Eijck, B. P.; Verwer, P.; Williams, D. E., *Acta Crystallogr.* 2000, *B56*, 697–714.

[18] Price, S. L., *Chem. Soc. Rev.* 2014, *43*, 2098–2111.

[19] Kitaigorodskii, A. I., *Molecular Crystals and Molecules*. Academic Press: New York and London, 1973; Vol. *29*, p 553.

[20] Hargittai, I.; Levy, J., *Struct. Chem.* 1999, *10*, 387–389.

[21] Warshel, A.; Huler, E.; Rabinovich, D.; Shakked, Z., *J. Mol. Struct.* 1974, *23*, 175–191.

[22] Colapietro, M.; Domenicano, A.; Portalone, G.; Schultz, G.; Hargittai, I., *J. Phys. Chem.* 1987, *91*, 1728–1737.

[23] Harris, N. J.; Lammertsma, K., *J. Am. Chem. Soc.* 1997, *119*, 6583–6589.

[24] Casalone, G.; Mariani, C.; Mugnoli, A.; Simonetta, M., *Mol. Phys.* 1968, *15*, 339–348.

[25] Almenningen, A.; Bastiansen, O.; Fernholt, L.; Cyvin, B. N.; Cyvin, S. J.; Samdal, S., *J. Mol. Struct.* 1985, *128*, 59–76.

[26] Dzyabchenko, A.; Scheraga, H. A., *Acta Crystallogr. B* 2004, *60*, 228–237.

[27] Cruz-Cabeza, A. J.; Liebeschuetz, J. W.; Allen, F. H., *CrystEngComm* 2012, *14*, 6797–6811.

[28] Day, G. M.; Motherwell, W. D. S.; Jones, W., *Phys. Chem. Chem. Phys.* 2007, *9*, 1693–1704.

[29] Kazantsev, A. V.; Karamertzanis, P. G.; Adjiman, C. S.; Pantelides, C. C.; Price, S. L.; Galek, P. T. A.; Day, G. M.; Cruz-Cabeza, A. J., *Int. J. Pharm.* 2011, *418*, 168–178.

[30] Karamertzanis, P. G.; Price, S. L., *J. Chem. Theory Comput.* 2006, *2*, 1184–1199.

[31] Kazantsev, A. V.; Karamertzanis, P. G.; Pantelides, C. C.; Adjiman, C. S., CrystalOptimizer: An Efficient Algorithm for Lattice Energy Minimization of Organic Crystals Using Isolated-Molecule Quantum Mechanical Calculations. In *Process Systems Engineering: Molecular Systems Engineering*, Volume *6*; Pistikopoulos, E. N., Georgiadis, M. C., Dua, V., Adjiman, C. S., Galindo, A., Eds; Wiley-VCH Verlag GmbH & Co. KGaA: Weinheim, 2011; pp 1–42.

[32] van Eijck, B. P., *J. Comput. Chem.* 2001, *22*, 816–826.

[33] van Eijck, B. P.; Mooij, W. T. M.; Kroon, J., *J. Phys. Chem. B* 2001, *105*, 10573–10578.

[34] Payne, R. S.; Roberts, R. J.; Rowe, R. C.; Docherty, R., *Int. J. Pharm.* 1999, *177*, 231–245.

[35] Cruz-Cabeza, A. J.; Bernstein, J., *Chem. Rev.* 2014, *114*, 2170–2191.

[36] Nangia, A., *Acc. Chem. Res.* 2008, *41*, 595–604.

[37] Thompson, H. P. G.; Day, G. M., *Chem. Sci.* 2014, *5*, 3173–3182.

[38] Baias, M.; Dumez, J.-N.; Svensson, P. H.; Schantz, S.; Day, G. M.; Emsley, L., *J. Am. Chem. Soc.* 2013, *135*, 17501–17507.

[39] Karamertzanis, P. G.; Day, G. M.; Welch, G. W. A.; Kendrick, J.; Leusen, F. J. J.; Neumann, M. A.; Price, S. L., *J. Chem. Phys.* 2008, *128*, 244708.

[40] Bruno, I. J.; Cole, J. C.; Kessler, M.; Luo, J.; Motherwell, W. D. S.; Purkis, L. H.; Smith, B. R.; Taylor, R.; Cooper, R. I.; Harris, S. E.; Orpen, A. G., *J. Chem. Inf. Comput. Sci.* 2004, *44*, 2133–2144.

[41] Groom, C. R.; Allen, F. H., *Angew. Chem. Int. Ed.* 2014, *53*, 662–671.

[42] Iijima, K.; Tanaka, K.; Onuma, S., *J. Mol. Struct.* 1991, *246*, 257–266.

[43] Cao, M.; Newton, S. Q.; Pranata, J.; Schäfer, L., *J. Mol. Struct. Theochem.* 1995, *332*, 251–267.

[44] Huang, Z.; Yu, W.; Lin, Z., *J. Mol. Struct. Theochem.* 2006, *758*, 195–202.

[45] Cooper, T. G.; Jones, W.; Motherwell, W. D. S.; Day, G. M., *CrystEngComm* 2007, *9*, 595–602.

[46] Cooper, T. G.; Hejczyk, K. E.; Jones, W.; Day, G. M., *J. Chem. Theory Comput.* 2008, *4*, 1795–1805.

[47] Day, G. M.; Cooper, T. G., *CrystEngComm* 2010, *12*, 2443–2453.

[48] Görbitz, C. H.; Dalhus, B.; Day, G. M., *Phys. Chem. Chem. Phys.* 2010, *12*, 8466–8477.

[49] van Eijck, B. P.; Kroon, J., *Acta Crystallogr. Sect. B: Struct. Sci.* 2000, *56*, 535–542.

[50] Brock, C. P.; Dunitz, J. D., *Chem. Mater.* 1994, *6*, 1118–1127.

[51] Dzyabchenko, A. V., *J. Struct. Chem.* 1984, *25*, 416–420.

[52] Verwer, P.; Leusen, F. J. J., *Rev. Comput. Chem.* 1998, *12*, 327–365.

[53] Karfunkel, H. R.; Leusen, F. J. J.; Gdanitz, R. J., *J. Comput. Aided Mater. Des.* 1993, *1*, 177–185.

[54] Beyer, T.; Day, G. M.; Price, S. L., *J. Am. Chem. Soc.* 2001, *123*, 5086–5094.

[55] Karamertzanis, P. G.; Pantelides, C. C., *J. Comput. Chem.* 2005, *26*, 304–324.

[56] Vasileiadis, M.; Kazantsev, A. V.; Karamertzanis, P. G.; Adjiman, C. S.; Pantelides, C. C., *Acta Crystallogr. Sect. B: Struct. Sci.* 2012, *68*, 677–685.

[57] Price, L. S.; McMahon, J. A.; Lingireddy, S. R.; Lau, S.-F.; Diseroad, B. A.; Price, S. L.; Reutzel-Edens, S. M., *J. Mol. Struct.* 2014, *1078*, 26–42.

[58] Vasileiadis, M.; Pantelides, C. C.; Adjiman, C. S., *Chem. Eng. Sci.* 2015, *121*, 60–76.

[59] Eddleston, M. D.; Hejczyk, K. E.; Bithell, E. G.; Day, G. M.; Jones, W., *Chem. Eur. J.* 2013, *19*, 7883–7888.

[60] Bygrave, P. J.; Case, D. H.; Day, G. M., *Faraday Discuss.* 2014, *170*, 41–57.

[61] Perrin, M.-A.; Neumann, M. A.; Elmaleh, H.; Zaske, L., Chem. Commun. 2009, 3181–3183.

[62] Steed, K. M.; Steed, J. W., *Chem. Rev.* 2015, *115*, 2895–2933.

[63] van de Streek, J.; Neumann, M. A., *CrystEngComm* 2011, *13*, 7135–7142.

[64] Williams, D. E., *Acta Crystallogr. Sect. A: Found. Crystallogr.* 1996, *52*, 326–328.

[65] Braun, D. E.; Orlova, M.; Griesser, U. J., *Cryst. Growth Des.* 2014, *14*, 4895–4900.

[66] D'Oria, E.; Karamertzanis, P. G.; Price, S. L., *Cryst. Growth Des.* 2010, *10*, 1749–1756.

[67] Eddleston, M. D.; Patel, B.; Day, G. M.; Jones, W., *Cryst. Growth Des.* 2013, *13*, 4599–4606.

[68] Della Valle, R. G.; Venuti, E.; Brillante, A.; Girlando, A., *J. Chem. Phys.* 2003, *118*, 807–815.

[69] Habgood, M.; Sugden, I. J.; Kazantsev, A. V.; Adjiman, C. S.; Pantelides, C. C., *J. Chem. Theory Comput.* 2015, *11*, 1957–1969.

[70] Wang, Y.; Lv, J.; Zhu, L.; Ma, Y., *Phys. Rev. B Condens. Matter* 2010, *82*, 094116/1–094116/8.

[71] Karfunkel, H. R.; Gdanitz, R. J., *J. Comput. Chem.* 1992, *13*, 1171–1183.

[72] Gdanitz, R. J., *Chem. Phys. Lett.* 1992, *190*, 391–396.

[73] Williams, D. E., *J. Comput. Chem.* 2001, *22*, 1154–1166.

[74] Coombes, D. S.; Price, S. L.; Willock, D. J.; Leslie, M., *J. Phys. Chem.* 1996, *100*, 7352–7360.

[75] Mitchell, J. B. O.; Price, S. L., *J. Phys. Chem. A* 2000, *104*, 10958–10971.

[76] Mitchell, J. B. O.; Price, S. L.; Leslie, M.; Buttar, D.; Roberts, R. J., *J. Phys. Chem. A* 2001, *105*, 9961–9971.

[77] Day, G. M.; Price, S. L., *J. Am. Chem. Soc.* 2003, *125*, 16434–16443.

[78] Misquitta, A. J.; Welch, G. W. A.; Stone, A. J.; Price, S. L., *Chem. Phys. Lett.* 2008, *456*, 105–109.

[79] Price, S. L., *CrystEngComm* 2004, *6*, 344–353.

[80] Stone, A. J., *The Theory of Intermolecular Forces*. 2nd ed.; Oxford University Press: Oxford, 2013.

[81] Stone, A. J., *Chem. Phys. Lett.* 1981, *83*, 233–239.

[82] Day, G. M.; Cooper, T. G.; Cruz-Cabeza, A. J.; Hejczyk, K. E.; Ammon, H. L.; Boerrigter, S. X. M.; Tan, J. S.; Valle, R. G. D.; Venuti, E.; Jose, J.; Gadre, S. R.; Desiraju, G. R.; Thakur, T. S.; van Eijck, B. P.; Facelli, J. C.; Bazterra, V. E.; Ferraro, M. B.; Hofmann, D. W. M.; Neumann, M. A.; Leusen, F. J. J.; Kendrick, J.; Price, S. L.; Misquitta, A. J.; Karamertzanis, P. G.; Welch, G. W. A.; Scheraga, H. A.; Arnautova, Y. A.; Schmidt, M. U.; van de Streek, J.; Wolf, A. K.; Schweizer, B., *Acta Crystallogr. Sect. B: Struct. Sci.* 2009, *65*, 107–125.

[83] Day, G. M.; Motherwell, W. D. S.; Ammon, H. L.; Boerrigter, S. X. M.; Della Valle, R. G.; Venuti, E.; Dzyabchenko, A.; Dunitz, J. D.; Schweizer, B.; van Eijck, B. P.; Erk, P.; Facelli, J. C.; Bazterra, V. E.; Ferraro, M. B.; Hofmann, D. W. M.; Leusen, F. J. J.; Liang, C.; Pantelides, C. C.; Karamertzanis, P. G.; Price, S. L.; Lewis, T. C.; Nowell, H.; Torrisi, A.; Scheraga, H. A.; Arnautova, Y. A.; Schmidt, M. U.; Verwer, P., *Acta Crystallogr. Sect. B: Struct. Sci.* 2005, *61*, 511–527.

[84] van Eijck, B. P.; Mooij, W. T. M.; Kroon, J., *J. Comput. Chem.* 2001, *22*, 805–815.

[85] Neumann, M. A., *J. Phys. Chem. B* 2008, *112*, 9810–9829.

[86] Oswald, I. D. H.; Allan, D. R.; Day, G. M.; Motherwell, W. D. S.; Parsons, S., *Cryst. Growth Des.* 2005, *5*, 1055–1071.

[87] Neumann, M. A.; van de Streek, J.; Fabbiani, F. P. A.; Hidber, P.; Grassmann, O., *Nat. Commun.* 2015, *6*, 7793.

[88] Habgood, M., *Cryst. Growth Des.* 2011, *11*, 3600–3608.

[89] Cruz-Cabeza, A. J.; Day, G. M.; Jones, W., *Phys. Chem. Chem. Phys.* 2011, *13*, 12808–12816.

[90] van Eijck, B. P., *Phys. Chem. Chem. Phys.* 2002, *4*, 4789–4794.

[91] Nyman, J.; Day, G. M., *CrystEngComm* 2015, *17*, 5154–5165.

[92] Anghel, A. T.; Day, G. M.; Price, S. L., *CrystEngComm* 2002, *4*, 348–355.

[93] Lewis, T. C.; Tocher, D. A.; Day, G. M.; Price, S. L., *CrystEngComm* 2003, *5*, 3–9.

[94] Day, G. M.; Price, S. L.; Leslie, M. L., *J. Phys. Chem. B* 2003, *107*, 10919–10933.

[95] Schmidt, M. U.; Ermrich, M.; Dinnebier, R. E., *Acta Crystallogr. Sect. B: Struct. Sci.* 2005, *61*, 37–45.

[96] Wu, H.; Habgood, M.; Parker, J. E.; Reeves-McLaren, N.; Cockcroft, J. K.; Vickers, M.; West, A. R.; Jones, A. G., *CrystEngComm* 2013, *15*, 1853–1859.

[97] Day, G. M.; van de Streek, J.; Bonnet, A.; Burley, J. C.; Jones, W.; Motherwell, W. D. S., *Cryst. Growth Des.* 2006, *6*, 2301–2307.

[98] Cruz-Cabeza, A. J.; Karki, S.; Fabian, L.; Friscic, T.; Day, G. M.; Jones, W., *Chem. Commun.* 2010, *46*, 2224–2226.

[99] Eddleston, M. D.; Hejczyk, K. E.; Bithell, E. G.; Day, G. M.; Jones, W., *Chem. Eur. J.* 2013, *19*, 7874–7882.

[100] Yates, J. R.; Pickard, C. J.; Mauri, F., *Phys. Rev. B Condens. Matter* 2007, *76*, 024401/1–024401/11.

[101] Baias, M.; Widdifield, C. M.; Dumez, J.-N.; Thompson, H. P. G.; Cooper, T. G.; Salager, E.; Bassil, S.; Stein, R. S.; Lesage, A.; Day, G. M.; Emsley, L., *Phys. Chem. Chem. Phys.* 2013, *15*, 8069–8080.

[102] Salager, E.; Day, G. M.; Stein, R. S.; Pickard, C. J.; Elena, B.; Emsley, L., *J. Am. Chem. Soc.* 2010, *132*, 2564–2566.

[103] Tremayne, M.; Grice, L.; Pyatt, J. C.; Seaton, C. C.; Kariuki, B. M.; Tsui, H. H. Y.; Price, S. L.; Cherryman, J. C., *J. Am. Chem. Soc.* 2004, *126*, 7071–7081.

[104] Bauer, J.; Spanton, S.; Henry, R.; Quick, J.; Dziki, W.; Porter, W.; Morris, J., *Pharm. Res.* 2001, *18*, 859–866.

[105] Chemburkar, S. R.; Bauer, J.; Deming, K.; Spiwek, H.; Patel, K.; Morris, J.; Henry, R.; Spanton, S.; Dziki, W.; Porter, W.; Quick, J.; Bauer, P.; Donaubauer, J.; Narayanan, B. A.; Soldani, M.; Riley, D.; McFarland, K., *Org. Process Res. Dev.* 2000, *4*, 413–417.

[106] Price, S. L., *Acta Crystallogr. Sect. B: Struct. Sci.* 2013, *69*, 313–328.

[107] Mooij, W. T. M.; van Eijck, B. P.; Price, S. L.; Verwer, P.; Kroon, J., *J. Comput. Chem.* 1998, *19*, 459–474.

[108] Vishweshwar, P.; McMahon, J. A.; Oliveira, M.; Peterson, M. L.; Zaworotko, M. J., *J. Am. Chem. Soc.* 2005, *127*, 16802–16803.

[109] Ouvrard, C.; Price, S. L., *Cryst. Growth Des.* 2004, *4*, 1119–1127.

[110] Lutker, K. M.; Matzger, A. J., *J. Pharm. Sci.* 2010, *99*, 794–803.

[111] Cruz Cabeza, A. J.; Day, G. M.; Motherwell, W. D. S.; Jones, W., *Cryst. Growth Des.* 2007, *7*, 100–107.

[112] Arlin, J.-B.; Johnston, A.; Miller, G. J.; Kennedy, A. R.; Price, S. L.; Florence, A. J., *CrystEngComm* 2010, *12*, 64–66.

[113] Leusen, F. J. J., *Cryst. Growth Des.* 2003, *3*, 189–192.

[114] Karamertzanis, P. G.; Price, S. L., *J. Phys. Chem. B* 2005, *109*, 17134–17150.

[115] McArdle, P.; Gilligan, K.; Cunningham, D.; Dark, R.; Mahon, M., *CrystEngComm* 2004, *6*, 303–309.

[116] Cruz-Cabeza, A. J.; Day, G. M.; Jones, W., *Chem. Eur. J.* 2008, *14*, 8830–8836.

[117] Issa, N.; Karamertzanis, P. G.; Welch, G. W. A.; Price, S. L., *Cryst. Growth Des.* 2009, *9*, 442–453.

[118] Karamertzanis, P. G.; Kazantsev, A. V.; Issa, N.; Welch, G. W. A.; Adjiman, C. S.; Pantelides, C. C.; Price, S. L., *J. Chem. Theory Comput.* 2009, *5*, 1432–1448.

[119] Chan, H. C. S.; Kendrick, J.; Neumann, M. A.; Leusen, F. J. J., *CrystEngComm* 2013, *15*, 3799–3807.

[120] Kuleshova, L. N.; Hofmann, D. W. M.; Boese, R., *Chem. Phys. Lett.* 2013, *564*, 26–32.

[121] Stepanovs, D.; Jure, M.; Kuleshova, L. N.; Hofmann, D. W. M.; Mishnev, A., *Cryst. Growth Des.* 2015, *15*, 3652–3660.

[122] Bucar, D.-K.; Day, G. M.; Halasz, I.; Zhang, G. G. Z.; Sander, J. R. G.; Reid, D. G.; MacGillivray, L. R.; Duer, M. J.; Jones, W., *Chem. Sci.* 2013, *4*, 4417–4425.

[123] Cruz Cabeza, A. J.; Day, G. M.; Motherwell, W. D. S.; Jones, W., Chem. Commun. 2007, 1600–1602.

[124] Cruz-Cabeza, A. J.; Day, G. M.; Jones, W., *Chem. Eur. J.* 2009, *15*, 13033–13040.

6

INTEGRATING COMPUTATIONAL MATERIALS SCIENCE TOOLS IN FORM AND FORMULATION DESIGN

JOSEPH F. KRZYZANIAK[1], PAUL A. MEENAN[1], CHERYL L. DOHERTY[2], KLIMENTINA PENCHEVA[2], SUMAN LUTHRA[3], AND AURORA CRUZ-CABEZA[4]

[1] *Pfizer Worldwide Research & Development, Groton, CT, USA*
[2] *Pfizer Worldwide Research & Development, Sandwich, Kent, UK*
[3] *Pfizer Worldwide Research & Development, Andover, MA, USA*
[4] *The School of Chemical Engineering and Analytical Science, The University of Manchester, Manchester, UK*

6.1 INTRODUCTION

Polymorphism of an active pharmaceutical ingredient (API) is an important area of focus in the development of life-saving medicines. Over the past three decades, multiple cases of clinical failures and drug product recalls have been related to a change in the API solid form, which raised concern for the overall safety and efficacy of the drug product. In an article by Lee et al. [1], such events were summarized in a timeline showing the impact that crystal form changes had on multiple solid oral dosage forms that include Tegretol (1988), Norvir (1998), Coumadin (2010), and Avalide (2010), to name a few. A further investigation was conducted by Lynn [2] revealing that approximately one injectable product recall has occurred each year from 2003 to 2011 due to precipitation/crystallization of the API [2]. Based on this information, it becomes critical to reflect on our scientific designs in the effort to overall improve our understanding of solid-state chemistry.

Computational Pharmaceutical Solid State Chemistry, First Edition. Edited by Yuriy A. Abramov.

Typically, polymorph screens have been designed to identify solid forms that would be present during the manufacturing of both API (crystallization, drying, milling, etc.) and drug product, as well as during long-term stability. Collaborations throughout the solid-state community have provided academic and industrial scientists the ability to better screen for polymorphs [3], control crystallization processes [4], and select particle properties to ensure robust drug product manufacture [5]. Utilizing scientific advancements in the understanding and control of nucleation [6], supersaturation [7], and hydrate screening has improved the ability to determine the most thermodynamically stable form for development. An additional approach, designated as high throughput screening (HTS), became popular over the past 15 years as it has the ability to test thousands of samples using multiple variables (temperature, solvents, water activities, excipients, etc.) to define the solid form landscape. Some success was made in identifying new crystal forms that are relevant in drug development [8]. Unfortunately, such traditional approaches, manual and HTS, in solid form screening have not reduced the occurrences of recalls due to solid form changes and/or precipitation as noted earlier. It has been recognized that there is a need to move away from this brute-force shotgun approach to a more proactive computer-aided design of API to impart the appropriate solid form characteristics for drug product development.

While the goal of the solid-state chemist is to select the thermodynamically stable form to be used in the development of a drug product that provides robust performance, stability, and manufacturability, it is important to consider the quote from Maria Kuhnert-Brandstatter (1975) "Probably every substance is potentially polymorphic. The only question is whether it is possible to adjust the external conditions in such a way that polymorphism can be realized or not." It is clear that the environmental conditions of an API solid form, at normal manufacturing and storage condition, are very different from the conditions that the API solid form is exposed to in the drug product. The formulation can alter the environment in such a way that the thermodynamically stable form becomes metastable. This is a common occurrence when thermodynamically stable anhydrous forms are formulated to prepare an aqueous-based dosage form, in which a precipitation of a hydrate with lower solubility is observed [9].

Over the past few years, materials scientists have designed and conducted research to determine the stable form in the conditions of the formulated dosage form. Luthra et al. identified that the thermodynamically stable form, anhydrous crizotinib free base, converted to a quaternary complex (API, phosphate, water, sucralose) in the formulated oral solution in which precipitation was observed [9]. This spurred the design of new solid form/polymorph investigation designated as the Formulation Form Screen. In this case as well as research published by Arora et al. [10, 11] and Bak et al. [12], unintended cocrystal formation occurred between the stable form of the API and the formulation components. As the number of variables and conditions exponentially increase as the stable API form is exposed to a formulated environment, there becomes a need to develop new approaches to test the reactivity of the API with its formulation components.

Computational tools have been accepted and applied within the pharmaceutical industry, typically in the bulk powder processing arena, where tools such as computational fluid dynamics (CFD) and discrete element methods (DEMs) have been utilized [13]. However, the field of computational materials science provides the opportunity to gain a fundamental understanding at the molecular level, connectivity between molecular structure, crystal structure, surface chemistries, and their relationship to bulk properties such as dissolution [14], particle shape [15, 16], and breakage [17, 18]. Although a more mature field within other industries, there has been increasing interest in this field from the pharmaceutical industry in an effort to seek better ways to optimize the bulk properties of the API for optimum performance, stability, and manufacturability. The intent of this chapter is to provide a brief summary of computational materials science approaches utilized to direct and guide experimental studies in assessing the risk of identifying a more stable API crystal form, while providing a rationale at the molecular level to justify the selection of the optimal form and particle for robust drug product development. In order to mitigate limitations of each of the methods described in this chapter, it is recommended that general computational support for solid form selection in the pharmaceutical industry should be based on a combination of these tools. Additionally, multiple computational approaches are covered in detail within this book, that is, structural informatics—Chapter 2, σ_{HB}-charges analysis—Chapter 3, crystal structure prediction (CSP)—Chapter 5, as well as in a recent review by Abramov [19].

6.2 FROM MOLECULE TO CRYSTAL STRUCTURE

Sulopenem was a broad-spectrum antibacterial drug candidate in development at Pfizer. In order to support the development of clinical supplies for Phase II studies, a polymorph screen was needed to identify the thermodynamically stable form to manufacture the drug substance. The experimental studies were challenging for this molecule due to its unique material properties. Chemical instability of the molecule in solution deterred the use of long-term slurries based on solvent-mediated phase transformations applying Ostwald's rule of stages, typically done in a conventional polymorph screen [20]. Also, specific impurities resultant due to chemical degradation can either inhibit nucleation of the stable form or stabilize metastable forms on the time scale of the low-energy polymorph screen experiment. Thus, there was the possibility that the polymorph screen could miss the thermodynamically stable form [21]. In addition to chemical reactivity, the compound was also known to be potent, which required special handling conditions resulting in more difficult and costly, safe experimental studies.

An abbreviated, nonconventional, polymorph screen was conducted on sulopenem free acid. A total of eight forms of sulopenem have been discovered and are consistent with three anhydrous forms (Forms A, B, and D), one hydrate (Form C), three solvates (Forms E, F, and G), and an amorphous form. Form B was confirmed as the most stable form in the short-term slurries at room temperature (RT) and 5°C. This form also resulted from competitive slurries between Form B and pairings of Forms A, C, and D. Since the isothermal slurries at 5°C and RT were conducted for limited time due to rapid chemical degradation, computational methods were investigated to complement the experimental results, that is, build more confidence to establish whether or not Form B is the most stable anhydrous form.

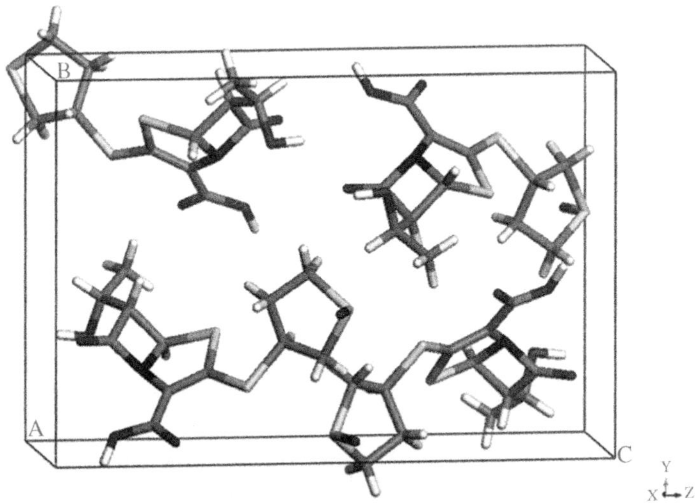

FIGURE 6.1 The unit cell of sulopenem Form B.

6.2.1 Single Crystal Structure

The crystal structure of the sulopenem Form B with the absolute chirality of the drug molecule was determined at RT by single crystal x-ray diffraction analysis. The unit cell for this crystal structure is shown in Figure 6.1. The molecular conformation confirmed attractive interactions between the oxygen and sulfur atoms [22].

6.2.2 Structural Analysis

Hydrogen bond (HB) propensity analysis was conducted for sulopenem, which contains two potential HB donors and five possible HB acceptors, as shown in Figure 6.2. COSMOtherm and Logit HB propensity (LHP) analyses were used to assess the competing pair-wise interactions, and confirm the ones that might be found in the most stable solid form.

6.2.2.1 HB Analysis: COSMOtherm Analysis of σ_{HB} Charges The COSMOtherm program [23] uses quantum mechanics to calculate screening (σ) HB charges for a molecule, hence giving a relative indication of the strengths and weaknesses of the HB donors and acceptors present. The best donor and acceptor pairings would be expected to be seen in the stable solid form [19]. The calculated sigma HB charges for the sulopenem are shown in Figure 6.2. The strongest acceptor and donor, that is, sulfonyl acceptor and carboxyl donor, respectively, are observed in the experimental crystal structure. While the sulfonyl acceptor is much stronger than all other acceptors, there is a noticeable competition between carboxyl, tertiary amide, and alcohol oxygens.

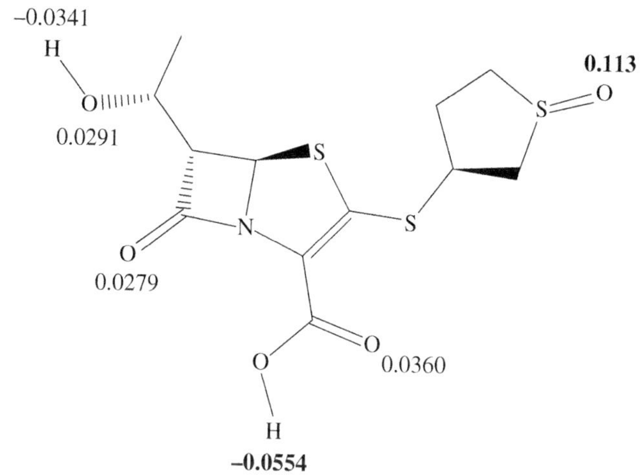

FIGURE 6.2 σ_{HB}-charges of the hydrogen bond (HB) centers. The highest donor and acceptor values are presented in bold.

6.2.2.2 ***HB Analysis: LHP Analysis*** The LHP model developed by Pfizer Institute for Pharmaceutical Materials Science (PIPMS) analyzes potential hydrogen bonding functional groups in order to make predictions on statistically likely pair-wise hydrogen bonding interactions [24]. These most likely interactions would be expected to be seen in the thermodynamically stable form.

The Cambridge Structural Database (CSD) was surveyed for crystal structures containing groups with similar functionality as is found in **sulopenem**, and the results were used to build and test a propensity model. The CSD search was performed within a subset of the crystal structures displaying the sulfonyl fragment (261 observations). In addition, these 32 observations of the combination of secondary alcohol and carboxylate fragments in the database were used in building the model. No unspecified fragments ("other") were included in the training set of the observed acceptor–donor pairs in order to improve esd's of the model parameters. The resulting model gives very high accuracy of predictions for the training set (831 acceptor–donor pairs) of 85.4%. The receiver operating characteristic (ROC) curve area under the curve (AUC) is 0.89. The hydrogen bonding probabilities predicted by the model for the sulopenem fragments are presented in the Table 6.1.

The Logit predictions are in a good agreement with the σ_{HB} charges analysis. The strongest donor–acceptor pair is the carboxyl–sulfonyl one, followed by the alcohol–sulfonyl pair. Probabilities of interactions between the fragments involving other than sulfonyl acceptor are much lower, displaying noticeable overlaps within a confidence interval of 80% (Table 6.1). Any of those fragments may form a HB in an actual crystal structure complementing the expected strongest interaction between the carbonyl and sulfonyl groups.

TABLE 6.1 HB Propensities Predicted by the Logit Model

Donor	Acceptor	HB Propensity	Propensity Lower Bound[a]	Propensity Upper Bound[a]
Carboxyl	Sulfonyl	0.784	0.706	0.846
Alcohol secondary	Sulfonyl	0.686	0.620	0.746
Carboxyl	Amide tertiary	0.233	0.113	0.422
Alcohol secondary	Amide tertiary	0.155	0.076	0.290
Carboxyl	Carboxyl	0.083	0.057	0.120
Carboxyl	Alcohol secondary	0.074	0.048	0.112
Alcohol secondary	Carboxyl	0.052	0.033	0.081
Alcohol secondary	Alcohol secondary	0.046	0.032	0.066
Carboxyl	Carboxyl	0.039	0.026	0.061
Alcohol secondary	Carboxyl	0.024	0.014	0.041
Carboxyl	Amide tertiary	0.007	0.002	0.022

[a] Propensity prediction range within confidence interval of 80%.

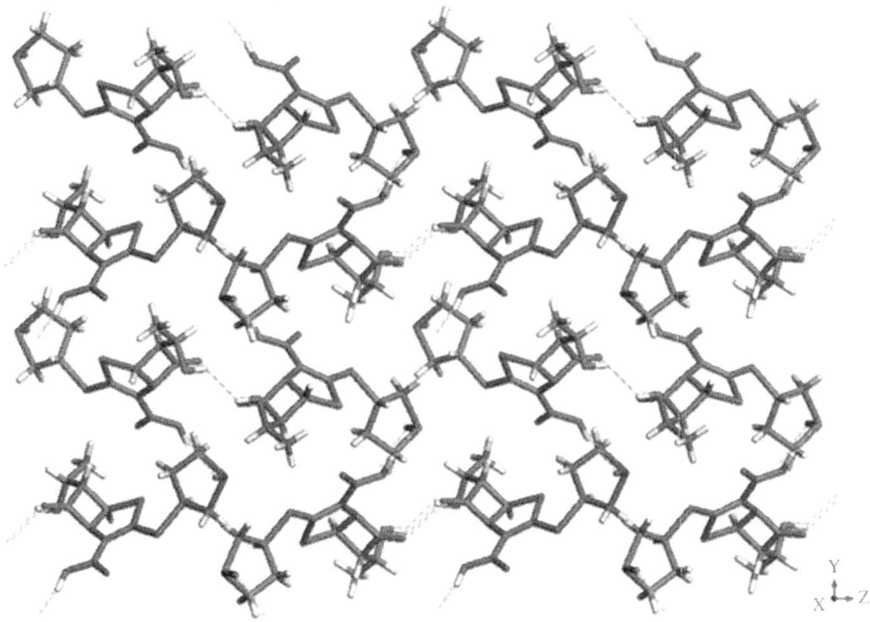

FIGURE 6.3 Hydrogen bonding and packing diagram in Form B sulopenem.

6.2.3 Molecular Packing and HB Geometry Analyses

The packing plot (Fig. 6.3) shows the two types of hydrogen bonds observed experimentally: (i) between the carboxyl and sulfonyl groups and (ii) between the alcohol and tertiary amide. This observation is in agreement with the hydrogen propensity assignment by both the σ_{HB} charges and Logit models. Moreover the strongest interaction, between carboxyl and sulfonyl functional groups, forms an extended network throughout the crystal, which is crucial for the polymorph stability.

6.2.4 Full Interaction Maps

The Full Interaction Map tool, developed at the Cambridge Crystallographic Data Centre (CCDC), allows for qualitative and quantitative analyses of the intermolecular environment to be carried out [25]. To prepare a full interaction map, the crystal structure conformation of a known polymorph is analyzed and compared with fragments exhibiting similar chemical features from the many published structures in the CSD. A map describing the environment around the molecule can then be prepared, which comes from combining the data obtained from the environment around the individual fragments analyzed from the CSD. Changes in both the chemical nature of the fragment and the conformation of the molecule are reflected in the resulting full interaction map.

A full interaction map [25] was prepared for sulopenem using data from fragments found in the CSD (v5.34) [26]. The results are shown in Figure 6.4a. The clouds indicate regions where HB acceptors (grey) and donors (gray) are most likely to be found, based on interaction data mined from the CSD. The most intensely clouds indicate a greater likelihood that an interacting group will be found in that region. The carboxyl and hydroxyl groups on sulopenem are seen to have distinct regions of interaction density located near them. Conversely, the region around the sulfoxide group exhibits a less dense and more diffuse cloud, indicating more flexibility in HB directionality is commonly observed around similar groups in the CSD.

A comparison of the fit of the calculated maps with the observed interactions allows the crystal structure to be assessed in terms of how well the intermolecular interactions are satisfied by the existing lattice. By overlaying the intermolecular interactions observed in the crystal structure of sulopenem, the carboxyl and hydroxyl donor groups are found to form hydrogen bonds in exactly the predicted locations, based on the full interaction map. The weaker groups fit the map less well. This analysis can be further quantified by creating hotspots, denoting the locations with the highest density for each cloud. The relative heights of these hotspots can be used to compare the interaction clouds. Figures 6.4a and b show these hotspots and their relative intensities for sulopenem, with observed hydrogen bonds overlaid.

It is clear that the regions with the strongest preference for acceptors to be located nearby are fully satisfied. Investigation around the strongest donor region, near the carboxyl group, shows that a long C—H···O interaction satisfies this acceptor group (Fig. 6.4a). The weaker directionality around the sulfoxide group is indicated by the lower density values on the hotspots in this area (Fig. 6.4b). This acceptor is satisfied by a hydrogen bonding interaction, but in a suboptimal geometry.

Overall the crystal structure of sulopenem matches the predicted interaction geometry as assessed by these full interaction maps. This supports the conclusion that the known

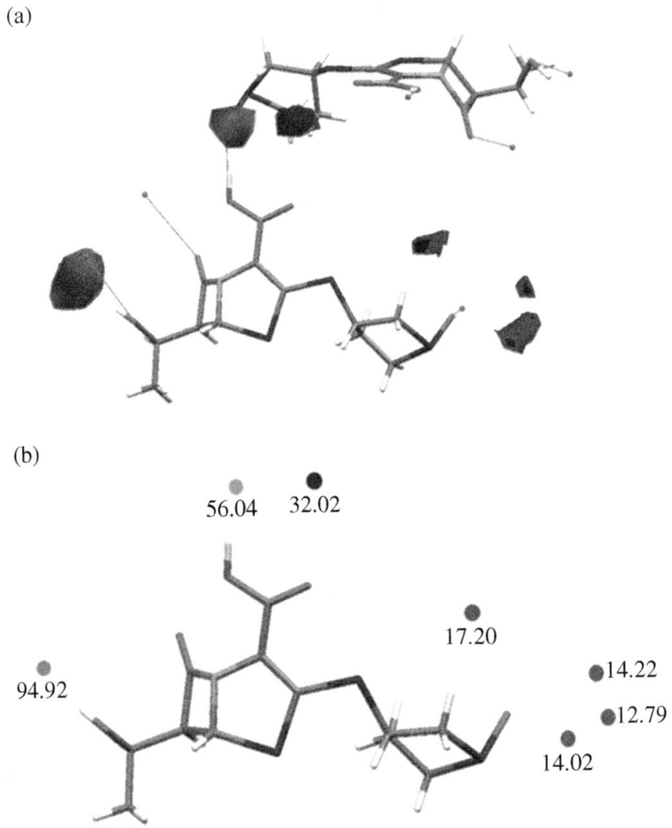

FIGURE 6.4 (a) Full interaction map for sulopenem. The location of neighboring molecules satisfying the carboxylic acid donor and acceptor group with full interaction map overlaid is shown. (b) Hotspots for sulopenem full interaction map. Relative heights are reported for each. (*see insert for color representation of the figure.*)

anhydrous form exhibits a satisfactory network of intermolecular interactions, and that a different packing scheme would not be likely to result in a more stable polymorphic form.

6.2.5 Crystal Structure Prediction

With five torsion angles (Fig. 6.5), **sulopenem** is a challenging compound for CSP calculations. Because of the flexibility, an initial exploration for likely conformations in the solid state needs to be conducted first before the computer generation of crystal structures.

6.2.5.1 Conformational Searches Rational selection of preferred conformations for the CSP study of sulopenem was performed according to a previously proposed approach [27]. For that, a combination of the OMEGA conformational search [28]

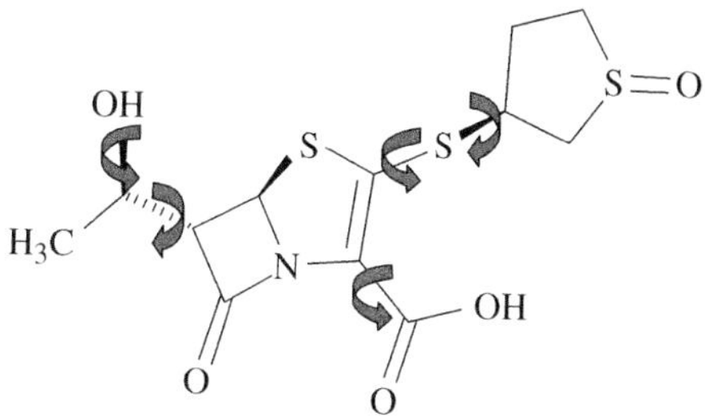

FIGURE 6.5 Sketch and main torsion angles in sulopenem.

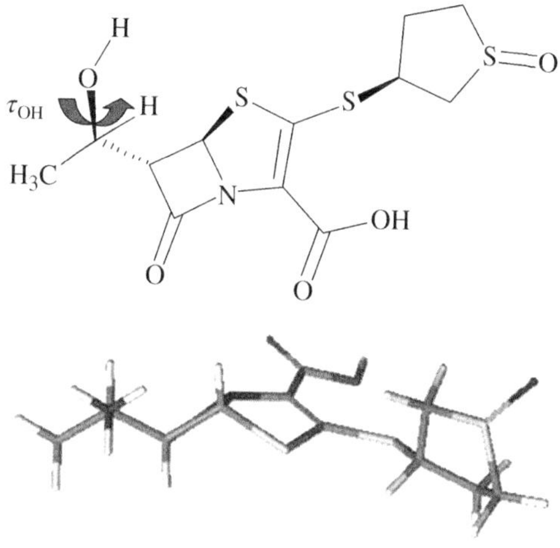

FIGURE 6.6 Staggered configurations used for the treatment of the —OH group flexibility.

and TURBOMOLE [29] calculations were used to effectively select conformations of high importance in the solid state. The OMEGA-generated conformations were reminimized at BP-TZVP-COSMO level of theory in an implicit solvation model for water [30]. This was followed by a COSMOtherm conformer population analysis in self-media, mimicking an amorphous solid-state environment [23]. Four conformers with the highest populations were selected for further study. Among those conformers the crystallographically observed one was the third top ranked.

This search method used for the selection of the conformers above did not consider the rotation of the hydroxyl group. The orientation of such group, however, can be of extreme importance for forming stable hydrogen bonds in the crystal structure. For each of the four conformers generated earlier, therefore, staggered conformations around the hydroxyl group were generated with $\tau_{OH}=180$, +60 and −60 (Fig. 6.6). This procedure resulted in 12 different conformers, which were taken as starting point in our CSP calculations.

6.2.5.2 CSP Methods The CSP methodologies used for this study, which are described in Chapter 4, have been developed at the PIPMS, Cambridge University [31]. The methodology can be divided into four main steps described (Fig. 6.7) as follows:

1. The 12 molecular models taken from the conformational search in the earlier section were partially geometry optimized using the code DMol^3 [32] (VWM-BP/DNP level of theory) implemented in the Material Studio software package [33]. For this, torsion angles were fixed and atomic point charges were derived to fit the molecular surface electrostatic potential (ESP) for these initial conformations.
2. Crystal structures were generated from the selected initial molecular conformations in the seven most popular chiral space groups using the Polymorph Predictor (PP) software implemented in Cerius2 [34]. Molecular models were treated as rigid during the whole PP process. Lattice energies were evaluated (i) using the empirically fitted parameters for the van der Waals contributions (Williams 99 or W99 force field) [35] and (ii) using the previously derived atomic ESP charges for the evaluation of the electrostatic terms. The generated crystal structures were clustered using the COMPACK algorithm also developed at the PIPMS [36].
3. Molecular conformation adjustments: The 50 most stable crystal structures for each family of conformations were reminimized allowing the torsion angles to relax and readjust to the effect of crystal forces. For this, the molecules were treated as partial rigid bodies (everything was kept rigid except for the torsion angles). The force field DREIDING with ESP charges was used in this step [37].
4. Accurate lattice and conformational energies were calculated using GAUSSIAN [38] and DMAREL [39]. For each of the low-energy structures, (i) the molecular model was taken from the crystal structure and was geometry optimized again constraining the six torsion angles to the optimized values in the crystal structure using GAUSSIAN03 (B3LYP/6-31G**), (ii) a single-point energy calculation was carried out for the optimized molecular model using a continuum polarization model with a dielectric constant of 2, (iii) atomic multipoles were derived from the molecular electron density in (ii) using a distributed multipole analysis (DMA) [40], and (iv) the crystal structures were lattice energy minimized with the GAUSSIAN optimized molecular models, the derived multipoles and the W99 intermolecular parameters and using the program DMAREL.

6.2.5.3 Choice of Force Field for the Molecular Conformation Adjustments The right selection of force fields is very important in CSP to ensure that the results of the model applied are meaningful. Rigid models tend to work well when using the W99 force fields and ESP charges; however, flexible force fields are not as accurate and can introduce important errors. In order to choose a suitable force field, experimental crystal structures were minimized with various models. Treating the molecules as

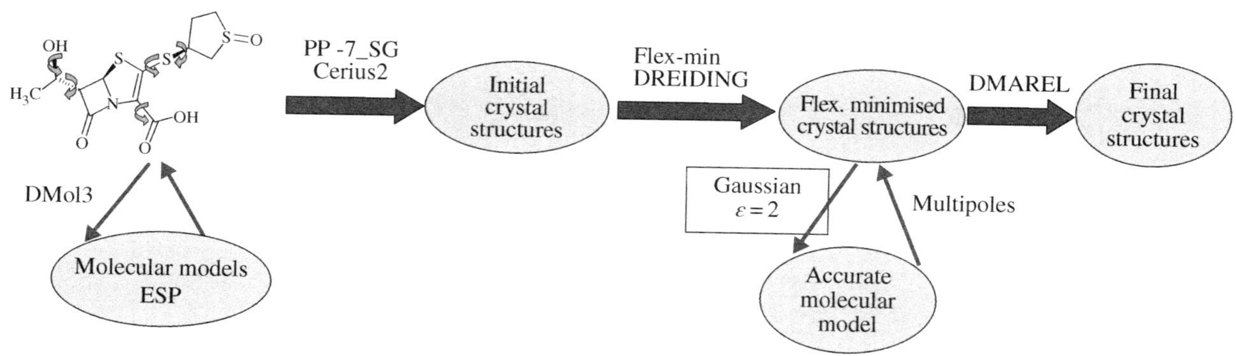

FIGURE 6.7 Crystal structure prediction (CSP) workflow.

fully flexible resulted in minimized structures, which were very far from the observed crystal structure. For example, using the DREIDING force field and a fully flexible molecular model with ESP charges, the minimization of the experimental crystal structure resulted in a completely different crystal structure. The RMS difference between structures was over 1.3Å.

A partial rigid body approximation was more appropriate in this case. Minimization of the observed crystal structures with partial rigid bodies and three different model potentials afforded the results given in Table 6.2. Although using COMPASS [41], the crystal structure RMS difference (observed/minimized crystal) was smaller, at this step of the calculations the key point was to get the conformations right. Therefore, the DREIDING+ESP model was selected as most appropriate for this system and used it in for the treatment of flexibility during the CSP calculations.

6.2.5.4 CSP Results The results obtained from the CSP calculations are plotted in Figure 6.8 in an energy versus density plot. Each of the circles in the plot corresponds to different hypothetical crystal structures generated computationally completely *ab initio*. The property "total energy" contains terms of conformational energy (intramolecular) and also lattice energy (intermolecular). The more negative the total energy, the more stable the crystal structures. The most stable predicted crystal structure (circled structure) corresponded to the crystal structure observed experimentally. The calculations also predict the observed structure to be the most stable form by almost 7 kJ/mol. This suggests that the potential of polymorphism for this system is unlikely, at least for other more stable polymorphs.

The experimental and hypothetical crystal structures are superimposed in Figure 6.9 using the COMPACK algorithm. The RMS deviation between these structures is fairly small. The W99 potential with the multipole description of the electrostatics, after a minimization of the structures using a partial rigid body model and the DREIDING potential, is a potential good model for this system.

From the lattice energy versus relative conformational energy plot (Fig. 6.10), it was observed that the experimental structure is the structure with the second lowest lattice energy. The lowest lattice energy structure has a very high relative conformational energy (>12 kJ/mol). The total energy of the observed structure is overall more stable than any other hypothetical polymorph.

In summary, a completely *in silico* screening was used to complement the experimental polymorphic screen of a challenging chemically reactive and potent compound. An effective molecular conformational sampling was successfully combined with CSP methodologies, which provided high-quality and reliable prediction results. The known anhydrous form of sulopenem free acid was predicted to be the most stable crystal structure of all the hundreds of polymorphs generated computationally. These results confidently indicated that the likelihood of finding a more stable anhydrous form for **sulopenem** is very low. Overall, this example illustrates the benefits of implementing integrated computational approaches as active tools in the identification of the thermodynamically stable form.

TABLE 6.2 Force Field and RMS Deviations between the Observed and the Crystal Optimized Conformations

Force Field Model	COMPASS + ESP	DREIDING + ESP	CVFF + ESP
Crystal structure RMS (Å)	0.597	0.914	0.922
Conformation RMS(Å)	0.381	0.244	0.385
Superimposed conformations (predicted/minimized)	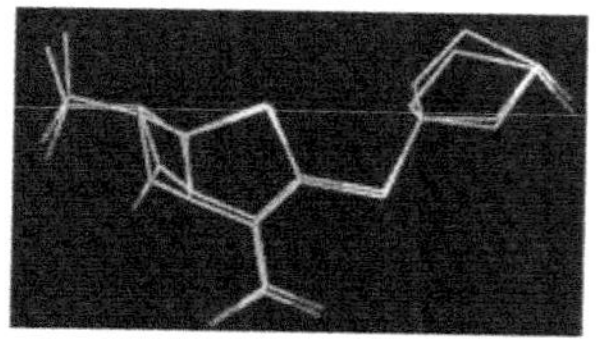	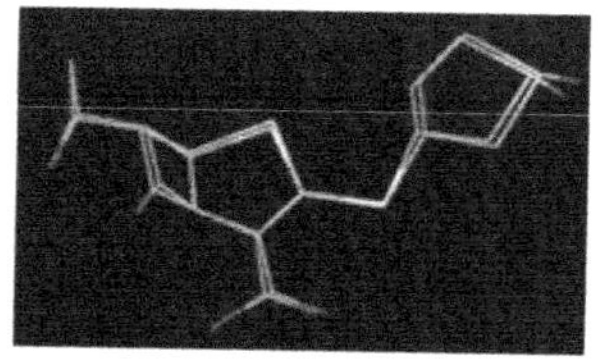	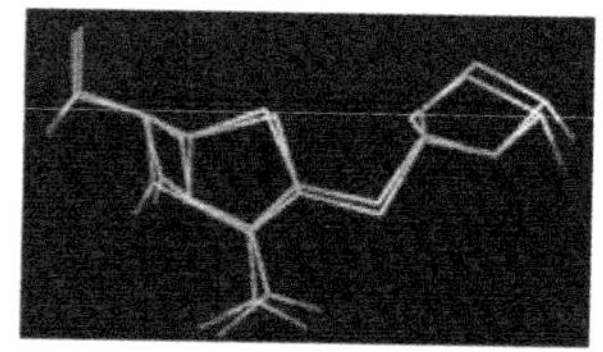

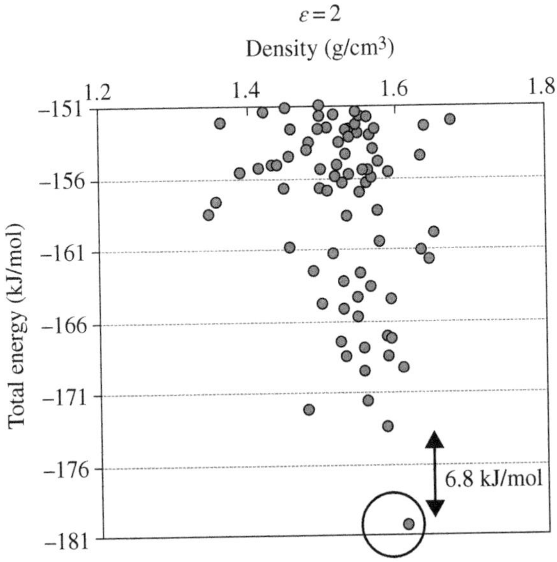

FIGURE 6.8 Crystal structure prediction (CSP) results for sulopenem. The circled structure matches the observed crystal structure.

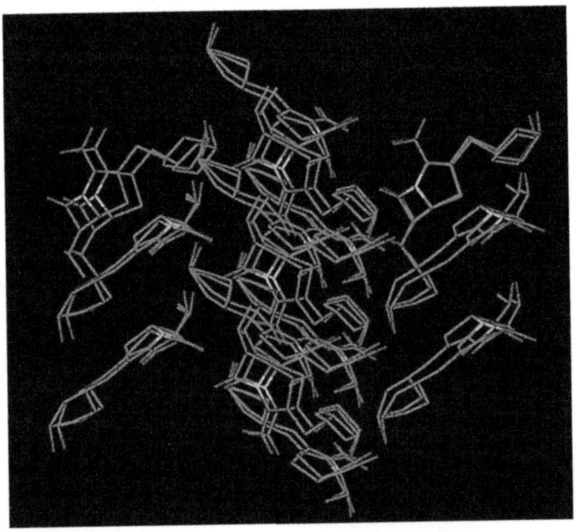

FIGURE 6.9 Superposition of the experimental and predicted crystal structures using the COMPACK algorithm (tolerance = 15%, RMS = 0.460). (*see insert for color representation of the figure.*)

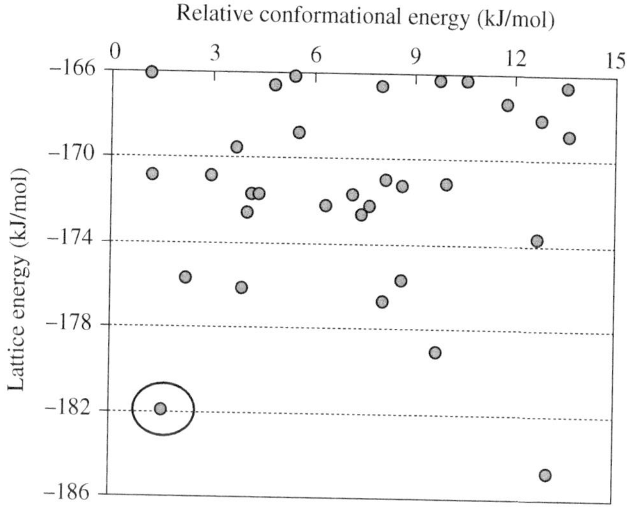

FIGURE 6.10 Lattice energy versus conformational energy plot.

6.3 FROM CRYSTALS TO PARTICLES

Crystal shape is an important property for pharmaceutical materials as it can influence the ease of API filtration, powder flow properties, stability, and quality. Therefore, the ability to control and engineer particle shape is a highly desirable goal. When crystals are obtained via crystallization from solution, the solvent choice is often considered as one of the primary parameters that can influence the crystal shape. The mechanism by which the solvent affects the crystal shape is typically considered to be via interaction of solvent molecules with growing crystal faces [42]. If solvent molecules interact strongly with a given crystal face, the growth of this face will be decreased resulting in crystal habit modification. An impurity formed during the synthetic route or a deliberately introduced additive will affect similarly the crystal growth and crystal habit. Although the effect of the crystal growing environment significantly influences the resultant habit, the shape of the crystal is also defined by the intermolecular forces defining the packing of the solid state and their degree of anisotropy. Crystal growth models that predict the crystal shape in isolation of growth-influencing factors, often referred to as vacuum morphology, and take into account only intermolecular interactions are very well understood and successfully applied [43]. These models rely on crystallographic structure data, which are very often available via Cambridge Structural Database. Therefore, clearly a combination approach taking into account the intermolecular interactions in the crystal bulk and surface interaction with solvent will help better understand crystal morphology and identify the potential for crystal shape alteration.

Methods for crystal morphology modeling from crystallographic data have become more common since Hartman and Perdok have proposed the periodic bond

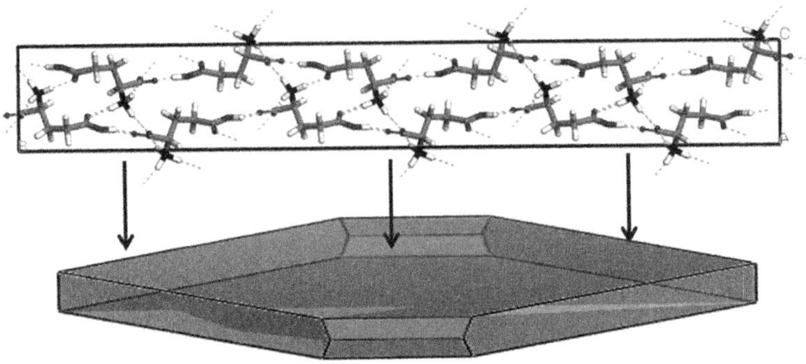

FIGURE 6.11 A growth slice attaching to the surface of the growing crystal utilized for attachment energy calculations.

chain model to define rate determining step that governs growth rate along the direction of a chain [44]. Advancements to this model were realized when crystal growth was viewed as attachment of layer of ordered structure to the existing crystal face. Hartman and Bennema then stated that relative growth rates defining crystal morphology can be calculated through attachment energy as the difference between bulk crystal energy and slice (surface-specific) energy [45]. The crystal and slice energy are calculated applying atom summation force field methods. The attachment energy is the energy released when a slice with thickness d_{hkl} is attaching to the crystal surface (Fig. 6.11). The algorithm for calculating attachment energies is available in commercial software package such as Materials Studio [33] as well as internal software developed in collaboration with academia.

A number of modeling techniques have been exploited to understand and predict the growth modifying effect of solvent on crystal surfaces and hence the resultant habit [42]. These computational approaches include molecular dynamics and surface docking algorithms to study the most favorable intermolecular interactions between crystal habit surfaces and solvents in order to establish the overall modification to crystal shape in terms of relative growth rates of different crystal faces. Typically Molecular Dynamics models are considered to be the most precise model to mimic the actual crystal–solvent interface. However, these models are computationally expensive due to the high number of molecules involved and requirements for longer simulation time in order to achieve reasonable statistics. Molecular docking approaches provide a less precise but quicker calculation approach for single solvent molecular–crystal face interactions, and could be quite timely if various solvents and more crystal surfaces need to be exploited. The work described here employs this alternative option for predicting the interaction of solvent media with crystal surface and uses this information to model the effect of solvent on the crystal shape grown from solution. This method adopts a quantum approach for the calculation of a dielectric continuum solvation, in combination with statistical thermodynamics [46]. Two commercial software packages—Materials Studio [34] and COSMOtherm [23]—are coupled to carry out these calculations. Typically solvation models are exploited mainly for prediction of thermodynamic physical–chemical properties such as solubility, vapor–liquid equilibrium (VLE), liquid–liquid equilibrium (LLE), and pK_a in various solvents and solvent mixtures. This methodology is further

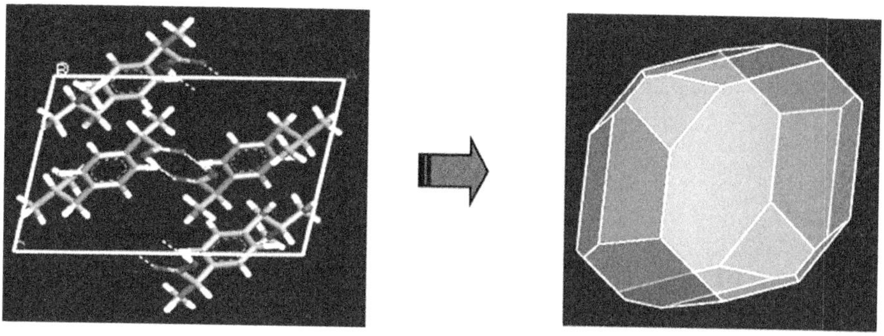

FIGURE 6.12 Crystal morphology of ibuprofen in isolation of solvent environment predicted using Materials Studio software.

extended in this work to the calculation of periodic systems, for example, crystal surfaces [47, 48] and applied to study solvent effects on morphology for the model compound ibuprofen. Combining the standard morphology prediction tools with the COSMO model provides a fast approach to explore the effect of various solvents and solvent mixtures on crystal growth. The main advantage of this method is that once the COSMO calculation has been carried out for all surfaces of interest, the effect of solvent can be quickly calculated due to the fast algorithms available in COSMOtherm software.

A crystal morphology simulation for ibuprofen was performed with Accelrys's Materials Studio software suite, utilizing the morphology module. The crystal structure was originally obtained from the CSD and then fully optimized using the COMPASS force field—a priori to morphological calculations. As shown in Figure 6.12, the unrelaxed vacuum slabs of crystal faces with thickness of 25 Å were constructed from the optimized crystal structure.

Surface charge densities for the most dominant crystal surfaces were calculated, applying the solvation model implemented in $DMol^3$ module in Materials Studio using PBE/DNP quantum level of theory. The calculated interaction energies using this approach were then used to modify the attachment energy ($E_{att\ mod}$) as predicted by the morphology simulations via using the following equation:

$$E_{\mathrm{att\,mod}} = \frac{\mathrm{Ab}}{\mathrm{As}} E_{\mathrm{slice}} - E_{\mathrm{bulk}}$$

where "E_{slice}" is the energy of a growth slice of thickness d_{hkl} in the solvent and "E_{bulk}" is the bulk crystal energy. The bulk energy was also calculated using $DMol^3$ in Materials Studio, applying the same level of theory as previously utilized for the energy of the slice. "Ab" and "As" are the number of atoms in the bulk unit cell and crystal surface, respectively.

The predicted morphology of ibuprofen from attachment energy considerations highlights four dominant faces: (100), (110), (001), and (002), as shown in Figure 6.13. These main faces were considered when modeling the solvent effect. Energies of the slices in solvent and modified attachment energies for six solvent systems are summarized in Table 6.3.

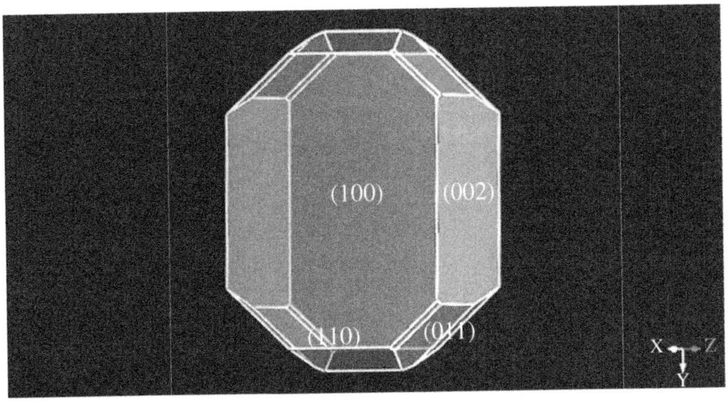

FIGURE 6.13 Predicted vacuum morphology of ibuprofen is dominated by four main crystal faces: (100), (110), (002), and (011).

The modified attachment energies were used to replot the modified morphologies. The resultant modified morphologies are then compared with experimental data [49]. It can be seen that there is very good agreement between simulated crystal shapes and experimentally observed morphologies crystallized from different solvent systems. The data highlight that nonpolar solvents such as hexane and toluene drive modification of the crystal morphology toward rod-like higher aspect ratio crystal habits. Polar solvents such as methanol and ethanol drive the morphology toward a more equant crystal shape (Table 6.4).

This work suggests that solvation modeling can be successfully applied to study the effects of solvent on crystal habit. A novel methodology has been tested on ibuprofen as a model compound. This compound has previously demonstrated a strong morphology dependence on solvent choice, with the simulated morphologies obtained demonstrating good agreement with experimental data. An obvious utility of this tool is to enhance crystal engineering workflows by simplifying the labor intensive and often time-consuming effort of solvent selection to aid API isolation with favorable crystal properties for consequent drug product development.

6.4 FROM PARTICLES TO DOSAGE FORMS

Obtaining API with the right solid form and consistent particle properties is critical not only from a drug substance manufacture standpoint but also from a drug product processing, performance, and stability perspective [50]. The importance of particle properties and material characterization previously well established in the bulk chemical industry is also now very well recognized in the pharmaceutical industry [51]. Establishing the right particle target attributes allowing for a robust and safe performance of the dosage form has become a key aspect in drug product development processes. Various experimental particle engineering technologies such as high shear wet milling, and sonocrystallisation are becoming more routine to ensure the most

TABLE 6.3 Slice and Modified Attachment Energy Results

	Faces							
	100	Ab/Asl = 1	−011	Ab/Asl = 1	002	Ab/Asl = 2	110	Ab/Asl = 1
Solvent	Slice (kcal/mol)	$E_{att\ mod}$ (kcal/mol)	Slice (kcal/mol)	$E_{att\ mod}$ (kcal/mol)	Slice (kcal/mol)	$E_{att\ mod}$ (kcal/mol)	Slice (kcal/mol)	$E_{att\ mod}$ (kcal/mol)
Hexane	−1646315.0	1.4	−1646284.7	31.7	−823157.4	1.7	−1646305.0	11.4
Toluene	−1646314.7	1.7	−1646287.6	28.9	−823157.6	1.2	−1646306.3	10.1
Ethyl acetate	−1646314.4	2.0	−1646298.4	18.0	−823157.4	1.6	−1646311.3	5.1
Acetonitrile	−1646312.6	3.9	−1646294.9	21.5	−823155.4	5.6	−1646307.4	9.0
Ethanol	−1646313.6	2.8	−1646300.2	16.2	−823156.5	3.4	−1646311.4	5.0
Methanol	−1646313.1	3.3	−1646299.8	16.6	−823155.9	4.6	−1646310.6	5.8

TABLE 6.4 Modified Attachment Energies were Used to Generate Solvent-Modified Crystal Shape

Solvent	Experimental Morphology	Predicted Morphology
Hexane	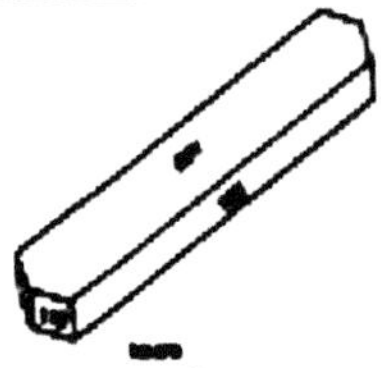	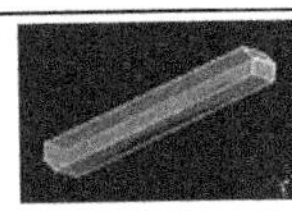
Toluene	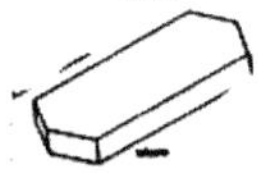	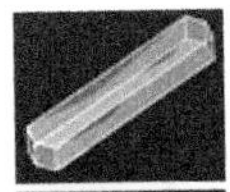
Ethyl acetate	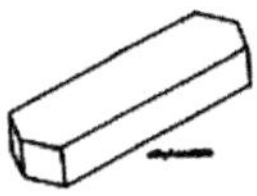	
Acetonitrile	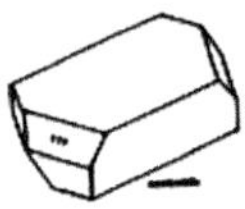	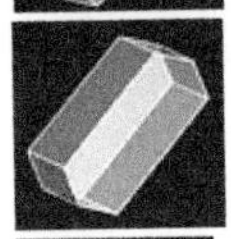
Ethanol	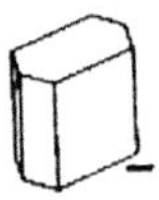	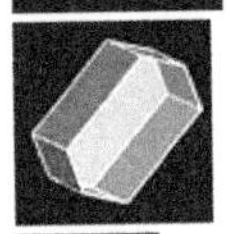
Methanol	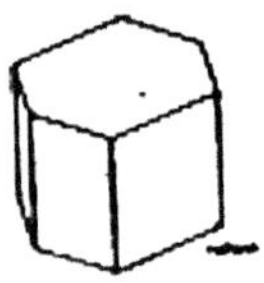	

Results are in good agreement with experimental observations reported in Bunyan et al. [49].

favorable particles will be introduced into pharmaceutical formulation and ultimately processed into a drug product [52]. Significant advances have already been made into defining the relationship between particle properties and formulation design and product performance. For example, the effect of particle size distribution on content uniformity, powder flow, and dissolution is well documented [53]. In addition, the challenges of processing needle and plate-like particles are well recognized and reported [54].

A growing area of research has focused on the characterization of surface properties of particles; however, this has typically being limited to the development of inhalation formulations [55]. Applications of surface techniques to study API/excipient interactions to understand better the target attribute space for API particle properties are less routine. An area that is often overlooked when considering and nominating critical particle attributes for formulation and process design is the surface structural characteristics of the crystallized particles. The surface properties of crystalline particles will be driven by the crystal structure and chemical surface topology that is allowed by particle morphology. Distinct particle morphology provides a definition of surface chemistry via a particular surface facet distribution, alteration of which could result in a change of overall surface properties and therefore the behavior of the powder in the formulation or process. Particle surface properties can also be subjected to change as a result of processing conditions, for example, particle size reduction and exposure to higher temperature induced by tableting. A structural investigation of possible surface structure changes can provide a clear advantage for understanding consequent drug product performance. This section will focus on drawing the attention to the wealth of information inherent in the structural features of the particles and highlight a potential utilization of such a molecular toolbox during drug product design.

6.4.1 Structural Investigation of Crystal Surfaces and Structure Dehydration

Since many pharmaceutical compounds can exist as hydrated forms, knowledge of the hydration and dehydration behavior of the drug substance is essential when developing a stable pharmaceutical formulation. Changes in the hydration state of the drug substance upon variation in processing conditions (temperature and humidity) could result in undesirable changes during drug product processing and affect the bioperformance of the dosage form. The importance of this has been well recognized by the pharmaceutical industry due to reported cases of failure of tablet formulations as a result of dehydration events [56]. However to date, few examples point to the potential of structural understanding of crystal surfaces related to potential challenges with formulation performance. The example of erythromycin B dihydrate is utilized here to demonstrate the potential advantage that could be gained by application of computational structural technologies. As detailed in the previous section, particle morphology may be derived from knowledge of the crystal structure. Modeling the particle morphology provides additional information about the surface chemistry and structure of the available particle facets. The structural position of the water and its environment within the exposed crystal surfaces can be examined through such modeling capabilities. The particle morphology of erythromycin was simulated from available single crystal data retrieved from the CSD (refcode NAVTEJ). The predicted vacuum morphology highlights that the particle shape is dominated by the larger (002) face and smaller (011) end face, respectively, which is shown in Figure 6.14. All the other predicted faces are currently not considered, but the approach could be extended to cover all potential crystal surfaces allowed by the predicted morphology.

An examination of the water position on the dominant (002) face (Fig. 6.15a) reveals that the water molecules are well protected by the molecule of erythromycin and are not exposed on the crystal surface of the particle. The crystal arrangement on this surface is very closely packed, binding the water molecule very tightly within the packing network. Due to this structural network, water molecules cannot be easily released from the structure through the surface. The (011) crystal face (Fig. 6.15b) reveals that the water molecules are very well exposed on this crystal surface.

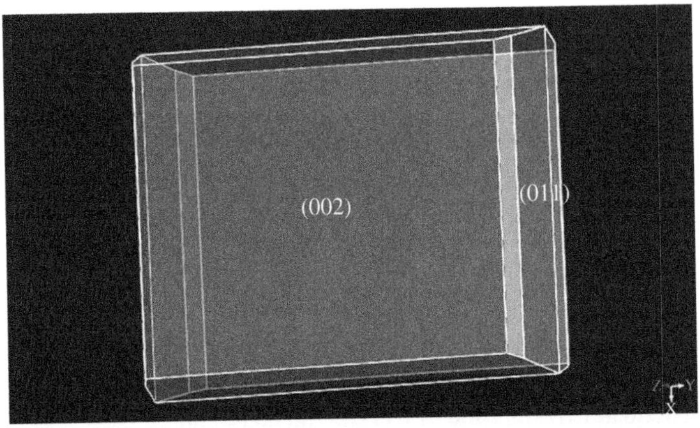

FIGURE 6.14 Predicted vacuum crystal morphology for erythromycin reveals that particle is dominated by the main (002) surface and the end (011) surface.

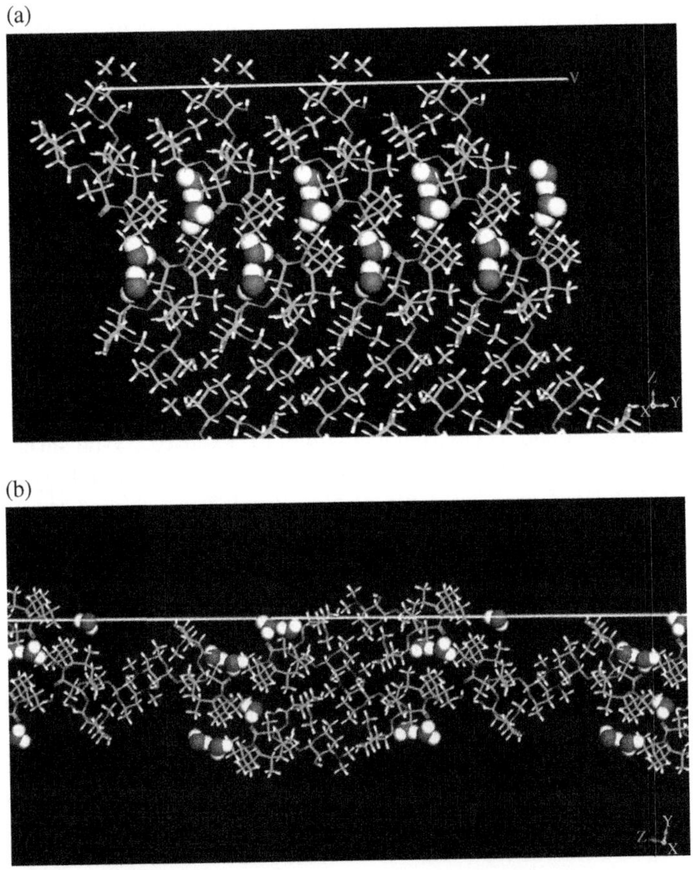

FIGURE 6.15 (a) Molecular structure of the dominant (002) surface of erythromycin. (b) Molecular structure of the end (011) surface of erythromycin. (*see insert for color representation of the figure.*)

The molecular packing on the surface is generally looser and this potentially allows for opportunity for free water molecules' movement and release from the structure through the surface as a result of change of conditions (e.g., increase in temperature or decrease in humidity). This analyses help build a hypothesis that if the relative exposure of the (011) face is increased with respect to (002) face via a thicker crystal habit, this could lead to more free and available water sites on the particle surface and thus change the typical dehydration behavior. A similar structural approach in combination with experimental techniques has been proven useful in exploring the dehydration mechanism of a pharmaceutical compound (7-methoxy-1-methyl-5-(4-(trifluoromethyl) phenyl)-[1,2,4] triazolo [4,3-a] quinolin 4-amine) as reported by Kang and co-workers [57]. These researchers demonstrated that the dehydration behavior of the aforementioned compound is highly influenced by the particle size and shape due to different surface accessibility to the outside environment of drug particles. This example illustrates how research within companies is becoming more open to utilize structural and computational technologies in synergy with standard experimental practices to design target attribute profile or troubleshoot challenges associated with drug development or manufacture.

6.4.2 Structural Investigations of Crystal Surfaces and Chemical Stability

Chemical stability of pharmaceutical materials is a key aspect to ensure product safety. Although the chemical pathway of API degradation is typically studied and understood during the process of drug product development, there are cases where an unexpected increase in degradation rate has been attributed to a change of API particles properties, formulation, or processing alterations.

Solid-state reactions usually require higher-activation energy compared to solution-state reactions due to the constraints of the crystal lattice. What is very often under-estimated is the negative effect of the bulk lattice on the crystal surface. It can be expected that the degradation-sensitive reaction sites in the solid state will be more vulnerable on the crystal surface, where particles are in direct contact with the external environment. Clearly environmental factors such as temperature, humidity, and interactions with excipient particles may have a negative effect on the rate of chemical degradation occurring on a crystal surface. Structural information on the particle surfaces can potentially provide additional information about the exposure of reaction-sensitive sites.

This effect has been demonstrated for the surface chemistry of aspirin. Hydrolysis is a well-known degradation chemical reaction occurring for aspirin (Fig. 6.16).

The reaction-sensitive atom is highlighted on the reaction scheme. These reaction-sensitive atoms can be visualized on the surface of dominant crystal surfaces of aspirin particles. First, crystal morphology was predicted using single crystal structure data for aspirin retrieved from the CSD (refcode ACSALA11). The resultant predicted crystal morphology reveals a particle with main (100) and (002) faces as shown in Figure 6.17. Second, the molecular topology and structural features of these two main surfaces adopting the molecular termination of the most stable slice were studied.

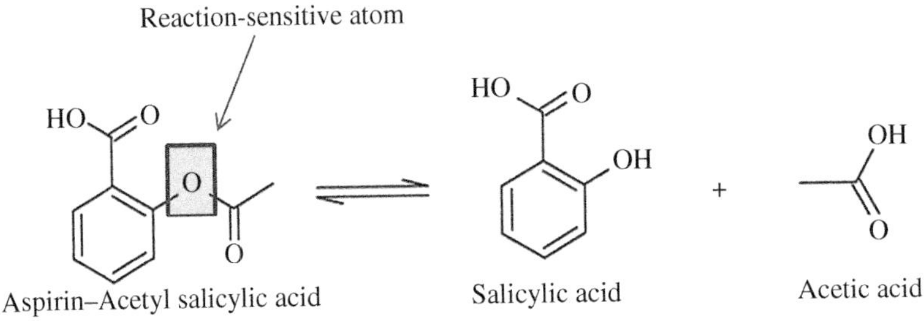

FIGURE 6.16 Degradation reaction of acetyl salicylic acid due to hydrolysis.

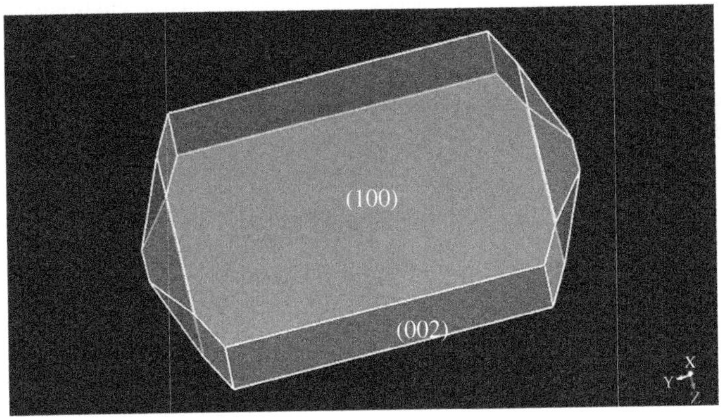

FIGURE 6.17 Predicted vacuum morphology for aspirin.

The reaction-sensitive oxygen is highlighted on the crystal surfaces of both (100) and (002). It can be seen that the (100) face (Fig. 6.18a) exposes the carboxylic end of the aspirin molecule, and the reaction-sensitive atom is buried below this functional group. The limited exposure of this oxygen atom on the crystal surface could potentially restrict the contact of this reactive site to the external environment, which could result in a decrease in degradation rate. In contrast to the (100) face, the (002) face (Fig. 6.18b) reveals a larger availability of the reaction site on the crystal surface, increasing the likelihood of this site to be in direct contact with the external environment, which could result in an increased degradation rate.

Such analysis could potentially suggest that a crystal morphology that minimizes the relative importance of (002) crystal face and maximizes the exposure of (100) via a thinner crystal habit would potentially have a reduced rate of degradation. Although direct experimental validation is not available for this case system, such a structural approach toward looking at surface-driven degradation process allows for a rapid and theoretically rationalized understanding of variation of product degradation rates often observed during product development.

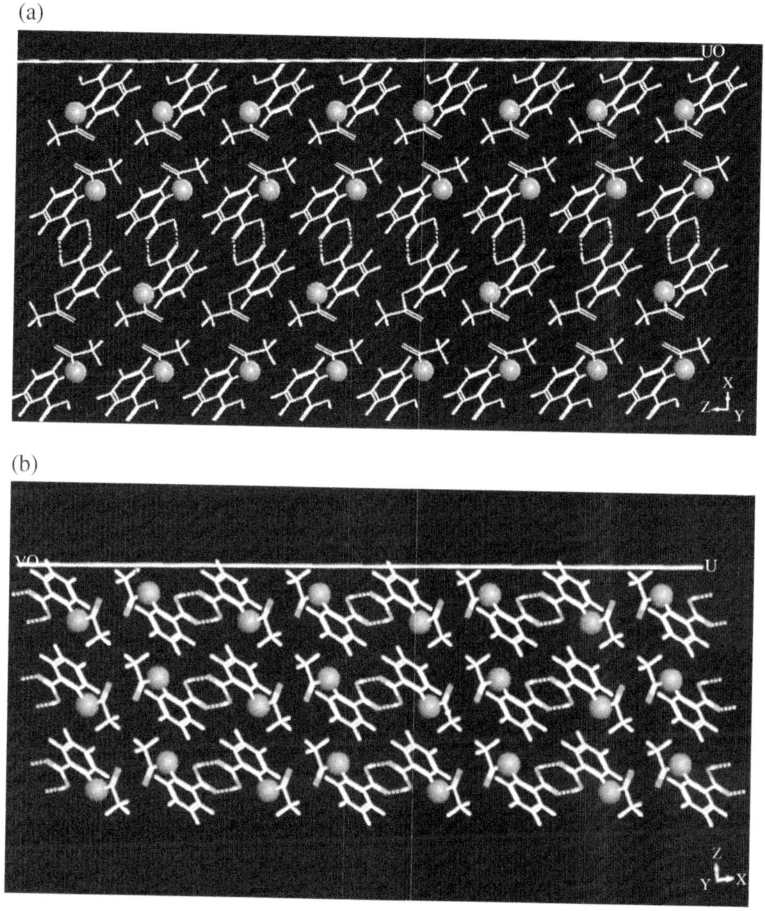

FIGURE 6.18 (a) Molecular structure of the (100) surface of aspirin. (b) Molecular structure of the (002) surface of aspirin.

6.5 CONCLUSION

In this chapter, the utility and importance of computational tools has been demonstrated to enhance the understanding of developing stable and robust drug products. Through the integration of these tools during development, a fundamental understanding of the relationship between molecular interactions, surface chemistry, and bulk properties makes it possible to identify the stable solid form as well as engineer the optimal crystal morphology for drug product processing and stability irrespective of the dosage form. Future advancements of our current computational tools will provide the framework for virtual API solid form, particle engineering, and drug product design, leading to a new era of drug product development. This new paradigm will create the cutting-edge knowledge needed to rapidly bring new drugs to patients with unmet medical needs.

ACKNOWLEDGMENTS

The authors would like to thank Dr. Neil Feeder and Dr. Yuriy A. Abramov for valuable discussions. We also would like to thank Dr. Yuriy A. Abramov for σ_{HB} charges analysis and CSP conformational prediction for sulopenem.

REFERENCES

[1] Lee, A. Y.; Erdemir, D.; Myerson, A. S., *Annu. Rev. Chem. Biomol. Eng.* 2011, *2*, 259–280.

[2] Lynn, D. Data presented by Steven Lynn, Chief, Recalls and Shortages, FDA/CDER Office of Compliance, Division of Manufacturing and Product Quality. Recalls, presentation to CASA, May 20, 2011, Baltimore, MD.

[3] Llinas, A.; Goodman Jonathan, M., *Drug Discov. Today* 2008, *13*, 198–210.

[4] Nagy, Z. K.; Fujiwara, M.; Braatz, R. D., *J. Process Cont.* 2008, *18*, 856–864.

[5] Perumalla, S. R.; Sun, C. C., *J. Pharm. Sci.* 2014, *103*, 1126–1132.

[6] Vekilov, P. G., *Cryst. Growth Des.* 2010, *10*, 5007–5019.

[7] Takiyama, H., *Adv. Powder Technol.* 2012, *23*, 273–278.

[8] Ware, E. C.; Lu, D. R., *Pharm. Res.* 2004, *21*, 177–184.

[9] Luthra, S. A.; Casteel, M.; Frericks-Schmidt, H.; Krzyzaniak, J., In-situ reactivity of API and excipients in an aqueous oral formulation resulting in formation of a quaternary supramolecular co-crystal complex. In *AAPS*, Washington, DC, October 25, 2011.

[10] Arora, K. K.; Tayade, N. G.; Suryanarayanan, R., *Mol. Pharm.* 2011, *8*, 982–989.

[11] Arora, K. K.; Thakral, S.; Suryanarayanan, R., *Pharm. Res.* 2013, *30*, 1779–1789.

[12] Bak, A.; Gore, A.; Yanez, E.; Stanton, M.; Tufekcic, S.; Syed, R.; Akrami, A.; Rose, M.; Surapaneni, S.; Bostick, T.; King, A.; Neervannan, S.; Ostovic, D.; Koparkar, A., *J. Pharm. Sci.* 2008, *97*, 3942–3956.

[13] Wassgren, C.; Curtis, J. S., *MRS Bull.* 2006, *31*, 900–904.

[14] Puri, V.; Dantuluri, A. K.; Kumar, M.; Karar, N.; Bansal, A. K., *Eur. J. Pharm. Sci.* 2010, *40*, 84–93.

[15] Hammond, R. B.; Pencheva, K.; Ramachandran, V.; Roberts, K. J., *Cryst. Growth Des.* 2007, *7*, 1571–1574.

[16] Dandekar, P.; Kuvadia, Z. B.; Doherty, M. F., *Annu. Rev. Mater. Res.* 2013, *43*, 359–386.

[17] Shariare, M. H.; Blagden, N.; de Matas, M.; Leusen, F. J. J.; York, P., *J. Pharm. Sci.* 2012, *101*, 1108–1119.

[18] Olusanmi, D.; Roberts, K. J.; Ghadiri, M.; Ding, Y., *Int. J. Pharm.* 2011, *411*, 49–63.

[19] Abramov, Y. A., *Org. Process Res. Dev.* 2013, *17*, 472–485.

[20] Gu, C.-H.; Young, V., Jr.; Grant, D. J. W., *J. Pharm. Sci.* 2001, *90*, 1878–1890.

[21] Gong, Y.; Collman, B. M.; Mehrens, S. M.; Lu, E.; Miller, J. M.; Blackburn, A.; Grant, D. J. W., *J. Pharm. Sci.* 2008, *97*, 2130–2144.

[22] Lupyan, D.; Abramov Yuriy, A.; Sherman, W., *J. Comput. Aided Mol. Des.* 2012, *26*, 1195–205.

[23] Eckert, F. K., *A COSMOtherm, C3.0; COSMOlogic.* GmbH & Co. KG: Leverkusen, 2012.

[24] Galek, P. T. A.; Fabian, L.; Motherwell, W. D. S.; Allen, F. H.; Feeder, N., *Acta Crystallogr., Sect. B: Struct. Sci.* 2007, *63*, 768–782.

[25] Wood, P. A.; Olsson, T. S. G.; Cole, J. C.; Cottrell, S. J.; Feeder, N.; Galek, P. T. A.; Groom, C. R.; Pidcock, E., *CrystEngComm* 2013, *15*, 65–72.

[26] Allen, F. H., *Acta Crystallogr., Sect. B: Struct. Sci.* 2002, *58*, 380–388.

[27] Abramov, Y. A.; Zell, M.; Krzyzaniak, J. F., Toward a rational solvent selection for conformational polymorph screening. In *Chemical Engineering in the Pharmaceutical Industry: R&D to Manufacturing*; am Ende, D. J., Ed.; John Wiley & Sons, Inc.: Hoboken, NJ, 2011; pp 491–504.

[28] Hawkins, P. C. D.; Skillman, A. G.; Warren, G. L.; Ellingson, B. A.; Stahl, M. T., *J. Chem. Inf. Model.* 2010, *50*, 572–584.

[29] TURBOMOLE, V6.0 2009, a development of University of Karlsruhe and Forschungszentrum Karlsruhe GmbH, 1989–2007, TURBOMOLE GmbH, since 2007. Available from http://www.turbomole.com (accessed October 30, 2015).

[30] (a) Schaefer, A.; Huber, C.; Ahlrichs, R., *J. Chem. Phys.* 1994, *100*, 5829–5835; (b) Becke, A. D., *Phys. Rev. A* 1988, *38*, 3098–3100; (c) Perdew, J. P., *Phys. Rev. B.* 1986, *33*, 8822–8824; (d) Perdew, J. P., *Phys. Rev. B.* 1986, *34*, 7406–7406.

[31] (a) Day, G. M.; Chisholm, J.; Shan, N.; Motherwell, W. D. S.; Jones, W., *Cryst. Growth Des.* 2004, *4*, 1327–1340; (b) Day, G. M.; Motherwell, W. D. S.; Jones, W., *Cryst. Growth Des.* 2005, *5*, 1023–1033; (c) Cooper, T.; Jones, W.; Motherwell, W. D. S.; Day, G. M., *CrystEngComm* 2007, *9*, 595–602; (d) Day, G. M.; Motherwell, W. D. S.; Jones, W., *Phys. Chem. Chem. Phys.* 2007, *9*, 1693–1704.

[32] Delley, B., *J. Chem. Phys.* 1990, *92*, 508–517.

[33] Accelrys Inc., *Materials studio, 6.0.* Accelrys Inc., San Diego, CA, 2013.

[34] Accelrys Inc., *Cerius2, version 4.6.* Accelrys Inc., San Diego, CA, 1997.

[35] (a) Coombes, D. S.; Price, S. L.; Willock, D. J.; Leslie, M., *J. Phys. Chem.* 1996, *100*, 7352–7360; (b) Williams, D. E.; Cox, S. R., *Acta Crystallogr., Sect. B: Struct. Sci.* 1984, *40*, 404–417; (c) Cox, S. R.; Hsu, L.-Y.; Williams, D. E., *Acta Crystallogr., Sect. A: Found. Crystallogr.* 1981, *37*, 293–301.

[36] Motherwell, S.; Chisholm, J. A., *J. Appl. Crystallogr.* 2005, *38*, 228–231.

[37] Mayo, S. L.; Olafson, B. D.; Goddard, W. A., III, *J. Phys. Chem.* 1990, *94*, 8897–909.

[38] Frisch, M. J., Trucks, G. W.; Schlegel, H. B.; Scuseria, G. E.; Robb, M. A.; Cheeseman, J. R.; Montgomery, Jr., J. A.; Vreven, T.; Kudin, K. N.; Burant, J. C.; Millam, J. M.; Iyengar, S. S.; Tomasi, J.; Barone, V.; Mennucci, B.; Cossi, M.; Scalmani, G.; Rega, N.; Petersson, G. A.; Nakatsuji, H.; Hada, M.; Ehara, M.; Toyota, K.; Fukuda, R.; Hasegawa, J.; Ishida, M.; Nakajima, T.; Honda, Y.; Kitao, O.; Nakai, H.; Klene, M.; Li, X.; Knox, J. E.; Hratchian, H. P.; Cross, J. B.; Bakken, V.; Adamo, C.; Jaramillo, J.; Gomperts, R.; Stratmann, R. E.; Yazyev, O.; Austin, A. J.; Cammi, R.; Pomelli, C.; Ochterski, J. W.; Ayala, P. Y.; Morokuma, K.; Voth, G. A.; Salvador, P.; Dannenberg, J. J.; Zakrzewski, V. G.; Dapprich, S.; Daniels, A. D.; Strain, M. C.; Farkas, O.; Malick, D. K.; Rabuck, A. D.; Raghavachari, K.; Foresman, J. B.; Ortiz, J. V.; Cui, Q.; Baboul, A. G.; Clifford, S.; Cioslowski, J.; Stefanov, B. B.; Liu, G.; Liashenko, A.; Piskorz, P.; Komaromi, I.; Martin, R. L.; Fox, D. J.; Keith, T.; Al-Laham, M. A.; Peng, C. Y.; Nanayakkara, A.; Challacombe, M.; Gill, P. M. W.; Johnson, B.; Chen, W.; Wong, M. W.; Gonzalez, C.; Pople, J. A., *Gaussian, 03*; Gaussian, Inc.: Wallingford, CT, 2004.

[39] Price, S. L. Willock, D. J.; Leslie, M.; Day, G. M., *DMAREL, 3.1.* CSE Department, CLRC Daresbury Laboratory: Warrington, UK, XX, 2001.

[40] Stone, A. J., *Chem. Phys. Lett.* 1981, *83*, 233–239.

[41] Sun, H., *J. Phys. Chem. B* 1998, *102*, 7338–7364.

[42] Schmidt, C.; Ulrich, J., *Chem. Eng. Technol.* 2012, *35*, 1009–1012.

[43] Clydesdale, G.; Roberts, K.; Docherty, R., The crystal habit of molecular materials: a structural perspective. In *Theoretical Aspects and Computer Modeling of the Molecular Solid State*; Gavezzotti, A., Ed.; John Wiley & Sons, Inc.: New York, 1997; p 203.

[44] Hartman, P.; Perdok, W. G., *Acta Crystallogr.* 1955, *8*, 521–524.

[45] Hartman, P.; Bennema, P., *J. Cryst. Growth* 1980, *49*, 145–156.

[46] Eckert, F.; Klamt, A., *AIChE J.* 2002, *48*, 369–385.

[47] Todorova, T.; Delley, B., *Mol. Simul.* 2008, *34*, 1013–1017.

[48] Luner, P. E.; Zhang, Y.; Abramov, Y. A.; Carvajal, M. T., *Cryst. Growth Des.* 2012, *12*, 5271–5282.

[49] Bunyan, J. M. E.; Shankland, N.; Sheen, D. B., *AIChE Symp. Ser.* 1991, *87* (284), 44–57.

[50] Vemavarapu, C.; Surapaneni, M.; Hussain, M.; Badawy, S., *Int. J. Pharm.* 2009, *374*, 96–105.

[51] Iacocca, R. G.; Burcham, C. L.; Hilden, L. R., *J. Pharm. Sci.* 2009, *99*, 51–75.

[52] Kougoulos, E.; Smales, I.; Verrier, H. M., *AAPS PharmSciTech* 2011, *12*, 287–294.

[53] (a) Yalkowsky, S. H.; Bolton, S., *Pharm. Res.* 1990, *7*, 962–966; (b) Johnson, K. C.; Swindell, A. C., *Pharm. Res.* 1996, *13*, 1795–1798; (c) Mullarney, M. P.; Leyva, N., *Pharm. Technol.* 2009, *33*, 126, 128–130, 132–134.

[54] Wood, W. M. L., *Powder Technol.* 2001, *121*, 53–59.

[55] Hooton, J. C.; Jones, M. D.; Harris, H.; Shur, J.; Price, R., *Drug Dev. Ind. Pharm.* 2008, *34*, 974–983.

[56] Bauer, J. F.; Dziki, W.; Quick, J. E., *J. Pharm. Sci.* 1999, *88*, 1222–1227.

[57] Kang, F.; Vogt, F. G.; Brum, J.; Forcino, R.; Copley, R. C. B.; Williams, G.; Carlton, R., *Cryst. Growth Des.* 2012, *12*, 60–74.

7

CURRENT COMPUTATIONAL APPROACHES AT ASTRAZENECA FOR SOLID-STATE AND PROPERTY PREDICTIONS

STEN O. NILSSON LILL, STAFFAN SCHANTZ, VIKTOR BROO, AND ANDERS BROO

Pharmaceutical Development, AstraZeneca, Göteborg, Sweden

7.1 INTRODUCTION

When a patient takes a tablet, it is probably likely that not many consider what the tablet really consists of. The active pharmaceutical ingredient (API) that is targeted for the disease is naturally the most important ingredient, but also excipients such as antioxidants, stabilizers, fillers, or sweeteners are components of the tablet that will give it the right properties related to, for example, dosage, dissolution rate, stability, and processability. The API itself may exist in different solid forms such as different polymorphs, salts, cocrystals, solvates, or amorphous material, and a particular form has therefore actively been chosen by the company for the tablet [1]. These different solid forms have different properties that can have a significant effect on a drug's bioavailability and processability. For example, different solid forms may have very different melting points, solubilities, water uptake, physical, or chemical stability. Therefore, it is of vital importance to have a detailed understanding of the different solid forms of the API and their properties as early as possible to be able to select the most suitable form of the API before it is taken further into being

Computational Pharmaceutical Solid State Chemistry, First Edition. Edited by Yuriy A. Abramov.

developed into a drug on the market. It is easy to imagine that the costs for drug project modifications increase as it emerges on the drug development time scale. Therefore, many pharmaceutical companies have made a significant investment into being able to predict solid state and material properties based on a computational approach and to use such information generated to support the solid form selection and minimize risks for late drug development surprises. This contribution highlights on how computational approaches currently are used at AstraZeneca for solid-state and property predictions.

7.2 POLYMORPHISM

A pharmaceutical compound may exist as different polymorphs, and it has been shown to be a general phenomenon. In a summary from 2006, it was described that about 80% of the drugs on the market show polymorphism under "normal" conditions [2]. For a pharmaceutical company polymorphism may be a serious problem and is one of the challenging issues for regulatory decisions since different polymorphs are considered different entities. It should though not be forgotten that having access to different polymorphs is not only negative; if you are able to control it, it may give a significant impact on the performance of a drug and give opportunities for explorations. The problem is that all systems are not controllable or at least not at the required level. Some examples with at first unknown more stable polymorphs that due to its impact have been given widespread attention in the community are ritonavir and rotigotine, resulting in withdrawal of products from the market with consequences not only in loss of sales and new nonexpected costs but also in scientific reputation. Traditionally, compounds are screened for finding polymorphs using experimental techniques under various conditions such as using different solvents, cooling and evaporation rates, temperature, and degree of supersaturation. Analysis is made by different techniques, for example, X-ray powder diffraction (XRPD), infrared (IR) spectroscopy, Raman spectroscopy, or nuclear magnetic resonance (NMR) spectroscopy, differential scanning calorimetry (DSC), and solubilities, and may also be done in a high-throughput fashion [3].

With the general development of computational capability in terms of hardware, software, and scientific advances, the scientific community has been taking a big step toward using rational approaches and techniques in predicting polymorphs and crystal structures during the last decades. We have seen a notable progress in the field of crystal structure predictions (CSPs) since the first computational blind test of predicting stable crystal structures was released in 2000 [4]. The more recent blind tests have shown an increased complexity with a higher degree of flexibility and elevated number of molecules in the asymmetric unit ($Z' = 2$) [5]. The field is constantly reviewed, and interested readers may look further into the details of these reviews [6, 7]. The approach that has shown the biggest success in CSP is probably the method developed by Neumann and co-workers where a molecule-specific force field is generated to reproduce dispersion-corrected density functional theory (DFT-D) structures and energies [8–10]. When the force field is fully parameterized, it can be used for rapid screening of both molecular conformers and crystal packing of the different conformers using the GRACE program. However, the development time for such a force field is long, typically in the time scale of weeks to months depending on the nature of the molecule of interest and the computer power available. Worth mentioning here is that incorporation of dispersion corrections in the description of both crystal structures and polymorph energies has been shown to be of utter importance to get reliable results. In this contribution, we do not wish to review or comment on the different ways of applying the dispersion correction, but rather remind the reader to incorporate it at any point. It should be remembered that differences in relative energies between polymorphs are very small, typically in the range 0–5 kJ/mol, and to calculate this energy difference correctly is very challenging in terms of computational accuracy although many sources of errors will cancel when comparing different spatial arrangements of the same molecule. The computational techniques that currently are mainly employed in CSP compare calculated energies at 0 K without zero-point energy correction, thus the vibrational or entropic contributions that would give a free energy are not encountered for. It is also often assumed that the thermodynamically most stable crystal is the one that is observed, although experimentally it may be just a kinetic phenomenon and the metastable form is formed more quickly and thus may be the subject of future transformations to a more stable polymorph. It should be noted though that it is possible to calculate free energies for crystal structures, although these require long calculation times.

As described earlier, the development time for generating a tailored force field used in CSPs is rather lengthy, and thus such an approach is not possible to use for each compound that a pharmaceutical company is interested in knowing the solid-state properties of [11–13]. Therefore, at AstraZeneca we have over the recent years developed a less time-consuming approach, which is here briefly described [14]. Rather than creating a new force field for a specific molecule, we have relied on the OPLS2005 force field and based on that created an in-house-modified force field—here called Solid State Prediction Tool (SSPT) for the treatment of the intermolecular interactions and for the creation of polymorph landscapes [15, 16]. Our calculations typically follow the procedures listed down:

1. Possible molecular conformers are searched via a mixed random torsional/low-mode sampling [17] and the OPLS2005 [18] or OPLS2.1 [19] force fields in a water continuum model as implemented in Macromodel (Schrödinger Inc.). The most stable conformers are then the subject of a refined geometry optimization using a DFT approach using either the B3LYP or M06 functionals. If we have access to an experimentally determined conformer from single-crystal XRD (SXRD) or high-resolution XRPD, we also use such structures into our forthcoming CSP protocol.
2. Crystal packings are then generated for each conformer using the in-house SSPT force field using the machinery of the GRACE program with structure solutions searched within the 13 most common space groups under the constraints of keeping the conformer rigid. This covers about 95% of all

structures found in the Cambridge Structural Database (CSD). If the compound of interest is chiral, the 12 most common chiral space groups are used in the search. Most searches are constrained to $Z' = 1$ although $Z' = 2$ investigations may also be performed. The polymorph space is searched via a Monte-Carlo (MC) parallel tempering method, and to ascertain statistical confidence we repeat the MC search 10 times. Structural duplicates are then removed based on energy and packing index criteria. Typically, the total time frame for such a CSP investigation is 1–6 h in comparison to weeks/months using the full Neumann approach described earlier. If different conformers are investigated they will run in parallel, thus the real time for calculating 2 or 20 conformers does not differ much for the end user although the computer time obviously is different. The workflow for performing all these polymorph searches and also for collection of results has been fully automated at AZ thus allowing for a streamlined, efficient and more error-free process. We can also allow the molecular conformation to be flexible in the polymorph search, thus without any other constraints except the quality of the underlying force field and space groups searched within. Such searches are more time consuming than the rigid conformer searches, and also require a good balance between the description of intra- and intermolecular interactions in the crystal structures.

3. The polymorph landscape is then analyzed, and a subset of solutions may be the subject of further DFT-D calculations. On a selection of structures, typically in the range 10–100, total energies are calculated, and based on those results a further subset of structures are extracted on which refinements of the molecular structure and cell parameters are performed. Depending on the final aim of the investigation, we may also extract conformers from such DFT-D optimizations. These conformers have been relaxed under the influence of the crystal packing forces and may therefore describe the "real" conformer differently than the conformer search performed in a water continuum or in gas phase. These conformers are then submitted into step 2, and the polymorph landscape investigation is reiterated with a slightly modified conformer, which could give bigger chances to find a crystal structure with more ideal intra- and intermolecular interactions.

As a test of the performance of the SSPT force field, we have done benchmark tests on its performance in describing crystal structures against some experimentally determined structures from CSD and in addition against some compounds from our internal crystal structure database. The conformer used in the crystal structure packing was taken as found in the databases, and the crystal packing was generated as described earlier in step. The analysis of the results of the investigation indicates that of the 63 small molecules taken from CSD the protocol was able to predict the observed crystal structure in about 75% of the cases. In about 20% of the cases, SSPT was able to directly rank the observed crystal structure as the most stable. In general, the polymorph landscapes for these fairly small molecules (mean MW = 110 g/mol) were very busy with an average of 350 unique solutions indicating that the interactions are weak, which allows for a multitude of possible packing arrangements making the predictions more challenging. The picture looks quite different when analyzing the 13 internal AZ-compounds. Now, the protocol was able to predict the observed crystal structure in 92% of the cases and in 85% of the cases it was ranked as no. 1. We also found the polymorph landscapes to be much simpler for these compounds with an average of only 92 unique solutions. This indicates the presence of stronger and more dominating intermolecular interactions which restrict access to other structural solutions. These results may seem contradictory that smaller molecules are more difficult to predict reliable polymorph landscapes of, but it is important to keep in mind that there is not only the conformational flexibility of the molecule but also the interaction flexibility that gives the final complexity of the system.

To illustrate this, we here provide an example of how a compound may be computationally analyzed when it comes to polymorphism, structural aspects of the crystal structures, and also example of an interaction analysis. As the example we have chosen a simple molecule, benzamide (Fig. 7.1).

A typical calculated polymorph landscape may look as in Figure 7.2 as generated for benzamide, with calculated interaction energies plotted versus the Kitaigorodsky packing index [20]. As a reminder to the reader, the most stable polymorph is the calculated crystal structure with the most negative interaction energy. However,

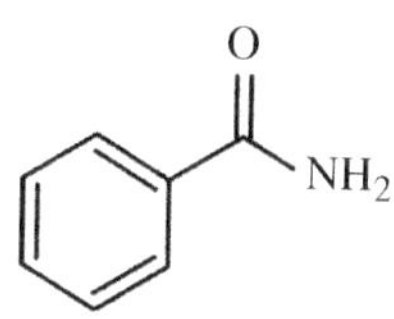

FIGURE 7.1 Schematic drawing of benzamide.

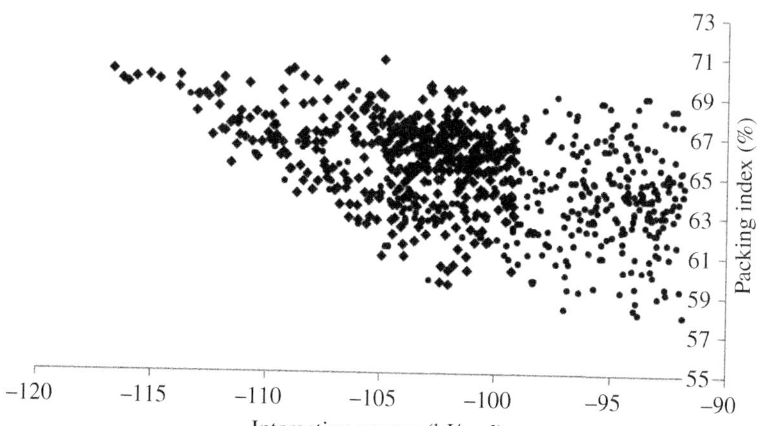

FIGURE 7.2 Predicted polymorph landscape for benzamide. The two conformers investigated are illustrated by diamonds and circles.

also predicted crystal structures with a high packing index may also be interesting in the evaluation of possible polymorphs. Thus, the upper left corner in our plots is where the most interesting structures will appear. It should also be reminded that the interaction energy does not say anything about the energy of the molecular conformer used in the study. In fact, a high-energy conformer may have a very beneficial packing interaction and thus low interaction energy. To include both the aspect of conformational penalty and interaction energy one therefore needs to compare total energies, as is calculated by DFT-D described already in step 3. In the plot, two different molecular conformers were used in separate crystal packing predictions, and it is seen that one of the conformers generally give solutions that have more favorable interaction energy and a higher packing index (diamonds vs. circles).

This is some of the information that this kind of polymorph landscape plots can provide. But what about the prediction of which polymorph of benzamide is the most stable? In the present case, the interaction energy shown was calculated based on the in-house SSPT force field originating from OPLS2005 [14]. The two starting molecular conformers were in this case found via a conformer search using OPLS2005 in the gas phase, in this case a trivial task since there are only two possible conformers. The crystal packing was generated under the constraints of having a rigid conformer and with $Z' = 1$. We then compared the 809 predicted crystal structures with the three known polymorphs I–III of benzamide available in the CSD database [21–23]. Polymorphs I and III both consist of one molecular conformer ($Z' = 1$), while polymorph II has two different conformers in the crystal structure ($Z' = 2$). Due to the restrictions used, we thus didn't expect to find polymorph II among our predicted structures. Polymorph I is experimentally determined to be the most stable [23]. Among the top predicted candidates we focused on the four most stable (**01–377**, **01–127**, **01–126**, and **01–85**) structures, all with the same molecular conformation. As a note, there were 13 predicted structures within only 4.1 kJ/mol. At the force field level of theory we predict **01–377** to be the most stable with the other three predicted to be within 1 kJ/mol. However, when comparing the structures between the predicted and experimentally determined crystal structures, we found that **01–126** is similar to polymorph I and **01–127** is similar to polymorph III. A standard approach for comparing similarities between different crystal structures is to compare the root-mean-square deviation RMSDx of interatomic distances based on a comparison of x interacting molecules in the crystal [24, 25]. Typically, such RMSD values are in the order 0.1–0.5 Å if there is a good match [13]. If $x < 15$, the predicted crystal structure is likely not a good enough candidate for representing the experimental structure and should therefore be regarded as another polymorph. Such a method is, for example, implemented in the program Mercury, which we have used at AZ. Further examples of tools from CCDC that we currently are using will be described in Section 7.6. Returning to the energy ranking of the different polymorphs, we then initially used CASTEP [26] to calculate DFT-D energies based on the force field-derived crystal structure. Now, structure **01–127** was found to be most stable, while **01–126** was ranked as no. 4. Clearly, this approach was not good enough in this case. By starting from the force field-generated crystal

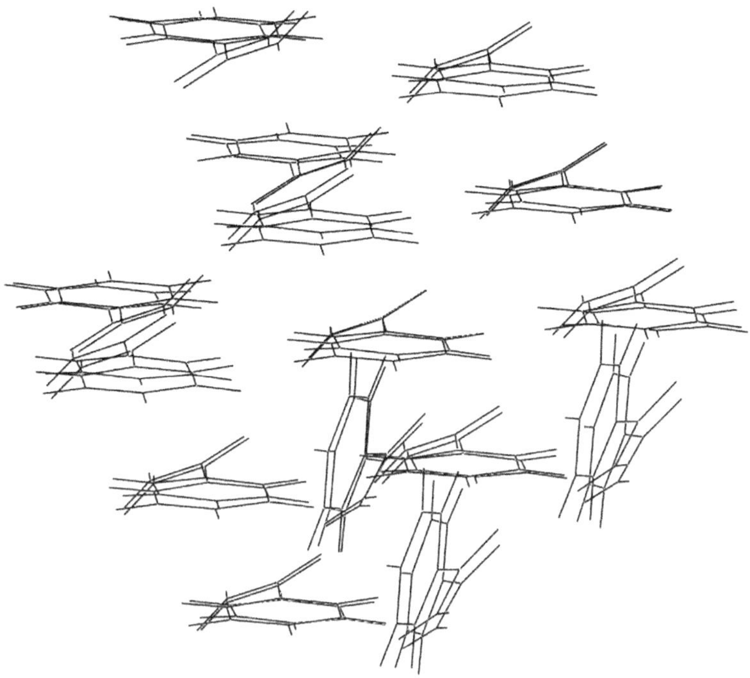

FIGURE 7.3 Overlay of experimental polymorph I and DFT-D optimized crystal structure **01–126** of benzamide. Fifteen molecules are overlaid to calculate the RMSD.

structure, but performing a reoptimizing with dispersion corrected DFT using Grimme's D2-correction [27], the relative energies of the different polymorphs could be more correctly determined. We now found candidate **01–126** (polymorph I) to be 0.67 kJ/mol more stable than **01–127** (polymorph III). As a comparison, we also calculated the relative energy of polymorph II starting from the experimental crystal structure and then reoptimizing using DFT-D. We found II to be approximately 21 kJ/mol less stable than I, in line with the experimental observations [23]. Interestingly, two other candidates that we investigated (**01–377** and **01–85**) were both predicted to be more stable than polymorph II and thus may be the subject of future structural determination under the right conditions. The final $RMSD_{15}$ value for DFT-D optimized polymorph I was determined to be 0.236 Å (Fig. 7.3), while it was 0.409 Å at FF-level, thus we see an improvement not only for the relative energies but also for the actual crystal structures (Table 7.1). It should be remembered though that temperature effects on the crystal cell are not included in the optimization, thus some structural deviation is expected.

A second way of comparing different structures is to compare their predicted XRPD patterns. The fingerprints of XRPD patterns are generally a very good way of quickly screening through a set of crystal structure candidates and thus allowing to focus on the most relevant structures. An example from the benzamide investigation is here shown in Figure 7.4.

TABLE 7.1 Relative Energies, $RMSD_{15}$, and XRPD Comparison, for Four Predicted Crystal Structures at Different Levels of Theory (FF, DFT-D)

	01–377	01–127	01–126	01–85
Rel. energy (kJ/mol)				
FF	0	0.42	0.65	1.01
DFT-D//FF	0.89	0	1.69	0.45
DFT-D//DFT-D	1.80	0.67	0	2.22
$RMSD_{15}$ (Å)				
FF		0.486[a]	0.409[b]	
DFT-D		0.308[a]	0.236[b]	
XRPD comparison				
FF		0.930[a]	0.950[b]	
DFT-D		0.950[a]	0.910[b]	

[a] Compared to experimental polymorph III.
[b] Compared to experimental polymorph I.

As can be immediately visualized, the predicted powder pattern for **01–126** (Fig. 7.4 middle) represents the predicted powder pattern of polymorph I (Fig. 7.4 top) better than what the predicted powder pattern of **01–127** (Fig. 7.4 bottom) does, especially noticeable at 2θ ~21°. In addition to comparing predicted powder pattern from an SXRD-determined structure, it is also common and useful to directly compare predicted powder patterns to experimentally determined powder patterns. This is a very valuable technique when there is no solution from single crystals available. Such an approach allows for rapid screening of several hundreds of predicted structural candidates to match an experimental powder pattern. Ideally the structure, both the molecular and crystal cell, from such a predicted structural crystal structure may be used as an input for further refinement of the powder pattern in a second iteration to completely match powder pattern intensities and peak positions.

A further and deeper analysis of crystal structures may be performed to understand why a group of predicted polymorphs are more stable than others and then group these together to possibly see common interaction patterns and packing motifs [28]. The crystal packing similarity tool in Mercury allows us to do such an analysis, and we will here show the "interaction tree" for the 10 most stable predicted crystal structures of benzamide as discussed earlier. The analysis is done in several steps by analyzing what crystal structures/motifs may be grouped together when the number of neighbors increase, that is, the interaction pattern get more complex and diverse and where possibly different crystal structures start to show differences. This is visualized by grouping crystal structures with the same intermolecular interactions together on a branch of a tree diagram (Fig. 7.5).

Here, it is shown that all 10 predicted crystal structures have the same dimer arrangement, being linked together by an amide hydrogen bond. For example, the four most stable predicted crystal structures have the same trimer arrangement by addition of another amide hydrogen bonds (Fig. 7.6). However, when tetramers and

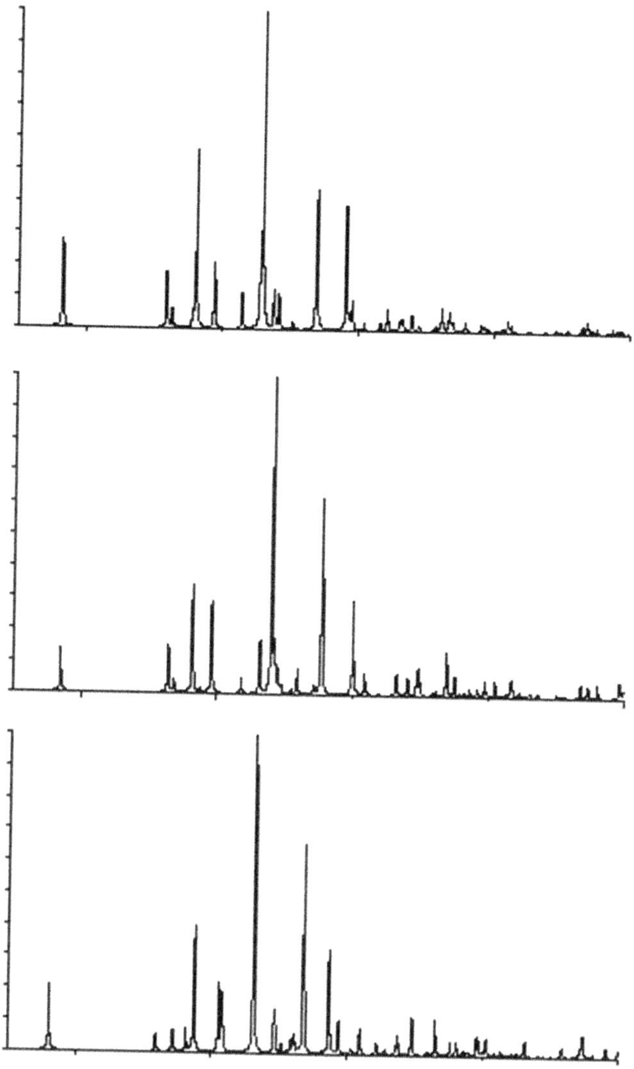

FIGURE 7.4 Top: Predicted XRPD pattern from experimental SXRD structure of Form I of benzamide Middle: Predicted XRPD pattern from predicted benzamide polymorph **01–126**. Bottom: Predicted XRPD pattern from predicted benzamide polymorph **01–127**.

pentamers are formed, different branches are generated with **01–377** and **01–85** on one branch, and **01–126** and **01–127** on another branch, the latter together with two other predicted structures (Figs. 7.7 and 7.8). In this way it is easy to visualize different interaction patterns, and such interaction pattern clusters are useful for building knowledge regarding possible cocrystal formation where specific interactions can be suppressed or enhanced. In the tree diagram in Figure 7.5, it is seen that polymorphs **01–377** and **01–85** share the same interaction pattern up to heptamers

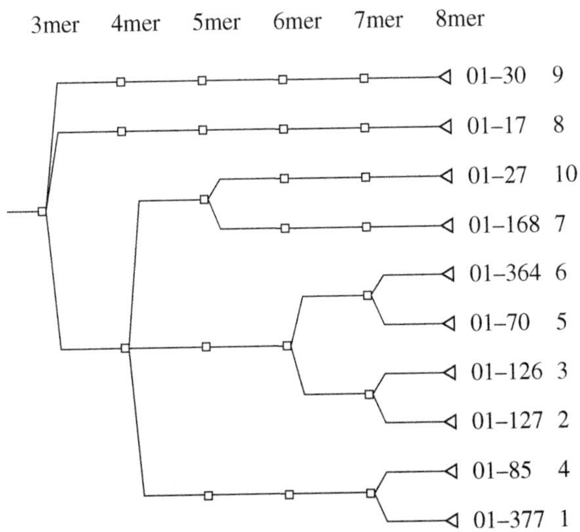

FIGURE 7.5 Interaction tree diagram for different polymorphs of benzamide. Polymorph IDs are highlighted in bold, and polymorph FF-energy ranks in numerals on the right-hand side.

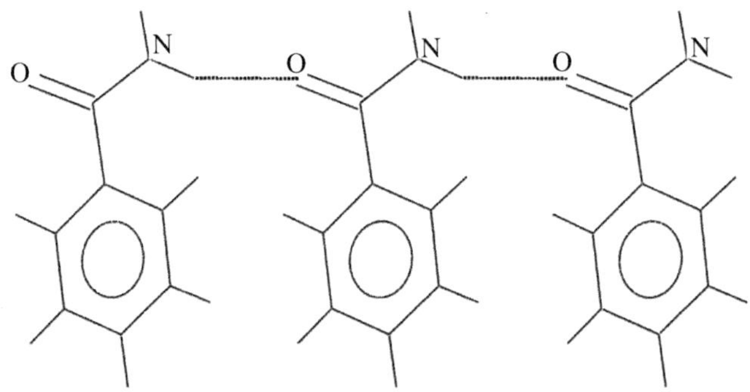

FIGURE 7.6 Trimer structural motif of polymorphs of benzamide. Hydrogen bonds indicated by dashed line.

(seven molecules interacting). Adding one more molecule from the respective crystal structure makes them separate and makes **01–377** the most stable polymorph at FF-level of theory. It is also seen that the two most stable DFT-D polymorphs **01–127** and **01–126** are closely related.

Similarly, in Figures 7.7 and 7.8 it is illustrated how tetramer cores of **01–126** and **01–377** differ in interaction patterns.

For a more detailed understanding of the intermolecular interactions involved in the crystal structures, we are also frequently employing the noncovalent interaction (NCI) plot analysis tool [29]. The tool allows visualizing NCIs such as hydrogen bonds,

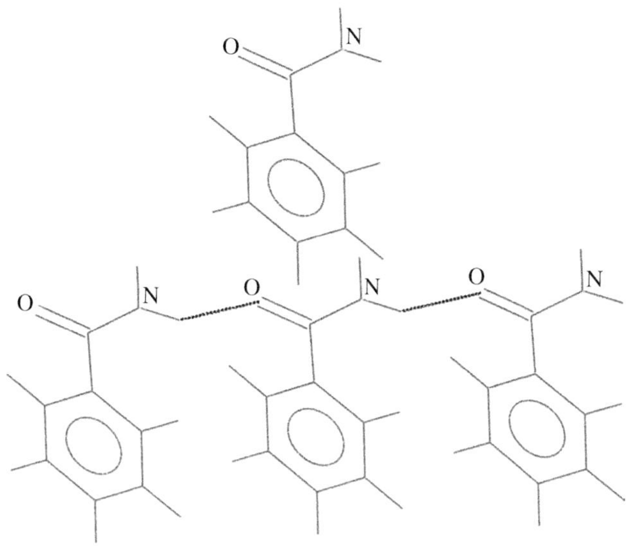

FIGURE 7.7 Tetramer structural motif of polymorph **01–126**.

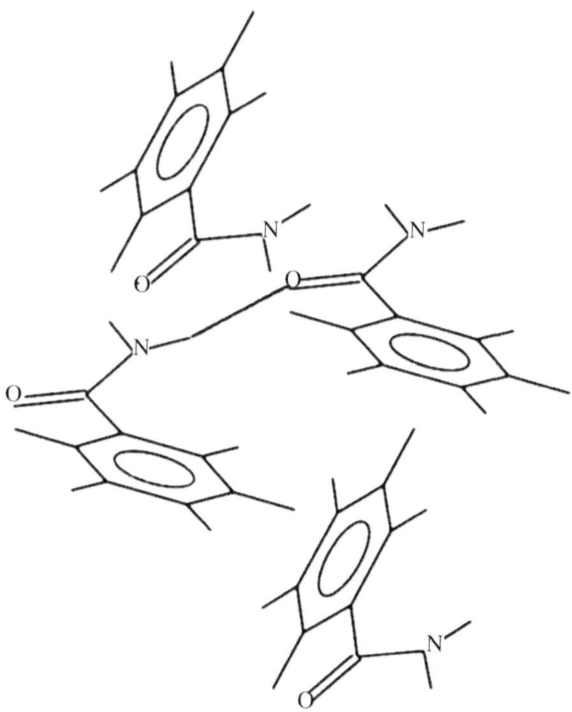

FIGURE 7.8 Tetramer structural component of polymorph **01–377**.

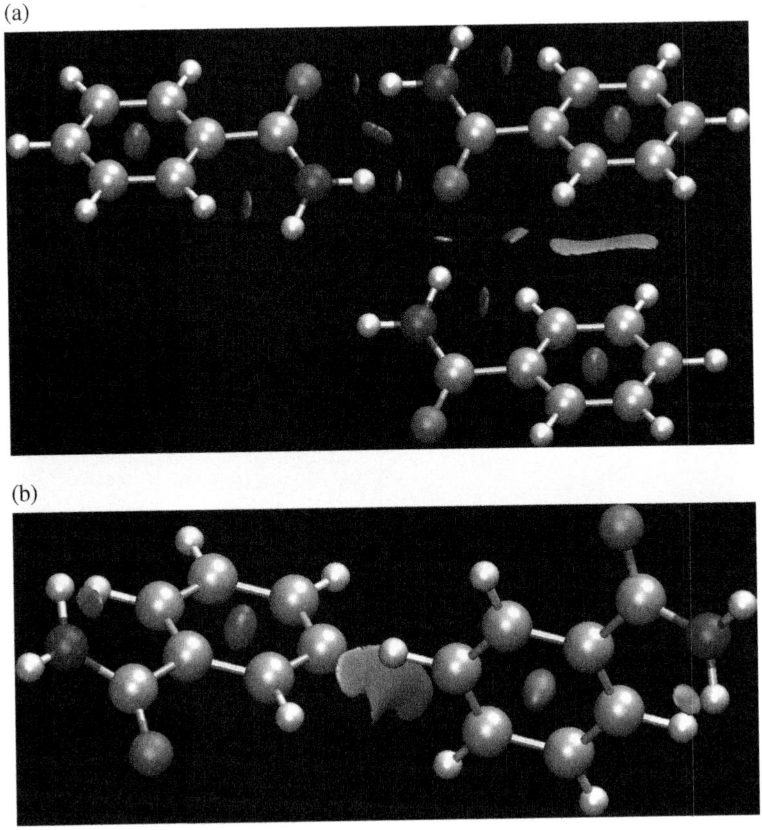

FIGURE 7.9 (a) Structure **01–126**. (b) Structure **01–127**. Noncovalent interaction (NCI) plot of predicted crystal structure. Reduced density gradient isosurface (IS) cutoff set to 0.25. Isosurfaces represent either unfavorable, weak, or strong favorable interactions. (*see insert for color representation of the figure.*)

aromatic ring stacking interactions, and CH-π interactions [30]. Examples of such plots are given in Figure 7.9 for two benzamide interaction types found for the structural candidates **01–126** and **01–127**. In Figure 7.9a, the hydrogen bonds between the amide functional groups are clearly visualized as surfaces and in addition the interaction between the aromatic rings is visualized as an interaction surface. In the **01–127** structure in Figure 7.9b, it can be seen that the aromatic rings interact differently.

One may now ask how polymorph landscape investigations are useful for risk assessments for drug projects. A valuable piece of information is the density of predicted polymorphs, as illustrated in Figure 7.10. In the top plot is shown a polymorph landscape with few solutions and the most stable polymorph clearly separated from the other. This is an ideal situation where the predictions indicate a low risk for other polymorphs to be formed. In the lower plot, on the other hand, we see many predicted polymorphs with similar energy, thus a high density in the most interesting area of

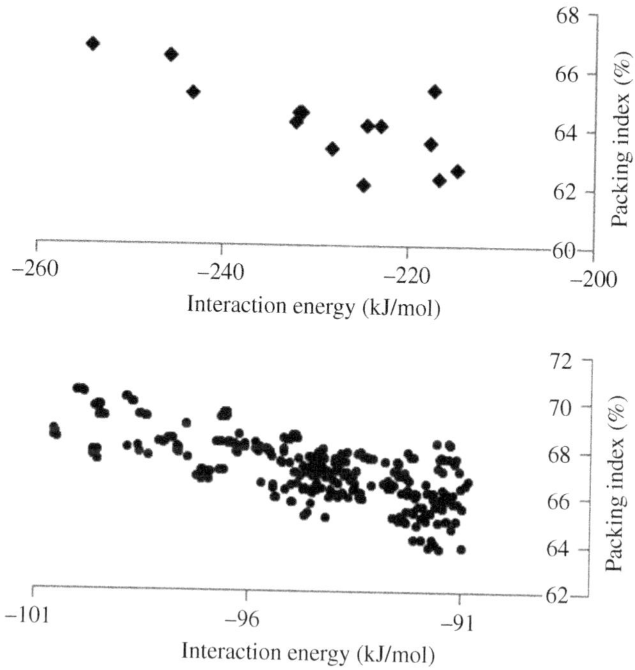

FIGURE 7.10 Predicted polymorph landscapes, illustrating different scenarios with low- (upper) and high risk (lower), respectively, for formation of several stable polymorphs.

the plot. In addition, the most stable polymorph is also not predicted to have the highest packing index. This would constitute a high risk for the project that other stable polymorphs may be formed.

7.3 CONFORMER SEARCH

As shown in Section 7.2, if we use the "correct" conformer in our crystal structure packing, we observe a good performance in predicting crystal structures and predicting which polymorph is the most stable. But the key question, what is the best way of finding the molecular conformer that is experimentally observed in crystal structures, is as yet unanswered in the scientific community [17, 31]. For some molecular systems, the global gas-phase minimum is the conformer also observed in the crystal structure while in most cases the observed conformer is a higher energy local minimum. How can one computationally find such a local molecular structure? CCDC has recently released a knowledge-based tool called Conformer Generator based on structural information from the CSD database to provide molecular geometries that are more likely to be observed. It is often observed that conformational searches in gas phase tend to give folded structures with several intramolecular interactions, while in experimental crystal structures elongated structures that rather use intermolecular interactions are more often found. An internal benchmark study was recently performed using varying force field, solvent treatment, and parameter settings in the actual conformational space search. The systems studied were taken from crystal structures in the CSD. In the study we found that performing the molecular conformational search in a continuum model environment simulating water enhances the chance of finding the conformer observed in the crystal structure compared to performing pure gas-phase calculations or in a continuum model of chloroform environment. Similarly, we could also manifest that the conformational space needs a thorough search and in general we generate up to 10,000 structures. The energy window within which we keep conformers is normally set to 63 kJ/mol. Among the different force fields tested (MMFF94s, OPSL2005, OPLS2.1) we found the most recent OPLS2.1 to give the best results.

An interesting approach of favoring extended conformations over folded ones was recently presented by Thompson and Day [32] who introduced a simple surface area–based bias energy to be added to the conformer energy. The bias energy was derived by plotting experimental sublimation enthalpies against calculated surface areas for a set of hydrocarbons where a linear relationship was found with a good correlation. By using the biased conformer energy, a clear enrichment in the ranking of observed conformers into the low-energy range of the conformational space was observed, thus giving us a higher probability in finding such conformers. We are currently exploring the use of such an approach for finding good conformational candidates to be used in polymorph searches.

7.4 MOLECULAR PERTURBATIONS TO ACHIEVE SOLUBILITY FOR GPR119 LIGANDS

Structural information and interaction analysis as provided by solving crystal structures, either experimentally or computationally, may also be used to give ideas and rationalize modifications of a lead compound to achieve an altered solubility. As an example of this, we want to highlight a drug project in which such information extracted from crystal structure of the ligand could be used to improve solubility, as recently reported by Scott et al. [33]. Initially the lead series was deemed to be unattractive due to the limited solubility of the first examples made in the series. During the lead generation phase, compounds with improved physicochemical properties were identified but it was challenging to combine good physicochemical properties with good potency. Consequently, the initial focus of our lead optimization campaign was rather to develop an alternate, more attractive series [34]. However, development of that series came to an end and the team rather picked up the series with solubility issues again and devised a strategy to address the shortcomings. It was hypothesized that strong interactions in the solid state as manifested by a rather high melting point (201–202°C) were limiting the intrinsic solubility. Ideas were generated from early structure property analysis and using the Solid-State Perturbation methodology that has been developed at AstraZeneca as described in more detail later [15, 16]. A single crystal of the lead compound, **GPR119-2** (Fig. 7.11), was grown and its crystal structure was determined. The structure was analyzed for interaction patterns and suggestions to new molecules that could perturb crystal interactions were manually generated. 3D coordinates of the new compounds were generated using the experimentally determined conformation of **GPR119-2**. These new compounds were subjected to CSPs using SSPT as described in Section 7.2. The final solutions were scored based on calculated lattice energy and density. Candidate molecules, with lower lattice energy, lower density, and similar $logD_{7.4}$-value than the reference molecule, were selected for synthesis consideration.

Eventually, the compound **GPR119-42** (Fig. 7.11) was made. **GPR119-42** had a significantly improved solubility as compared to **GPR119-2**, and the crystal structure of **GPR119-42** confirmed the initial hypothesis that it was possible to perturb interactions in the solid state to improve solubility, in this case by introducing a methyl group in a key position in the piperazine ring. Furthermore, **GPR119-42** showed a better potency than **GPR119-2** and was selected for further development. The process of generating molecular substitutions to the API and calculation of individual predicted polymorph landscape for each substitution has now been fully automated and also made available to nonexpert users through an in-house web interface. In this way, a CSP screening of a multitude of ligands can be performed automatically just by providing the structure of the parent ligand candidate.

To further validate our modeling tool, SSPT and the modified force field, we set out to predict the crystal structure of **GPR119-2** without using any information from the observed crystal structure. Of course this is not a true blind test as we have access to the "correct answer" beforehand. Nevertheless, it is an illustrative example on the performance of our simplified process using predetermined conformations in a crystal packing prediction. First, we confirmed that we get the right crystal structure as the highest-ranked solution if we use the 3D coordinates from the conformer of **GPR119-2** extracted from the crystal structure, see Figure 7.12. The highest ranked solution had an $RMSD_{15}$ of 0.19 Å. Furthermore, only very few unique solutions were found in the MC packing prediction step, thus validating the approach.

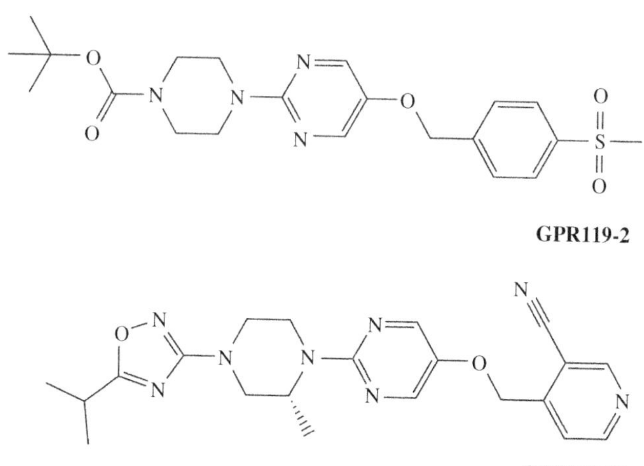

GPR119-2

GPR119-42

FIGURE 7.11 Schematic drawing of **GPR119-2** and **GPR119-42**.

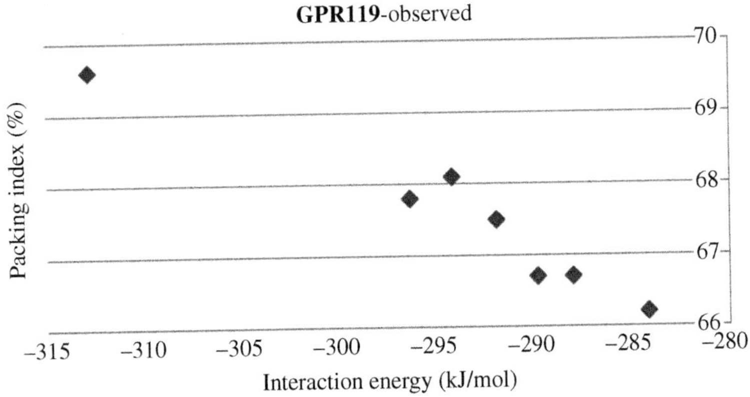

FIGURE 7.12 Polymorph landscape of **GPR119-2** generated with the SSPT CSP methodology. Plotted is the interaction energy (EI in kJ/mol) versus packing index (ck in %). The crystal structure in the left-hand upper corner is the energetically most favorable structure.

Next, starting from a 2D sketch of **GPR119-2** we generated 3D coordinates using CORINA [35] and relaxing the coordinates using the SZYBKI program [36]. We then performed a conformer search according to step 1 as described in Section 7.2 to generate a conformation ensemble. We then followed the method suggested by Thompson and Day as described earlier, and added the energy bias based on total surface area to favor elongated conformations over folded conformations. The top four conformations were then selected for further geometry optimization using B3LYP/6-31G** level of theory. At this stage we could compare the theoretical molecular conformations with the observed crystal structure rather than performing the crystal packing prediction. In Figures 7.13a–d, overlays of the geometry optimized top four conformations on the observed conformation are displayed.

All the four top-ranked conformations overlay fairly well with the observed structure, but there are small differences. The rank 3 conformation (Fig. 7.13c) is closest to the observed conformation; there are two small differences. The first difference is the torsion angle between the two aromatic ring systems. In the observed structure, the torsion angle is 0.1° and in the rank 3 structure the angle is 27.8°. The second difference is the conformation of the sulfone group. On the other hand, in the rank 1 conformation (Fig. 7.13a), the two aromatic rings are in the same plane, but the piperidine–carbamate system is not in the correct conformation. It is possible to retrieve the observed conformation by rotating 180° around the pyrimidine–piperazine bond, hereafter called rank1-rot. However, if this rotation is applied followed by a B3LYP geometry optimization, we retrieve the rank 4 conformation (Fig. 7.13d) rather than the observed conformation. In Figure 7.14, the polymorph landscape obtained with SSPT using the rank 1 conformation of **GPR119-2** is shown. Clearly, a much busier plot with more unique solutions than when the observed conformation was used is found.

Now, if the rank1-rot conformation is used in a rigid CSP, the polymorph landscape is still busy, but a few solutions are clearly separated from the main group, see

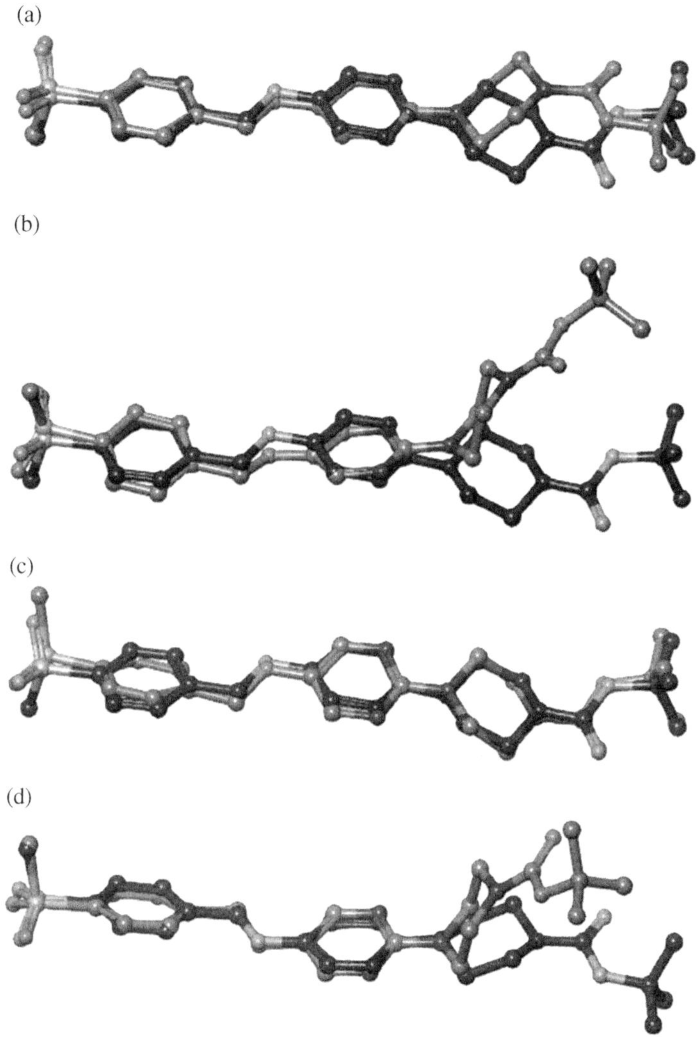

FIGURE 7.13 (a) Rank 1 conformation in gray overlaid on the observed conformation in black. (b) Rank 2 conformation in gray overlaid on the observed conformation in black. (c) Rank 3 conformation in gray overlaid on the observed conformation in black. (d) Rank 4 conformation in gray overlaid on the observed conformation in black.

Figure 7.15. As our second-ranked solution, the best fit to the observed structure is found. The crystal structure similarity tool gives an $RMSD_8$ value of 0.66 Å.

At this point, we can conclude that the fully unbiased method fails as only small differences between the theoretical conformations and the observed will translate into a too complex polymorph landscape to allow for systematic reranking using DFT-D calculations on the tentative crystal structures. However, in this case we know the answer, and from all the predicted crystal structures we can pick

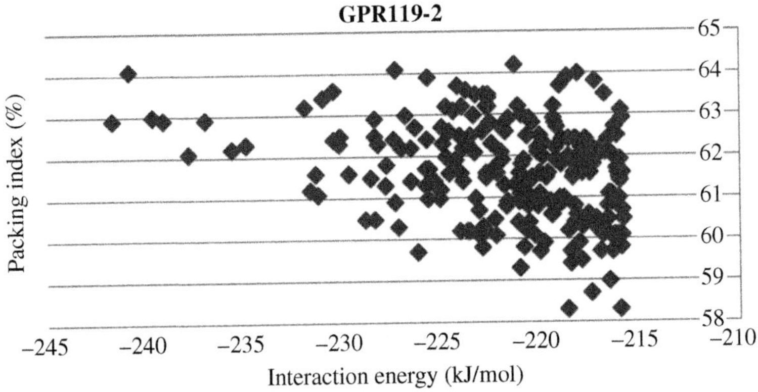

FIGURE 7.14 Polymorph landscape of **GPR119-2** rank 1 conformation generated with the SSPT CSP methodology. Plotted is interaction energy (EI in kJ/mol) versus packing index (ck in %).

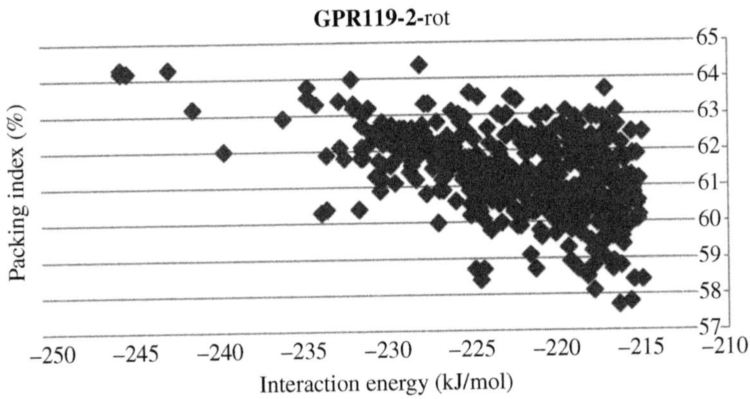

FIGURE 7.15 Polymorph landscape of **GPR119-2** rank1-rot conformation generated with the SSPT CSP methodology. Plotted is interaction energy (EI in kJ/mol) versus packing index (ck in %).

out the crystal structure showing the best overlap with the experimental structure provided by the use of the "crystal packing similarity tool" in Mercury as described in Section 7.2. We then optimized the crystal structure of this match using the PBE functional with a Tkatchenko–Scheffler (TS) dispersion correction added [37], as implemented in the plane wave code CASTEP [26]. The fully optimized crystal structure was compared with the observed crystal structure and the final $RMSD_{15}$ value was 0.576 Å, see Figure 7.16.

By this exercise we have demonstrated that it is possible to use an unbiased conformational search followed by a rigid-body crystal packing prediction to reproduce the observed structure, but the reranking step becomes too expensive in terms of computational time as we need to screen through the full set of unique solutions

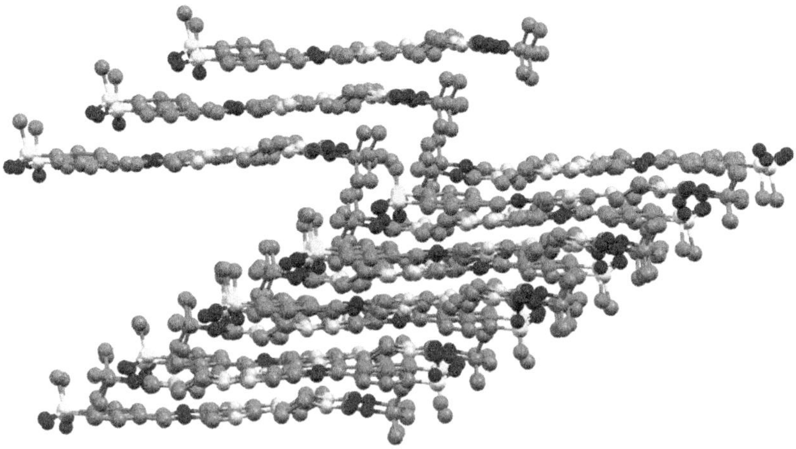

FIGURE 7.16 Overlay of the observed and the predicted crystal structure of **GPR119-2**. The observed structure is rendered in light gray and the predicted structure is in dark gray.

using DFT-D geometry optimization. Further development is needed to get the "right" solid-state conformation directly from the conformation search step. Research on this is ongoing in our lab and will be communicated in due time.

7.5 SOLID-STATE NUCLEAR MAGNETIC RESONANCE AND AZD8329 CASE STUDY

Solid-state nuclear magnetic resonance (NMR) is nowadays an established technique in the pharmaceutical industry, used mainly as a tool to distinguish different polymorphs. Its advantages are high versatility and resolution, which allow for studies of all the materials in a formulation. Compared to, for example, powder XRPD and Raman scattering, spectral overlap is most often much less of a problem in NMR. Also, the primary parameter, the resonance frequency or the chemical shift, is very sensitive not only to the intramolecular structure but also to intermolecular interactions and spatial arrangement, which is the basis for polymorph selectivity. A range of nuclei can be studied for complementary information, for example, ^{1}H, ^{13}C, ^{15}N, and ^{19}F. These occur at different natural abundances that, together with their nuclear properties, give different sensitivities at standard conditions.

More recently, solid-state NMR has seen enormous progress in absolute structure determination of solids, and methods have been introduced that complement the more traditional powder diffraction methods based on X-rays or neutrons. Many recent studies have also shown that plane wave DFT calculations can accurately reproduce measured NMR chemical shifts in solids. This complementary approach has been used to validate and refine a number of crystal structures [38–42]. The DFT structure validation requires a structural starting point for chemical shift calculations, so therefore it must be coupled with CSP methodology for generation of such candidates.

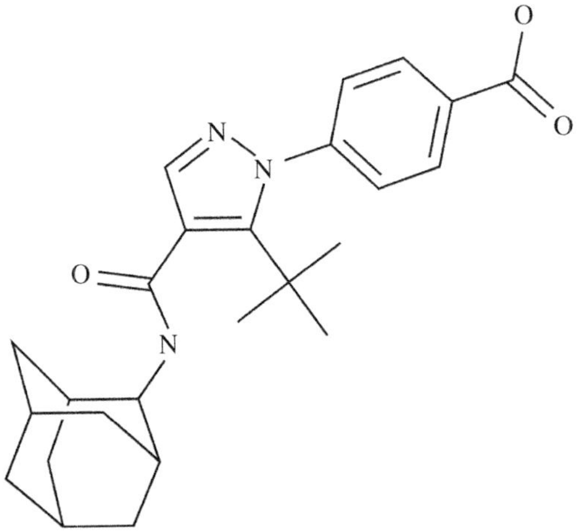

FIGURE 7.17 Chemical structure of **AZD8329**.

In a university collaboration we have recently applied a protocol developed by Emsley and Gray that combines CSP, measured ^{1}H NMR chemical shifts, and DFT chemical shift calculations to determine *de novo* the crystal structure of a polymorph of a relatively large pharmaceutical compound (422 g/mol) with previously unknown structure, 4-[4-(2-adamantylcarbamoyl)-5-ter-butyl-pyrazol-1-yl] benzoic acid (Fig. 7.17) (AstraZeneca-ID: **AZD8329**) [43, 44]. **AZD8329** is an active drug compound with potential for treatment of metabolic syndrome such as Type 2 diabetes through inhibition of 11β-hydroxysteroid dehydrogenase type 1 (11β-HSD1), an NADPH dependent reductase that converts cortisone to cortisol [45], which in turn could reduce intracellular glucocorticoid concentrations [46–48]. Several polymorphs of the compound have been found to exist, at least seven ansolvate/solvate forms are known today. Of the anhydrous forms **1–4**, the two polymorphs **1** and **4** were the most interesting since they had been chosen for development due to their suitable material properties. The structure of Form **1** had been determined by single crystal diffraction; however, the structure of Form **4** was still unsolved and was therefore the target for the investigation described here. Starting from only the chemical formula of compound **AZD8329** (Fig. 7.17), crystal structures were predicted in an unbiased way by exploring the polymorph landscape for the most stable local molecular conformers. The conformational variety of **AZD8329** was searched by generating torsional energy profiles about all exocyclic single bonds and combining these to generate an ensemble of probable conformers. *Cis–trans* isomerization commonly occurs in organic molecules [49] although one of the two isomers usually has a much lower energy. For **AZD8329**, we calculated *trans*-amide isomers to be 40–50 kJ/mol more stable than *cis*-amide isomers. However, crystal structures were nevertheless generated for both *cis*- and *trans*-isomers in case the crystallization conditions had "locked" the molecule in one or the other isomer or if strong intermolecular interactions could stabilize the less stable *cis*-form, a phenomenon observed frequently in crystal structures. Predicted crystal structures within 30 kJ/mol of the most stable one, in either the *cis*- or the *trans* sets, were then the subject for a computational NMR analysis and for more advanced DFT geometry optimizations.

For both Form **1** and **4**, ^{13}C chemical shifts were assigned with a natural abundance ^{13}C—^{13}C INADEQUATE [50] NMR spectrum, which gives the connectivity between carbons that are bonded, see Figure 7.18. Proton chemical shifts were obtained from a ^{1}H—^{13}C HETCOR NMR spectrum by their connection to the carbon nuclei previously assigned (Fig. 7.18).

For each predicted structure, chemical shifts were generated with the GIPAW approach described in detail elsewhere [39]. For the set of *cis*-isomers this included 14 structures, and for the *trans*-isomers this involved 20 structures. For each structure, measured and calculated shifts were then compared and the corresponding RMSD was obtained.

In Figure 7.19, the lowest RMSD values for ^{1}H shifts determined in this way for Form **4** of **AZD8329** are shown. Based on its low RMSD value we predicted structure 21 (Figs. 7.19 and 7.20), which is the crystal structure based on the *cis*-isomer with the lowest predicted total energy, to be the correct crystal structure for this polymorph. This structural candidate was one that gave predicted calculated chemical shifts in best agreement with the experimental data, having an RMSD between calculated and experimental ^{1}H chemical shifts of 0.48 ppm. This is within the expected errors for ^{1}H chemical shift calculations. None of the other candidate structures were in satisfaction for the criteria for agreement.

Furthermore, we were also able to solve a structure from powder X-ray diffraction for Form **4**. The heavy atom positions of the two structures agree to an all-atom RMSD of 0.284 Å, which confirms the framework of the structure. However, the XRPD data were not sufficient to determine the positions of the protons, in particular which oxygen in the carboxylic acid group that is protonated. Given that the structure is not symmetric, and that a carboxylic acid dimer is not formed, this is a significant point. In contrast, the NMR method is highly sensitive to such features. The hydrogen bonding network in Form **4**, as determined from the combined CSP/NMR structure solution formed by the candidate 21, is shown in Figure 7.21. It is observed that the carboxylic acid group forms a double hydrogen bond to the amide group of a neighboring molecule; the acid group carbonyl oxygen accepts a hydrogen bond from the NH group, while the OH proton donates a hydrogen bond to the amide carbonyl O atom forming hydrogen-bonded chains running along the crystallographic *c* axis. As a second verification, we calculated chemical shifts for a structure where the acidic proton was switched on to the second oxygen of the carboxylic acid group, and gratifyingly we found a very poor agreement with the experimental observations. Thus, we conclude that the predicted CSP/NMR structure 21 has the correct position for the carboxylic acid proton.

To our knowledge, Form **4** is the first example of a *de novo* CSP/NMR structure determination of a molecular solid of previously unknown structure. Furthermore, with a molecular weight of 422 g/mol **AZD8329** is the largest molecule so far tackled by such a protocol for NMR powder crystallography.

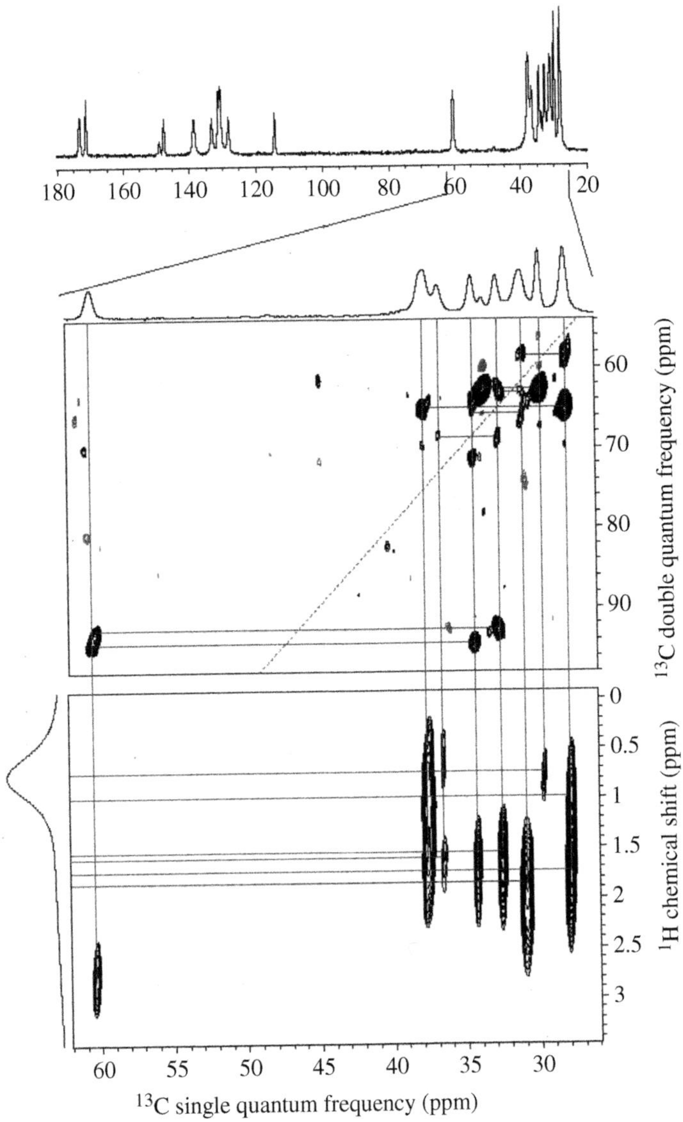

FIGURE 7.18 ^{13}C CPMAS NMR spectrum (upper), expansions of the aliphatic regions of the ^{13}C—^{13}C INADEQUATE NMR spectrum (middle) and ^{1}H—^{13}C HETCOR spectrum (lower) of **AZD8329** Form **4**. Source: Baias et al. [43]. Reprinted with permission of American Chemical Society.

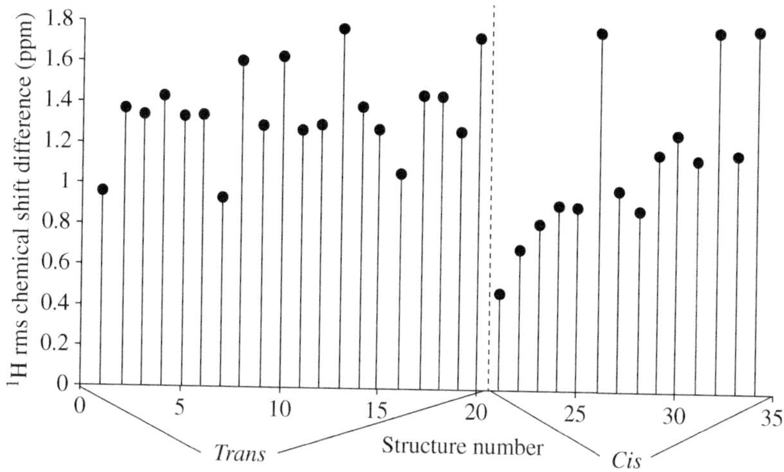

FIGURE 7.19 Comparison between experimental ^{1}H chemical shifts recorded for powdered Form **4** of **AZD8329** and calculated shifts for the predicted structures. Predicted structures are ordered by increasing calculated lattice energies (decreasing predicted stability). The first 20 structures correspond to the predicted most stable *trans*-isomers.

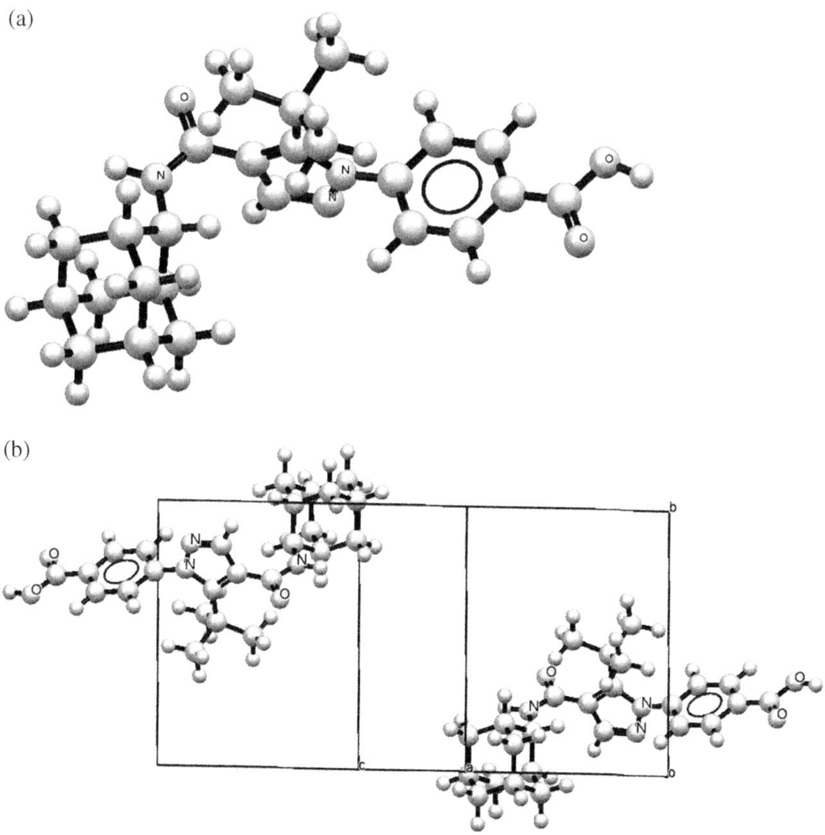

FIGURE 7.20 (a) The single-molecule conformation of **AZD8329** Form **4** determined by powder ^{1}H NMR and computational modeling. (b) Molecular packing in the unit cell of **AZD8329** Form **4** determined by powder ^{1}H NMR and computational modeling.

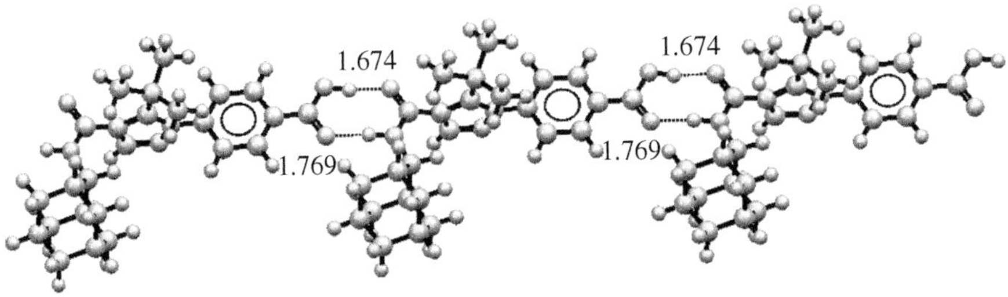

FIGURE 7.21 Illustration of the intermolecular hydrogen bonding network in Form **4** determined from the NMR structure. Note that if the carboxylic acid proton is permuted to the other oxygen, then the H-bond network cannot form.

7.6 CCDC TOOLS

In our work in form assessment, we employ some tools developed by CCDC. One of these tools is MOGUL [51], which allows for a rapid "control" of generated crystal structures. It analyzes structural parameters and compares with the data available in the CSD database. In this way it is possible to identify if the structure analyzed falls within a reasonable range of the parameter analyzed and helps to identify unusual or false structures. We also analyze structures in terms of hydrogen bonds and other noninteraction patterns if catemers are formed or if there are structural features such as dimers and sheets involved. A few years ago, CCDC released a new tool, hydrogen bond propensity (HBP), aimed at being used in solid-state form design and as a complement in risk assessment of polymorphs [52, 53]. The theory of HBP is described elsewhere, and it has been shown to be a useful approach [54, 55]. We here very briefly describe the merits of the tool, but at the same time we wish to stress that it should be used as a diagnostic tool rather than in an absolute manner. HBP uses the information of known hydrogen bond interactions provided in the CSD database. For a given molecular structure it analyzes the functional groups and for donor and acceptor combinations it predicts by probabilities which hydrogen bonds are likely or less likely to be formed. In this way, a risk assessment of polymorphism is created. A crystal structure with high HBP indicates that the more likely hydrogen bond alternatives are satisfied, while a low HBP could be a warning that there possibly are other structures with a more favorable hydrogen bond interaction pattern and a higher thermodynamic stability. In this way, it gives a decision basis for if it is worth pursuing and investing in additional polymorph screening. It also helps to set focus on specific hydrogen bond patterns among predicted structures. The tool may also be used for matching hydrogen-bond possibilities with coformers and solvents, thus it may be used for not only creating ideas for design of cocrystals and/or solvates to enhance physicochemical properties but also for IP reasons. With respect to solvates, it is also useful for evaluation of possible hydrate formation. Hydrates are generally not the preferred solid form in a project as they may be the subject of hydration/dehydration and therefore considered less desired, and HBP is therefore a useful tool in understanding where such forms may be formed.

Based on an approach already used for identification of interactions in proteins, the CCDC team has developed the full interaction maps analysis tool [56]. It relies on a library consisting of information of interaction between pairs of functional groups. Around a central functional group is plotted the interaction "densities," allowing for an identification of the directionality of interactions to be performed. Hydrogen bonds, hydrophobic interactions, as well as halogen bonds may be visualized. The screening of a structure is done rapidly and this allows for investigation of series of conformers/molecules/solvents/coformers for easy identification whether the more likely interaction patterns are fulfilled or not.

7.7 TAUTOMERISM

As a final example, we also illustrate how a computational approach was used in a study on possible tautomerism. Molecule **VI**, studied in the second CCDC blind test [57] may form three different stable polymorphs, all observed as sulfonimides. However, in one of these forms, a slight shift in the proton positions within a dimer core of the crystal structure gives the sulfonamide tautomer. Therefore, we initiated a study to elucidate the energy difference between sulfonimide and sulfonamide tautomers of **VI** (Fig. 7.22).

We studied monomers, dimers both in gas phase and in solvent environments, and also periodic solid state crystal structures [14]. It was found that monomers of molecule **VI** prefer to exist as sulfonamide tautomers while the equilibrium is shifted towards sulfonimides for larger aggregates due to stronger hydrogen bonds and stronger polarization effects. Also, we could determine activation barriers for proton transfer between sulfonamide and sulfonamide aggregates indicating this to be a rapid process, which might indicate this to be a general phenomenon for sulfonamides. Similar investigations have, for example, been performed for other imine/amine tautomerisms in the solid state [38].

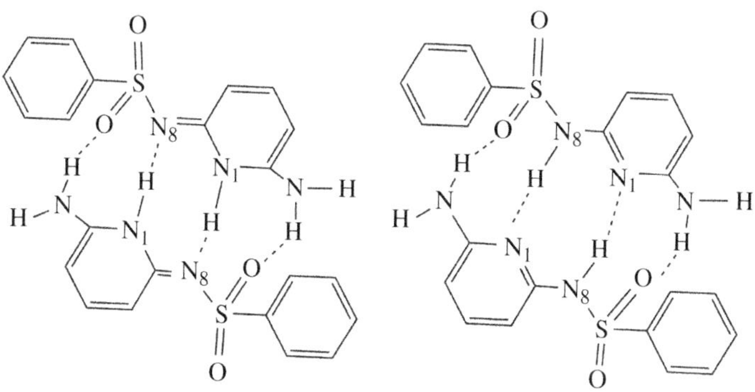

FIGURE 7.22 Sulfonimide (left) and sulfonamide (right) dimers of molecule **VI**.

7.8 CONCLUSIONS

In this chapter, we have highlighted a few important areas within pharmaceutical development where predictive science can play an important role for making scientifically informed project decisions. We have shown examples on the usefulness of predicting risks for polymorph formation and some of the underlying tools behind our current approaches in the field. Some discussed case studies illustrate where CSPs have been used in drug projects and on the direct impact of such studies, and how they can be used in connection to an early discovery phase (Section 7.4), or at a later phase for helping determining the right form of a drug candidate (Section 7.5). Stimulated by the scientific achievements and the application of predictive science in pharmaceutical solid-state chemistry, we are also looking forward at applying computational techniques for predicting materials science properties of interest for drug development.

ACKNOWLEDGMENTS

We wish to acknowledge Maria Baias, Jean-Nicolas Dumez, Per Svensson, Graeme Day, and Lyndon Emsley for the fruitful collaboration in the study of **AZD8329**. We also acknowledge Per Svensson and Jan Rosdahl for the initial development of SSPT.

REFERENCES

[1] G. P. Stahly, *Cryst. Growth Des.* 2007, *7*, 1007–1026.

[2] Lohani, S.; Grant, D. J. W. Thermodynamics of Polymorphs. In *Polymorphism: in the Pharmaceutical Industry*; Hilfiker, R., Ed.; Wiley-VCH Verlag GmbH & Co. KGaA: Weinheim, FRG, 2006, pp. 21–42.

[3] R. Storey, R. Docherty, P. Higginson, C. Dallman, C. Gilmore, G. Barr, W. Dong, *Crystallogr. Rev.* 2004, *10*, 45–56.

[4] J. P. M. Lommerse, W. D. S. Motherwell, H. L. Ammon, J. D. Dunitz, A. Gavezzotti, D. W. M. Hofmann, F. J. J. Leusen, W. T. M. Mooij, S. L. Price, B. Schweizer, M. U. Schmidt, B. P. van Eijck, P. Verwer, D. E. Williams, *Acta Crystallogr., Sect. B: Struct. Sci.* 2000, *56*, 697–714.

[5] D. A. Bardwell, C. S. Adjiman, Y. A. Arnautova, E. Bartashevich, S. X. M. Boerrigter, D. E. Braun, A. J. Cruz-Cabeza, G. M. Day, R. G. Della Valle, G. R. Desiraju, B. P. van Eijck, J. C. Facelli, M. B. Ferraro, D. Grillo, M. Habgood, D. W. M. Hofmann, F. Hofmann, K. V. J. Jose, P. G. Karamertzanis, A. V. Kazantsev, J. Kendrick, L. N. Kuleshova, F. J. J. Leusen, A. V. Maleev, A. J. Misquitta, S. Mohamed, R. J. Needs, M. A. Neumann, D. Nikylov, A. M. Orendt, R. Pal, C. C. Pantelides, C. J. Pickard, L. S. Price, S. L. Price, H. A. Scheraga, J. van de Streek, T. S. Thakur, S. Tiwari, E. Venuti, I. K. Zhitkov, *Acta Crystallogr., Sect. B: Struct. Sci.* 2011, *67*, 535–551.

[6] J. D. Dunitz, *Chem. Commun.* 2003, 545–548.

[7] S. L. Price, *Chem. Soc. Rev.* 2014, *43*, 2098–2111.

[8] M. A. Neumann, M.-A. Perrin, *J. Phys. Chem. B* 2005, *109*, 15531–15541.

[9] J. van de Streek, M. A. Neumann, *Acta Crystallogr., Sect. B: Struct. Sci.* 2014, *70*, 1020–1032.

[10] J. van de Streek, M. A. Neumann, *Acta Crystallogr., Sect. B: Struct. Sci.* 2010, *66*, 544–558.

[11] S. Z. Ismail, C. L. Anderton, R. C. B. Copley, L. S. Price, S. L. Price, *Cryst. Growth Des.* 2013, *13*, 2396–2406.

[12] D. E. Braun, J. A. McMahon, L. H. Koztecki, S. L. Price, S. M. Reutzel-Edens, *Cryst. Growth Des.* 2014, *14*, 2056–2072.

[13] A. V. Kazantsev, P. G. Karamertzanis, C. S. Adjiman, C. C. Pantelides, S. L. Price, P. T. A. Galek, G. M. Day, A. J. Cruz-Cabeza, *Int. J. Pharm.* 2011, *418*, 168–178.

[14] S. O. Nilsson Lill, A. Broo, *Cryst. Growth Des.* 2014, *14*, 3704–3710.

[15] L.-E. Briggner, R. Hendrickx, L. Kloo, J. Rosdahl, P. H. Svensson, *ChemMedChem* 2011, *6*, 60–62.

[16] L.-E. Briggner, L. Kloo, J. Rosdahl, P. H. Svensson, *ChemMedChem* 2014, *9*, 724–726.

[17] I. J. Chen, N. Foloppe, *Bioorg Med. Chem.* 2013, *21*, 7898–7920.

[18] J. L. Banks, H. S. Beard, Y. Cao, A. E. Cho, W. Damm, R. Farid, A. K. Felts, T. A. Halgren, D. T. Mainz, J. R. Maple, R. Murphy, D. M. Philipp, M. P. Repasky, L. Y. Zhang, B. J. Berne, R. A. Friesner, E. Gallicchio, R. M. Levy, *J. Comput. Chem.* 2005, *26*, 1752–1780.

[19] D. Shivakumar, E. Harder, W. Damm, R. A. Friesner, W. Sherman, *J. Chem. Theory Comput.* 2012, *8*, 2553–2558.

[20] A. I. Kitaigorodsky, *Molecular Crystals and Molecules*. Academic Press: New York, 1973.

[21] Q. Gao, G. A. Jeffrey, J. R. Ruble, R. K. McMullan, *Acta Crystallogr., Sect. B: Struct. Sci.* 1991, *47*, 742–745.

[22] K. Kobayashi, A. Sato, S. Sakamoto, K. Yamaguchi, *J. Am. Chem. Soc.* 2003, *125*, 3035–3045.

[23] J. Thun, L. Seyfarth, C. Butterhof, J. Senker, R. E. Dinnebier, J. Breu, *Cryst. Growth Des.* 2009, *9*, 2435–2441.

[24] C. F. Macrae, I. J. Bruno, J. A. Chisholm, P. R. Edgington, P. McCabe, E. Pidcock, L. Rodriguez-Monge, R. Taylor, J. van de Streek, P. A. Wood, *J. Appl. Crystallogr.* 2008, *41*, 466–470.

[25] J. A. Chisholm, S. Motherwell, *J. Appl. Crystallogr.* 2005, *38*, 228–231.

[26] J. Clark Stewart, D. Segall Matthew, J. Pickard Chris, J. Hasnip Phil, I. J. Probert Matt, K. Refson, C. Payne Mike, *Z. Kristallogr.* 2005, *220*, 567.

[27] S. Grimme, *J. Comput. Chem.* 2006, *27*, 1787–1799.

[28] S. L. Childs, P. A. Wood, N. Rodríguez-Hornedo, L. S. Reddy, K. I. Hardcastle, *Cryst. Growth Des.* 2009, *9*, 1869–1888.

[29] J. Contreras-García, E. R. Johnson, S. Keinan, R. Chaudret, J.-P. Piquemal, D. N. Beratan, W. Yang, *J. Chem. Theory Comput.* 2011, *7*, 625–632.

[30] S. O. Nilsson Lill, *Phys. Chem. Chem. Phys.* 2011, *13*, 16022–16027.

[31] E. Perola, P. S. Charifson, *J. Med. Chem.* 2004, *47*, 2499–2510.

[32] H. P. G. Thompson, G. M. Day, *Chem. Sci.* 2014, *5*, 3173–3182.

[33] J. S. Scott, A. M. Birch, K. J. Brocklehurst, A. Broo, H. S. Brown, R. J. Butlin, D. S. Clarke, Ö. Davidsson, A. Ertan, K. Goldberg, S. D. Groombridge, J. A. Hudson, D. Laber,

A. G. Leach, P. A. MacFaul, D. McKerrecher, A. Pickup, P. Schofield, P. H. Svensson, P. Sörme, J. Teague, *J. Med. Chem.* 2012, *55*, 5361–5379.

[34] K. J. Brocklehurst, A. Broo, R. J. Butlin, H. S. Brown, D. S. Clarke, Ö. Davidsson, K. Goldberg, S. D. Groombridge, E. E. Kelly, A. Leach, D. McKerrecher, C. O'Donnell, S. Poucher, P. Schofield, J. S. Scott, J. Teague, L. Westgate, M. J. M. Wood, *Bioorg. Med. Chem. Lett.* 2011, *21*, 7310–7316.

[35] J. Sadowski, J. Gasteiger, G. Klebe, *J. Chem. Inf. Comput. Sci.* 1994, *34*, 1000–1008.

[36] S. Wlodek, A. G. Skillman, A. Nicholls, *J. Chem. Theory Comput.* 2010, *6*, 2140–2152.

[37] A. Tkatchenko, M. Scheffler, *Phys. Rev. Lett.* 2009, *102*, 073005/1–073005/4.

[38] X. Li, A. D. Bond, K. E. Johansson, J. Van de Streek, *Acta Crystallogr. Sect. C: Cryst. Struct. Commun.* 2014, *70*, 784–789.

[39] J. R. Yates, S. E. Dobbins, C. J. Pickard, F. Mauri, P. Y. Ghi, R. K. Harris, *Phys. Chem. Chem. Phys.* 2005, *7*, 1402–1407.

[40] A. L. Webber, L. Emsley, R. M. Claramunt, S. P. Brown, *J. Phys. Chem. A* 2010, *114*, 10435–10442.

[41] A. L. Webber, B. Elena, J. M. Griffin, J. R. Yates, T. N. Pham, F. Mauri, C. J. Pickard, A. M. Gil, R. Stein, A. Lesage, L. Emsley, S. P. Brown, *Phys. Chem. Chem. Phys.* 2010, *12*, 6970–6983.

[42] C. Ochsenfeld, S. P. Brown, I. Schnell, J. Gauss, H. W. Spiess, *J. Am. Chem. Soc.* 2001, *123*, 2597–2606.

[43] M. Baias, J.-N. Dumez, P. H. Svensson, S. Schantz, G. M. Day, L. Emsley, *J. Am. Chem. Soc.* 2013, *135*, 17501–17507.

[44] J. S. Scott, J. de Schoolmeester, E. Kilgour, R. M. Mayers, M. J. Packer, D. Hargreaves, S. Gerhardt, D. J. Ogg, A. Rees, N. Selmi, A. Stocker, J. G. Swales, P. R. O. Whittamore, *J. Med. Chem.* 2012, *55*, 10136–10147.

[45] C. R. W. Edwards, R. Benediktsson, R. S. Lindsay, J. R. Seckl, *Steroids* 1996, *61*, 263–269.

[46] J. W. Tomlinson, E. A. Walker, I. J. Bujalska, N. Draper, G. G. Lavery, M. S. Cooper, M. Hewison, P. M. Stewart, *Endocr. Rev.* 2004, *25*, 831–866.

[47] R. Thieringer, A. Hermanowski-Vosatka, *Expert Rev. Cardiovasc. Ther.* 2005, *3*, 911–924.

[48] M. Wamil, J. R. Seckl, *Drug Discov. Today* 2007, *12*, 504–520.

[49] J. Bauer, S. Spanton, R. Henry, J. Quick, W. Dziki, W. Porter, J. Morris, *Pharm. Res.* 2001, *18*, 859–866.

[50] A. Lesage, P. Charmont, S. Steuernagel, L. Emsley, *J. Am. Chem. Soc.* 2000, *122*, 9739–9744.

[51] I. J. Bruno, J. C. Cole, M. Kessler, J. Luo, W. D. S. Motherwell, L. H. Purkis, B. R. Smith, R. Taylor, R. I. Cooper, S. E. Harris, A. G. Orpen, *J. Chem. Inf. Comput. Sci.* 2004, *44*, 2133–2144.

[52] M. Majumder, G. Buckton, C. F. Rawlinson-Malone, A. C. Williams, M. J. Spillman, E. Pidcock, K. Shankland, *CrystEngComm* 2013, *15*, 4041–4044.

[53] E. Nauha, J. Bernstein, *Cryst. Growth Des.* 2014, *14*, 4364–4370.

[54] P. T. A. Galek, L. Fabian, W. D. S. Motherwell, F. H. Allen, N. Feeder, *Acta Crystallogr., Sect. B: Struct. Sci.* 2007, *63*, 768–782.

[55] P. T. A. Galek, J. A. Chisholm, E. Pidcock, P. A. Wood, *Acta Crystallogr., Sect. B: Struct. Sci.* 2014, *70*, 91–105.

[56] P. A. Wood, T. S. G. Olsson, J. C. Cole, S. J. Cottrell, N. Feeder, P. T. A. Galek, C. R. Groom, E. Pidcock, *CrystEngComm* 2013, *15*, 65–72.

[57] W. D. S. Motherwell, H. L. Ammon, J. D. Dunitz, A. Dzyabchenko, P. Erk, A. Gavezzotti, D. W. M. Hofmann, F. J. J. Leusen, J. P. M. Lommerse, W. T. M. Mooij, S. L. Price, H. Scheraga, B. Schweizer, M. U. Schmidt, B. P. van Eijck, P. Verwer, D. E. Williams, *Acta Crystallogr., Sect. B: Struct. Sci.* 2002, *58*, 647–661.

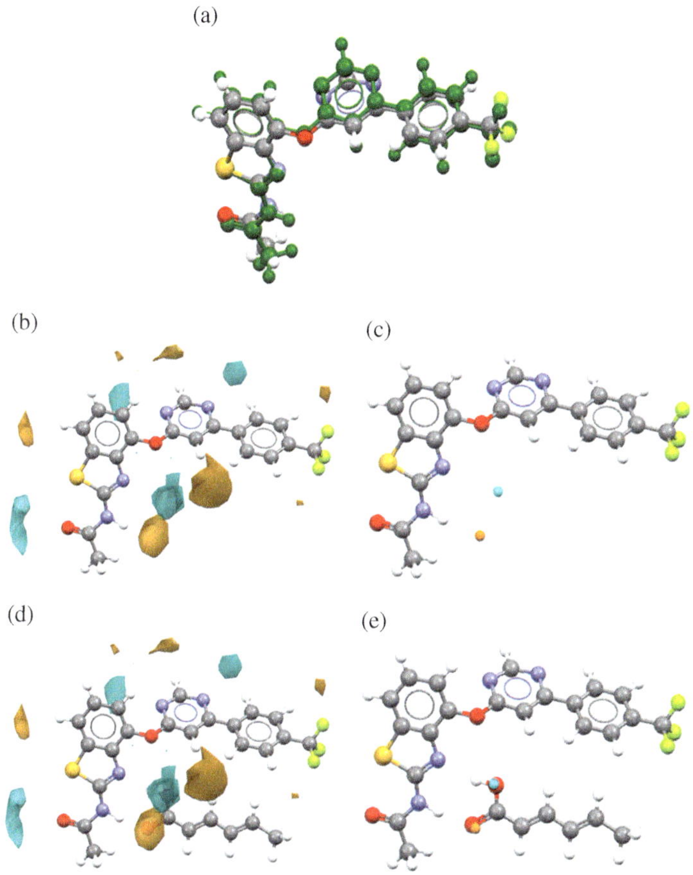

FIGURE 2.8 Conformation and Full Interaction Maps of AMG 517. (a) Overlay of the AMG 517 molecule from the sorbic acid cocrystal with a likely conformation suggested by Mogul analysis (dark). Overlay RMSD = 0.392Å. (b) Contour maps for hydrogen-bond donor (cyan) and acceptor (orange) probes, (c) maximum hot spots in the donor (cyan) and acceptor (orange) maps, (d) overlay of a sorbic acid molecule onto the contour maps and (e) overlay of a sorbic acid molecule onto the hot spots (figures prepared using *Mercury* [18]).

Computational Pharmaceutical Solid State Chemistry, First Edition. Edited by Yuriy A. Abramov.

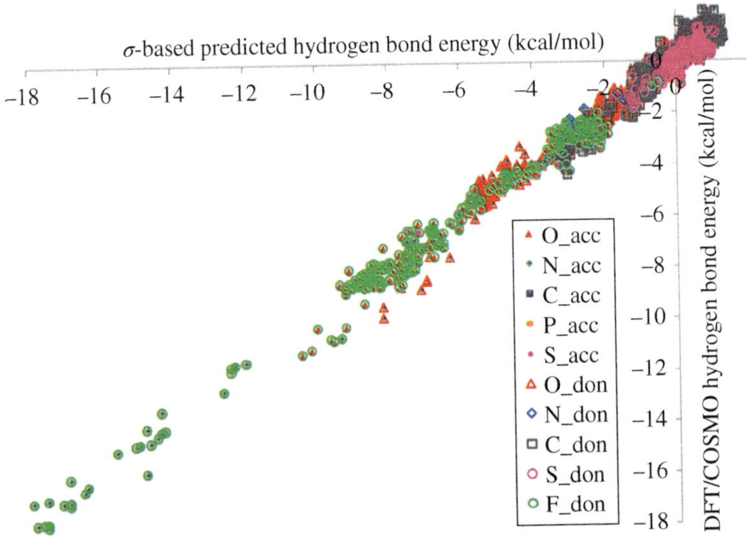

FIGURE 3.2 Scatter plot of the DFT/COSMO H-bonding energy versus the predicted HB energy based on a product of donor and acceptor polarization charge densities. Source: Adapted from Klamt et al. [55]. Reproduced with permission of Royal Society of Chemistry.

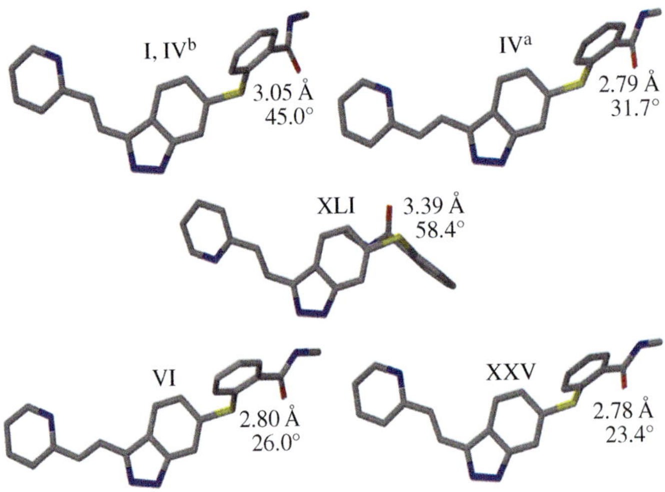

FIGURE 4.8 Five experimentally observed crystallographic conformations of axitinib. The CSD refcodes for forms I, IV, VI, XXV, and XLI are VUSDIX06, VUSDIX05, VUSDIX03, VUSDIX, and VUSDIX04, respectively. Form IV has two molecules in the asymmetric unit cell. Form IV conformation "b" (IV^b) is nearly identical to the conformation of form I. S···O distance and SCCC=O motif planarity are shown for each conformation. Source: Adapted from Lupyan et al. [37]. Reproduced with permission of Springer.

(a)

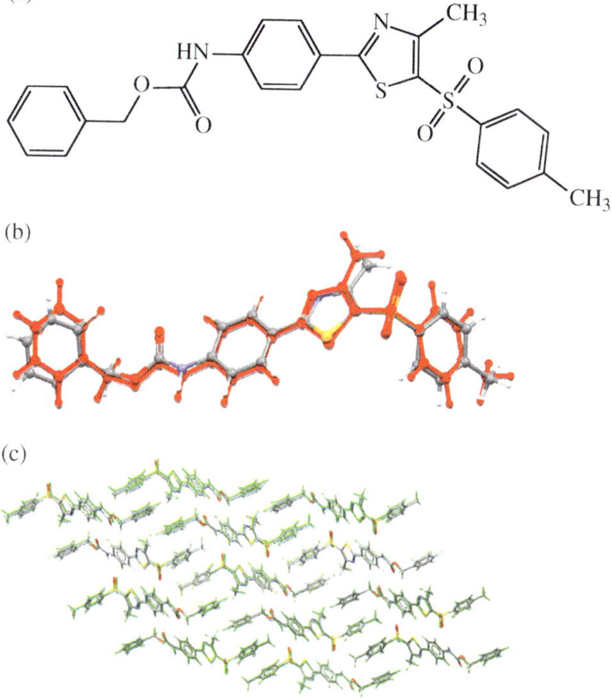

(b)

(c)

FIGURE 5.4 Molecule XX (benzyl-(4-(4-methyl-5-(*p*-tolylsulfonyl)-1,3-thiazol-2-yl) phenyl)carbamate) from the fifth blind test of crystal structure prediction. (a) Chemical diagram, (b) overlay of one of the 48 database generated conformations (red) with the conformation in the observed crystal structure, (c) overlay of the CSP global minimum in lattice energy (green) with the observed structure from X-ray diffraction. Source: Kazantsev et al. [29]. Reprinted with permission of Elsevier.

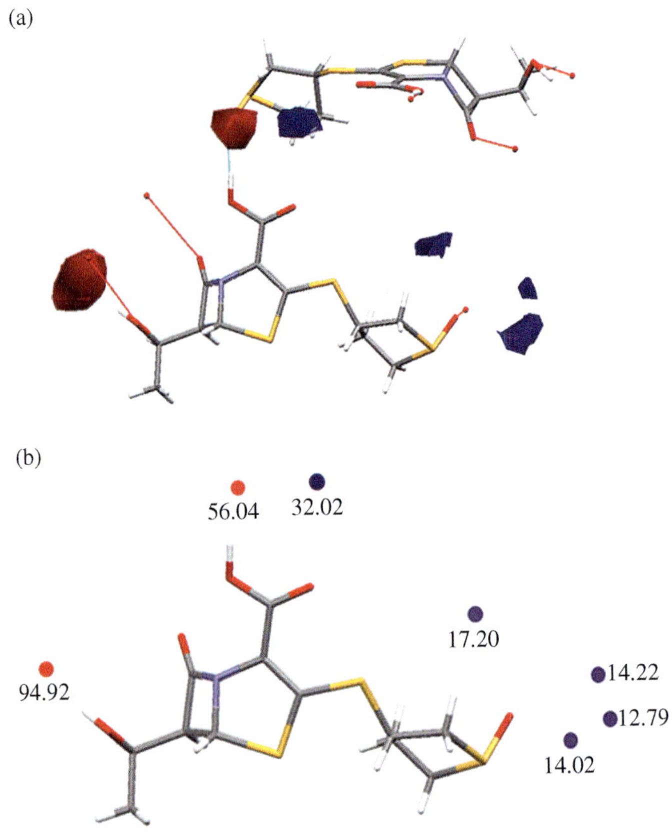

FIGURE 6.4 (a) Full interaction map for sulopenem. The location of neighboring molecules satisfying the carboxylic acid donor and acceptor group with full interaction map overlaid is shown. (b) Hotspots for sulopenem full interaction map. Relative heights are reported for each.

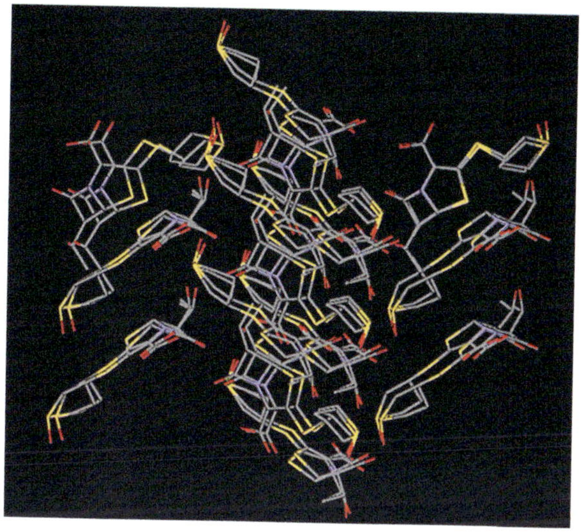

FIGURE 6.9 Superposition of the experimental and predicted crystal structures using the COMPACK algorithm (tolerance = 15%, RMS = 0.460).

(a)

(b)

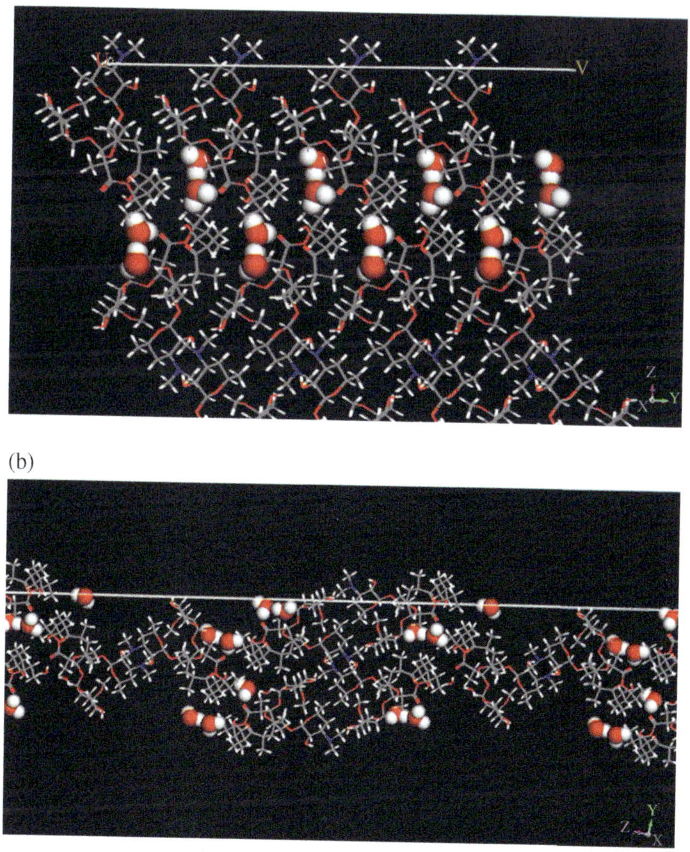

FIGURE 6.15 (a) Molecular structure of the dominant (002) surface of erythromycin. (b) Molecular structure of the end (011) surface of erythromycin.

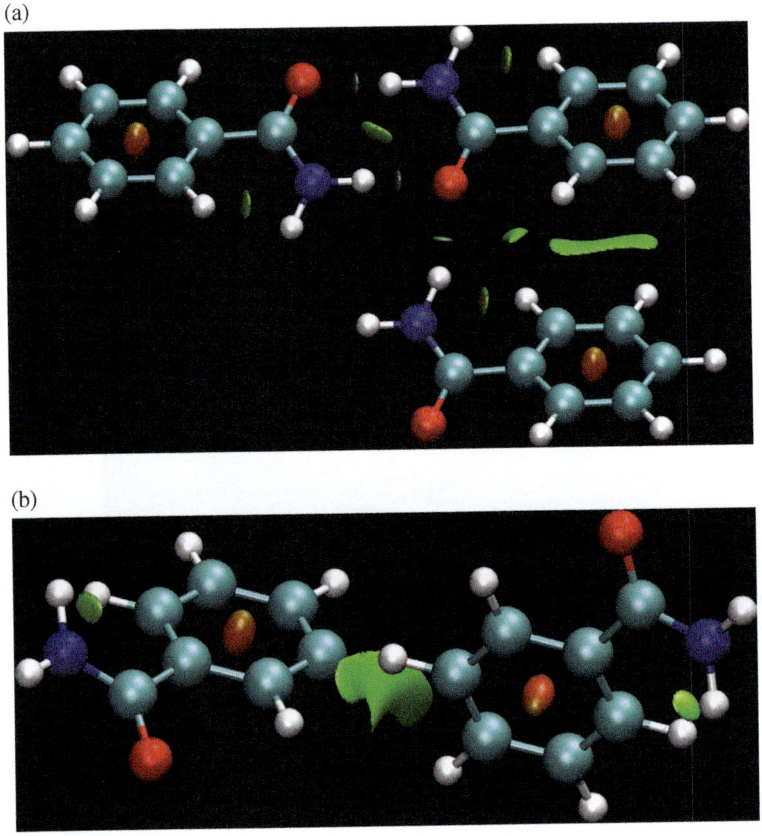

FIGURE 7.9 (a) Structure **01–126**. (b) Structure **01–127**. Noncovalent interaction (NCI) plot of predicted crystal structure. Reduced density gradient isosurface (IS) cutoff set to 0.25. Isosurfaces represent either unfavorable, weak, or strong favorable interactions.

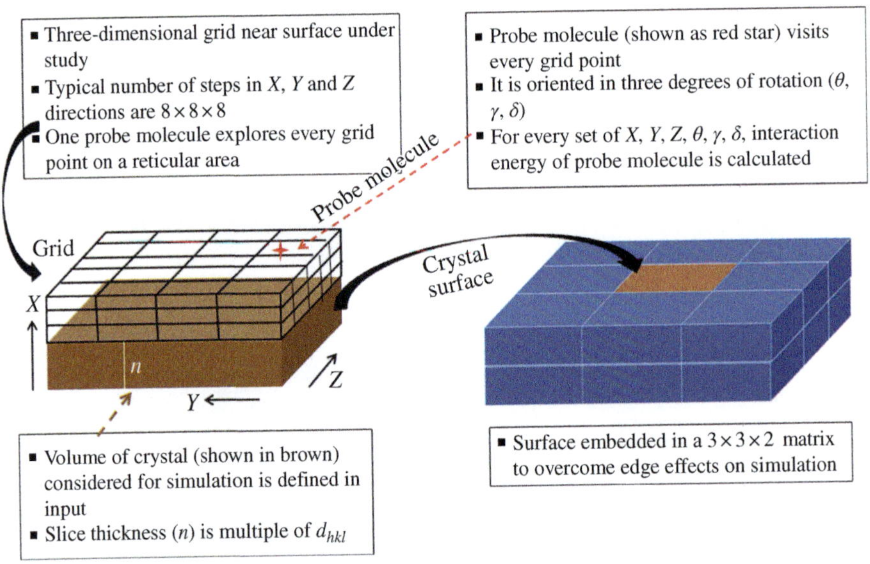

FIGURE 8.5 Schematic representation highlighting the main computational methodology associated with the grid search; the probe molecule in the context of the surface and the virtual grids. Source: Adapted from Ramachandran et al. [63]. Reproduced with permission of American Chemical Society.

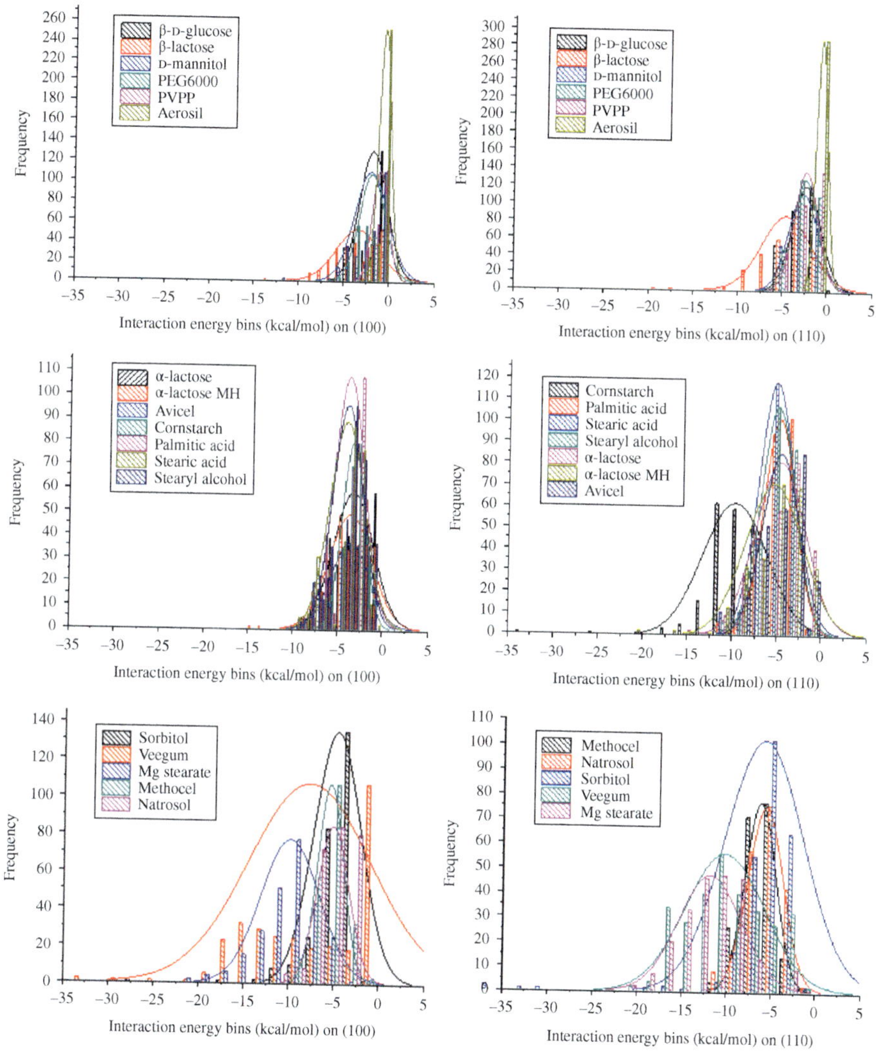

FIGURE 8.10 Comparisons of interaction energy distributions of 17 excipients with ketoprofen on (100) surface (left-hand side) and on (110) surface (right-hand side). The plots are split into six separate plots for clarity. Source: Adapted from Ramachandran et al. [81].

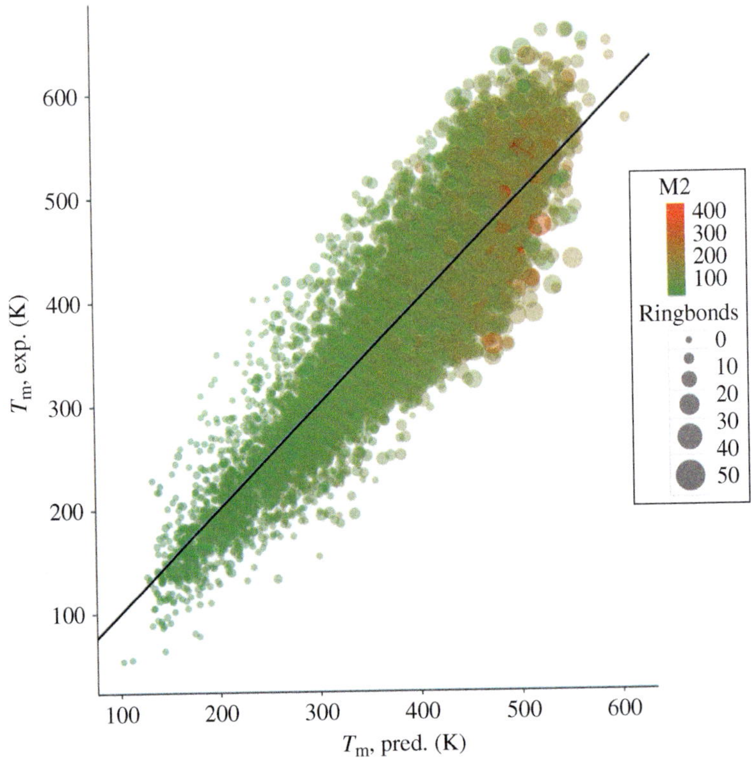

FIGURE 9.5 Random forest–based QSPR model for the prediction of melting points. Additionally, the number of ring bonds and second σ-moment (M2) are shown via point size and color gradient, respectively.

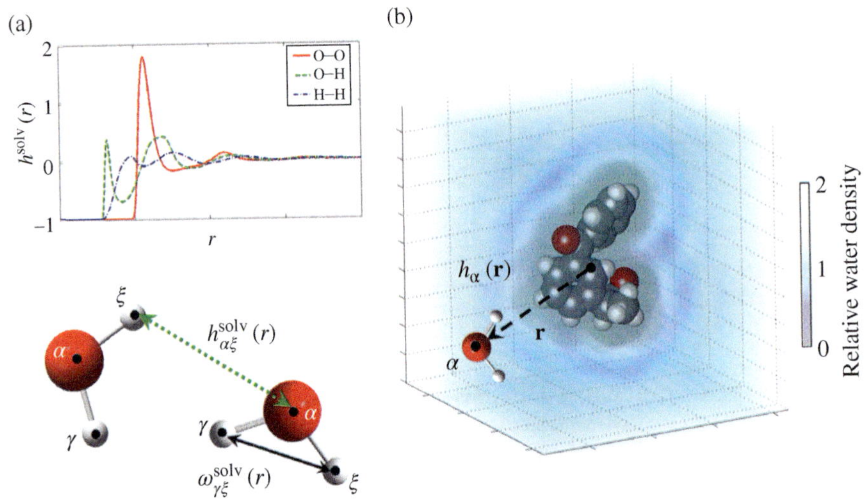

FIGURE 11.4 Correlation functions in the 3D RISM approach. (a) Site–site intramolecular ($\omega^{solv}_{\gamma\xi}(\mathbf{r})$) and intermolecular ($h^{solv}_{\alpha\xi}(\mathbf{r})$) correlation functions between sites of solvent molecules. The graph shows the radial projections of water solvent site–site density correlation functions: oxygen–oxygen (O–O, solid line), oxygen–hydrogen (O–H, dashed line), and hydrogen–hydrogen (H–H, dashed-dotted line). (b) Three-dimensional intermolecular solute–solvent correlation function $h_\alpha(\mathbf{r})$ around a model solute (diclofenac). Source: Palmer et al. [94]. Reprinted with permission of American Chemical Society.

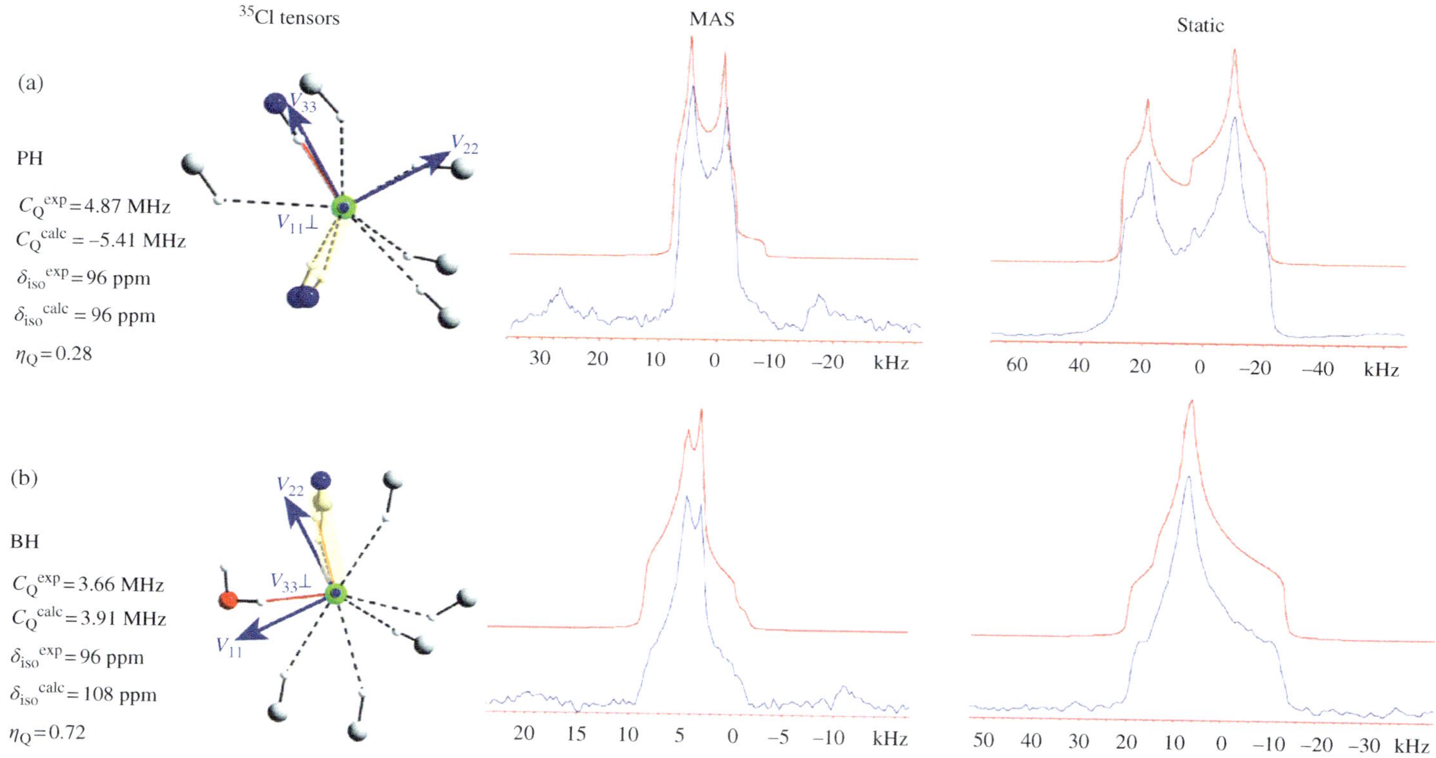

FIGURE 12.3 ^{35}Cl EFG tensor orientations and ^{35}Cl SSNMR spectra at 21.1 T for (a) Procaine HCl and (b) Bupivacaine HCl Monohydrate. Simulations for the static and MAS spectra are shown on top (red line) of the experimental lines (in blue). The yellow-shaded region represented on the tensor orientation diagrams highlights the N–H···Cl interactions. Experimental and calculated C_Q and δ_{iso} values are also shown for each compound. Adapted with permission from Ref. [39]. Copyright 2008, American Chemical Society.

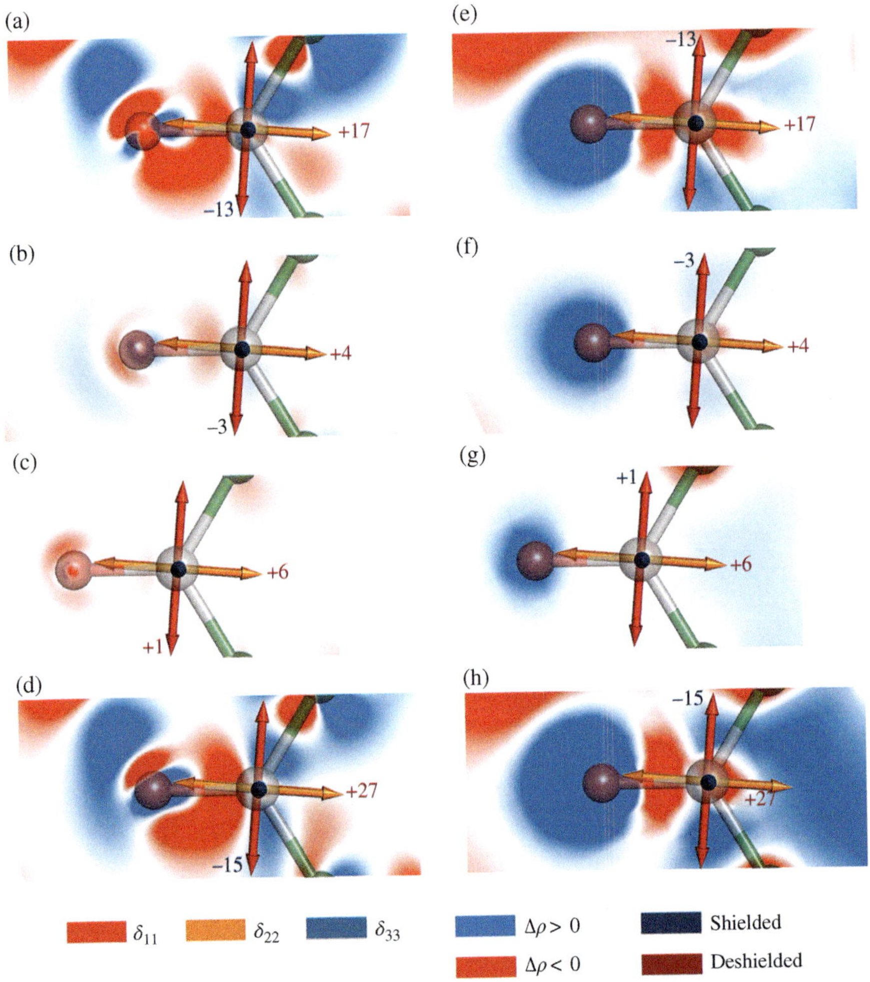

FIGURE 12.12 Visualization of the in-plane ($\delta_{11}|\delta_{22}$) slices of the EDD (left) and SDD (right) for atom C2 in theobromine. (a and e) HB2; (b and f) cluster composed of molecules 1 and 6 (weak H-bonding, see Fig. 12.11); (c and g) S7; and (d and h) HBS13. The arrows schematically represent the calculated relative orientations of the in-plane principal components of the ^{13}C CST, and the numbers denote the calculated changes in the magnitude of these components upon formation of the clusters. Adapted with permission from Ref. [58]. Copyright 2013, American Chemical Society.

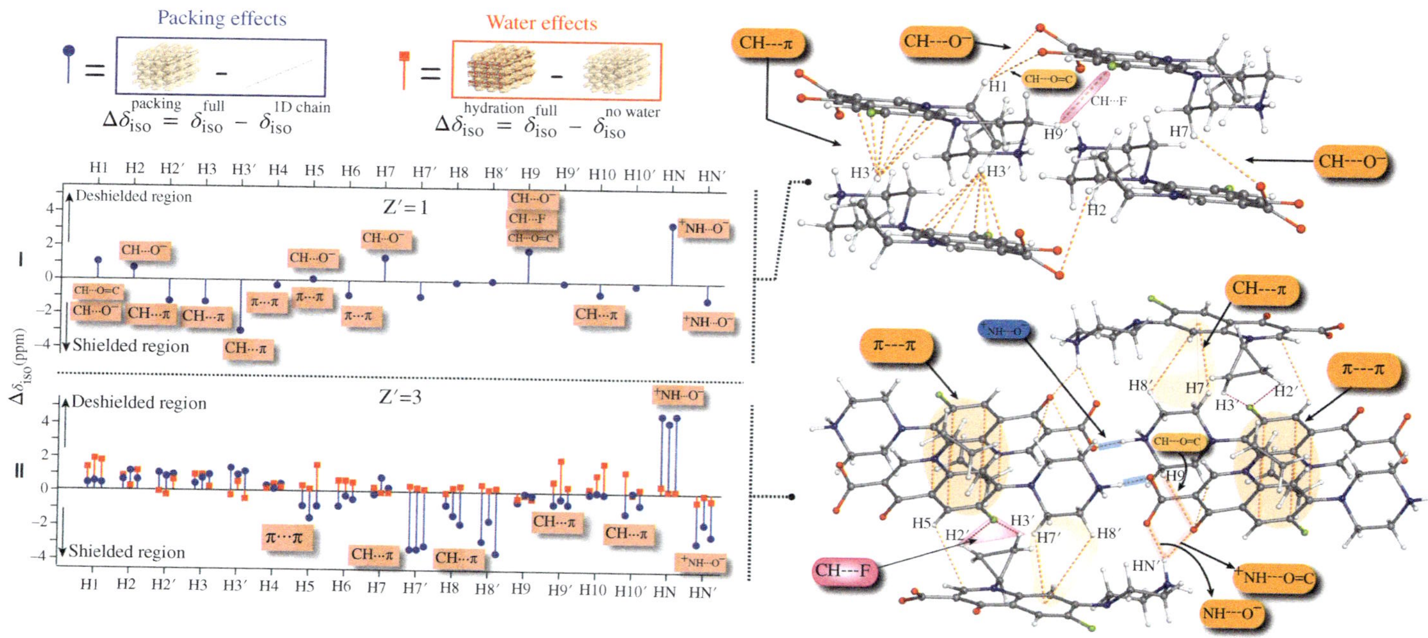

FIGURE 12.14 (Left) Δδ(ppm) plot showing crystal packing contributions (blue stems) and hydration contributions (red stems) to the GIPAW calculated ¹H ICS of ciprofloxacin Form I and Form II. In II, each of the three stems per proton corresponds to a crystallographic unique molecule ($Z' = 3$). Right: Detailed view of intermolecular interactions in the ciprofloxacin Forms I and II structures. Reprinted with permission from Ref. [20]. Copyright 2012, American Chemical Society.

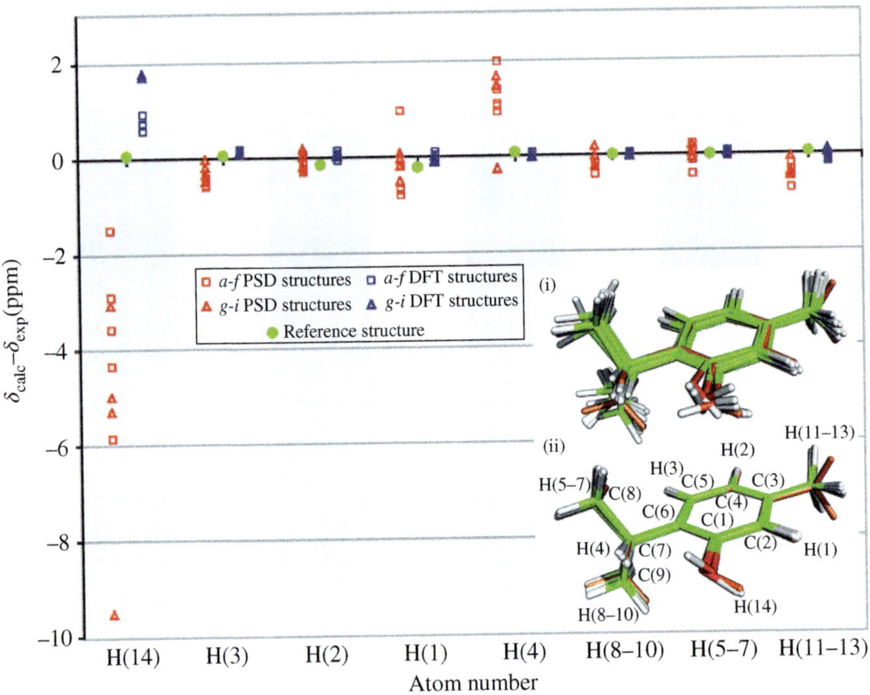

FIGURE 12.17 Comparison between calculated and experimental ^{1}H CSs for each resonance of thymol. Red (left) corresponds to the structures before the DFT-optimization (the nine structures are displayed in i) and blue (right) after optimization (same nine structures after periodic DFT optimization represented in ii, where the orange structure is the reference structure). Squares and triangles indicate structures where the orientation of the hydroxyl proton H(14) is respectively similar to (structures a–f) or different from (structures g–i) the reference structure (represented by green circles), respectively. Adapted with permission from Ref. [71]. Copyright 2009, RSC Publishing.

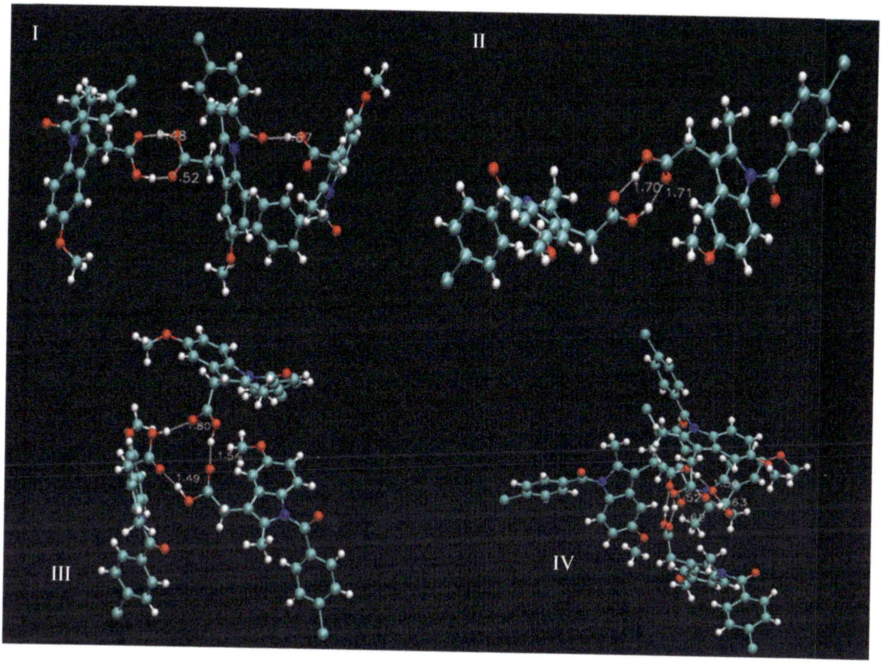

FIGURE 13.9 Representative H-bonding patterns in a simulated IMC glass: (I), H-bonded trimer similar to that in the unit cell of the α-crystalline polymorph; (II) H-bonded dimer similar to that in the unit cell of the γ-polymorph; (III) H-bonded cyclic trimer; and (IV) H-bonded tetramer. Source: Xiang and Anderson [24b]. Reprinted with permission of American Chemical Society.

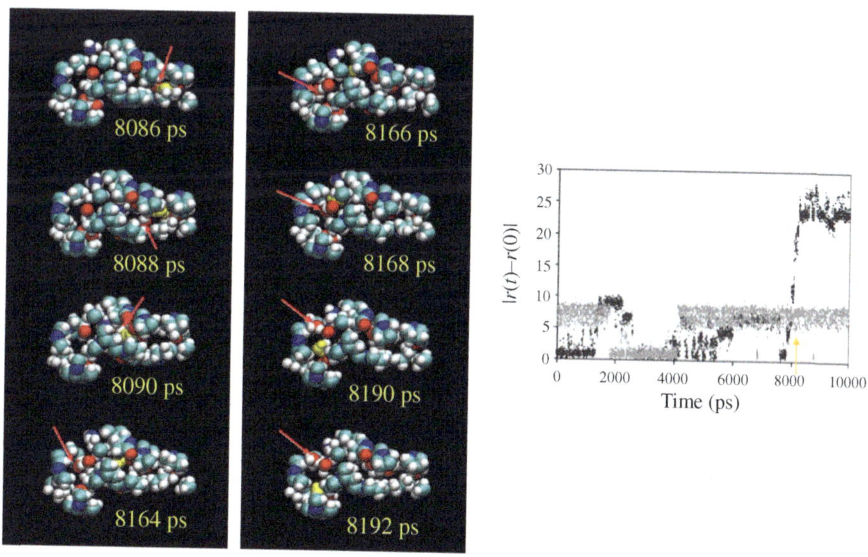

FIGURE 13.13 (Left): Snapshots of a water hopping event in simulated PVP containing 10% water over a approximately 100 ps time span. The yellow-tagged water molecule underwent a rapid displacement from one free volume pocket to another approximately 20 Å to the left. The red arrow tracks a second water molecule that co-migrated with the first water molecule. Right: The displacement of the yellow-tagged water molecule versus time is marked with an arrow.

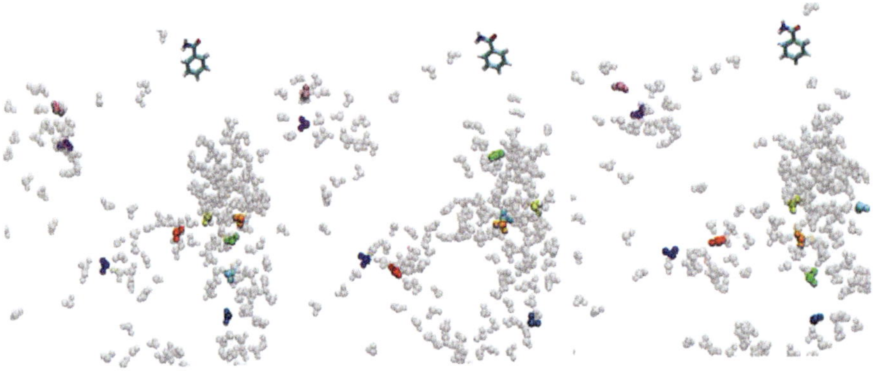

FIGURE 13.14 Three snapshots of water clusters at different times in a simulation of a water saturated 60% tricaprylin/40% monocaprylin lipid mixture. A few water molecules are color coded to allow tracking of their facile movement within water clusters. A single molecule of benzamide remains nearly stationary throughout this time frame.

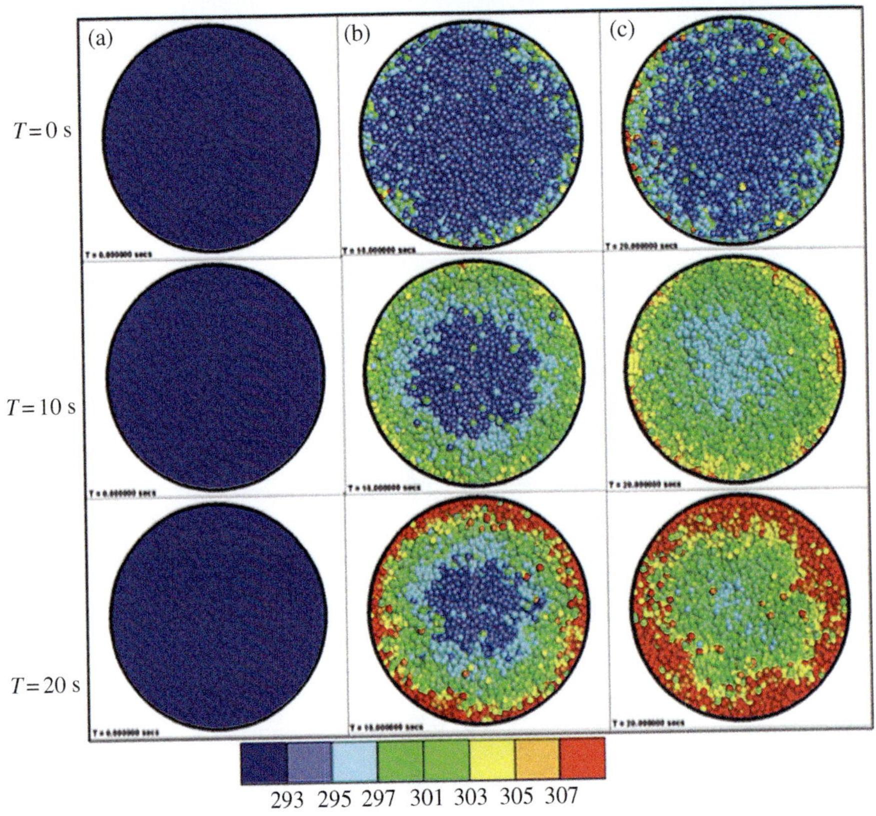

FIGURE 14.3 Shows the snapshots comparing the temperature profile for the effect of wall temperature (a) 318 K, (b) 328 K, and (c) 338 K, for $T = 0$, 10, and 20 s; $\omega = 5$ rpm and $f = 45\%$. Adapted from Ref. [12] with permission. Copyright 2013, Elsevier.

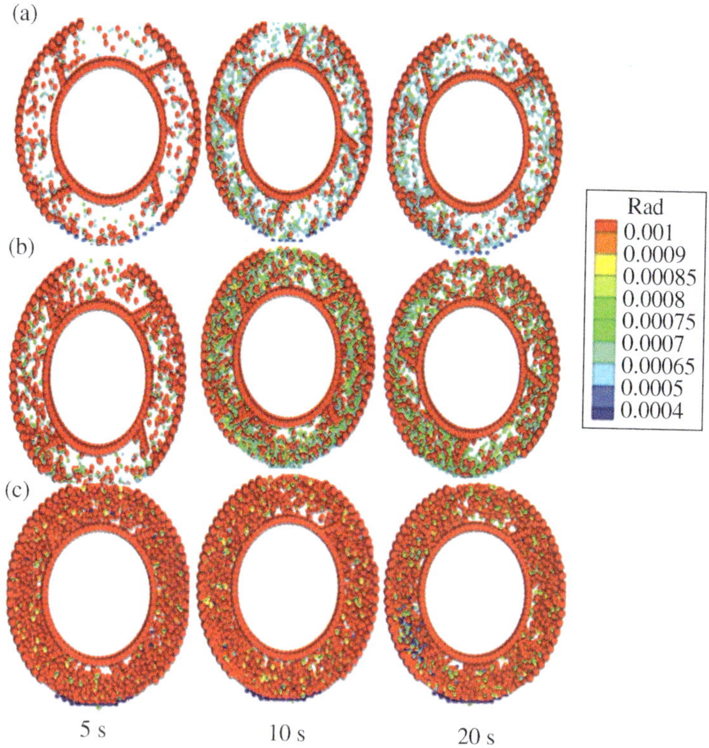

FIGURE 14.11 Temporal distribution of particle size under the impeller speed of 600 rpm at feed rates of (a) 60 g/min, (b) 120 g/min, and (c) 200 g/min. Adapted from Ref. [18] with permission. Copyright 2013, Elsevier.

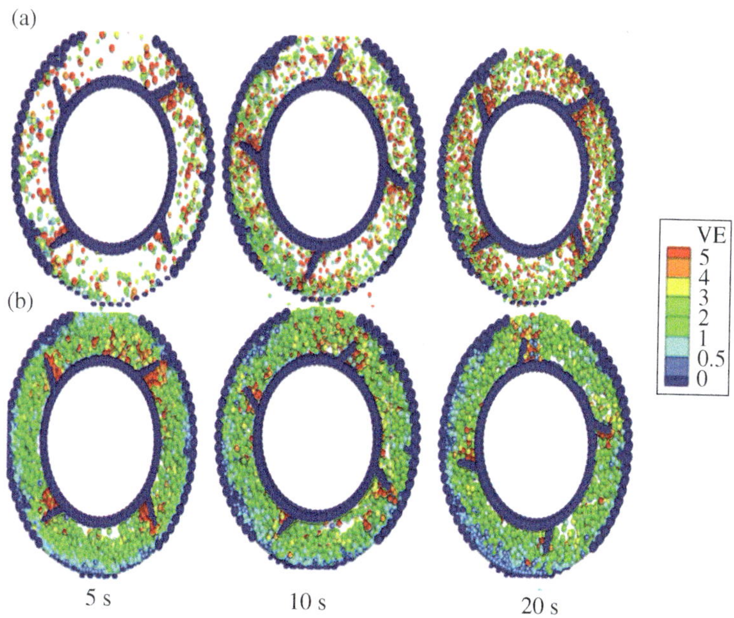

FIGURE 14.13 Temporal distribution of velocity under impeller speed of 600 rpm at feed rates (a) 60 and (b) 200 g/min. Adapted from Ref. [18] with permission. Copyright 2013, Elsevier.

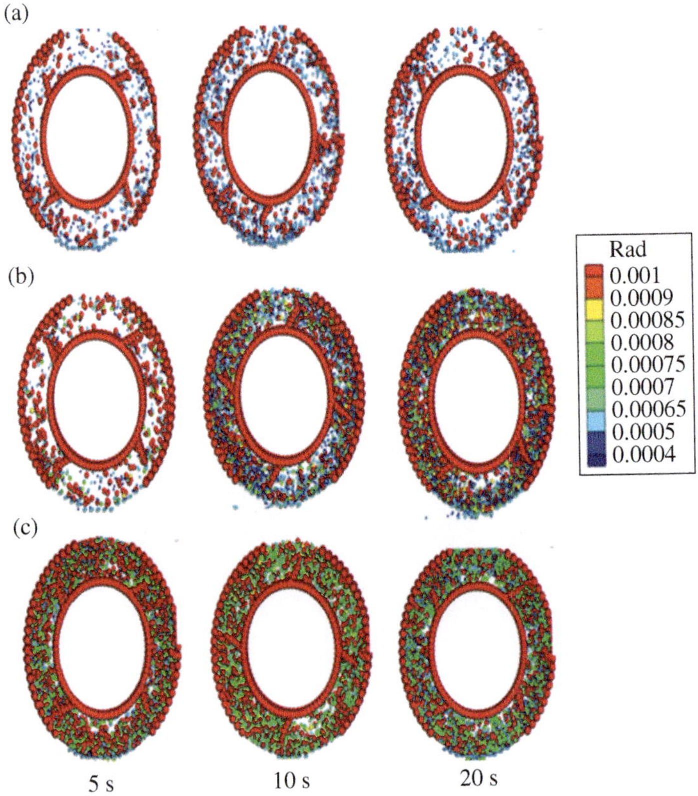

FIGURE 14.14 Temporal distribution of particle size under impeller speed of 1140 rpm at feed rates of (a) 60, (b) 120, and (c) 200 g/min. Adapted from Ref. [18] with permission. Copyright 2013, Elsevier.

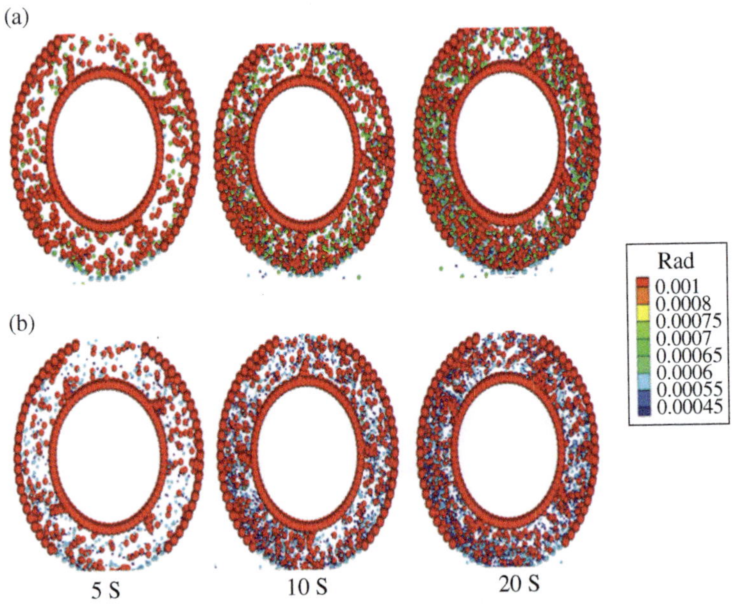

FIGURE 14.15 Temporal distribution of particle size for impeller wall tolerance 3.9 mm at different impeller speeds of (a) 600 rpm and (b) 1140 rpm. Adapted from Ref. [18] with permission. Copyright 2013, Elsevier.

8

SYNTHONIC ENGINEERING: FROM MOLECULAR AND CRYSTALLOGRAPHIC STRUCTURE TO THE RATIONAL DESIGN OF PHARMACEUTICAL SOLID DOSAGE FORMS

KEVIN J. ROBERTS[1], ROBERT B. HAMMOND[1], VASUKI RAMACHANDRAN[1], AND ROBERT DOCHERTY[2]

[1] *Institute of Particle Science and Engineering, and Institute of Process Research & Development, School of Chemical and Process Engineering, University of Leeds, Leeds, UK*
[2] *Pfizer Worldwide Research & Development, Sandwich, Kent, UK*

8.1 INTRODUCTION

Selection of the commercial solid-form and the associated crystallisation process is one of the key milestones in the development of any new drug molecule. It is critical not only from a drug substance manufacturing standpoint but also from a drug product processing, performance and stability perspective. The regulatory landscape associated with the solid-form and particle attributes of the active pharmaceutical ingredient (API) and dosage form development has been described previously [1, 2]. The issues associated with the emergence of an unexpected solid-form and the importance of intellectual property around crystallisation

Computational Pharmaceutical Solid State Chemistry, First Edition. Edited by Yuriy A. Abramov.

process design and polymorphs have also been well documented [3]. The progress of automation and structural informatics technologies, which allow development scientists to search and identify the solid-form with optimal properties, has also been reported [4–6].

In 1987, the Nobel Prize for chemistry was awarded to Cram, Lehn and Pedersen for their work on supra-molecular chemistry. Since then publication(s), for example [7], have charted the evolution of pharmaceutical materials science. More recently the importance of the materials science tetrahedron in depicting the relationships between internal structure, particle properties and the processing and performance of a drug product has been described [8]. The industrial perspective on engineering pharmaceutical materials has also been highlighted [9]. Pharmaceutical materials science has emerged as a foundation of quality by design (QbD) [10] with solid-form, crystallisation and particle engineering being core elements linking the drug product attributes to the final steps of the synthetic pathway of the API.

Whilst increasing interest in the crystallisation of pharmaceutical entities within academia has resulted in substantial progress over the past decade, the challenge for the pharmaceutical scientist in tackling the crystallisation of highly complex, new chemical entities remains a significant one. These complex organic structures are exacting because of the following reasons:

- Increasing molecular complexity results in a complicated solid-form space (salts, cocrystals, polymorphs, hydrates and solvates).
- Different crystal faces exhibit different surface chemistry and interactions with solvents, inherent process impurities and excipients.
- Different solid-forms may have different chemical and physical stabilities, biopharmaceutical properties and drug product processing behaviour.

Whilst traditionally, the solid-form selection process has focused on achieving an appropriate degree of bioavailability, increasing emphasis is now being placed on selection of solid-forms at the pre-formulation stage which have optimal downstream properties such as chemical and physical stability, mechanical behaviour, surface properties and particle shape. Given such a perspective, this chapter outlines some recent progress on the application of emerging computational technologies as foundation elements of the modern QbD strategy for the development and manufacture of advanced particulate products (Fig. 8.1). This chapter highlights the opportunity for computational tools to build the bridge across the chemical, analytical and formulation disciplines. These relationships, when combined with institutionalised corporate knowledge of formulation design practices [12], can provide a clear route map for a fully integrated, product design process. This holistic process is consistent with the emerging QbD philosophy [10] and can be realised through the use of increasingly sophisticated particle and surface chemistry design tools.

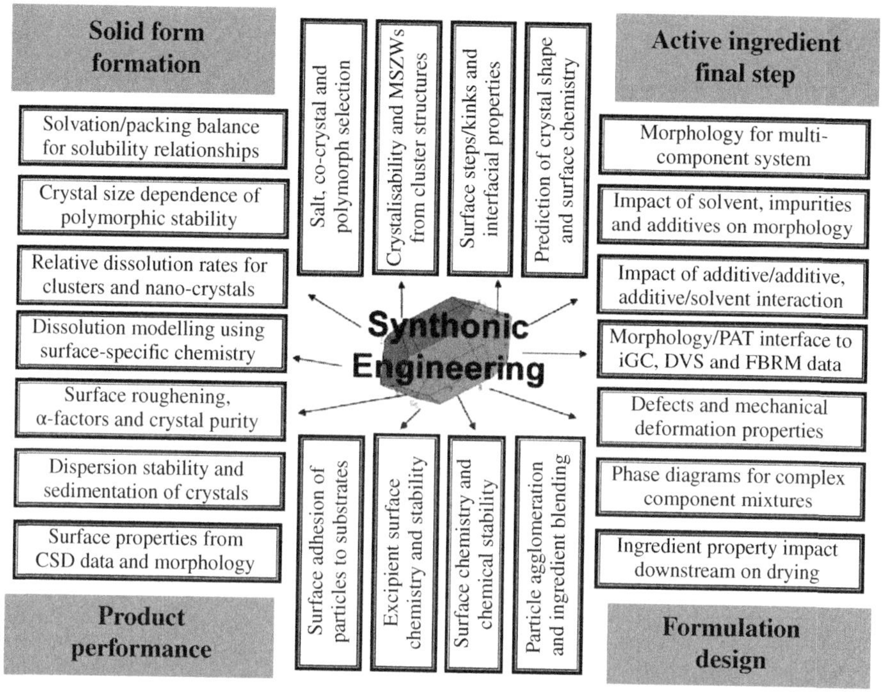

FIGURE 8.1 The potential application of synthonic engineering tools at the active ingredient and drug product interface. Source: Adapted from Roberts et al. [11].

8.2 THE CRYSTAL

8.2.1 Crystallography

Crystals may be considered simply as three-dimensional (3D) repeating patterns of atoms or molecules. As with any other pattern, they can be described by defining the following:

1. The item to be repeated (the motif)
2. The way in which it is repeated (symmetry operations).

Extending this general concept to crystal structures, the motif is an atom, molecule or a collection of molecules and/or ions. The lattice describing the scheme of repetition is now a 3D array of points, and the unit cell is the smallest repeating unit within this 3D structure. The unit cell is fully described by six lattice parameters, comprising three lengths of the unit cell (**a, b** and **c**) with the three inter-axial angles (**α, β** and **γ**). Consideration of the relative magnitude of these parameters gives rise to the definition of the seven crystal systems. Consideration of having extra lattice points

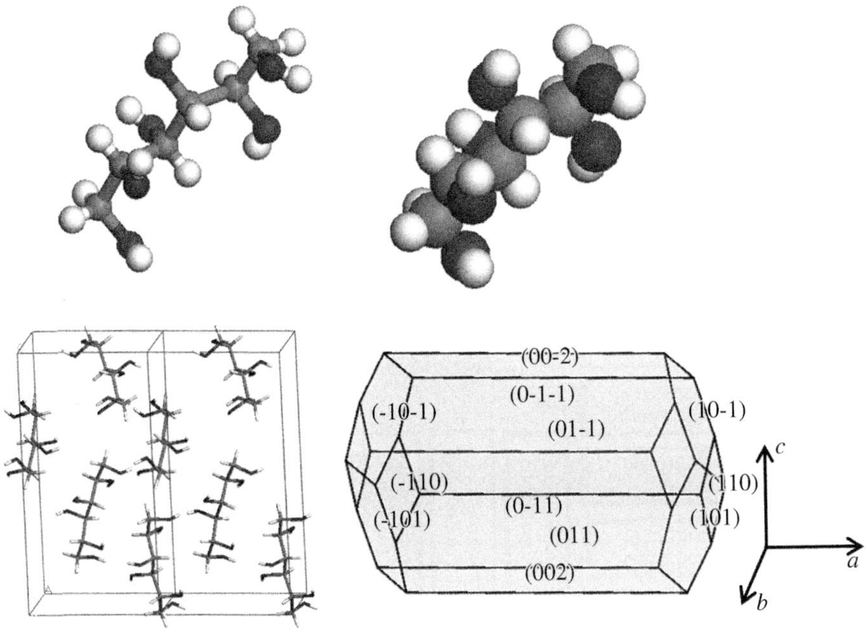

FIGURE 8.2 The molecule, the crystal chemistry and the crystal morphology of D-mannitol (CSD refcode: DMANTL09) [15]. The molecular structure (top) is shown in ball and stick and space-fill. The crystal packing of mannitol (bottom left) contains one unit cell in the *a* and *c* directions with two unit cells of D-mannitol along the *b*-direction of the unit cell. Each unit cell contains four D-mannitol molecules. The crystal morphology (bottom right) shows the Miller indices.

in face- or body-centred sites produces the fourteen unique Bravais lattice types spread over these seven systems. Examination of these seven crystal systems reveals that the unit cells become progressively less symmetrical upon moving from cubic through orthorhombic to triclinic crystal systems [13, 14].

So far we have considered the unit cells in terms of their basic shape and relative dimensions. It is also possible to refine the seven crystal systems in terms of the symmetry elements which they possess. These elements represent various combinations of rotation, translation and inversion and together with the Bravais lattice type define the full 3D arrangement of atoms or molecules within a given structure. The symmetry exhibited by a unit cell is thus reflected in both the arrangement of atoms in the internal crystal structure and in the physical and chemical properties of the resulting macroscopic crystal. Symmetry is evident in properties such as crystal growth rates and crystal shape, surface chemistry and mechanical properties. Exactly the same principles are involved in describing the structure of drug molecules. However, the greater number of atoms typically involved can make visualisation of the structure a little more difficult. Figure 8.2 (bottom left) shows two unit cells of D-mannitol along the crystallographic *b*-direction. Each unit cell contains four mannitol molecules [15].

The crystallographic planes that define the external growth morphology of the 'as grown' crystal can be described by the law of rational indices [16, 17] as expressed through their Miller indices (*hkl*). The former enables the 3D nature of the crystal lattice to be expressed through integer variables where the indices normally have only small values (usually 0 or 1 but sometimes higher) of either parity. Crystal planes are thus defined through their Miller indices as the reciprocals of the fractional intercepts which the plane makes with the crystallographic axes. When the fractional intercept is negative, the resulting negative Miller index is indicated by placing a bar above the corresponding integer. The convention of brackets in crystallography is as follows:

- Round brackets are used to denote a specific plane – (*hkl*).
- Curly brackets are used for a set of planes related by symmetry – {*hkl*}.
- Square brackets are used to denote a direction which is the vector – [*uvw*].
- Pointed brackets are used for a set of directions related by symmetry – <*uvw*>.

Miller indices are important to the crystallisation scientist as they provide a link between the modern structural crystallography of X-ray diffraction and classical morphological crystallography of shape and habit. This allows the process chemist or pharmaceutical scientist to link the internal molecular structure to the chemical functionality of the external surface structure. Figure 8.2 (bottom right) shows the observed morphology for D-mannitol with the Miller indices labelled.

Reticular area is the projection of the unit cell on to a surface plane and can effectively be thought of as the 2D unit cell at the surface. Reticular area (S_{hkl}) is calculated from the ratio of the unit cell volume (V_{cell}) to the *d*-spacing (d_{hkl}) as shown in Equation 8.1. The concept will be utilised later when describing the surfaces of particles.

$$S_{hkl} = \frac{V_{\text{cell}}}{d_{hkl}} \tag{8.1}$$

8.2.2 Crystal Chemistry and Crystal Packing of Drug Molecules

Molecules can essentially be regarded as impenetrable systems whose shape and volume characteristics are governed by the molecular conformation and the radii of the constituent atoms. The atomic radii are essentially exclusion zones in which no other atom may enter except under special circumstances, such as to form a covalent bond. Figure 8.2 (top) shows a comparison between a ball and stick and van der Waals (space-fill) representation of D-mannitol.

The structures and crystal chemistry of molecular materials are often classified into different categories according to the type of inter-molecular forces present [18]. A number of factors are of particular importance in assessing the influence of

TABLE 8.1 Space Groups with Highest Ranking for Population from 587,899 Structures in the Cambridge Structural Database on 1 January 2011

Space Group Number	Hermann–Mauguin Symbol	Crystallographic System	Ranking
14	$P2_1/c$	Monoclinic	1
2	P-1	Triclinic	2
15	C2/c	Monoclinic	3
19	$P2_12_12_1$	Orthorhombic	4
4	$P2_1$	*Monoclinic	5
61	Pbca	Orthorhombic	6
33	$Pna2_1$	Orthorhombic	7
62	Pnma	Orthorhombic	8
9	Cc	Monoclinic	9
1	P1	*Triclinic	10
60	Pbcn	Orthorhombic	11
5	C2	*Monoclinic	12
29	$Pca2_1$	Orthorhombic	13
148	R-3	Trigonal	14
13	P2/c	Monoclinic	15
11	$P2_1/m$	Monoclinic	16
12	C2/m	Monoclinic	17
18	$P2_12_12$	*Orthorhombic	18
7	Pc	Monoclinic	19
56	Pccn	Orthorhombic	20

An asterisk (*) means that the group is a chiral space group, that is it is one of the 65 space groups in which a chiral molecule can crystallize.

inter-molecular bonding on the physico-chemical properties of organic solids [19, 20]. These include the following:

- The size and shape of the molecular entities that make up the structure
- The strength of the interaction
- The distance over which the interaction exerts an influence
- The extent to which the interaction is directional or not

Organic molecules in general and drug molecules in particular are usually expressed as solids in only a limited number of low-symmetry crystal systems. The lengths of the three principal axes describing the shape of a molecule are often very unequal which implies a non-spherical shape. In turn, this is reflected in crystal structures in which the unit cell edge lengths are unequal. A further consequence of their non-equant shape is that organic molecules prefer to adopt space groups which have translational symmetry elements, as this allows the most efficient spatial packing of the protrusions of one molecule into the gaps left by the packing arrangements of its neighbours. These tendencies are reflected in an analysis of the Cambridge Crystallographic Database (CCD) [14, 21]. The vast majority of the organic structures reported prefer the triclinic, monoclinic and orthorhombic crystal systems [14, 22, 23]. Table 8.1 shows a list of the 20 most populated space groups from the database.

To understand the principles which govern the wide variety of solid-state properties and structures of drug molecules, it is important to describe both the energy and direction of interactions between molecules. As a result of the pioneering work in the development of atom–atom inter-molecular potentials [24–29], it is now possible to interpret inter-molecular packing effects in organic crystals in terms of their interaction energies [23, 25, 27]. The basic assumption of the atom–atom method is that the interaction between two molecules can be considered to simply consist of the sum of the interactions between the constituent atoms considered in pairs. The lattice energy E_{latt} (often referred to as the crystal binding or cohesive energy) for molecular materials can be calculated by summing up all the interactions between a designated, central molecule and all the surrounding molecules. Hence, if there are n atoms in the central molecule and n' atoms in each of the N surrounding molecules, then the lattice energy can simply be calculated by the Equation 8.2 [30, 31]. The model is based on 0 K simulation.[1]

$$E_{latt} = \frac{1}{2}\sum_{k=1}^{N}\sum_{i=1}^{n'}\sum_{j=1}^{n} V_{kij} \tag{8.2}$$

Each atom–atom interaction pair (V_{kij}) consists of a short-range repulsive and attractive dispersive interaction which can be described by Lennard-Jones potential together with an electrostatic interaction describing monopole–monopole interactions, and in some cases (particularly for pharmaceuticals) a hydrogen-bonding (H-bonding) potential [24, 28, 29]. The former two are, broadly speaking, undirected interactions, whilst the latter is not. On a per atom basis, the hydrogen bond (H-bond) is much stronger than, say, a dispersive interaction but in an organic crystal the latter can involve many more pairs of atomic interactions and so, for molecules such as pharmaceuticals, where the molecular weight is relatively high, contributions from the undirected van der Waals interactions can tend to dominate the lattice energy. The interactions that make up the main molecular coordination sphere in the case of D-mannitol (CSD refcode: DMATL09) [15] calculated using the computer programme, HABIT [32], are listed in Table 8.2.

This table gives the interaction type, the multiplicity of the interaction, the energy value and the relative contributions to the overall lattice energy. It can be seen that there are eight key inter-molecular interactions with a calculated interaction energy greater than 1 kcal/mol (neglecting the negative sign) which surround the central molecule, and these interactions contribute about 71% of the total lattice energy. This analysis shows that the top two contributions in D-mannitol are stronger contributing more than 22% each to the lattice energy.

8.2.3 Deconstructing the Supra-Molecular Interactions in Bulk – Intrinsic Synthons

The use of these atom–atom potentials has been validated extensively by comparing the theoretical values against the known crystal structures and experimentally measured lattice energies derived from sublimation enthalpies [23]. A particular

[1] This depends on the potential used, which could be standard state.

TABLE 8.2 Eight Strongest Inter-Molecular Interactions from D-mannitol (CSD refcode: DMANTL09 [15])

Bond	Type of Interaction	Multiplicity	Distance (Å)	Inter-Molecular Energy (kcal/mol)	% Contribution to Lattice Energy
A	O–H···O hydrogen bond	2	5.54	−3.80	12.68
B	van der Waals	2	5.44	−3.75	12.51
C	van der Waals	2	7.86	−1.90	6.34
D	van der Waals	2	6.79	−1.30	4.34
Total		8		−21.50	71.74

Top eight interactions make up over 70% of the lattice energy therefore majority of important inter-molecular interactions.

advantage of the lattice energy calculated in this way is that it can be broken down into constituent inter-molecular interactions and related to particular crystallographic directions. It can also be further partitioned onto the constituent atom–atom and/or group contributions. An analysis of this type provides a complete description of the supra-molecular synthons within solid-state structures, and the lattice energy can effectively be de-convoluted into the following:

- Molecular coordination both in terms of number of interactions and total energy.
- The lattice energy as a collection of inter-molecular interactions – intrinsic synthons aligned along specific crystallographic directions.
- The lattice energy deconstructed onto molecular fragments and even individual atomic contributions.
- The surface as a collection of unsaturated interactions due to the surface termination of the bulk crystal structure –extrinsic synthons. This will be discussed further in Section 8.3.4.

8.2.3.1 Molecular Coordination: Number and Energy Table 8.3 summarises an examination of the crystal chemistry and key inter-molecular interactions, highlighting how they contribute to the lattice energy in a selection of aromatic compounds (benzoic acid (BA), terephthalic acid, para-amino benzoic acid (PABA) and ortho-amino benzoic acid (OABA)) all of which have at least one carboxylic acid group. The crystal structures were optimised within Materials Studio [38] using Dreiding potential [39] prior to the lattice energy calculation. The lattice energy was calculated using the empirical force-field parameters of Momany et al. [29] and by assigning atomic point charges calculated using the semi-empirical molecular orbital programme, MOPAC [40], for each form and compared to experimentally measured sublimation enthalpies. In the case of para-amino benzoic acid [37], restrained electrostatic potential (RESP) charges based on the *ab initio* MP2/aug-cc-pVDZ method/basis set [41, 42] and derived from the Antechamber within Ambertools were used. The lattice energy is found to converge at a limiting radius of 12–15 Å in all the

TABLE 8.3 Summary of Crystallography, and Lattice Energy Contributions for Aromatic Compounds[a]

Attribute	Benzoic Acid	Terephthalic Acid		Para-Aminobenzoic Acid		Ortho-Aminobenzoic Acid
	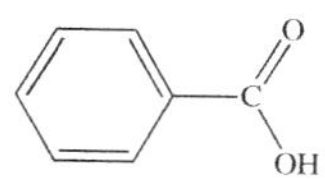	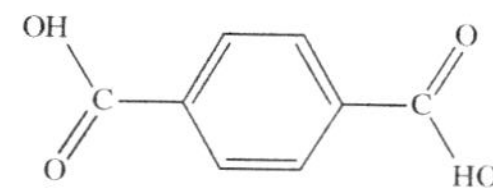				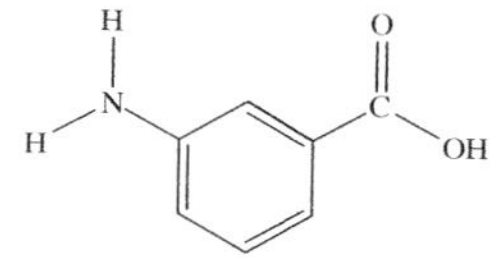
Refcode	BENZAC12 (M. Nieger, Private Communication, 2012.)	TEPHTH [33]	TEPHTH01 [33]	AMBNAC06 [34]	AMBNAC04 [35]	AMBACO03 [36]
Polymorph		α	β	α	β	form II
a (Å)	5.29	7.73	9.54	18.57	6.27	16.45
b (Å)	6.14	6.44	5.34	3.84	8.58	12.01
c (Å)	19.98	3.75	5.02	18.63	12.36	7.29
α (°)	90.00	92.75	86.95			90.00
β (°)	87.20	109.15	134.65	93.56	100.13	90.00
γ (°)	90.00	95.95	104.90			90.00
Space group	$P2_1/n$	$P\bar{1}$	$P\bar{1}$	$P2_1/n$	$P2_1/n$	P b c a
Z: Z':	Z: 4 Z': 1	Z: 1 Z': 1	Z: 1 Z': 1	Z: 8 Z': 2	Z: 4 Z': 1	Z: 8 Z': 1
Lattice energy (kcal/mol)	−20.81	−37.19	−35.29	−24.45	−22.73	−26.27
C_6H_4 (%)	62.3	32.32	32.40	39.40	42.50	25.64
R1	34.82% (COOH)	33.83% (COOH)	33.77% (COOH)	48.30% (COOH)	33.70% (COOH)	47.87% (COOH)
R2	2.88% (H)	33.83% (COOH)	33.77% (COOH)	12.30% (NH_2)	23.80% (NH_2)	26.51% (NH_2)

[a] Source: Adapted from Rosbottom et al. [37]. Reproduced with permission of Royal Society of Chemistry.
The space groups in this small selection reflect the key space groups identified in Table 8.1.

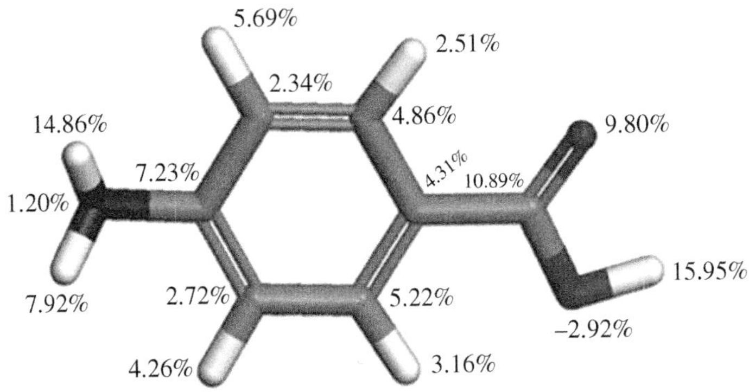

FIGURE 8.3 Molecular structure of para-amino benzoic acid (PABA) highlighting the percentage contribution of each atom to the total lattice energy per molecule. The group contributions for C_6H_4, COOH and NH_2 groups are listed in Table 8.3. Source: Adapted from Rosbottom et al. [37]. Reproduced with permission of Royal Society of Chemistry.

cases. It takes a 30-molecule cluster for the α form, and a 35-molecule cluster for the β form of PABA to obtain a fully converged lattice energy (not shown here). These larger clusters satisfy the OH···O interactions between carboxylic acid groups in the α-form and also express the supporting NH···O H-bonding interactions. The NH···O and OH···N H-bond formation appears to be important for lattice energy convergence in the β-form.

Contributions to the lattice energy of the two independent molecules in the asymmetric unit of the α-form are broadly similar due to the similar environments, whereas the β-form manifests a contribution from the amino hydrogen atoms and hydroxyl hydrogen, and a decrease in the contribution from the carbonyl oxygen. This reflects the formation of centrosymmetric, H-bonded carboxylic acid dimers in the α-form and NH_2 donor and acceptor interactions in the β-form [37]. Figure 8.3 shows the lattice energy deconstructed onto the molecular fragments and individual atoms for the β polymorph of PABA. This approach sort of allows the pharmaceutical scientist to look at the group contributions of various intrinsic synthons and understand the relative importance in defining the solid-state stability and crystal chemistry.

8.2.3.2 The Lattice Energy Deconstructed onto Molecular Fragments Table 8.3 demonstrates how the different inter-molecular packing in each crystal structure affects the respective contribution of the functional groups to the overall lattice energy. When a hydrogen atom in the aromatic ring is replaced by a functional group with a H-bonding donor or acceptor capability, there is a relative decrease in the contribution of C_6H_4 ring structure to the overall lattice energy. Comparing the relative contributions of the moiety C_6H_4, it is 62% in BA; whereas, it ranges from 25 to 40% in the other compounds which have two functional groups substituted at R1 and R2. The positions of substitution in the BA molecule, that is in OABA and PABA have an impact on the relative amount the C_6H_4 moiety contribution to the lattice energy, which is 26% for the meta substituents compared to 40% for the para substituents. This could be due to the fact that stronger inter- and intra-molecular H-bonds form between OH···N in OABA. The contribution of the substituent groups to the lattice energy is proportional to their size and H-bonding potential of the particular functional group, for example the contributions from the –COOH group versus $-NH_2$ group are 48 and 12%, 34 and 24% and 48 and 27%, respectively, in α-PABA, β-PABA and OABA. For PABA, the NH_2 group contributes significantly more to the lattice energy of the β form than the α-form, as in the β structure the NH_2 acts as both a H-bonding donor and acceptor, whilst in the α structure the NH_2 acts only as a donor. The strong H-bonds formed between the COOH groups in the α-structure result in a larger contribution from the COOH group in α-PABA.

Interaction between molecules and their associated energy is the key link between the intrinsic molecular structure and the crystal packing, allowing a profile of the important interactions to be built up within families of compounds. This will be highlighted in Section 8.4. This approach has recently been used to examine the relative importance of the solvation and crystal packing effects on low-solubility pharmaceutical materials [37], thus facilitating discussions between medicinal chemists and the pharmaceutical scientists about optimising the design of molecular features and, thereby, the physical properties for the intended dosage form.

8.3 MORPHOLOGY AND SURFACE STRUCTURE

8.3.1 Nucleation and the Crystal Growth Process

Solution crystallisation can be viewed as a two-step process involving first the dissolution of a material, and then changing some attribute of the crystallising system, such as temperature, solubility or solvent content to induce crystallisation. At a given temperature and pressure there is a fixed amount of solute that can dissolve in a given amount of solvent to achieve chemical equilibrium. When this amount of solute is added, the solution is said to be saturated. The amount of solute required to make a saturated solution at a specified temperature and pressure is equivalent to the solubility [43].

During crystallisation the supersaturation drives the solute molecules which segregate from their solvated state and self-assemble, aligning certain structural elements such as conformation and inter-molecular packing to produce, a stable 3D, ordered, crystallographic array of molecules. This highly time-dependent process means that crystallisation is essentially kinetically driven. The time required for crystallisation to proceed depends on the supersaturation driving force. Supersaturation is the ratio of actual solute concentration (or strictly speaking activity) to the equilibrium solubility at a specified temperature and pressure. Supersaturated solutions are metastable, implying that crystallisation will ultimately occur, albeit after a certain amount of time has elapsed, but that the process is inhibited by a kinetic barrier. As the rate of supersaturation generation increases, say due to solution cooling, a given solution will manifest an upper limit in supersaturation before it becomes unstable and crystallisation occurs spontaneously [43, 44].

In the nucleation stage, small clusters of solute molecules are formed; some of these clusters may grow sufficiently to form stable nuclei and subsequently form crystals. Others fail to reach adequate dimensions before they dissolve again. Within the metastable zone width (MSZW), the induction time to the onset of crystallisation has an inverse relationship with the supersaturation [44–47].

Following their nucleation, crystals grown from solution typically exhibit regular, planar facets characterised by their Miller indices. Although appearing flat to the naked eye, these crystalline surfaces are rarely so at the molecular level. The various features which make up the nanoscale surface topography of crystal faces are intimately involved in the mechanisms by which crystals grow [48].

A molecule from solution which adsorbs onto a flat part of a crystal face is relatively weakly bound to the surface as the molecule is bound on only one side. There is, therefore, a high probability that the molecule will be desorbed again in a relatively short period of time. To remain on the flat surface, molecules must form stable clusters or nuclei. This consecutive birth of new nuclei and spread to a new layer (B and S) on the crystal face is the 2D analogue of the 3D nucleation process. The 2D nucleation is controlled by the competition between the free energy reduction produced by molecules moving from solution to the crystal and the increase in free energy due to the formation of a new surface [43, 49]. At low supersaturation, growth by the 2D nucleation is difficult since a large number of molecules are required to form a stable nucleus. Detailed analysis reveals that at low supersaturation, the rate of growth by 2D nucleation is negligibly small. Experimental studies, however, show that crystals do in fact grow at appreciable rates even at very low supersaturation. This apparent contradiction between theory and experiment was resolved by Burton, Cabrera and Frank (BCF) [50] with the proposition that imperfections in the crystal packing, due to the presence of lattice defects such as screw dislocations, provide the permanent surface steps needed to promote crystal growth. This step provides a site where a molecule can have multiple binding interactions with a surface, thereby removing the need for 2D nucleation as part of the growth process [48].

8.3.2 Particle Morphology and Surface Structure

The external shape of a crystal is referred to as the crystal morphology. Crystals are bounded by the slowest growing faces and the crystal habit is determined by the relative growth rates of the various faces. The crystal habit is traditionally described using a variety of qualitative terms such as plate-like, prismatic and needle-like. In the current vision we should be moving to more quantitative descriptions of the shape utilising Miller indices to allow a greater understanding of the different surface chemistry being exposed. Early crystallographers were fascinated by the flat and symmetry-related external faces in both natural and synthetic crystallised solids. This led them to postulate that the ordered external arrangement was a result of an ordered internal arrangement [51, 52].

Morphological simulations based on crystal lattice geometry were initially proposed [52–55]. Subsequent work focused on quantifying the crystal morphology in terms of the interaction energies between crystallising units [31, 56]. Attachment and slice energies can be calculated directly from the crystal structure by partitioning the lattice energy in certain crystallographic directions [31, 54]. The calculated attachment energies for crystal faces can be used as a measure of relative growth rates, and so a theoretical morphology may be computed by determining the smallest polyhedron that can be enclosed by these faces given their relative growth rates [31, 54].

The external shape of crystals not only has a considerable influence on the properties of the solid but also has an impact on the handling of particulate materials and their interactions with the excipients used in product design. The shape of the growing crystals is influenced by a wide variety of factors as follows:

- Supersaturation
- Nature of the solvent
- Impurities
- Nature, number and distribution of crystallographic defects

Given the complex nature of crystal growth, it is not surprising that a variety of mechanisms exist through which impurities can influence the growth process. The dominant effect of impurities on the growth process is to interact with a crystal surface, thereby retarding the growth of that face. As we have seen, the structure and chemistry of crystal surfaces varies with crystallographic orientation. A given impurity will, therefore, interact differently with symmetrically non-equivalent crystal faces and retard their growth to different relative extents. The action of a tailor-made additive, benzoic acid, on the crystal growth of a structurally similar compound, benzamide, suggests a mechanism is which a benzoic acid molecule is readily incorporated into the growing crystal face by virtue of its molecular similarity to the host molecule and its capacity for H-bonding [48]. Once incorporated, the addition of further benzamide molecules is, however, strongly impacted by the termination of the infinite chain of H-bonds in the host crystal lattice by the presence of the benzoic acid impurity. This leads to a repulsion between the lone pairs of electrons on the oxygen of the benzoic acid hydroxyl group and the carbonyl group of benzamide. Overall, this leads to a retardation of growth of the {0 1 1} form [31, 57].

The crystallisation solvent is also an impurity and is present at very high concentrations around the crystal. Any of the mechanisms described for other impurities may also apply to the solvent and can produce similar changes in crystal habit. Finally, the interaction between solvent and crystal surface can change the surface energetics of the crystal faces due to the differential binding energies of solute and solvent on the different surfaces [58].

The modern process chemist has at their disposal tools that allow a link to be made between the bulk crystal chemistry, the surface chemistry of the crystal faces and their potential interactions with solvents and impurities. Examples of industrial application include aspirin [58], the agrochemical product paclobutrazol [59]. The subsequent sections in this chapter provide the details of how such an understanding of the intrinsic and extrinsic synthons can be used in the design of pharmaceutical products in particular addressing the challenges in form selection, particle formation and formulation design.

8.3.3 Crystal Morphology Prediction

Early crystallographers were fascinated by the flat, symmetry-related external faces in both natural and synthetic-crystallised solids. This led them to postulate that the ordered, external arrangement was a result of an ordered internal arrangement. Hauy [16] showed that classical polyhedral shapes could be made up by stacking cubes (basic building blocks, now referred to as unit cells) in a variety of 3D sequences. Morphological simulations based on crystal-lattice geometry were proposed by Bravais Friedel [51], Donnay and Harker [52], and this work spans over seventy years but is often quoted collectively by modern workers as the BFDH law (see e.g. Ref. [55]). This law can be summarised

> taking into account sub-multiples of the interplanar spacing *d*(*hkl*) due to space group symmetry, the most important crystallographic forms will be those with the greatest interplanar spacings

Using the BFDH principle and given the unit cell dimensions and space group symmetry conditions, it is possible to identify the most important morphological forms. For molecular systems this approach shows good agreement with experiment and can be used with confidence as an initial prediction of the crystal shape before refinement by further, more sophisticated energy calculations.

Hartman and Perdok [53] were the first to extensively quantify the crystal morphology in terms of the interaction energies between crystallising units. They used the assumption that the surface energy is directly related to the inter-molecular interaction energies and identified by chains of 'strong' inter-molecular interactions called periodic bond chains (PBCs). The strength of a PBC is determined by the weakest link in that chain. They also characterised flat, stepped and kinked faces according to the number of PBCs that are present in these faces. PBC analyses can be used to determine the slice (E_{slice}) and attachment (E_{att}) energies. The slice energy is defined as the energy released on the formation of a growth slice of a particular thickness [52, 55]. The attachment energy is defined as the energy released on the attachment of a particular growth slice onto the crystal. Faces with the lowest attachment energies will be the slowest growing and therefore will be the morphologically most important.

Attachment and slice energies can be calculated directly from the crystal structure by partitioning the inter-molecular interactions contributing to the lattice energy as calculated from each symmetrically independent molecule in the unit cell into slice and attachment energies. Slice (intrinsic synthons) and attachment energies (extrinsic synthons) can be calculated by summing all the interactions between a central molecule and all the molecules within a defined slice (E_{slice}) and all the molecules outside the slice (E_{att}). The thickness of the slab for the slice energy calculations is defined by the interplanar spacing d_{hkl}. This process is shown schematically in Figure 8.4 where *C* is the central molecule, *P* is a molecule outside the slice, *S* is a molecule inside the slice, *N*+ is the growth normal to the planes (*hkl*), *N*− is the growth normal to the planes (-h-k-l) and θ is the angle between the growth normal

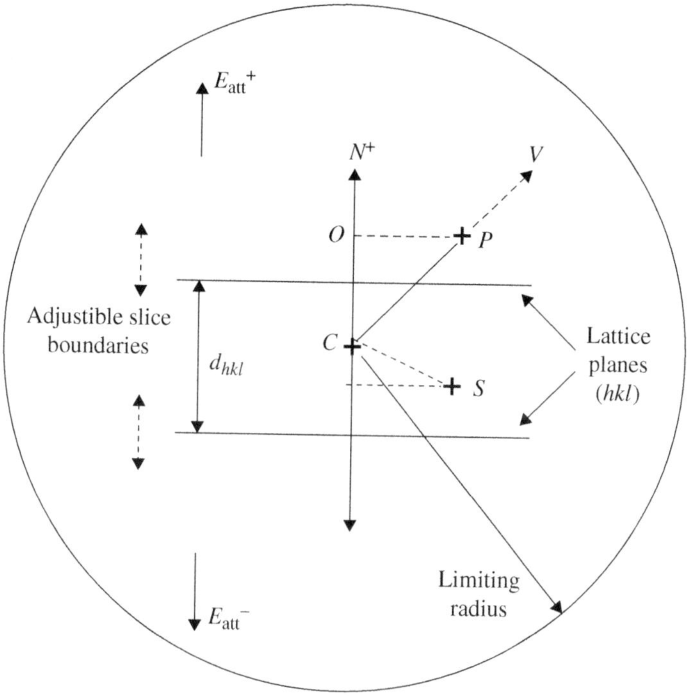

FIGURE 8.4 Two dimensional representation of the inter-molecular interactions using atom–atom method in calculating attachment and slice energies within a sphere of limiting radius. *C* is the central molecule, *P* is a molecule outside the slice and *S* is a molecule inside the slice. Source: Adapted from Roberts et al. [48] with permission. © 1996 John Wiley & Sons, Inc.

and the bonding vector CP. The slice boundaries are adjusted along the growth normal to maximise the absolute magnitude of E_{slice} and hence obtain the energetically most stable slice. A polar Wulff plot having a radius proportional to E_{att} is constructed to obtain the predicted morphology.

A computer program – HABIT – has been developed to allow such calculations to be carried out. The calculated attachment energies can be used as a measure of relative growth rates and so the theoretical morphology may be computed by determining the smallest polyhedron that can be enclosed by the specified forms given their relative growth rates. The faces on which to carry out the energy calculations are usually those identified from a BFDH calculation. This approach has proved successful for a number of compounds and applied to a number of challenges during drug product design and these will be highlighted in the following sections. The attachment energy model assumes that the surface is a perfect termination of the bulk and little or no surface relaxation takes place. Surface relaxation has been shown to play an important role for inorganic systems such as alumina and haematite [60]. Calculations of this type for acids and amides [31, 61], non-linear optical materials and agrochemicals [59] have shown, in general, for molecular materials that these surface relaxation effects can be neglected without introducing significant errors in the calculated crystal morphology.

8.3.4 Deconstructing the Supra-Molecular Interactions at Surfaces – Extrinsic Synthons

The lattice energy can be partitioned into slice and attachment energy per surface as defined by specified Miller planes (*hkl*). The magnitude of the attachment energy per face can be taken to predict the relative growth rate and hence morphological importance and surface area of the surface. For stable crystal growth it is the slowest growing (lowest attachment energy) surfaces that dominate the external morphology of a material. Face-specific information such as which of the bulk intrinsic synthons are unsaturated (broken) due to surface termination can be outputted for subsequent analysis. These unsaturated interactions are known as 'extrinsic synthons'. The nature and strength of these interactions, combined with constructions of the predicted surfaces using molecular visualisation software, can reveal detailed information on the surface chemistry of the important faces and how the solute molecule binds and incorporates into the lattice. This information can then be directly related to the relative growth rate and size of the crystal face. The following case studies will highlight the application of these concepts to interparticle interactions, dissolution and solvent binding.

8.3.5 Grid Searching – Probing Inter-molecular Interactions at Surfaces and Environments

Inter-molecular grid search suite [58, 62] is a tool designed to characterise the nature and behaviour of the unsaturated supra-molecular extrinsic synthons with the external environment. In this the inter-molecular interactions between a crystal surface, with a well-defined termination, are probed which represent/describe an external environment. This approach calculates the interaction energies of a probe molecule with all the molecules at and near the crystal surface within a sphere of pre-defined radius. Similar to lattice energy calculations, an atomistic approach is utilised to calculate the inter-molecular interaction energy. A 3D virtual grid as shown in the schematic in Figure 8.5 is created above the crystal surface whose in-plane dimensions are periodic (multiples of reticular area) whereas the plane normal to the crystal surface is non-periodic. The search area is further divided into grids whose sizes are defined by the step sizes which are expressed in fractional coordinates in the *y*- and *z*-directions (in-plane directions) and in Cartesian coordinates in the *x*-direction. The probe molecule visits every grid point and is subjected to sequential orientations. The technique has the advantage of being robust and easy to use; however, the approach is based on some of the following assumptions:

- The probe molecule is treated as a rigid body which is optimised in the gas phase prior to the grid search.

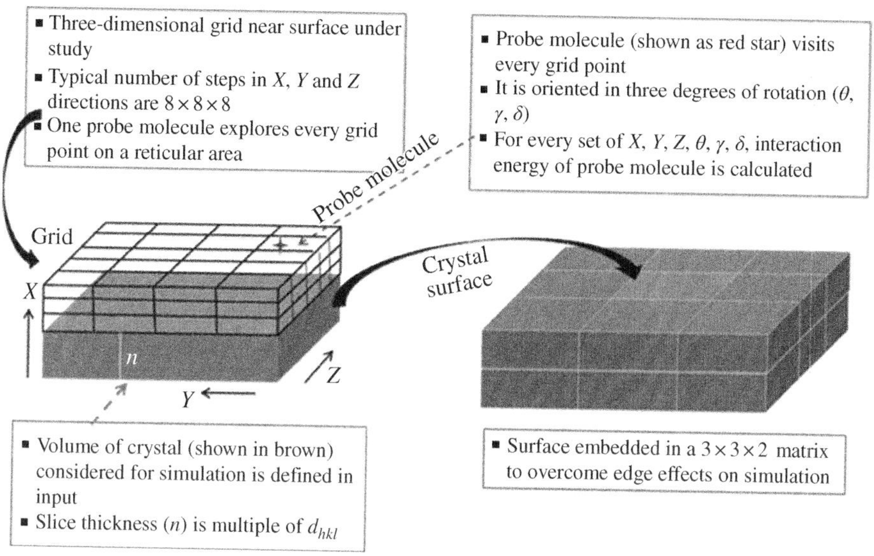

FIGURE 8.5 Schematic representation highlighting the main computational methodology associated with the grid search; the probe molecule in the context of the surface and the virtual grids. Source: Adapted from Ramachandran et al. [63]. Reproduced with permission of American Chemical Society. (*see insert for color representation of the figure.*)

- The surface is optimised prior to the grid search; however, it is not relaxed further during the grid search.
- The molecular vibrations are not taken into account.

The technique has applications in understanding the effects of solvent or impurity binding on surfaces that will lead to changes in morphology, in cohesion/adhesion balance of API/excipient, or in surface adsorption studies. The grid search or systematic search is further described in the schematic shown in Figure 8.5.

8.4 THE CRYSTALLISATION PERSPECTIVE

8.4.1 Nucleation, Surface Energies and Directed Polymorphism

Understanding the early stages of the crystal growth process is important due to its role in directing the crystal form and hence physical properties. However, characterising the detailed structural nature of such nanosized crystals associated with the formation of a distinct polymorph post-nucleation is experimentally challenging due to the small particle sizes involved. Hence, molecular modelling techniques are now being used to provide an understanding of the inter-relationship between crystal form and surfaces at the nanoscale size with those prevalent in the bulk material properties. However, at small particle size, for example the size of a crystal nucleus, the predominating surface structure may differ from the bulk structure and therefore a significant change of the properties of such particles could be expected. Drawing upon Ostwald's Rule of Stages [64], it has been stated that in the emergence of a solid phase from solution a metastable phase should be the first to appear and that this would subsequently transform to the most stable phase. Metastable forms have often been referred to as kinetic forms as they are created when non-ideal crystallisation conditions prevail such as fast cooling rates associated with high levels of the supersaturation in the solution and hence a small size of the critical nucleation cluster. Therefore, by controlling the degree of supersaturation during crystallisation, a control of the critical cluster size and hence the relative stability of polymorphic forms might be expected. The hypothesis that polymorphic phase stability might be size-dependent was suggested by Keller et al. [65]. They proposed that the development of a new phase from a nucleus to macroscopic size could be coupled with the possibility that, during this process, the stability relationship between competing polymorphs could become inverted.

This concept of a size-dependent phase stability in polymorphic materials is addressed here by confronting this model and testing it against the well-characterised H-bonded compound L-glutamic acid (L-GA) which is a material with two monotropically related polymorphic modifications: the metastable α-form that transforms into the stable β-form through a solvent-mediated transformation [66].

The observed crystal structures (LGLUAC03 [67], LGLUAC11 [68]), taken from the CSD [21], were optimised allowing the hydrogen atoms to relax. The morphologies were simulated using the technologies described in Section 8.3 with specific details of the calculations described elsewhere [66]. The slowest-growing face in the α-crystal was calculated to be the (002) with smaller (101), and (111), (−111) faces present in the morphology in excellent agreement with experimental observation. The simulated morphology for β-form crystals revealed dominant (020) crystal faces and smaller (101); (021) smaller faces were also in good agreement with experimental results and previous observations from the literature. The 'metastable' α-form crystals showed a prismatic shape whilst the stable β-form was predicted to have elongated needle-like shape [69].

Molecular clusters of different sizes were created by overlaying the predicted morphological shape and the optimised crystal structure. Figure 8.6 shows the facetted molecular clusters for the α- and β-forms.

The energies of the facetted molecular clusters of different sizes were plotted as a function of cluster size for both polymorphic forms of L-GA (Fig. 8.6c). The results were fitted with a power law function which enabled calculation of cluster energy for any molecular cluster size. The results revealed that the metastable α-form is a more thermodynamically stable form than the 'stable' β-form at smaller cluster sizes with a crossover point of 240 molecules.

Therefore, it is concluded that the first nucleating form (when the cluster size is small), according to Ostwald rule, would be the 'metastable' α-form which has become surface stabilised due to its small size hence making it more stable than the 'stable' β-form. In this respect 'stable' and 'metastable' as used here refer to the bulk crystallographic structures.

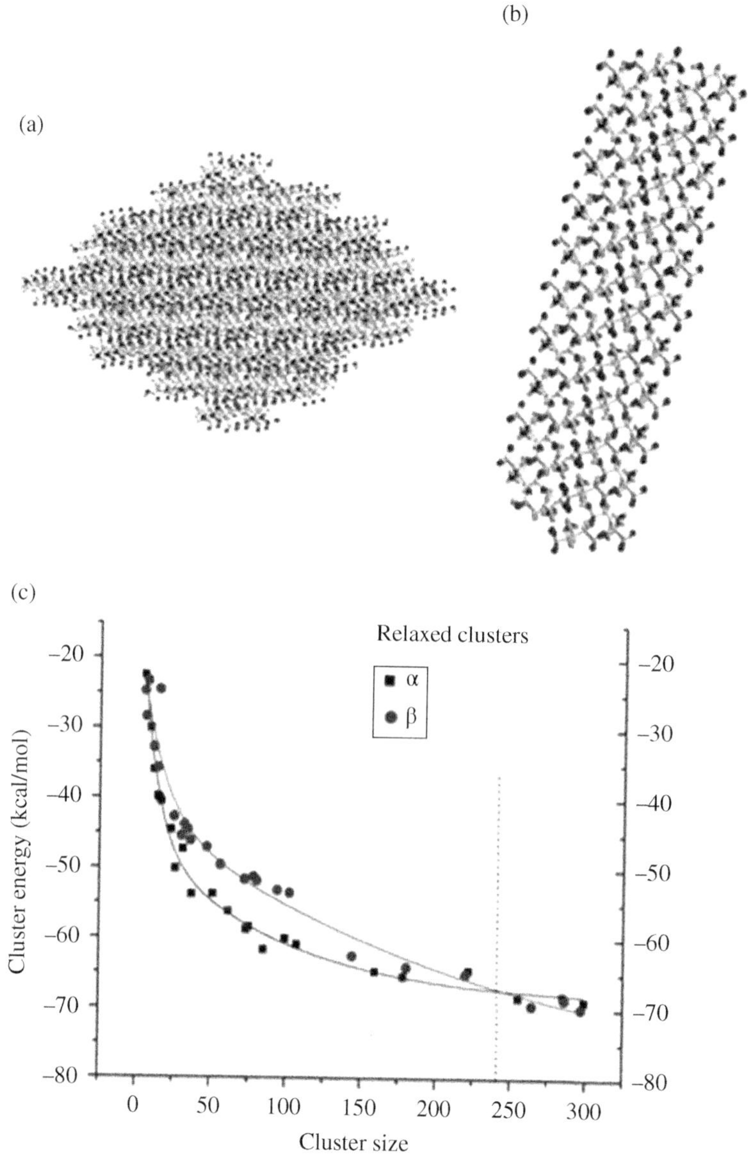

FIGURE 8.6 Facetted molecular clusters of α- (a) and β-forms (b) and energy minimisation (c) of relaxed facetted clusters of L-GA. Source: Adapted from Hammond et al. [70]. Reproduced from Elsevier.

This work [71] also examines the impact of facetted versus non-facetted clusters as well as the variation in the conformation of the glutamic acid molecule as a function of distance from the centre of the nanoparticle. This shows the impact of relaxation and illustrates the possibility of mapping the molecular conformation from solution to surface and eventually bulk conformations.

8.4.2 The Impact of Solvent on Morphology

The strong influence of the growth solvent on crystal morphology (and habit) has been well documented [72]. It is known that the underlying mechanism involves face-specific, solvent–solute interactions resulting in different degrees of solvent binding which inhibits crystal growth on selective crystal habit planes. A number of researchers have studied the impact of solvent choice on the resultant crystal morphology by relating the mechanistic aspects associated with habit modification to the specificity of the inter-molecular interactions between crystallographically ordered crystal habit surfaces and solvent and/or impurity molecules [73]. An excellent review covers the current and emerging modelling approaches in terms of practical applicability, from the standpoint of process engineering [74]. Recently a different approach has been suggested to study the growth of crystals from solution using kinetic Monte Carlo simulations whose parameters are derived from atomistic, molecular dynamics simulations [75]. This use of Monte Carlo techniques enabled the simulation of length scales up to 1 μm and of timescales up to a millisecond.

Recently, we have successfully demonstrated the application of a surface-specific, grid-based search method, as outlined schematically in Figure 8.7, to predict both the solute and solvent binding to crystal-habit surfaces of aspirin [58] associated with crystallisation from aqueous ethanol solution. In this approach, the relevant crystal

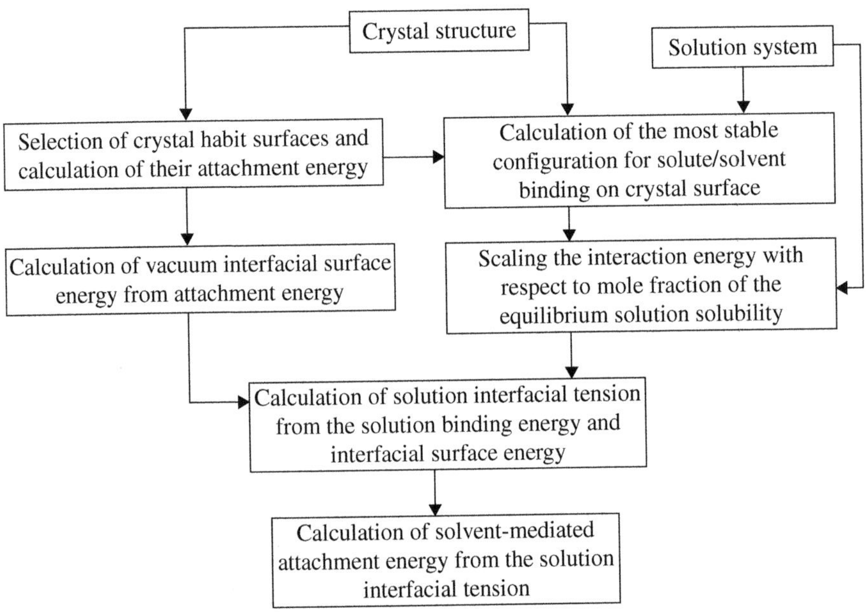

FIGURE 8.7 Schematic diagram summarising an overall approach used to calculate specific surface energies (interfacial tensions) of crystal surfaces in the presence of solution. Source: Adapted from Hammond et al. [58]. Reproduced with permission of American Chemical Society.

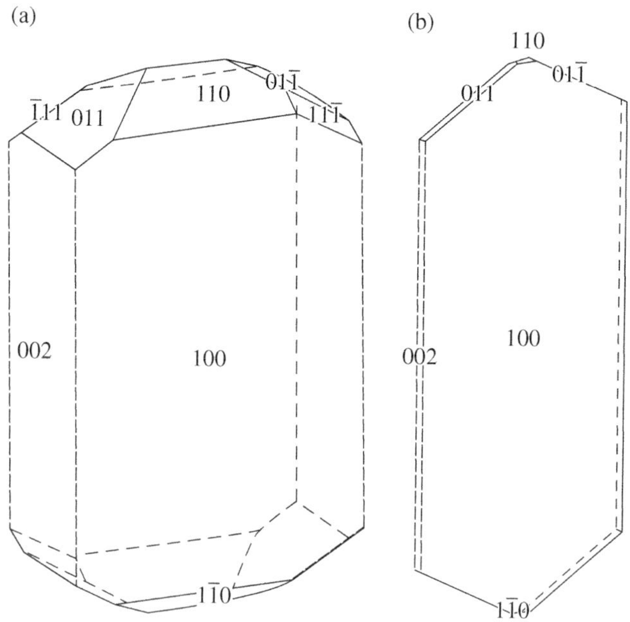

FIGURE 8.8 Crystal habit for aspirin as predicted (a) via the attachment energy model and as derived (b) from solution surface energies (38% ethanol–water, soluble at 50°C). Source: Adapted from Hammond et al. [58]. Reproduced with permission of American Chemical Society.

surfaces were modelled individually to construct a molecular-scale model with each surface having a particular crystallographic orientation (*hkl*), being cleaved. In the subsequent systematic grid-based search calculations, the binding energetics of the crystal habit surfaces (see Fig. 8.7) for every configuration of the probe molecule were calculated. In addition, these configurations were subjected to energy minimisation, treating the probe molecule as a rigid body. The most favourable interaction energy identified in this way was used to adjust the specific (vacuum) surface energy for each form identified as being present in the experimentally observed crystal morphology for aspirin.

Figure 8.8 summarises the results from these simulations with representative, experimentally observed crystal habits for aspirin crystals grown from ethanol and water. A comparison of Figures 8.8a and b reveals that the unadjusted attachment energy model provides a good general match to the experimental morphology but, that the crystal habit, as predicted, manifests a much thicker and more tabular shape when compared to the experimentally observed shape of the crystals grown from solution. The crystal habit prediction, as adjusted to allow for the effect of surface wetting by the contacting solution (Fig. 8.8b), reveals, in contrast, a much better match to the experimental data consistent with this model providing a more appropriate method for predicting the crystal habit of solution-grown crystals.

8.4.3 The Impact of Impurities on Morphology

Earlier in the chapter (Section 8.3.2) the case of incorporation of an impurity during growth was described in concept. Here molecular modelling has been applied to study the mechanism through which L-alanine, acting as an impurity, disrupts the growth and modifies the morphology of α-glycine crystals. L-alanine was selected for this case study because of its structural similarity to glycine differing only by the substitution of a methyl group for a hydrogen atom.

An approach employing the atom–atom formalism has been extended to model the effects of the chiral molecule, L-alanine, when it becomes incorporated as an impurity in αglycine (achiral) crystals [73]. In this, a quantitative measure of the changes in lattice-site energy within the host α-glycine (CSD refcode: GLYCIN02 [76]) lattice accompanying incorporation of an L-alanine molecule [77] is provided.

Slice and attachment energies were calculated for the significant growth forms including {020}, {011} and {110} in the presence of L-alanine. A new approach was implemented which employs a weighted average of attachment energies calculated for the pure host, and host modified by additive, to predict crystal morphology and its dependence on the proportion of lattice sites at which host molecules are replaced by additive molecules. The results show that whereas L-alanine is able to attach both at the growing (020)

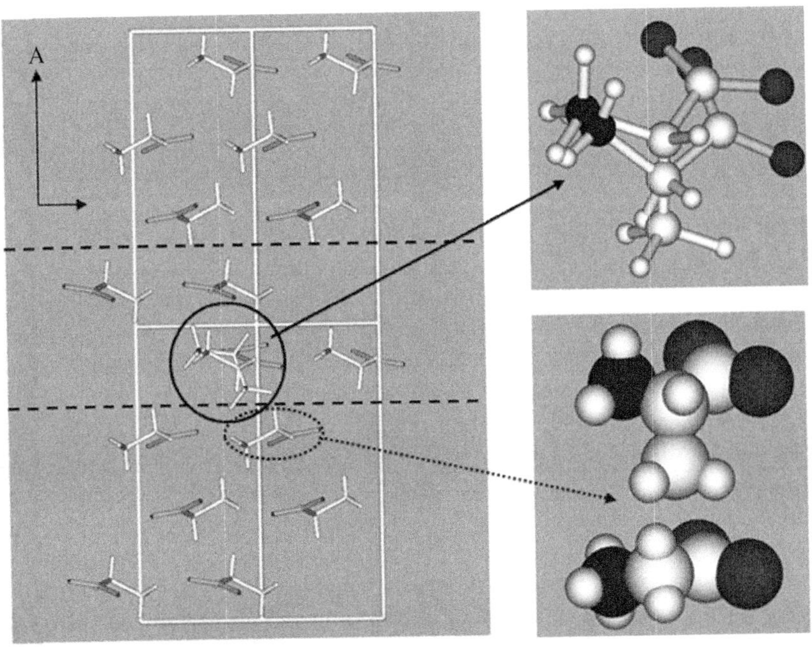

FIGURE 8.9 Illustration of the substitution of a host α-glycine molecule by an additive L-alanine molecule at lattice position [0003] so that the methyl group of the L-alanine molecule is aligned with the pro-chiral S hydrogen atom of the glycine molecule. The enlargement, top right, shows the comparative positions of the host and additive molecules occupying that specific lattice position. The enlargement bottom right shows the most significant intermolecular ‘clash’ between the additive and a host molecule. Source: Adapted from Hammond et al. [77]. Reproduced with permission of American Chemical Society.

(which has the pro-chiral R hydrogen atom, bonded to the α carbon, orientated normal to the surface) and $(0\bar{2}0)$ crystal surface of α-glycine (which has the pro-chiral S hydrogen, atom bonded to the α carbon, orientated normal to the surface), the reduction in the rate of growth perpendicular to the $(0\bar{2}0)$ surface, ascribed from the reduction in the associated relative attachment energy, is greater. This is summarised in Figure 8.9.

8.5 THE DRUG PRODUCT PERSPECTIVE

8.5.1 Excipient Compatibility

A key challenge in product design is that during drug formulation one or more active pharmaceutical ingredients (APIs) are combined with non-active excipients to produce a marketable product. The excipients provide a range of features from bulking and flavouring to optimising the mechanical and processing properties of the active ingredient. Excipients can also have a major impact on the stability of the product and so appropriate selection is critical to the success of product design.

To achieve this, a thorough understanding of the components, their interactions with the API and proposed manufacturing process is required. Grid search methods allied to surface chemistry modelling have been utilised to define particle properties in terms of surface chemistry examining dissolution [78], mechanical properties [79], agglomeration [66, 70, 80] and sticking tendency. In this section we highlight recent work using computer simulation of API excipient interactions to predict adhesive strength at the API/excipient interface.

Ketoprofen and 17 commonly used excipients were chosen as the probe systems for this study [81]. The molecular modelling has two parts: prediction of crystal morphology followed by the prediction of binding energy of a probe molecule on the surfaces of the ketoprofen crystal. Three conformers for each of these excipients (probe molecules) were taken into consideration. For those excipient molecules which have more than one molecule in the asymmetric unit (e.g. sorbitol and α-lactose monohydrate), both the complete asymmetric unit as well as three conformers of a single molecule were considered as the probe conformations for the systematic search study. For those excipients that are polymers, a representative monomer unit was used as a probe molecule.

The systematic search results include the minimum interaction energy and the distribution of interaction energies. The results are summarised in Table 8.4. The minimum interaction energies on the two most morphologically important surfaces increase with an increasing number of atoms of the probe (excipient) molecule as expected gives an increased pairwise additive of the interatomic interactions. The interaction energy distributions of the three conformers of each excipient were compared with each other (Fig. 8.10). The peak position refers to the mean and mode of the interaction energy between an excipient as the probe molecule and a surface of the API. Similarly, the minimum interaction energy of every conformer was compared one with another. The most stable conformer was chosen from the energy distribution and the minimum interaction energy, and its values were compared with those of the other excipients in order to assess the order of preference of these excipients to bind with ketoprofen.

TABLE 8.4 Minimum Interaction Energy (kcal/mol) of All the Excipients on the Ketoprofen (100) and (110) Surfaces[a]

	Minimum Interaction Energy (kcal/mol)	
Excipient Name	(100)	(110)
Ketoprofen molecule	**−6.3506**	**−9.1831**
Aerosil	**−3.44**	**−2.50**
Avicel	−8.71	−11.56
Corn starch	−14.28	−37.19
β-D-glucose (GLUCSE02)	**−5.84**	**−6.30**
α-Lactose	−13.72	−15.15
α-Lactose monohydrate (LACTOS11)	−16.30	−20.92
β-Lactose (BLACTO02)	−13.74	−19.35
Magnesium stearate	−21.70	−21.95
Magnesium stearate	−20.14	−22.17
D-mannitol (DMANTL09)	−11.05	**−7.01**
Methocel (hydroxy methyl cellulose)	−9.1082	−11.696
Natrosol (hydroxyl ethyl cellulose)	−10.501	−11.24
Palmitic acid (YEFWEM01)	−10.051	−10.67
PEG6000	−5.8839	**−6.7986**
PVPP	**−3.3135**	**−6.0176**
Sorbitol	−11.748	**−10.59**
Sorbitol ASU (GLUCIT03)	−26.755	−36.531
Stearic acid (STARAC05)	−10.978	−11.898
Stearyl alcohol	−10.371	−11.867
Veegum	−40.194	−20.395

[a] Source: Adapted from Ramachandran et al. [81].
Colour code: dark grey – interaction weaker than ketoprofen; clear – interaction stronger than ketoprofen; and light grey – similar binding to ketoprofen (within 1 kcal/mol of the ketoprofen interaction).

It was found from this study that all the seventeen excipients gave a favourable interaction. Compared with the base line value for ketoprofen itself, Table 8.4 shows that four excipients have a poorer binding energy on the (100) face and five excipients were found to have a poorer binding energy on the (110) face. It was observed that the excipients bound more favourably on (110) than (100) which could be due to the availability of H-bonding donors and acceptors on the former surface. Generally, aerosil, glucose, PEG6000 and PVP have relatively poor binding energies on both faces. Mannitol exhibits good binding on the (100) face but poorer binding on (110). The opposite was found to be true for sorbitol. One of the interesting observations is the differences between glucose and lactose for both (100) and (110) faces: lactose has stronger binding than ketoprofen but for glucose the binding on (110) is significantly poorer.

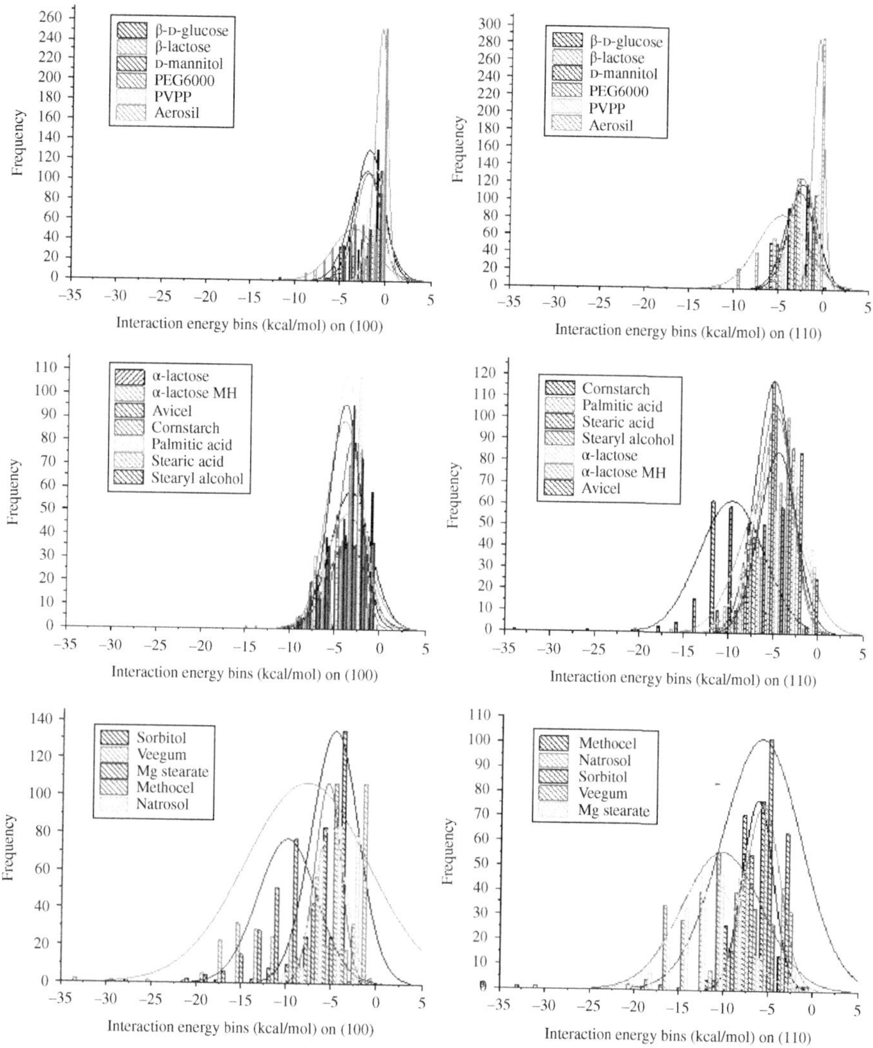

FIGURE 8.10 Comparisons of interaction energy distributions of 17 excipients with ketoprofen on (100) surface (left-hand side) and on (110) surface (right-hand side). The plots are split into six separate plots for clarity. Source: Adapted from Ramachandran et al. [81]. (*see insert for color representation of the figure.*)

8.5.2 Inhaled Drug Delivery Design

A challenge in inhalation formulations is that they are comprised of micronised particles of an active ingredient and excipient(s) which tend to form agglomerates. These agglomerates need to be dispersed effectively upon delivery for an effective drug deposition to the lungs. Therefore, for studying the deagglomeration and aerosolisation behaviour, it is essential to have a complete understanding of the

TABLE 8.5 Predicted Cohesive Energy of Inhalation Formulation Components (LMH, FP, BUD, SB)[a]

FP		BUD		SB		LMH	
Forms	Strongest (Minimum) Interaction Energy (kcal/mol)	Forms	Strongest (Minimum) Interaction Energy (kcal/mol)	Forms	Strongest (Minimum) Interaction Energy (kcal/mol)	Forms	Strongest (Minimum) Interaction Energy (kcal/mol)
{2 0 0}	−12.96	{0 0 2}	−8.69	{2 0 0}	−7.75	{0 2 0}	−17.32
{1 1 0}	−9.26	{1 0 1}	−10.75	{0 0 2}	−7.62	{0 0 1}	−15.64
{1 0 1}	−13.22	{0 1 1}	−9.27	{1 1 1}	−7.73	{0 1 1}	−14.24
{0 1 1}	−10.61	{1 1 0}	−11.11	{1 0 2}	−8.34	{1 0 0}	−16.81
				{2 1 0}	−7.44	{1 0 −1}	−14.51
						{1 1 0}	−16.23
						{0 −2 0}	−14.66
						{0 −1 1}	−14.66
Average	*−11.5*		*−9.9*		*−7.8*		*−15.8*
Range	*3.7*		*2.42*		*0.72*		*3.08*

[a] Source: Adapted from Ramachandran et al. [63]. Reproduced with permission of American Chemical Society.

inter-particle interactions between the active ingredient and excipients or, in other words, the cohesive and adhesive forces between particles of homogeneous and heterogeneous species, respectively. Grid search methods have been successfully applied to predict the cohesive behaviours of an excipient, α-lactose monohydrate (LMH) and three APIs: salbutamol (SB), fluticasone propionate (FP) and budesonide (BUD) [63]. The prediction has been validated through independent experimental de-agglomeration studies using laser diffraction techniques [82].

The interaction energies between a probe molecule (of LMH, SB, FP or BUD) and all the surfaces of their respective crystal morphology were calculated. The cohesive energies are the average of the minimum interaction energy of the probe molecule on every crystal surface of its respective morphology as shown in Table 8.5. The cohesive strength is stronger if the interaction energy is more negative. On that basis, it is predicted that FP is the most cohesive followed by BUD and SB. One of the interesting features in Table 8.6 is the binding energy range for SB binding is very tight, across the different faces being within 1 kcal/mol, whilst for FP the range is over 3 kcal/mol thus reflecting a different range of binding modes and/or different surface chemistry.

Thus the prediction showed that FP (−11.5 kcal/mol) has the highest cohesive strength when compared to BUD (−9.9 kcal/mol) or SB (−7.8 kcal/mol) and this ranking correlated well to the laser diffraction measurements where the airflow pressure required for complete dispersion (CPP) was 3.5, 2.0 and 1.0 Bar for FP, BUD and SB, respectively. This case study demonstrates that the technologies also have the potential to be used as predictive tools for assessing the cohesive–adhesive strength balance for inhaled drug formulations.

TABLE 8.6 Analysis of the Structural Factors Related to Likely Slip Due to Mechanical Deformation[a]

(*hkl*)	Surface Energy mJ/mm^2	Slip Plane Rugosity	Interlocking of Slip Planes?	Slip System with $b = [010]$	Slip Involving Hydrogen Bond Breaking	Surface Cleavage Likely?	Plastic Deformation Likely?
(100)	0.0829	Very low	N	Y	N	Y	Y
(001)	0.0904	Low	N	Y	N	Y	Y
(011)	0.1550	High	Y	N	Y	N	N
(110)	0.1430	High	Y	N	Y	N	N

[a] Source: Adapted from Olusanmi et al. [79]. Reproduced with permission of Elsevier.

8.5.3 Mechanical Properties

Solid-form selection is important in identifying a suitable route for tablet processing. Direct compression can be chosen over dry granulation or wet granulation only if the solid-form has high compressibility and high flowability. The mechanical strength of the API, which is dominant in the formulation, is detrimental to the strength of the tablet. From the crystallographic structure, mechanical deformation properties of aspirin were predicted successfully [79].

Organic crystals in general and drug molecules in particular are soft materials due to the presence of weak inter-molecular interactions. The mechanical deformation properties of these materials can often be highly anisotropic, reflecting the anisotropy in a number of factors such as elastic constants, dislocation Burgers vector, $\bar{b}$ $[u\ v\ w]$ and available slip planes (*hkl*). Crystal deformation under mechanical stress takes place elastically and then plastically. Plastic deformation requires dislocation motion along the low index slip planes subject to the requirement as given in Equation 8.3:

$$hu + kv + lw = 0 \tag{8.3}$$

The relative lengths of the crystallographic unit cell edge parameters together with the inter-molecular packing pattern of aspirin as projected down the three crystallographic axes are shown in Figure 8.11. Analyses suggest that the most likely slip plane is (100) followed by (001). Slip on the (010) plane would not seem to be very likely given that this would involve breaking the dimer H-bonds. Slip on the lower index surfaces such as the (101) plane would appear to be feasible in terms of the inter-molecular packing but would perhaps be unlikely given that this would require breaking of the dimer bond. This analysis is summarised in Table 8.6 together with the calculated values of the surface energies. The most likely Burgers vector for dislocation slip is parallel to the shortest unit cell axis [010], this being the shortest lattice translation, and given that the Bravais lattice is primitive there would be no halving of this distance. Given that the other two lattice constants are nearly twice the magnitude of $\vec{b}$, it can be speculated that this lattice dislocation would be the most

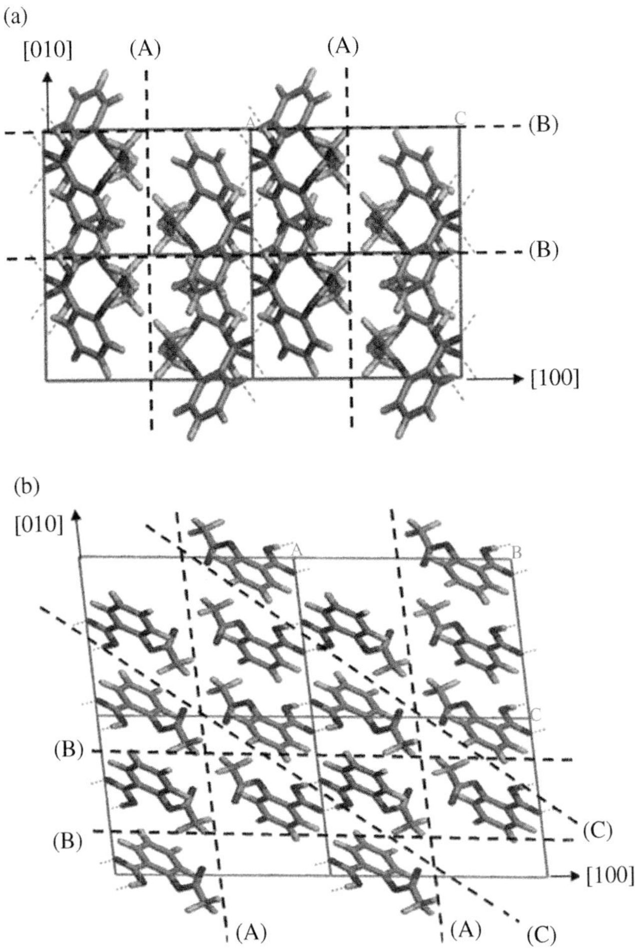

FIGURE 8.11 (a) (1 0 0) projection showing slip possibilities on (0 0 1) (A) planes. Note that slip on (0 1 0) (B) would require breaking the dimer hydrogen bonds. The black and grey parts of the molecule represent oxygen and hydrogen, respectively. (b) (0 1 0) projection showing comparatively easy slip possibilities on (1 0 0) (A) crystal planes and less so at (0 0 1) (B). Note that slip on (1 0 1) (C) would involve breaking dimer hydrogen bonds. Source: Adapted from Olusanmi et al. [79]. Reproduced with permission of Elsevier.

likely and the only Burgers vector that needs to be considered. Consideration of the combination of slip plane and Burgers vector would suggest that the most likely active slip systems would be in the (100) plane along the $\vec{b}$ direction giving the slip system, (100)[010] and in the (001) plane along the $\vec{b}$ direction giving the slip system, (001)[010].

Observations of the damaged morphology of the product particles after impact tests are in agreement with the breakage mechanisms observed during nanoindentation

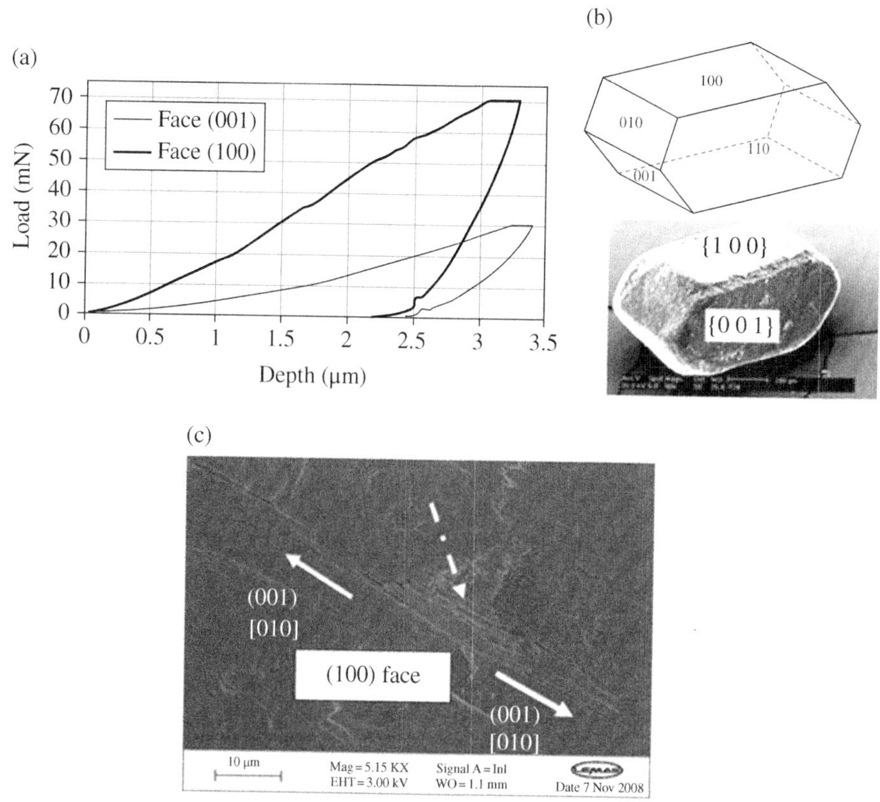

FIGURE 8.12 (a) Load–displacement curves from indentation on faces (0 0 1) and (1 0 0) of aspirin carried out at loading rates of 5 mN/s and at similar depths. 'Pop-ins' can be observed on both curves, indicated by the arrows. (b) Morphological sketch of aspirin drawn using SHAPE and scanning electron micrograph of a representative aspirin particle. (c) SEM image of an indent of aspirin (1 0 0) face. Source: Adapted from Olusanmi et al. [79]. Reproduced with permission of Elsevier.

studies with the deformation and fracture of aspirin occurring on the preferential slip planes as shown in Figure 8.12.

8.5.4 Dissolution

Selection of solid-form and appropriate particle attributes is one of the foundation elements of drug product design. One challenge is to deliver consistent dissolution rates of API particles used in formulations. Additionally as the chemistry route is optimised and the final step isolation is refined, there is a need to be able to define the potential impact of changes of API particle size and shape on product efficacy. The model used to predict the dissolution kinetics of the monodisperse crystalline

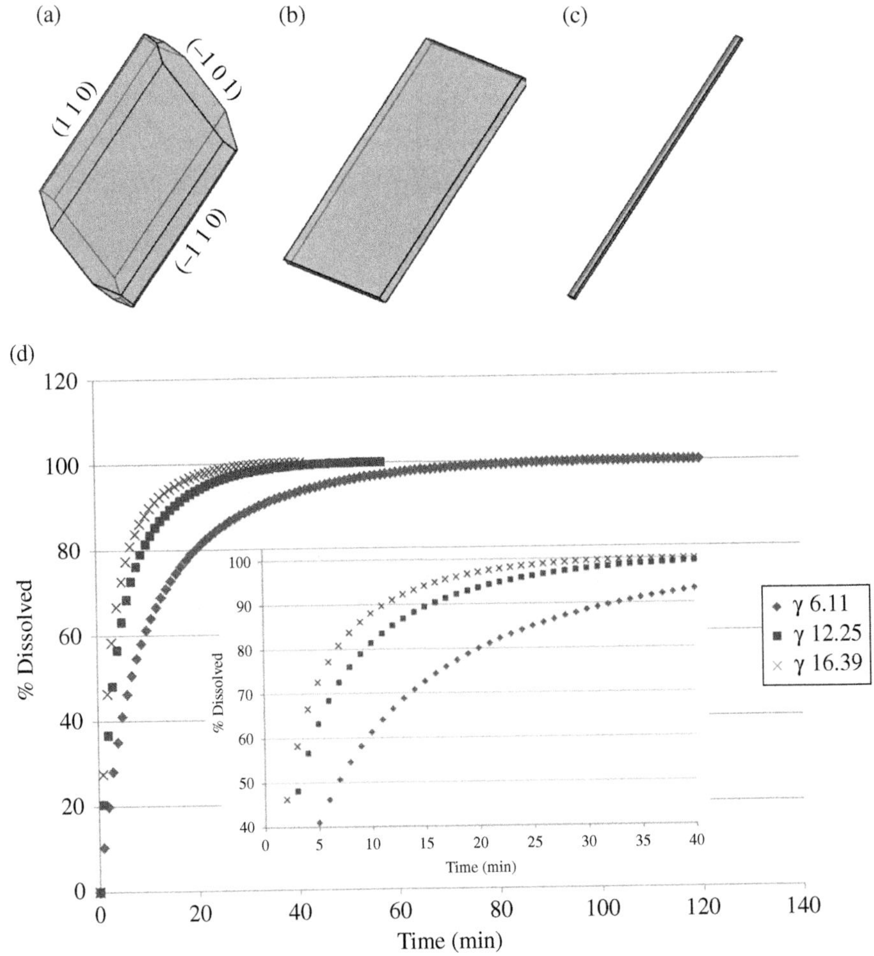

FIGURE 8.13 Shapes of celecoxib (form C) crystal for different shape factors (a) attachment energy, (b) plate-like morphology, (c) needle-like morphology and (d) the predicted dissolution profiles of celecoxib crystals with different shapes (needle shape dissolves faster than the plate-like crystal and the attachment energy model takes the longest to dissolve) with an inset showing the magnification of 0–40 min and from 40 to 100% dissolution. Source: Adapted from Ramachandran [62].

TABLE 8.7 Shape Factors of the Three Different Celecoxib Crystals and the % Surface Area of the Forms[a]

	% Surface Area of Particle						
Shape Factor (Γ)	{0 1 0}	{1 1 0}	{1 0 0}	{0 0 1}	{0 1 −1}	{1 0 −1}	{1 −1 0}
6.11	35.0	31.1	3.2	0.6	2.1	20.8	7.8
12.25	84.4	0	10.6	1.6	3.3	0	0
16.39	63.2	0	36.0	0	0.8	0	0

[a] Source: Adapted from Ramachandran [62].

particles is based on Noyes–Whitney equation (Eq. 8.4) [83]. In this, the surface area S of the particle is dependent on the morphology of the crystal which can be expressed in terms of the shape factor, Γ (Eq. 8.5) which can be predicted from morphological simulations.

$$-\frac{dX_s}{dt} = \frac{DS}{h}\left(C_s - \frac{X_d}{V}\right) \tag{8.4}$$

$$\Gamma = \frac{P_a}{P_v^{2/2}} \tag{8.5}$$

where, X_s is the mass of solid drug, t is time, D is the diffusion coefficient, S is the surface area at time t, h is the boundary layer (stagnant layer) thickness, C_s is the aqueous solubility, X_d is the mass of dissolved drug, V is the volume of the dissolution medium, and P_a and P_v are the particle's surface area and volume, respectively.

This approach has been applied to predict the morphology of celecoxib (CSD refcode: DIBBUL) [84], as described in Section 8.3.3 together with its associated shape factor. The shape factor was used subsequently in estimating the dissolution of API particles, celecoxib [62]. In order to see the effect of the dissolution rate on celecoxib crystals with varying morphology (i.e. a range of shape factors), the centre to face distance that is proportional to the attachment energies was manually adjusted to produce three different morphologies for the celecoxib crystal for which new shape factors were obtained [62, 85]. These morphologies are shown in Figure 8.13a–c with their respective shape factors shown in Table 8.7. The corresponding dissolution rates, as predicted by the dissolution model, are given in Figure 8.13d.

Although not yet taking into account the structural chemistry of the surfaces, this case study shows that the crystalline morphology can have an effect on the rate of dissolution, that is dissolution rate is faster when the shape factor which is greater. This model is robust and can be applied to compare the dissolution rates of different morphologies and extended to examine different polymorphs. The approach can also take into account the polydisperse nature of particles. Ultimately, this will allow the formulator and process chemist to examine the link between the evolving chemical route and API particles and the dissolution rate.

8.6 SUMMARY AND FUTURE OUTLOOK: SYNTHONIC ENGINEERING PARTICLE PASSPORT AND THE FUTURE OF THE DRUG PRODUCT DESIGN

Over the past decade, through embracing both academic advances and technology initiatives, significant progress has been made in defining relationships between the API properties of new and existing chemical entities and the formulation design aspects of new products. Examples of established progress include the following:

- API particle size distributions and content uniformity [86]
- API particle size distributions and flow [87]

- API particle size distributions, surface area and mechanical properties [88]
- API particle size and dissolution [89]
- Crystal brittleness and milling behaviour [90]

Whilst, traditionally, the solid-form selection process has focused on two main factors, that is achieving an appropriate degree of product stability and bioavailability, increasing emphasis is also being focused on selection of solid-forms at the pre-formulation stage which have optimal physical properties such as mechanical behaviour, surface properties and particle shape. Given this perspective, this chapter has, briefly, outlined some recent research on the application of emerging computational technologies as foundation elements of the modern paradigm:

- Understanding inter-particle interactions associated with polymorphic transformation [66]
- Predicting crystal surface/solvent interactions and on the solvent mediation of the crystal habit [58]
- Understanding the effect of impurity on the crystal growth and therefore modifying its morphology [77]
- Screening and selection of excipients through understanding of their interactions with the API as a critical step in product design [81]
- Predicting the deagglomeration and aerosolisation behaviour of inhaled formulation from the inter-particle interactions between the API and excipients [63]
- Visualising the impact of crystal packing motifs on the potential for mechanical deformation processes [91]
- Predicting the dissolution of API particle as a function of its shape and size [62]

Models have been built that allow the pharmaceutical scientists to explore the impact of particle variation on dissolution rate and bioavailability. The Biopharmaceutics Classification System (BCS) [89] is used do define classes of compounds based on the solubility and permeability of the compounds. Permeability is a molecular property, but solubility and dissolution rates are related to the internal structure (salt and polymorph) and particle size distribution/surface area. More recently this concept was built upon where the interplay between permeability and solubility and the impact of using simulated gastric fluids in measuring solubility have brought greater definition to the impact of API particle attributes on drug efficacy [92]. Additionally, a recent cross-industry attempt to develop a Manufacturing Classification System (MCS) [93] highlights the intense interest in this area from the manufacturing sector. It is envisaged that an MCS would aid product and process development facilitating the transition between clinical and commercial manufacturing sites providing a common understanding of risk. A further goal of an MCS would be to provide regulatory relief for the development and manufacturing of dosage forms with well-controlled API and excipient properties targeted/suited to particular manufacturing process trains.

These relationships, combined with institutionalised corporate knowledge of formulation design practices [12, 94] plus computational methodologies of the sort highlighted in this chapter open up the potential of a fully integrated holistic product design process consistent with the emerging QbD philosophy [10, 94].

ACKNOWLEDGEMENTS

Synthonic Engineering research has been funded at Leeds for a number of years through EPSRC, TSB and industrial research grants. More recently, the development of the methods in this research to provide modelling tools suitable for use in industrial process and product design has been funded through an EPSRC follow-on grant (EP/I028293/1) in collaboration with CCDC, Infineum, Malvern Instruments and Nexia Solutions, together with support funding from Pfizer, Novartis, Syngenta and Boehringer Ingelheim to whom we are most grateful.

REFERENCES

[1] DeCamp, W. H., *Am. Pharm. Rev.* 2001, *4*, 70–77.

[2] Byrn, S. R.; Pfeiffer, R.; Stowell, J. G., *Am. Pharm. Rev.* 2002, *5*, 92–99.

[3] Chemburkar, S. R.; Bauer, J.; Deming, K.; Spiwek, H.; Patel, L.; Morris, J.; Henry, R.; Spanton, S.; Dziki, W.; Poter, W.; Quick, J.; Bauer, P.; Donaubauer, J.; Narayanan, B. A.; Soldani, M.; Riley, D.; McFarland, K., *Org. Process Res. Dev.* 2000, *4*, 413–417.

[4] Storey, R.; Docherty, R.; Higginson, P. D., *Am. Pharm. Rev.* 2003, *6*, 100–105.

[5] Ticehurst, M.; Docherty, R., *Am. Pharm. Rev.* 2006, *9*, 32–36.

[6] Thayer, A., *Chem. Eng. News* 2007, *85*, 11–29.

[7] Desiraju, G. R., *The Crystal as a Supramolecular Entity*. John Wiley & Sons, Ltd.: Chichester, 1997.

[8] Sun, C. C., *J. Pharm. Sci.* 2009, *98*, 1671–1687.

[9] Chow, K.; Tong, H. Y. H.; Lum, S.; Chow, A. H. L., *J. Pharm. Sci.* 2008, *97*, 2855–2877.

[10] Federal Register. 2005; Vol. *70*, p 134.

[11] Roberts, K. J.; Docherty, R.; Marshall, A. Cross-Industry Synthonic Engineering Workshop, Institute of Physics, London, 2012.

[12] Shekunov, B. Y.; Chattopadhyay, P.; Tong, H. Y. T.; Chow, A. H. L., *Pharm. Res.* 2007, *24*, 203–227.

[13] Phillips, F. C., *An Introduction to Crystallography*. Longmans: London, 1963.

[14] Bladon, P.; Gorton, J.; Hammond, R. B., *Molecular Modelling: Computational Chemistry Demystified*. RSC Publishing: Cambridge, UK, 2011.

[15] Fronczek, F. R.; Kamel, H. N.; Slattery, M., *Acta Crystallogr. Sect. C: Cryst. Struct. Commun.* 2003, *59*, o567–o570.

[16] Hauy, R. J., *Essai d'ue theorie sur la structure des cristaux*, 1784.

[17] Hauy, R. J., *Traite de Mineralogie*. Paris, 1801.

[18] Wright, J. G., *Molecular Crystals*. Cambridge University Press: Cambridge, 1987.

[19] Etter, M. C., *Acc. Chem. Res.* 1990, *23*, 120–126.

[20] Taylor, R.; Kennard, O., *Acc. Chem. Res.* 1984, *17*, 320–326.

[21] Cambridge Structural Database System (CCDC). Mercury CSD 3.1.1 In *Cambridge Structural Database System 2013 Release, v 5.34*. CCDC: Cambridge, UK, 2013.

[22] Taylor, R.; Kennard, O., *J. Chem. Soc.* 1982, *104*, 5063–5070.

[23] Docherty, R.; Jones, W. Theoretical Methods for Crystal Structure Determination. In *Organic Molecular Solids: Properties and Applications*; Jones, W., Ed.; CRC Press: London, 1997; pp. 113–148.

[24] Williams, D. E., *J. Chem. Phys.* 1966, *4545*, 3770–3778.

[25] Gavezzotti, A.; Filippini, G., *J. Phys. Chem.* 1994, *98*, 4831–4837.

[26] Brock, C. P.; Dunitz, J. D., *Chem. Mater.* 1994, *6*, 1118–1117.

[27] Kitaigorodscky, A. I., *Molecular Crystals and Molecules*. Academic Press: New York, 1973.

[28] Lifson, S.; Hagler, A. T.; Dauber, P., *J. Am. Chem. Soc. 101*, 5111–5121.

[29] Momany, F. A.; Carruthers, L. M.; McGuire, R. F.; Scheraga, H. A., *J. Phys. Chem.* 1974, *78*, 1595–1620.

[30] Docherty, R.; Roberts, K. J.; Saunders, V.; Black, S. N.; Davey, R. J., *Faraday Discuss.* 1993, *95*, 11–25.

[31] Clydesdale, G.; Docherty, R.; Roberts, K. J., Computational Studies of the Morphology of Molecular Crystals Through Solid State Intermolecular Force Calculations Using the Atom-Atom Method. In *Colloid and Surface Engineering: Controlled Particle, Droplet and Bubble Formation*; Wedlock, D. J., Ed.; Butterworth Heinman: London, 1994; pp. 95–135.

[32] Clydesdale, G.; Roberts, K. J.; Docherty, R., *J. Cryst. Growth* 1996, *166*, 78–83.

[33] Bailey, M.; Brown, C. J., *Acta Crystallogr.* 1967, *22*, 387–391.

[34] Athimoolam, S.; Natarajan, S., *Acta Crystallogr. Sect. C: Cryst. Struct. Commun.* 2007, *63*, o514–o517.

[35] Gracin, S.; Fischer, A., *Acta Crystallogr., Sect. E: Struct. Rep. Online* 2005, *61*, o1242–o1244.

[36] Boone, C. D. G.; Derissen, J. L.; Schoone, J. C., *Acta Crystallogr., Sect. B Struct. Sci.* 1977, *33*, 3205–3206.

[37] Rosbottom, I.; Roberts, K. J.; Docherty, R., *CrystEngComm* 2015, *17*, 5768–5788.

[38] Accelrys. *Materials Studio, 6.0*. Accelrys Software Inc.: San Diego, 2015.

[39] Mayo, S. L.; Olafson, B. D.; Goddard, W. A., *J. Phys. Chem.* 1990, *94*, 8897–8909.

[40] Ferenczy, G. G.; Reynolds, C. A.; Richards, W. G., *J. Comp. Chem.* 1990, *11*, 159–169.

[41] Vanquelef, E.; Simon, S.; Marquant, G.; Garcia, E.; Klimerak, G.; Delepine, J. C.; Ciepak, P.; Dupradeau, F.-Y., *Nucleic Acids Res.* 23 May 2011, *39*, 1–7.

[42] Anisimov, V. M.; Lamoureux, G.; Vorobyov, I. V.; Huang, N.; Roux, B.; MacKerell, A. D. Jr., *J. Chem. Theory Comput.* 2005, *1*, 153–168.

[43] Davey, R. J.; Garside, J., *From Molecules to Crystals*. Oxford University Press: Oxford, 1998.

[44] Roberts, K. J.; Docherty, R.; Taylor, L. S., Materials Science: Solid Form Design and Crystallisation Process Development. In *Pharmaceutical Process Development: Current Chemical and Engineering Challenges*; Blacker, J., Williams, M. T., Eds.; Royal Society of Chemistry: Cambridge, UK, 2011; pp. 286–316.

[45] Nyvlt, J., *J. Cryst. Growth* 1968, *3–4*, 377–383.

[46] Mullin, J. W., *Crystallization*. Butterworth-Heinmann: Oxford, 2001.

[47] Myerson, A. S., *Handbook of Industrial Crystallisation*. Butterworth-Heinmann: Woburn, 2002.

[48] Roberts, K. J.; Walker, E. M.; Clydesdale, G., The Crystal Habit of Molecular Materials: A Structural Perspective. In *Theoretical Aspects and Computer Modelling of the Molecular Solid State*. Gavezzotti, A., Ed.; John Wiley & Sons, Inc.: New York, 1997; pp. 203–232.

[49] Roberts, K. J.; Hendriksen, B. A., *J. Cryst. Growth* 1993, *128*, 1218–1224.

[50] Burton, W. K.; Cabrera, N.; Frank, F. C., *Phil. Trans. R. Soc. Lond. A* 1951, *243*, 299–358.

[51] Freidel, G., *Bull. Soc. Fr. Mineral.* 1907, *30*, 326–455.

[52] Donnay, J. D. H.; Harker, D., *Am. Minerol.* 1937, *11*, 467–468.

[53] Hartman, P.; Perdok, W. G., *Acta Crystallogr.* 1955, *8*, 49–52.

[54] Berkovitch-Yellin, Z., *J. Am. Chem. Soc.* 1985, *107*, 8239–8254.

[55] Docherty, R.; Clydesdale, G.; Roberts, K. J.; Bennema, P., *J. Phys. D: Appl. Phys.* 1991, *24*, 89–99.

[56] Hartman, P.; Bennema, P., *J. Cryst. Growth* 1980, *49*, 145–156.

[57] Roberts, K. J.; Sherwood, J. N.; Yoon, C. S.; Docherty, R., *Chem. Mater.* 1994, *6*, 1099–1101.

[58] Hammond, R. B.; Pencheva, K.; Ramachandran, V.; Roberts, K. J., *Crys. Growth Des.* 2007, *7*, 1571–1574.

[59] Black, S. N.; Williams, L. J.; Davey, R. J.; Moffat, F.; McEwan, D. M.; Sadler, D. E.; Docherty, R.; Williams, D. J., *J. Phys. Chem.* 1990, *94*, 3223–3226.

[60] Mackrodt, W. C.; Davey, R. J.; Black, S. N.; Docherty, R., *J. Cryst. Growth* 1987, *80*, 441–446.

[61] Docherty, R.; Roberts, K. J., *J. Cryst. Growth* 1988, *88*, 159–167.

[62] Ramachandran, V., Synthonic Engineering Tool – Prediction of Dissolution – Case Study Report on Celecoxib. Unpublished work, University of Leeds: Leeds, UK, 2012.

[63] Ramachandran, V.; Murnane, D.; Hammond, R. B.; Pickering, J.; Roberts, K. J.; Soufian, M.; Forbes, B.; Jaffari, S.; Martin, G.; Collins, E.; Pencheva, K., *Mol. Pharm.* 2015, *12*, 18–33.

[64] Ostwald, W., *Z. Phys. Chem.* 1897, *22*, 289–330.

[65] Keller, A.; Goldbeck-Wood, G.; Hokosaka, M., *Faraday Discuss.* 1993, *95*, 109–128.

[66] Hammond, R. B.; Pencheva, K.; Roberts, K. J., *Cryst. Growth Des.* 2007, *7*, 875–884.

[67] Lehmann, M. S.; Nunes, A. C., *Acta Crystallogr., Sect B: Struct. Sci.* 1980, *36*, 1621–1625.

[68] Lehmann, M. S.; Koetzle, T. F.; Hamilton, W. C., *J. Cryst. Mol. Struct.* 1972, *2*, 225–233.

[69] Ferrari, E. S.; Davey, R. J., *Cryst. Growth Des.* 2004, *4*, 1061–1068.

[70] Hammond, R. B.; Pencheva, K.; Roberts, K. J., *J. Phys. Chem. B* 2005, *109*, 19550–19552.

[71] Hammond, R. B.; Pencheva, K.; Roberts, K. J., *CrystEngComm* 2012, *14*, 1069–1082.

[72] Weissbuh, I.; Leiserowitz, L.; Lahav, M. "Tailor-Made Additives" and Impurities. In *Crytallization Technolgy Handbook*; Mersmann, A., Ed.; Marcel Dekker: New York, 1995; pp. 563–612.

[73] Lahav, M.; Leiserowitz, L., *Chem. Eng. Sci.* 2001, *56*, 2245–2253.

[74] Winn, D.; Doherty, M. F., *AIChE J.* 2000, *46*, 1348–1367.

[75] Reilly, A. M.; Briesen, H., *J. Cryst. Growth* 2012, *354*, 34–43.

[76] Marsh, R. E., *Acta Crystallogr.* 1958, *11*, 654–663.

[77] Hammond, R. B.; Ramachandran, V.; Roberts, K. J., *CrystEngComm* 2011, *13*, 4935–4944.

[78] Hammond, R. B.; Pencheva, K.; Roberts, K. J., *Cryst. Growth Des.* 2006, *6*, 1324–1334.

[79] Olusanmi, D.; Roberts, K. J.; Ghadiri, M.; Ding, Y., *Int. J. Pharm.* 2011, *411*, 49–63.

[80] Hammond, R. B.; Jeck, S.; Ma, C. Y.; Pencheva, K.; Roberts, K. J.; Auffret, T., *J. Pharm. Sci.* 2009, *98*, 4589–4602.

[81] Ramachandran, V.; Galet, L.; Hammond, R. B.; Roberts, K. J.; Pickering, J.; Jones, M.; Pencheva, K.; Docherty, R., Molecular Pharmaceutics, (to be submitted), 2016.

[82] Jaffari, S.; Forbes, B.; Collins, E.; Barlow, D. J.; Martin, G. P.; Murnane, D., *Int. J. Pharm.* 2013, *447*, 124–131.

[83] Noyes, A.; Whitney, W. R., Sol. Solid Subs. *J. Am. Chem. Soc.* 1897, *19*, 930–934.

[84] Vasudev, R.; Shashi Rekha, K.; Vyas, K.; Mohanti, S. B.; Kumar, R. P.; Reddy, G. O., *Acta Crystallogr. Sect. C: Cryst. Struct. Commun.* 1999, *55*, 9900161–9900169.

[85] Modi, S. R.; Dantuluri, A. K. R.; Perumalla, S. R.; Sun, C. C.; Bansal, A. K., *Cryst. Growth Des.* 2014, *14*, 5283–5292.

[86] Yalkowsky, S. H.; Bolton, S., *Pharm. Res.* 1990, *7*, 962–966.

[87] Narayan, P.; Hancock, B. C., *Mater. Sci. Eng. A* 2003, *A355*, 24–36.

[88] Sun, C. C.; Hou, H.; Gao, P.; Ma, C.; Medina, C.; Alvarez, J., *Pharm. Sci.* 2009, *98*, 239–247.

[89] Amidon, G. L.; Lennernas, H.; Shah, V. P.; Crison, J. R. A., *Pharm. Res.* 1995, *12*, 413–420.

[90] Johnson, K.; Swindell, A. C., *Pharm. Res.* 1996, *13*, 1795–1798.

[91] Reddy, C. M.; Krishna, G. R.; Ghosh, S., *CrystEngComm* 2010, *12*, 2296–2314.

[92] Butler, J. M.; Dressman, J. B., *J. Pharm. Sci.* 2010, *99*, 4940–4954.

[93] Leane, M.; Pitt, K.; Reynolds, G., *Pharm. Dev. Technol.* 2014, *20*, 12–21.

[94] Docherty, R.; Kougoulos, R.; Horspool, K., *Am. Pharm. Rev.* 2009, *12*, 39–46.

9

NEW DEVELOPMENTS IN PREDICTION OF SOLID-STATE SOLUBILITY AND COCRYSTALLIZATION USING COSMO-RS THEORY

CHRISTOPH LOSCHEN[1] AND ANDREAS KLAMT[1,2]

[1] *COSMOlogic GmbH & Co. KG, Leverkusen, Germany*

[2] *Faculty of Chemistry and Pharmacy, Institute of Physical and Theoretical Chemistry, University of Regensburg, Regensburg, Germany*

9.1 INTRODUCTION

One of the major problems of modern drug development consists of the increasing prevalence of poorly water-soluble drugs, which is most often accompanied by a low bioavailability [1]. Computational predictions can help address this issue either by solubility estimation of new compounds at the early stage of drug design or at a later stage by supporting the drug development process, for instance by screening of suitable solvents. Among the more commonly used solubility prediction schemes are the nonrandom two-liquid segment activity coefficient (NRTL-SAC) method [2], the Hansen solubility parameter approach [3], the perturbed-chain statistical associating fluid theory (PC-SAFT) equation of state [4], and the UNIQUAC functional-group activity coefficient (UNIFAC) method [5]. All those conventional methods have in common that they are strongly parameterized and rely on a set of molecule or functional group–specific parameters, which are usually determined by fitting to experiments.

Computational Pharmaceutical Solid State Chemistry, First Edition. Edited by Yuriy A. Abramov.

They usually perform well in their core region of parameterization, but extrapolation should always be done with care. This poses a problem if nonstandard or unusual chemistry is to be explored where those experiments are not yet available, which may well be the case for a newly synthesized drug. In contrast, the conductor-like screening model for realistic solvation (COSMO-RS) is based on first-principles calculations. In order to predict the property of a new compound, it is sufficient to carry out a density functional theory (DFT) computation, followed by an efficient procedure routed in statistical thermodynamics to take into account intermolecular interactions. Only the molecular structure is necessary as an input and the first, sometimes costly, quantum chemical part can be abbreviated using databases containing precomputed results. Although COSMO-RS also contains a non-negligible number of parameters, they are element specific, that is, no specific interactions are parameterized for each molecule and thus there is no need for any reparameterization for new drugs. Due to this unique predictive capability COSMO-RS theory and its software implementation, COSMO*therm* [6] are nowadays used by many groups in chemical and pharmaceutical research.

In the following paragraphs, some application examples will be presented, starting with a short introduction to COSMO-RS (Section 9.2), followed by solubility predictions in pure and mixed solvents (Section 9.3). A modification using several reference solubilities is shown in Section 9.4 whereas Section 9.5 is about quantitative structure–property relationship (QSPR) models of the melting point and the enthalpy of fusion. The final Sections 9.6 and 9.7 deal with COSMO-RS-based coformer selection for cocrystal screening and the related issue of solvent selection to avoid solvate formation.

9.2 COSMO-RS

COSMO-RS is a fluid phase thermodynamics method for property prediction of the liquid phase that was first introduced in 1995 by A. Klamt [7]. A concise review of the methodology was published in 2011 [8], and a detailed introduction can be found in Ref. [9], thus only a short summary of the most important aspects of the approach will be given here.

COSMO-RS was originally created as an improvement to the conductor-like screening model COSMO [10]. COSMO is a quantum chemical solvation model that takes into account solvent effects implicitly and thus belongs to the group of dielectric continuum solvation models (DCMs). Within COSMO, the solute is placed inside a conductor (dielectric constant $\varepsilon = \infty$), which results in a perfect screening of the electric field of the molecule. The polarization of the conductor and the subsequent response of the molecular electrostatic field have to be computed iteratively by the quantum chemical method until self-consistency is reached. For finite dielectrics the solute potential in the conductor is corrected by a simple dielectric screening function $f(\varepsilon)$. This simplification gives a significant improvement in efficiency and numerical stability as compared to other DCMs based on a finite dielectric constant. Though COSMO is nowadays one of the most often applied solvation models, Klamt pointed out the shortcomings of the underlying DCM approach [8, 9]: the continuum description of a DCM will be inadequate in particular for polar solvents, that is, for solvents that possess a non-negligible dipole moment. In such solvents the largest part of the dielectric response originates from the dipole reorientation of the solvent molecules (for water with $\varepsilon = 78.4$ the contribution of the dipole reorientation amounts to 99% of the polarization) and implicit solvation cannot give the full picture.

This finding leads to the concept of COSMO-RS, which combines the DCM approach with the statistical thermodynamic treatment of interacting surface segments. The interactions of the segments are derived from the screening charge density of the DCM calculation, whereas the 3-dimensional information is projected into a histogram of equally binned charge densities—the so-called σ-profile. A σ-profile represents the amount of surface area $p(\sigma)$ as a function of the polarization charge density σ. It is now possible to formulate intermolecular interactions based just on those surface segments, which introduces the novel concept of an ensemble of interacting surface segments in solution. Compared to the usual approach of an ensemble of molecules like in molecular dynamics, this leads to a dramatic increase in efficiency for the evaluation of the thermodynamic statistics. COSMO-RS intermolecular interactions take into account van der Waals interactions, electrostatic interactions, and hydrogen bonding (HB). At the same time the solute or solvent embedded in the virtual conductor corresponds to a well-defined energetic reference state.

The van der Waals energy for a solute X is given as a surface proportional term with element-specific coefficients $\tau_{\mathrm{el}(\alpha)}$:

$$E_{\mathrm{VdW}}^{\mathrm{X}} \cong \sum_{\alpha} \tau_{\mathrm{el}(\alpha)} A_{\alpha}^{\mathrm{X}} \tag{9.1}$$

The summation is over all atoms α, and A_{α}^{X} is the COSMO surface area on atom α.

The so-called electrostatic misfit term arises from the mismatch of the surface segments of two neighboring molecules if intermolecular interactions are introduced by leaving the grounded conductor reference state. The misfit term equals zero for two interacting surface segments of the same size and opposite charge but leads to an electrostatic energy penalty for unlike segment pairs proportional to the square of the net charge $\sigma + \sigma'$:

$$\Delta E_{\mathrm{misfit}} \cong a_{\mathrm{contact}} c_{\mathrm{misfit}} \left(\sigma + \sigma'\right)^2 \tag{9.2}$$

c_{misfit} is a coefficient, which can be derived from basic electrostatics, and a_{contact} is the surface segment contact area. Finally, the intermolecular HB interaction is obtained via the following functional form:

$$\Delta E_{\mathrm{HB}} \cong a_{\mathrm{contact}} c_{\mathrm{HB}} \min\left(0, \sigma\sigma' - \sigma_{\mathrm{HB}}^2\right) \tag{9.3}$$

Due to the threshold σ_{HB} only contributions from sufficiently polar surface segment pairs contribute to the HB term. The term $\sigma\sigma'$ ensures that the hydrogen bond term increases with increasing polarity of the donor and the acceptor. Recently it could be demonstrated that this product of polarization charge densities indeed reflects the HB energy of experimentally determined hydrogen bond strengths over a wide range of acceptor molecules to within an accuracy of 2 kJ/mol [11].

With the functional form of intermolecular interactions at hand, they have now to be evaluated for a liquid ensemble of interacting segments. A detailed derivation of the necessary statistical thermodynamics and a proof of thermodynamic consistency have been given by Klamt et al. [12]. This yields the chemical potential as a function of the polarization charge density $\mu_S(\sigma)$ and one finally obtains the chemical potential of solute X in solvent S μ_S^X by integrating the solvent chemical potential $\mu_S(\sigma)$ over the binned solute surface weighted by the σ-profile:

$$\mu_S^X = \int d\sigma p^X(\sigma)\mu_S(\sigma) + RT\ln\{x\gamma_{\text{comb.,S}}^X\} \tag{9.4}$$

The logarithmic term contains the ideal entropic contribution depending on the molar fraction x of the solute X and a combinatorial contribution to the activity coefficient due to the relative sizes of solute and solvent. This approach is extended trivially from pure solvents to mixtures because the σ-profile of the mixture can be additively composed out of the mixture ingredients:

$$p_S(\sigma) = \sum_i \frac{x_i p_S^{X_i}(\sigma)}{x_i A^{X_i}} \tag{9.5}$$

By means of the previous two equations COSMO-RS gives access to the chemical potential of any solute in arbitrary pure solvents or solvent mixtures as a function of temperature and concentration. This opens the field to nearly the entire fluid phase equilibrium thermodynamics, and allows for the computation of any kind of liquid phase properties, that is, all kinds of activity coefficients, partition coefficients, solubilities, excess quantities, solvation energies, and many more. If quantum chemical gas phase energies are incorporated, this also grants access to vapor–liquid equilibria like vapor pressures and Henry's law constants. Accordingly, in addition to solubility prediction [13] and solvent screening [14], COSMO-RS theory has been applied successfully in numerous fields such as for pK_a predictions [15], ionic liquids [16], partitioning coefficients [17], cocrystal or solvate formation [18], and excipient ranking [19] to name just some important applications. In addition, a σ-profile-based algorithm for quantifying ligand–receptor interactions in proteins has been introduced [20].

The standard COSMO-RS workflow starts with a conformer search at a low-level method like AM1 or a molecular mechanics force-field. High-energy conformers are then subsequently filtered out by going to higher-level density functional methods. Clustering according to the chemical potential in solvents with different polarity ensures that no relevant conformers are left out. The resulting σ-profiles of the low-energy conformers are stored, for example, in a database and can then be used as an input to the COSMO*therm* program. Next COSMO*therm* carries out the statistical thermodynamics computation including Boltzmann weighting of the conformers to quantify the intermolecular interactions. Most of the COSMO*therm* functionality is accessible via the graphical user-interface COSMO*therm* X.

At a somewhat reduced accuracy it is also possible to circumvent the sometimes costly quantum chemical calculations and to generate σ-profiles on-the-fly from fragments of precomputed COSMO files stored in a database. This approach is implemented in the software COSMO*quick* and is particularly useful for solubility prediction using one or several reference solvents (see also Section 9.4).

If not stated otherwise the following results have been generated with recent COSMO*therm* [6] and COSMO*quick* [21] releases and their respective COSMO-RS parameterizations, BP_TZVP_C30_1301.ctd (TZVP level of theory) and BP_SVP_AM1_C30_1301.ctd (SVP level of theory). QSPR calculations have been done using the *R* statistics package [22].

9.3 PREDICTION OF DRUG SOLUBILITY USING COSMO-RS

Because the solubility prediction capabilities of COSMO-RS have been reviewed before [14, 23], here we summarize just shortly the theoretical foundations and then focus on some recent results.

The mole fraction solubility of the compound X in solvent S is readily obtained from the condition of equal chemical potentials at the solid–liquid equilibrium:

$$\log_{10}\left(x_S\right) = \frac{\mu_X^X - \mu_S^X - \Delta G_{fus}^X}{RT\ln(10)} \tag{9.6}$$

The Gibbs free energy of fusion, ΔG_{fus}, is the free energy that has to be expended to bring the solid into the supercooled liquid state at a given temperature and thus by convention $\Delta G_{fus} \geq 0$. This quantity can be expressed by the melting enthalpy ΔH_{fus}, the melting point T_m, and the heat capacity change at the solid–liquid phase transition ΔCp_{fus}:

$$\Delta G_{fus} = \Delta H_{fus} - T\Delta S_{fus} - \Delta Cp_{fus}\left(T_m - T\right) + T\Delta Cp_{fus}\ln\left(\frac{T_m}{T}\right) \tag{9.7}$$

Often the approximation $\Delta Cp_{fus} = 0$ is made and usually gives decent results as long as the temperature range under consideration ($T_m - T$) is not too large. From a computational point of view, ΔH_{fus} and the melting point T_m require at least knowledge about the crystal structure of the solute and are difficult to predict. A typical procedure for solubility prediction is to compute the liquid chemical potentials only and to derive the missing free energy of fusion from experimentally determined melting point and fusion enthalpy. Figure 9.1 shows the chemical potential of paracetamol in CCl_4 at different concentrations (in mole fractions) at 298 K as computed by COSMO*therm*. The two intersections correspond to the equilibrium state, where the chemical potential of the pure liquid equals the potential of the solvated species and the equilibrium state, where the chemical potential of the solid equals the potential of the solvated species. As paracetamol is solid at room temperature ($\Delta G_{fus} > 0$), the state belonging to the first intersection is not thermodynamically stable but corresponds to the supercooled drug in equilibrium with its solution.

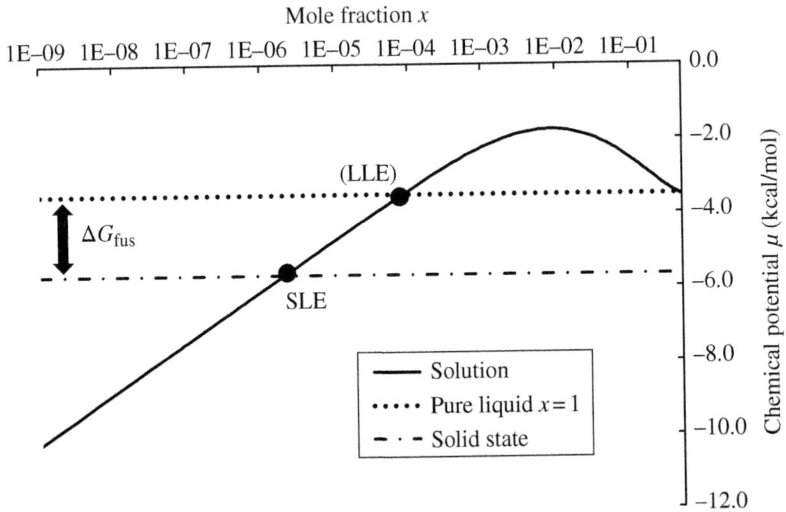

FIGURE 9.1 The chemical potential of paracetamol in CCl_4 at varying concentrations (in mole fractions x) at 298 K. The intersection with the solid-state chemical potential defines the solid–liquid equilibrium (SLE) and corresponds to the solubility of the drug in this particular solvent. The difference between the pure liquid and the solid-state chemical potential is the free energy of fusion ΔG_{fus}. Also shown is the liquid–liquid equilibrium (LLE) where the virtually supercooled liquid is at equilibrium with the dissolved drug.

For higher solubilities (e.g., concentration >10 volume%), Equation 9.6 should be solved iteratively, because the dependence of the chemical potential of the solute in the solvent, μ_S^X, on the mole fraction x cannot be neglected anymore. Equation 9.6 is solved iteratively by default within COSMO*therm*. If the ΔG_{fus} term is not known or cannot be derived from an experiment, COSMO*therm* provides a QSPR estimate, however leading to an increase in the overall prediction error.

In Figure 9.2 COSMO*therm* predictions for the solubility of some drugs in different solvents are presented. Experimental data for the hypolipidemic agent lovastatin [24–27], the nonsteroidal anti-inflammatory drug meloxicam [28], the antimalarial artemisinin [29], and the sulfonamide antibiotic sulfamethoxypyridazine [30] have been taken from the literature. ΔG_{fus} values for the drugs have been derived from the melting point and the enthalpy of fusion where available, or computed by using Equation 9.6 with ethanol as a reference solvent. All of the drugs have a very low solubility in water and except for sulfamethoxypyridazine the aqueous solubility is even the lowest among all solvents, which is another manifestation of the low bioavailability of current drugs.

Table 9.1 compares the overall deviation of the predictions from the experiment. The overall prediction accuracy for the five drugs amounts to an RMSE=0.49 (root mean squared error of the decimal logarithm of the mole fraction solubility), which is about the typical error bar for COSMO-RS solubility predictions. Due to our experience this is already close to the usually achieved experimental accuracy.

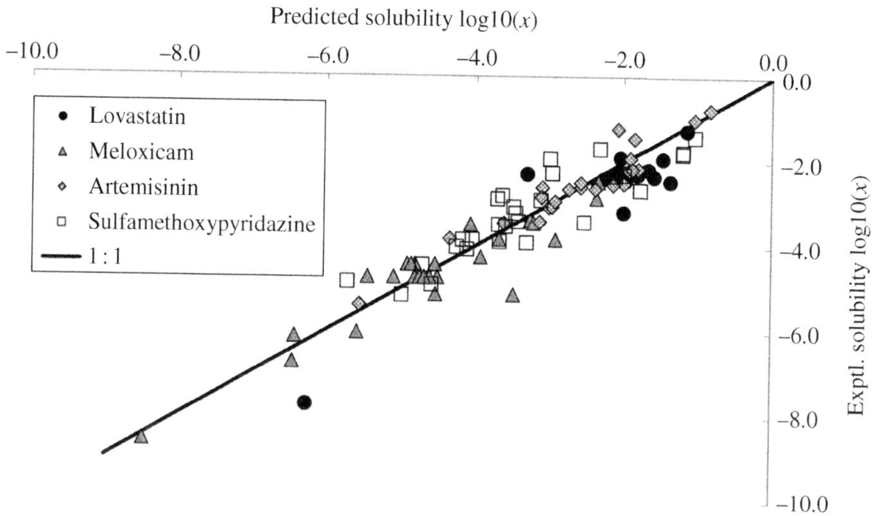

FIGURE 9.2 COSMO-RS predicted versus experimental solubilities for several drugs in logarithmic units (mole fraction based).

TABLE 9.1 Summary of COSMO-RS Solubility Predictions of Some Drugs, Computed at the TZVP Level of Theory

Drug	RSME	Experimental Data Used	References
Lovastatin	0.62	$T_m = 446$ K, $\Delta H_{fus} = 10.3$ kcal/mol	[24–27]
Meloxicam	0.49	$T_m = 530$ K, $\Delta H_{fus} = 17.1$ kcal/mol	[28]
Artemisinin	0.32	$\Delta G_{fus} = 2.7$ kcal/mol	[29]
Sulfamethoxypyridazine	0.53	log10(x),ethanol = −3.42	[30]
Mean (stdev)	0.49 (0.11)		

RMSE, root mean squared error; stdev, standard deviation.

Areas where COSMO-RS-based predictions are particularly useful are for instance the low solubility regime where it is difficult to measure accurately or for hazardous substances where a measurement is time consuming and costly.

Furthermore, COSMO-RS is known to perform well for predictions of activity coefficients and solubilities in solvent mixtures [14, 23] and in most cases reproduces solubility maxima correctly. A challenging problem case, however, the solubility of paracetamol in a water–dioxane mixture, is shown in Figure 9.3. Experimental data for the system paracetamol/water–dioxane have been taken from three different sources [31–33] and compared with COSMO*therm* computations at the TZVP and the newer FINE parameterization [6].

There is a tremendous difference between the different experimental measurements, which amounts to more than 0.5 log units at intermediate mole fractions. Such large noise is not unusual but rather a typical problem that one has to deal with when experimental data is being reproduced. In spite of this scatter all of the experiments

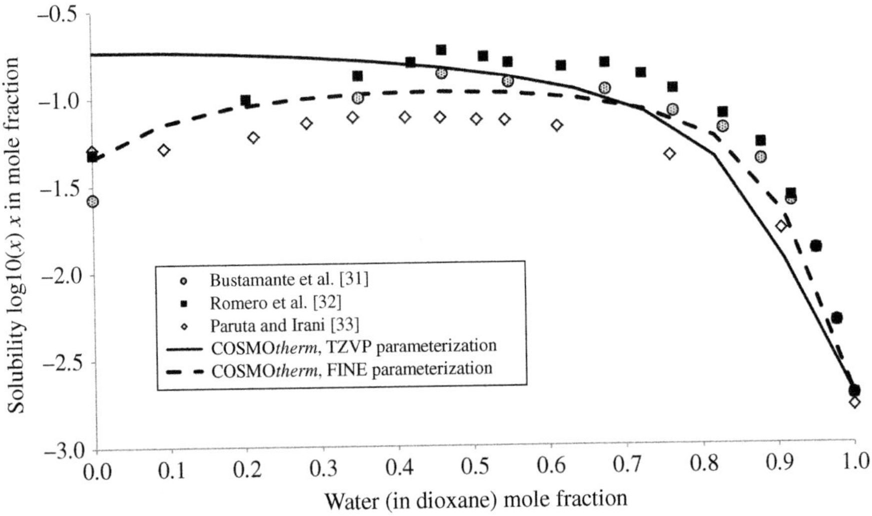

FIGURE 9.3 Predicted and experimental logarithmic mole fraction solubility of paracetamol in a water–dioxane mixture. Three different experiments have been taken from the literature. Solubilities are predicted with COSMO*therm*. The FINE and the standard COSMO*therm* TZVP level are shown.

exhibit a solubility maximum. Though the standard level (TZVP) gives decent results for the absolute solubility, it does not reproduce this solubility maximum qualitatively in this particular case. The COSMO-RS FINE level is an enhancement that uses a finer grid for the COSMO cavity construction and an extended basis set (TZVPD) with an additional diffuse basis function for the DFT calculation. The FINE parameterization was introduced in 2012 into COSMO*therm* and is still being improved continuously. Using the same reference solvent as the standard TZVP the new approach is able to reproduce the maximum in the solubility curve nicely.

In general, quantitative solubility prediction for solids relying on liquid phase thermodynamics is hampered by the fact that in addition to the activity coefficients also accurate solid-state information like melting point and fusion data have to be available. In the preceding part this issue could be mitigated by the use of an experimental solubility as a point of reference. In the following sections two additional workarounds addressing this problem are presented. First, the use of several, instead of only one, experimental reference points is shown and second, a QSPR for melting point and enthalpy of fusion is introduced that can be used to determine the enthalpy of fusion.

9.4 SOLUBILITY PREDICTION WITH MULTIPLE REFERENCE SOLVENTS

Solubility predictions with COSMO*therm* can be applied without the need for reparameterization for most chemical systems. Once the quantum chemical computations to generate the σ-surface are done, they are also quite fast and usually finished within a fraction of a second. Nevertheless, highly parameterized methods like NRTL-SAC and

PC-SAFT sometimes perform better for interpolation, as long as one stays within the core region of their parameterization. A compromise between those two different procedures represents the solubility prediction scheme implemented recently in the software package COSMO*quick* ("COSMO*quick* approach") [34]. The idea behind the COSMO*quick* approach is to correct the chemical potential in solution μ_S^X (see Eq. 9.6) by using several reference solubilities in different solvents. Formally, different references should yield the same ΔG_{fus}. In practice, of course, this is not the case due to the approximations that come along with the theoretical model. The deviations from the mean $\langle \Delta G_{fus} \rangle$ are used to correct the chemical potential in solution based on the similarity of a specific solvent with the reference solvents. Usually solubilities from three different solvents are sufficient for accurate predictions. In addition, the COSMO*quick* approach does not need any costly quantum-chemical computation. Rather it generates σ-profiles on-the-fly from a simplified molecular input (e.g., SMILES notation) by carrying out a quick analysis of an extensive database containing precomputed COSMO files. Of course, that leads to somewhat less accurate σ-profiles and thus to a reduced accuracy concerning property prediction. However, using multiple references one is able to compensate the partially reduced σ-profile quality to some extent. The COSMO*quick* approach takes into account solubility in a noniterative fashion making it somewhat less reliable for solutes with high solubilities, in practice however this is not of relevance for most drugs, which typically have low solubility.

Table 9.2 shows some results of the COSMO*quick* approach in comparison with the popular solubility estimation tool NRTL-SAC. NRTL-SAC is a methodology that

TABLE 9.2 Comparison of NRTL-SAC and COSMO*quick* Solubility Prediction Results on Different Solutes

Solute	Method	No. Reference Solvents	Total No. of Solvents	RMSE[a]
Sulfadiazine	NRTL-SAC	7	19	1.24
	COSMO*quick*	3	19	0.42
Paracetamol	NRTL-SAC	8	23	0.60
	COSMO*quick*	4	23	0.48
Cimetidine	NRTL-SAC	6	11	0.82
	COSMO*quick*	4	11	0.49
Fluorenone	NRTL-SAC	5	21	0.39
	COSMO*quick*	3	21	0.34
Xanthene	NRTL-SAC	5	19	0.53
	COSMO*quick*	3	19	0.27
Monuron	NRTL-SAC	5	24	0.25
	COSMO*quick*	3	24	0.25
Cinchonidine	NRTL-SAC	5	23	0.94
	COSMO*quick*	3	23	0.87
Saccharin	NRTL-SAC	5	9	0.95
	COSMO*quick*	3	9	0.42
<mean>	NRTL-SAC	5.8	18.6	0.72
	COSMO*quick*	3.3	18.6	0.44

[a]Root mean squared error (RMSE) of logarithmic solubility log10(x), x in mole fractions. Table partially reproduced from Ref. [34].

was introduced by Chen and coworkers in 2004 [2], which uses segment-specific activity coefficients, whereas the segment contributions have to be determined by a fit to the experiment for each drug or solvent.

The drugs sulfadiazine, paracetamol, and cimetidine and their respective solvents in Table 9.2 have been chosen according to the original NRTL-SAC publication [35], the other solute/solvent datasets have been taken from the available literature. NRTL-SAC results have been obtained with our own implementation of the method. Additional details on the parameters used and original literature references can be found in Ref. [34]. From Table 9.2 it becomes obvious that although a smaller number of reference solvents are used, the COSMO*quick* results are significantly more accurate; on the set of eight molecules the average score for COSMO*quick* amounts to RMSE = 0.44, where NRTL-SAC gives a RMSE = 0.72.

Moreover, in contrast to NRTL-SAC and most other highly parameterized prediction models, relative solubilities for new drugs can be computed even if no experimental data is available. If just one reference is used, the approach reduces to the standard COSMO*therm* approach for solubility prediction.

This is shown in Table 9.3, which illustrates the difference between relative solubility predictions and predictions using one or more references.

Relative solubility prediction, that is, setting ΔG_{fus} to an arbitrary fixed value in Equation 9.6 will give already a good trend, as seen in the squared correlation coefficient R^2. Including one experimental reference allows now for quantitative predictions maintaining the same trend as it only shifts the results. Optimal results concerning R^2 and RMSE are usually obtained using three to four reference solvents, going further gives only minor improvements. The reference solvents should be chosen to reflect different solvent characteristics. Ideally they contain a nonpolar solvent like hexane, a polar-aprotic solvent as acetone, and a solvent like ethanol with mixed donor and acceptor features.

There exist a few restrictions for the instant generation of σ-profiles as used by the COSMO*quick* approach: currently no conformational effects can be taken into account and the support of stereochemistry is only limited. The underlying COSMO database contains more than 100,000 compounds containing rare functional groups and ionic species and covers a very broad but of course limited chemical space. Thus, for very rare functional groups sometimes the generation of σ-profiles may fail, which can be alleviated, however, by adding a suitable molecule to the database. Again, no refitting is necessary in such a case.

TABLE 9.3 Comparison of COSMO*quick* Solubility Predictions for the Paracetamol Dataset [36] Using Runs without Any Reference (Relative), with One to Six Experimental Solubilities

n, references	R^2	RMSE
0 (relative)	0.83	—
1	0.83	0.61
4	0.86	0.48
6	0.87	0.45

In summary, this workflow constitutes an efficient and probably one of the most accurate ways to predict solubilities. Moreover, it is applicable also for nonexperts in the field of computational chemistry.

9.5 MELTING POINT AND FUSION ENTHALPY QSPR MODELS

COSMO-RS itself is a liquid phase theory but, according to Equation 9.6, for a quantitative solubility prediction some sort of information about the solid state is indispensable. In many cases the problem of a missing free energy of fusion or equivalently of unknown melting point and fusion enthalpy can be circumvented by referring to an experimental solubility, as seen in the preceding section. But sometimes even this information may not be available. The full computational prediction of fusion data requires at least a crystal structure prediction of the solute, which is not yet practically feasible for most cases. QSPR methods offer a simple and efficient alternative, but have mostly the drawback of missing a sound physical model underneath, and are often not able to provide much insights, especially when a multitude of unthoughtful descriptors are used. Nevertheless, in order to predict the enthalpy of fusion and the melting point in a reasonable amount of time a QSPR approach seems to be without an alternative. Moreover, statistically robust models using sound descriptors will not only allow for readily available quantitative solubilities but may also take into account temperature dependency according to Equation 9.7. Therefore, two QSPR models will be presented in the following, concerning the enthalpy of fusion and the melting point, respectively.

First, we have created a model for the fusion enthalpy using the dataset of Chickos and coworkers [37, 38], which contains about 2300 organic molecules. A set of about 30 descriptors were computed using COSMO*quick*, using only structural information by starting from SMILES strings and the subsequently generated σ-profiles. To the latter belong the so-called σ-moments, which build the basis for a linear expansion of the σ-potential of a compound and have been used successfully for instance for describing partition coefficients [39]. Then, a multivariate linear regression was fitted and descriptors were selected by an iterative forward search taking into account the full set. Model stability was checked by fivefold cross-validation yielding an accuracy of about 2.2 kcal/mol (root mean squared error, RMSE), no significant improvement could be obtained using a nonlinear model. Among the variables that were considered as significant are the chemical potential of the pure compound (mu_self), the number of intramolecular hydrogen bonds (h_int), the second σ-moment and the hydrogen bond donor moment (M2 and Mdon2), the number of alkylatoms (alkylatoms), and the number of rotatable bonds of linear chains (nbr11).

In order to develop a melting point model a dataset was created from prior published sources. Melting points from the PHYSPROP database [40] and from the work of Karthikeyan and coworkers [41] were combined. However, we had to find out that this combined dataset contains a significant amount of noise, plainly speaking basically questionable entries. Thus, suspect data points were removed or corrected resulting in a dataset containing the melting point and SMILES of more than 12,000 compounds.

As a test set 277 compounds were kept aside, corresponding to the data used by Karthikeyan and coworkers as a test set.

Probably due to the fact that the melting point is a highly nonlinear property, it was not possible to build a model using linear regression having an accuracy of less than 55 K. Instead, a random forest regression was used to cope with the nonlinearity. Random forests consists of an ensemble of several hundred unpruned decision trees, where a subset of the training samples are held back in each iteration to obtain an accurate generalization error. To build each binary split a random subset of the variables is used. The finally predicted value is obtained by just averaging the prediction from each tree. Since their introduction in 2002 by Leo Breiman [42], random forests experienced strongly increasing popularity, which is mostly due to their statistically robustness (i.e., practically no overfitting, see Table 9.4), their high accuracy also with default parameter sets, and their simplicity of implementation and ease of code parallelization. As tree-based algorithm, they are also invariant under monotone transformation of the variables. Disadvantageous is their lack of direct interpretability; however, they allow for a straightforward determination of variable importances [43].

For this specific melting point model 14 variables have been selected by iterative forward selection (Fig. 9.4), yielding an overall accuracy (RMSE) after fivefold cross-validation of 41.2 K (Table 9.4).

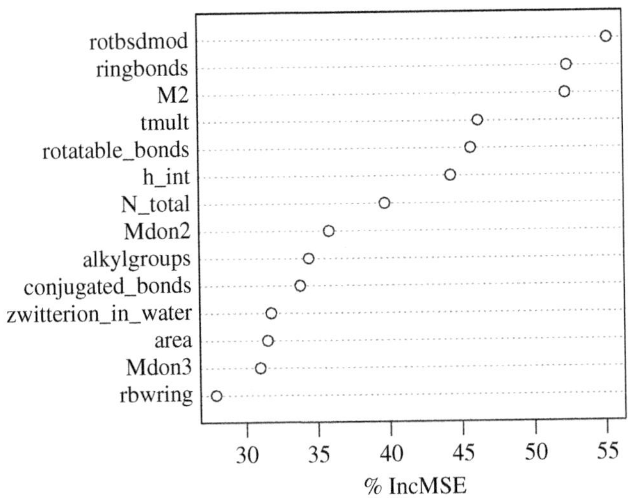

FIGURE 9.4 Variable importances due to the change in mean squared error (MSE) by random permutation of variables in the random forest. Variable acronyms: alkylgroups, number of alkylgroups; area, molecular surface area; conjugated_bonds, number of conjugated bonds; h_int, internal enthalpy; M2, second σ-moment; MDon2 and MDon3, 2nd and 3rd σ–donor moment; N_total, number of nitrogen atoms; rbwring, ring flexibility parameter; ringbonds, number of ringbonds; rotatable_bonds, number of freely rotatable bonds in the molecule; rotbsdmod, a general flexibility parameter including rings; tmult, topological (2D) symmetry; zwitterion_in_water, binary variable, that is, 1 if compounds presumably forms a zwitterion in water, according to its σ-profile, otherwise 0.

TABLE 9.4 Comparison of Train Set (Cross-Validation) and Test Set Results of Different Melting Point QSPR Models

Model	n, descriptors	n,train	RMSE,train	n,test	RMSE,test
NN (Karthikeyan et al. [41])	203	4173	48.0	277	41.4
MLR	14	12096	59.8[a]	277	52.6
RF	14	12096	41.2[a]	277	39.9
GBM	14	12096	40.8[a]	277	40.6

GBM, gradient boosting models; MLR, multivariate linear regression; NN, artificial neural net; RF, random forest. n, descriptors, number of descriptors used; n, train, size of training set, n, test, size of test set.
[a]Fivefold cross-validated results.

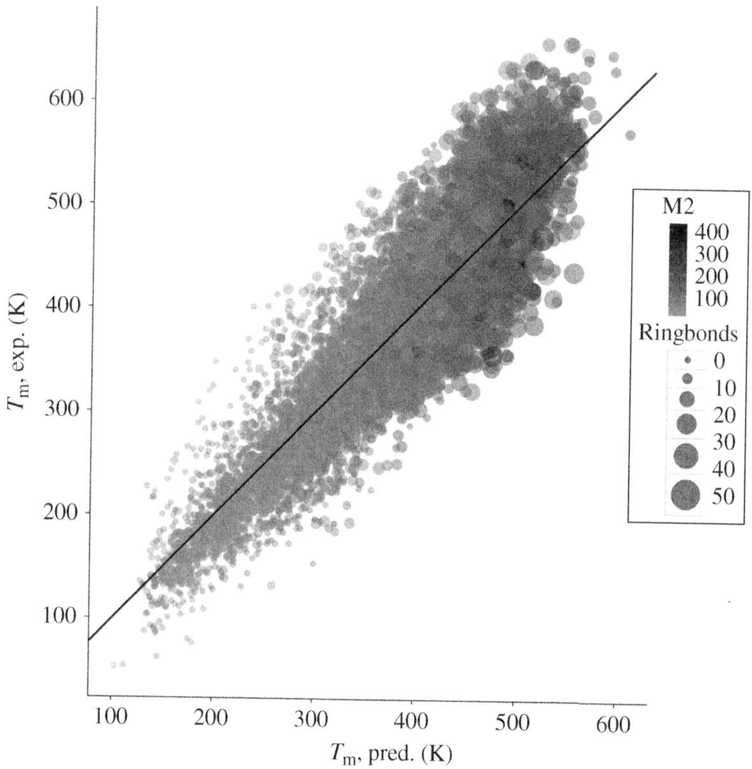

FIGURE 9.5 Random forest–based QSPR model for the prediction of melting points. Additionally, the number of ring bonds and second σ-moment (M2) are shown via point size and color gradient, respectively. (*see insert for color representation of the figure.*)

The results of the model prediction are shown in Figure 9.5, which contains a comparison of experimental and predicted melting points.

Most of the strong deviations between predictions and experiment are due to less common functional groups like silanes or fluorocompounds or due to high symmetry like adamantane and related compounds, which is difficult to cope with via a non3D approach. Furthermore, the dataset probably still contains experimental inaccuracies. Models exceeding a general accuracy of 40 K based only on the 2D structure are currently probably out of reach; see, for example, the overview of recent melting point models in Ref. [44].

Using other nonlinear regression techniques than random forests may give slightly better results but affords parameter tuning and thus opens the door for potential overfitting. Gradient tree boosting as introduced by Friedman [45] also is a decision tree–based technique that is fitting iteratively the residuals of the predictions from a so-called weak learner, for instance a shallow decision tree. Moreover, it uses only a subset of the training instances for each tree (subsampling) and scales the contribution of each tree by a constant (shrinkage) parameter. The approach needs some parameter tweaking, but thoroughly tuned it does not overfit and is known for excellent prediction accuracy. Table 9.4 compares results obtained by using different regression techniques including random forests and gradient boosting using internal validation (cross-validation) and the test set of Karthikeyan et al. [41]. Random forests and gradient boosting yield models with a similar accuracy, with an RMSE approximately 41 K. Concerning model storage the gradient boosting model is to be preferred as its decision trees are not as deep as for a random forest and thus afford less hard disc space. Please note the significant difference in the number of variables used for building the models, whereas the prediction accuracy is about the same. Although Karthikeyan and coworkers have carried out principal component analysis prior to model building, this only reduces the dimensionality of the feature space but does not reduces the number of variables effectively used. It is obvious, that descriptors based on COSMO*quick* and COSMO*therm* are highly informative in a chemical sense and allow for the generation of QSPR models using only small feature sets. This is important in order to avoid overfitting and additionally increases the efficiency of the models.

The combination of the melting point and the enthalpy of fusion model allows for a temperature-dependent estimation of the free energy of fusion ΔG_{fus} from scratch. Thus, Equation 9.6 can be used for quantitative predictions even in the case of missing experimental data, some results using this procedure are compiled in Table 9.5.

Table 9.5 shows that using the just introduced QSPR models gives reasonable results that are comparable to NRTL-SAC (see Table 9.2, though it should be noted,

TABLE 9.5 Solubility Prediction Using QSPR Models in Comparison with Using Multiple Reference Solubilities

Drug	RMSE, multiple ref.	RMSE, QSPR	R^2, multiple ref.	R^2, QSPR
Sulfadiazine	0.42	1.15	0.93	0.86
Paracetamol	0.48	0.62	0.86	0.82
Cimetidine	0.49	0.89	0.86	0.94

Root mean squared error (RMSE) and squared correlation coefficient (R^2) are shown for the COSMO*quick* approach using multiple reference solvents (multiple reference) and using a free energy of fusion estimate via the melting point and enthalpy of fusion quantitative structure–property relationship (QSPR) models.

that in contrast to NRTL-SAC no specific experimental input has been used now). Nevertheless, if possible, using (multiple) experimental solubilities is to be preferred when a high accuracy is desired.

9.6 COCRYSTAL SCREENING

Although cocrystals have been known for a long time, it was realized only very recently that they provide an additional leverage to modify physicochemical properties like the solubility and hence the bioavailability of traditional drugs. Consequently, nowadays the development of cocrystalline structures of a drug and an excipient, also called coformer, has become of high relevance to pharmaceutical industry. The rational design via computational approaches helps focussing on the relevant chemical space of potential coformers and thus may serve to speed up such a development. Several attempts have been made so far to assess possible cocrystal formation via computational methods.

Crystal structure prediction can give insights into molecular packing and the relevant solid-state interactions of cocrystals [46, 47] but is yet too time consuming to be of practical use for flexible, multicomponent systems, or even for the screening over several coformers.

Fabian analyzed cocrystal structures stored within the Cambridge Structural Database, statistically establishing a QSPR-like approach [48], which was reported to yield likely coformers in the case of artemisinin [49].

Musumeci and coworkers used the molecular electrostatic potential (MEP) of drug and coformer derived from a DFT calculation to identify potential cocrystals by a hierarchical mapping of complementary donor and acceptor sites [50].

A study by Seaton relates cocrystal formation with Hammett constants, which is limited to compounds where such constants are available or measurable [51].

Considering those attempts, a favorable cocrystal screening approach should satisfy several important aspects: it should be comparatively fast to be able to scan molecular libraries in a decent amount of time, accurate, broadly applicable, and ideally based on sound physicochemical principles.

Those criteria are met by a cocrystal screening based on COSMO-RS theory, using the mixing enthalpy H_{mix} (or equivalently the excess enthalpy H_{ex}) of a supercooled cocrystal mixture [18]. The main reason for the predictive capability of this theory concerning cocrystal formation is its accurate description of intermolecular interactions. The idea behind a COSMO-RS-based screening can be best understood by having a look at the thermodynamic cycle shown in Scheme 9.1.

COSMO-RS, being a liquid phase theory, predicts readily ΔG_{mix}, the free energy obtained by mixing the virtual subcooled components A and B into the subcooled cocrystal AB. In order to complete the thermodynamic cycle and to get the free energy of cocrystal formation ΔG°_{cc} the free energy of fusion for the reactants and the cocrystal is needed. Interestingly, by neglecting the lower part of the cycle, the strength of ΔG_{mix} and in particular ΔH_{mix} alone seems to be already an excellent quantity to predict for cocrystal formation [18]. The somewhat better performance of

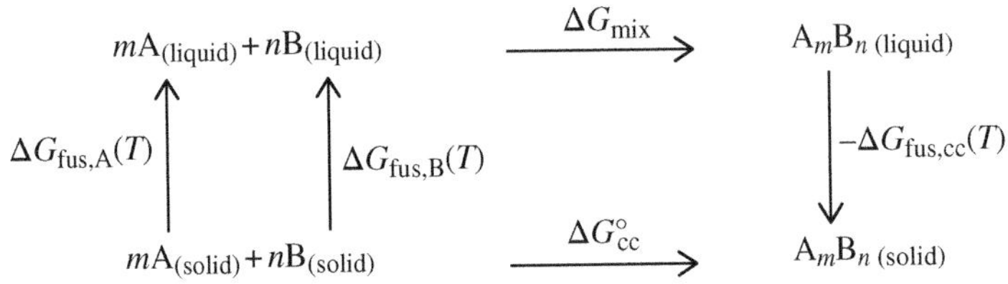

SCHEME 9.1 Thermodynamic cycle for the standard Gibbs free energy of formation (ΔG°_{cc}) for a cocrystal A_mB_n out of its solid components A and B. $\Delta G_{fus}(T)$ is the free energy of fusion for bringing a subcooled liquid into the solid state at the temperature T. Please note the sign convention for the free energy of fusion adopted here, that is, solid systems below the melting point having a positive free energy of fusion. ΔG_{mix} is the Gibbs free energy for the hypothetical process of mixing the liquid components A and B into a cocrystal A_mB_n, in its (supercooled) liquid state.

the enthalpy in comparison to the Gibbs free energy may be explained by the fact that the entropy change along the thermodynamic cycle, that is, starting from solid reactants A and B and ending with solid cocrystal AB should be negligible. Additionally, the contributions by the fusion free energy of reactants and cocrystal seem to be to some extent contrary, leaving the mixing enthalpy as most influential quantity in the cycle of Scheme 9.1:

$$\Delta G^{\circ}_{cc} = \Delta H_{mix} - T\Delta S_{mix} - \Delta\Delta G_{fus} \approx \Delta H_{mix} \tag{9.8}$$

One has to conclude, that a large part of the decisive intermolecular interactions that make up the cocrystal are probably already dominating in solution and are well described by the ΔH_{mix} term as calculated by COSMO-RS theory. Of course Equation 9.8 is a considerable approximation and the total neglect of the $\Delta\Delta G_{fus}$ term will be the main source for deviations from experimental findings (see also Table 9.6). In practice, the mixing enthalpy or equivalently the excess enthalpy for mixing the drug A and coformer B is computed by COSMO*therm* and ΔH_{mix} is used as a measure for the propensity of the system AB to form a cocrystal. Hence, sorting the list of potential coformers with respect to ΔH_{mix} yields an enrichment of the likely best coformers at the top. Statistically, such an ordered screening list can be evaluated most comprehensively with the area under curve (AUC) of a receiver operating characteristics (ROC) diagram. Such a curve plots the true-positive rate (sensitivity) against the false-positive rate (1-specificity), an AUC value of 1.0 corresponds to perfect prediction, whereas an AUC value of 0.5 corresponds to a random result.

Unfortunately, as long as it is not possible to develop any predictive model for the free energy of fusion of cocrystals, a systematic improvement beyond simply using ΔG_{mix} or ΔH_{mix} is out of scope. However, we found an empirical, but significant improvement of the screening function by including the number of rotatable bonds in the form of the following screening function (f_{screen})

$$f_{screen} \sim H_{mix} + a\left(\max(1, n_{drug}) + \max(1, n_{coformer})\right) \tag{9.9}$$

TABLE 9.6 Results of COSMO-RS-Based Virtual Cocrystal Screenings

API	AUC, ΔH_{mix}	AUC, f_{screen}	AUC, COSMO*quick*	n_{cf}	n_{cc}	References
Itraconazole	1.00	0.88	0.75	8	4	[52]
3-Cyanophenol	0.98	0.93	0.93	18	3	[53]
4-Cyanophenol	1.00	1.00	1.00	18	4	[53]
3-Cyanopyridine	0.88	0.82	0.82	18	1	[53]
4-Cyanopyridine	0.96	0.93	0.93	18	3	[53]
Bicalutamide	0.94	0.97	0.75	18	2	[53]
Nicotinamide	0.67	0.68	0.70	16	10	[54–56]
Paracetamol	0.53	0.81	0.75	13	4	[57]
Meloxicam	0.73	0.93	0.87	17	15	[58]
Benzamide	0.75	0.75	0.58	12	6	[59]
Indomethacin	0.49	0.79	0.81	21	4	[60]
Indomethacin	0.48	0.70	0.81	45	5	[61]
Diclofenac	0.61	0.57	0.49	18	11	[62]
Artemisinin	0.95	0.95	0.97	67	2	[49]
Praziquantel	0.88	1.00	1.00	11	8	[63]
Average	0.79 (±0.19)	0.85 (±0.13)	0.81 (±0.14)			

Computed AUC scores as obtained by comparison with different experimental cocrystal screening sets as taken from the literature, number of coformers in the set (n_{cf}), the number of observed cocrystals (n_{cc}), and references to the experiments are presented. In the last row, the averaged AUC score and the standard deviation are given. Coformers were ranked according to their mixing enthalpy (AUC, ΔH_{mix}) and Equation 9.9 (AUC, f_{screen}). In addition results using σ-profiles approximated by the COSMO*quick* approach in combination with approximation 9.9 are shown (AUC, COSMO*quick*).

where a is a fit parameter to be determined on a set of experimental results, and n is the total number of rotatable bonds of drug and coformer. Effectively, a function is added to the mixing enthalpy that just penalizes floppy molecules, whereas molecules with zero or one rotatable bonds are being treated equally. Thermodynamically, the number of rotatable bonds should not have a dramatic influence on cocrystal formation, as this should affect reactants and cocrystal equally. Most probably some of the crystallization kinetics has been captured empirically by Equation 9.9 as an increasing flexibility renders the crystallization of a cocrystal less probable. Table 9.6 summarizes the statistics based on the AUC value for COSMO-RS-based cocrystal screening on a set of different experimental sets taken from the literature. Using Equation 9.9 an overall mean AUC = 0.85 is achieved over the 15 screenings. As an example, AUC = 0.85 corresponds to a hypothetical scenario of 25 potential coformers including only three hits, and having those hits in position 2, 5, and 9 of the list.

Figure 9.6 contains two challenging cases, paracetamol [57] and indomethacin, based on the experimental data from reference [60], for which the original screening via ΔH_{mix} gave the worst results with AUC ~ 0.5. The figure demonstrates the obtained enrichment, that is, how many hits are found successively if one follows the ΔH_{mix} ordered list of drug–conformer pairs.

Taking into account the number of rotatable bonds for both cases leads to a significant overall improvement with AUC = 0.81 and AUC > 0.7, respectively.

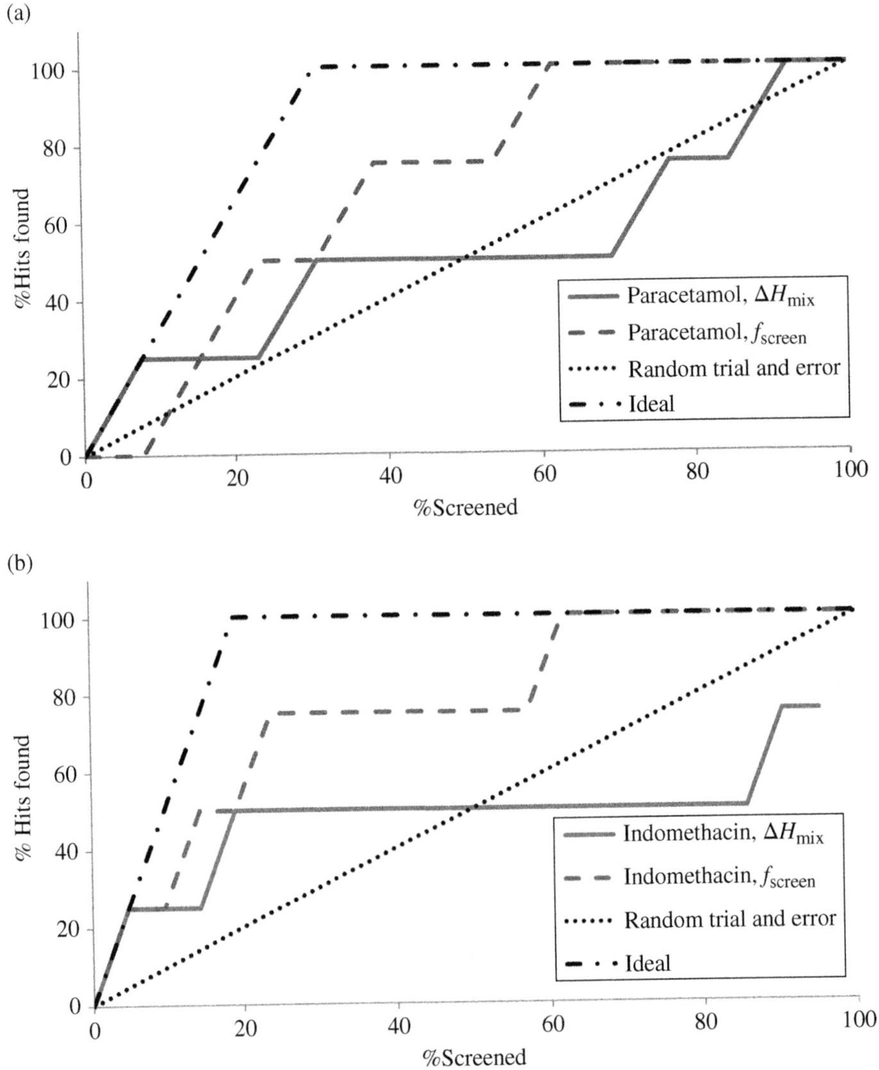

FIGURE 9.6 Cocrystal virtual screening results for paracetamol (a) and indomethacin (b) computed based on the mixing enthalpy only (ΔH_{mix}) and taking into account molecular flexibility (f_{screen}). For reference, a random trial-and-error experiment (dotted line) and a theoretical perfect screening giving immediately all cocrystals (dashed-dotted line) are given.

In addition to the results (AUC, ΔH_{mix} and AUC, f_{screen}) from full COSMO*therm* calculations, Table 9.6 also contains data obtained by using approximate σ-profiles via the COSMO*quick* approach (AUC, COSMO*quick*) in combination with approximation (Eq. 9.9). At the sake of a somewhat reduced accuracy, COSMO*quick*-based cocrystal screenings can easily handle large molecular libraries and identify within minutes the most promising compounds among thousands from possible candidates [18].

9.7 SOLVATE FORMATION

During drug development it is often desirable to identify solvents, which do not form solvates with a specific drug. Thermodynamically, solvate formation and cocrystal formation are strongly related and differ only by the fact that for solvates one of the reactants is liquid at the temperature of the experiment and thus having by definition a free energy of fusion $\Delta G_{fus} = 0$. Therefore, the just presented workflow for cocrystals can as well be applied to assess the potential or risk of solvents to form solvates.

COSMO-RS-based solvate screening has been applied successfully to the cases of the drugs axitinib, thiophanate-methyl (TM), and thiophanate-ethyl (TE) by Abramov et al. [18]. Axitinib is a small-molecule tyrosine kinase inhibitor developed by Pfizer being a particular challenging polymorphic system with at least 5 anhydrous forms and more than 60 solvated forms [64, 65]. The fungicides TM and TE are each close analogues, however, showing different molecular arrangements in the solid state. They both possess several polymorphic forms, additionally 14 solvates have been reported for TM and 7 for TE, respectively [66, 67].

Figures 9.7 and 9.8 present the results of the COSMO*therm*-based solvate screening for those three drugs whereas drug-solvent pairs with a more positive excess enthalpy are predicted to have a lower solvate formation tendency. For the case of axitinib (Fig. 9.7) heptane was found to have the highest H_{ex} value among all considered solvents and indeed is one of the few solvents, which does not form solvates with this drug. In the original dataset [18] positive H_{ex} values were predicted for 24 of

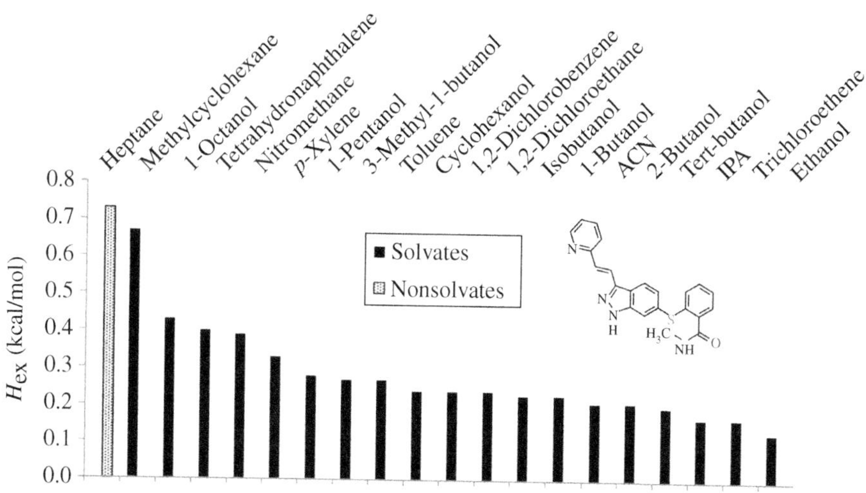

FIGURE 9.7 Results of COSMO*therm* solvate screening for the drug axitinib [18]. Enthalpies H_{ex} are calculated at the BP-TZVP-COSMO level of theory and are presented in kcal/mol. Solvate (black) and nonsolvate (gray) forming solvents are indicated. Experimental stoichiometries were used in the calculations whenever available, otherwise 1 : 1 mixtures were considered. Only the first 20 solvents from Ref. [18] are presented.

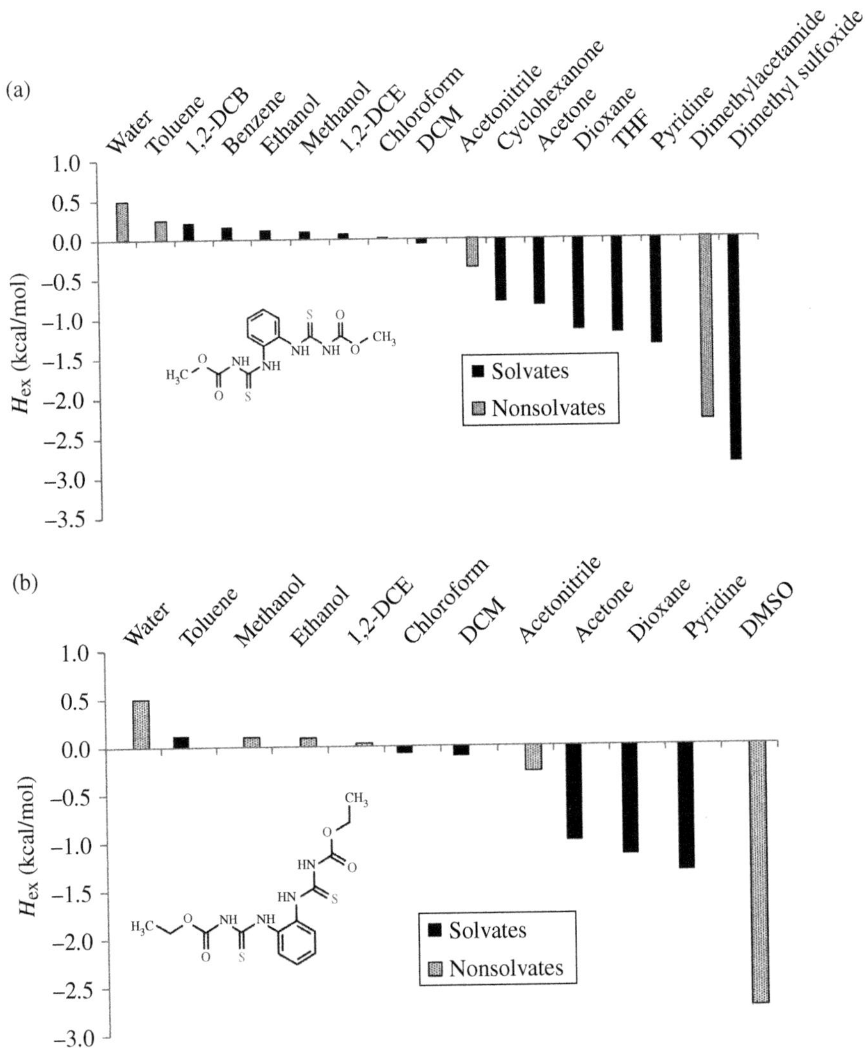

FIGURE 9.8 Results of COSMO*therm* solvate screening for the drugs TM (a) and TE (b) [18]. Enthalpies H_{ex} are calculated at the BP-TZVP-COSMO level of theory and are presented in kcal/mol. Solvate (black) and nonsolvate (gray) forming solvents are indicated. Experimental stoichiometries were used in the calculations whenever available, otherwise 1 : 1 mixtures were considered.

the 46 solvents under investigation [18]. That indicates that low miscibility of axitinib with these 23 solvents in the supercooled liquid phase is counterbalanced by lattice packing contributions in the solid state, which are ignored in current calculations.

Concerning TM and TE (Fig. 9.8a and b), a few solvents having low predicted excess enthalpy are not known to form solvates. This could be due to unfavorable packing in the crystal or just due to fact that they have been experimentally missed to date. Nevertheless, for both drugs the workflow yields nonsolvate formers at the top of the set of screened solvents and thus allows to readily identify such solvents from a heap of possible candidates. Because one is usually not interested in finding all but only one or a few nonsolvate formers, this renders the COSMO*therm* (non)solvate screening as a highly useful method from a practical point of view.

9.8 SUMMARY

COSMO-RS's unique combination of quantum chemistry and surface segment–based statistical thermodynamics allows for computing nearly any thermodynamic property in the liquid phase. For the prediction of a solid phase–related property like solubility either experimental data or QSPR models have to be incorporated. This enables a straightforward extension to the computation of solid–liquid equilibria. Usage of multiple reference solubilities as implemented in the COSMO*quick* approach allows for high prediction accuracy. Unlike most of the available solubility prediction methods, COSMO-RS is not limited to certain functional groups and does not need any reparameterization prior to use. Due to its predictive character it has become a valuable tool in pharmaceutical research in various fields such as solubility and partitioning coefficient prediction, solvent screening, and more recently cocrystal and solvate screening.

REFERENCES

[1] Blagden, N.; de Matas, M.; Gavan, P. T.; York, P. *Adv. Drug Deliv. Rev.* 2007, *59*, 617–630.

[2] Chen, C.-C.; Song, Y. *Ind. Eng. Chem. Res.* 2004, *43*, 8354–8362.

[3] Hansen, C. M. *J. Paint Tech.* 1967, *39*, 104–117.

[4] Gross, J.; Sadowski, G. *Ind. Eng. Chem. Res.* 2001, *40*, 1244–1260.

[5] Fredenslund, A.; Jones, R. L.; Prausnitz, J. M. *AIChE J.* 1975, *21*, 1086–1099.

[6] Eckert, F.; Klamt, A. COSMO*therm, Version C3.0-Revision 13.01*; COSMO*logic* GmbH&Co.KG, Leverkusen, Germany, 2013. See also http://www.cosmologic.de, accessed November 4, 2015.

[7] Klamt, A. *J. Phys. Chem.* 1995, *99*, 2224–2235.

[8] Klamt, A. *WIREs Comput. Mol. Sci.* 2011, *1*, 699–709.

[9] Klamt A. COSMO-*RS: From Quantum Chemistry to Fluid Phase Thermodynamics and Drug Design*. Elsevier Science Ltd., Amsterdam, The Netherlands: 2005.

[10] Klamt, A.; Schüürmann, G. *J. Chem. Soc., Perkin Trans. 2* 1993, 799–805.

[11] Klamt, A.; Reinisch, J.; Eckert, F.; Graton, J.; Le Questel, J.-Y. *Phys. Chem. Chem. Phys.* 2013, *15*, 7147–7154.

[12] Klamt, A.; Krooshof, G. J. P.; Taylor, R. *AIChE J.* 2002, *48*, 2332–2349.

[13] Klamt, A.; Eckert, F.; Hornig, M.; Beck, M. E.; Bürger, T. *J. Comput. Chem.* 2002, *23*, 275–281.

[14] Eckert, F.; Klamt, A. *AIChE J.* 2002, *48*, 369–385.

[15] Eckert, F.; Klamt, A. *J. Comput. Chem.* 2006, *27*, 11–19.

[16] Diedenhofen, M.; Eckert, F.; Klamt, A. *J. Chem. Eng. Data* 2003, *48*, 475–479.

[17] Klamt, A.; Eckert, F.; Hornig, M. *J. Comput. Aided Mol. Design* 2001, *15*, 355–365.

[18] Abramov, Y. A.; Loschen, C.; Klamt, A. *J. Pharm. Sci.* 2012, *101*, 3687.

[19] Pozarska, A.; da Costa Mathews, C.; Wong, M.; Pencheva, K. *Eur. J. Pharm. Sci.* 2013, *49*, 505–511.

[20] Klamt, A.; Thormann, M.; Wichmann, K.; Tosco, P. *J. Chem. Inf. Model.* 2012, *52*, 2157.

[21] Loschen, C.; Hellweg, A.; Klamt A. COSMO*quick, Version 1.2*; COSMO*logic* GmbH & Co. KG, Leverkusen, Germany, 2013.

[22] R core team. *R: A Language and Environment for Statistical Computing*, 2013. See also: http://www.r-project.org/, accessed November 4, 2015.

[23] Eckert F. Prediction of Solubility with COSMO-RS. In *Developments and Applications in Solubility*; Letcher T., Ed.; RSC, Cambridge, UK; 2007, pp 188–198.

[24] Nti-Gyabaah, J.; Chiew, Y. C. *J. Chem. Eng. Data* 2008, *53*, 2060–2065.

[25] Nti-Gyabaah, J.; Chmielowski, R.; Chan, V.; Chiew, Y. C. *Int. J. Pharm.* 2008, *359*, 111–117.

[26] Sun, H.; Gong, J.-B.; Wang, J.-K. *J. Chem. Eng. Data* 2005, *50*, 1389–1391.

[27] Tung, H.-H.; Tabora, J.; Variankaval, N.; Bakken, D.; Chen, C.-C. *J. Pharm. Sci.* 2008, *97*, 1813–1820.

[28] Babu, P. S.; Subrahmanyam, C.; Thimmasetty, J.; Manavalan, R.; Valliapan, K. *Pak. J. Pharm. Sci.* 2007, *20*, 311–316.

[29] Lapkin, A. A.; Peters, M.; Greiner, L.; Chemat, S.; Leonhard, K.; Liauw, M. A.; Leitner, W. *Green Chem.* 2010, *12*, 241–251.

[30] Martin, A.; Bustamante, P.; Escalera, B.; Sellés, E. *J. Pharm. Sci.* 1989, *78*, 672–678.

[31] Bustamante, P.; Romero, S.; Peña, A.; Escalera, B.; Reillo, A. *J. Pharm. Sci.* 1998, *87*, 1590–1596.

[32] Romero, S.; Reillo, A.; Escalera, B.; Bustamante, P. *Chem. Pharm. Bull.* 1996, *44*, 1061–1064.

[33] Paruta, A. N.; Irani, S. A. *J. Pharm. Sci.* 1965, *54*, 1334–1338.

[34] Loschen, C.; Klamt, A. *Ind. Eng. Chem. Res.* 2012, *51*, 14303–14308.

[35] Chen, C.-C.; Crafts, P. A. *Ind. Eng. Chem. Res.* 2006, *45*, 4816–4824.

[36] Granberg, R. A.; Rasmuson, Å. C. *J. Chem. Eng. Data* 1999, *44*, 1391–1395.

[37] Chickos, J. S.; Acree, W.E. Jr.; Liebman, J. F. *J. Phys. Chem. Ref. Data* 1999, *28*, 1535–1673.

[38] Chickos, J. S.; Acree W. E. *Thermochim. Acta* 2002, *395*, 59–113.

[39] Zissimos, A. M.; Abraham, M. H.; Klamt, A.; Eckert, F.; Wood, J. *J. Chem. Inf. Comput. Sci.* 2002, *42*, 1320–1331.

[40] Syracuse Research Corporation (SRC) 2010. PhysProp Database. See also: http://www.srcinc.com/what-we-do/environmental/scientific-databases.html, accessed October 10, 2013.

[41] Karthikeyan, M.; Glen, R. C.; Bender, A. *J. Chem. Inf. Model.* 2005, *45*, 581–590.

[42] Breiman, L. *Mach. Learn.* 2001, *45*, 5–32.

[43] Hastie, T.; Tibshirani, R.; Friedman, J. *The Elements of Statistical Learning: Data Mining, Inference and Prediction*. Springer, New York, 2011.

[44] Salahinejad, M.; Le, T. C.; Winkler, D. A. *J. Chem. Inf. Model.* 2013, *53*, 223–229.

[45] Friedman, J. H. *Comput. Stat. Data Anal.* 2002, *38*, 367–378.

[46] Issa, N.; Karamertzanis, P. G.; Welch, G. W. A.; Price, S. L. *Cryst. Growth Des.* 2009, *9*, 442–453.

[47] Karamertzanis, P. G.; Kazantsev, A. V.; Issa, N.; Welch, G. W. A.; Adjiman, C. S.; Pantelides, C. C.; Price, S. L. *J. Chem. Theory Comput.* 2009, *5*, 1432–1448.

[48] Fabian, L. *Cryst. Growth Des.* 2009, *9*, 1436–1443.

[49] Karki, S.; Friscic, T.; Fabian, L.; Jones, W. *CrystEngComm* 2010, *12*, 4038–4041.

[50] Musumeci, D.; Hunter, C. A.; Prohens, R.; Scuderi, S.; McCabe, J. F. *Chem. Sci.* 2011, *2*, 883–890.

[51] Seaton, C. C. *CrystEngComm* 2011, *13*, 6583–6592.

[52] Remenar, J. F.; Morissette, S. L.; Peterson, M. L.; Moulton, B.; MacPhee, J. M.; Guzmán, H. R.; Almarsson, Ö. *J. Am. Chem. Soc.* 2003, *125*, 8456–8457.

[53] Bis, J. A.; Vishweshwar, P.; Weyna, D.; Zaworotko, M. J. *Mol. Pharm.* 2007, *4*, 401–416.

[54] Berry, D. J.; Seaton, C. C.; Clegg, W.; Harrington, R. W.; Coles, S. J.; Horton, P. N.; Hursthouse, M. B.; Storey, R.; Jones, W.; Friščić, T.; Blagden, N. *Cryst. Growth Des.* 2008, *8*, 1697–1712.

[55] Báthori, N. B.; Lemmerer, A.; Venter, G. A.; Bourne, S. A.; Caira, M. R. *Cryst. Growth Des.* 2011, *11*, 75–87.

[56] Fabian, L.; Hamill, N.; Eccles, K. S.; Moynihan, H. A.; Maguire, A. R.; McCausland, L.; Lawrence, S. E. *Cryst. Growth Des.* 2011, *11*, 3522–3528.

[57] Karki, S.; Friščić, T.; Fábián, L.; Laity, P. R.; Day, G. M.; Jones, W. *Adv. Mater.* 2009, *21*, 3905–3909.

[58] Cheney, M. L.; Weyna, D. R.; Shan, N.; Hanna, M.; Wojtas, L.; Zaworotko, M. J. *Cryst. Growth Des.* 2010, *10*, 4401–4413.

[59] Seaton, C. C.; Parkin, A. *Cryst. Growth Des.* 2011, *11*, 1502–1511.

[60] Mohammad, M. A.; Alhalaweh, A.; Velaga, S. P. *Int. J. Pharm.* 2011, *407*, 63–71.

[61] Kojima, T.; Tsutsumi, S.; Yamamoto, K.; Ikeda, Y.; Moriwaki, T. *Int. J. Pharm.* 2010, *399*, 52–59.

[62] Aakeröy, C. B.; Grommet, A. B.; Desper, J. *Pharmaceutics* 2011, *3*, 601–614.

[63] Espinosa-Lara, J. C.; Guzman-Villanueva, D.; Arenas-García, J. I.; Herrera-Ruiz, D.; Rivera-Islas, J.; Román-Bravo, P.; Morales-Rojas, H.; Höpfl, H. *Cryst. Growth Des.* 2013, *13*, 169–185.

[64] Campeta, A. M.; Chekal, B. P.; Abramov, Y. A.; Meenan, P. A.; Henson, M. J.; Shi, B.; Singer, R. A.; Horspool, K. R. *J. Pharm. Sci.* 2010, *99*, 3874–3886.

[65] Chekal, B. P.; Campeta, A. M.; Abramov, Y. A.; Feeder, N.; Glynn, P. P.; McLaughlin, R. W.; Meenan, P. A.; Singer, R. A. *Org. Process Res. Dev.* 2009, *13*, 1327–1337.

[66] Nauha, E.; Saxell, H.; Nissinen, M.; Kolehmainen, E.; Schafer, A.; Schlecker, R. *CrystEngComm* 2009, *11*, 2536–2547.

[67] Nauha, E.; Ojala, A.; Nissinen, M.; Saxell, H. *CrystEngComm* 2011, *13*, 4956–4964.

10

MODELING AND PREDICTION OF SOLID SOLUBILITY BY GE MODELS

LARISSA P. CUNICO[1], ANJAN K. TULA[1], ROBERTA CERIANI[2], AND RAFIQUL GANI[1]

[1] *SPEED, Department of Chemical & Biochemical Engineering, Technical University of Denmark, Kongens Lyngby, Denmark*
[2] *State University of Campinas, Cidade Universitária Zeferino Vaz, Campinas-SP, Brazil*

10.1 INTRODUCTION

The information of solid solubility and solid–liquid equilibrium (SLE) data of chemicals is necessary in product and process design, for example, in the extraction, adsorption, and/or crystallization processes. These processes are important steps in recovery and purification of chemicals. Here the knowledge of solvents and solid saturation diagrams play an important role. For chemicals involved in the pharmaceutical and related industries, a common problem encountered is the low solubility of active ingredients (or active pharmaceutical ingredient, API) in water—this makes the solubility analysis an important issue for API efficacy [1]. The chemical systems encountered are usually complex, consisting of the API that is usually a large complex molecule dissolved in single solvents or solvent mixtures together with additives such as lipids and surfactants [2]. Since the measurement of experimental data is often time-consuming, expensive and may not be even feasible in some cases, a predictive model-based approach for solubility calculation procedure would be desirable, at least in the early stages of product-process evaluation. In this chapter, model-based prediction methods for solid solubility using excess Gibbs energy

Computational Pharmaceutical Solid State Chemistry, First Edition. Edited by Yuriy A. Abramov.

models (GE models) are highlighted together with analysis of data, model parameter estimation, and calculations of solid solubilities.

An important first step in any model-based calculation procedure is the analysis and type of data used. Here, the accuracy and reliability of the measured data sets to be used in regression of model parameters is a very important issue. It is clear that reliable parameters for any model cannot be obtained from low-quality or inconsistent data. However, for many published experimentally measured solid solubility data, information on measurement uncertainties or quality estimates are unavailable. Also, pure component temperature limits and the excess GE models typically used for nonideality in vapor–liquid equilibrium (VLE) may not be reliable for SLE (or solid solubility). To address this situation, an alternative set of consistency tests [3] have been developed, including a new approach for modeling dilute solution SLE, which combines solute infinite dilution activity coefficients in the liquid phase with a theoretically based term to account for the nonideality for dilute solutions relative to infinite dilution. This model has been found to give noticeably better descriptions of experimental data than traditional thermodynamic models (nonrandom two liquid (NRTL) [4], UNIQUAC [5], and original UNIversal Functional group Activity Coefficient (UNIFAC) [6]) for the studied systems.

In this chapter, the solid solubility of chemicals in different organic solvents is reviewed and evaluated. In particular, the solubilities of API and lipid systems using different property (GE models) models such as NRTL [4], UNIQUAC [5], original UNIFAC [6], and NRTL-SAC [7, 8] implemented in Computer-Aided Process Engineering Center (CAPEC) software are evaluated. Also implemented in the CAPEC, the Perturbed-Chain Statistical Associating Fluid Theory (PC-SAFT) equation of state [9] with group contribution (GC) prediction of the pure component parameters [10] was selected to evaluate solid solubility predictions in this work. In the case of original UNIFAC [6] model, which is based on the concept of GC, the prediction of missing group interaction parameters to describe the complex systems (such as API solvents) is also considered. A computer-aided model-based framework for solid solubility calculations and for solvent selection and design, called SolventPro, is also presented. The framework integrates different methods and tools needed in the management of the complexity of such chemicals in an efficient, flexible, and robust way. The application of SolventPro together with the NRTL segment activity coefficient (NRTL-SAC) and the PC-SAFT models for solubility predictions, multilevel property estimation, and solution of pharmaceutical industry problems is highlighted.

10.2 FRAMEWORK

To predict the solid solubility, in addition to model-based property models, databases and numerical solvers are necessary. To better illustrate each step of the solid solubility calculation, the necessary workflow and dataflow are highlighted in Figure 10.1, starting with the necessary pure component properties and ending with the phase diagram generation. It is important to note that when the experimental values of the

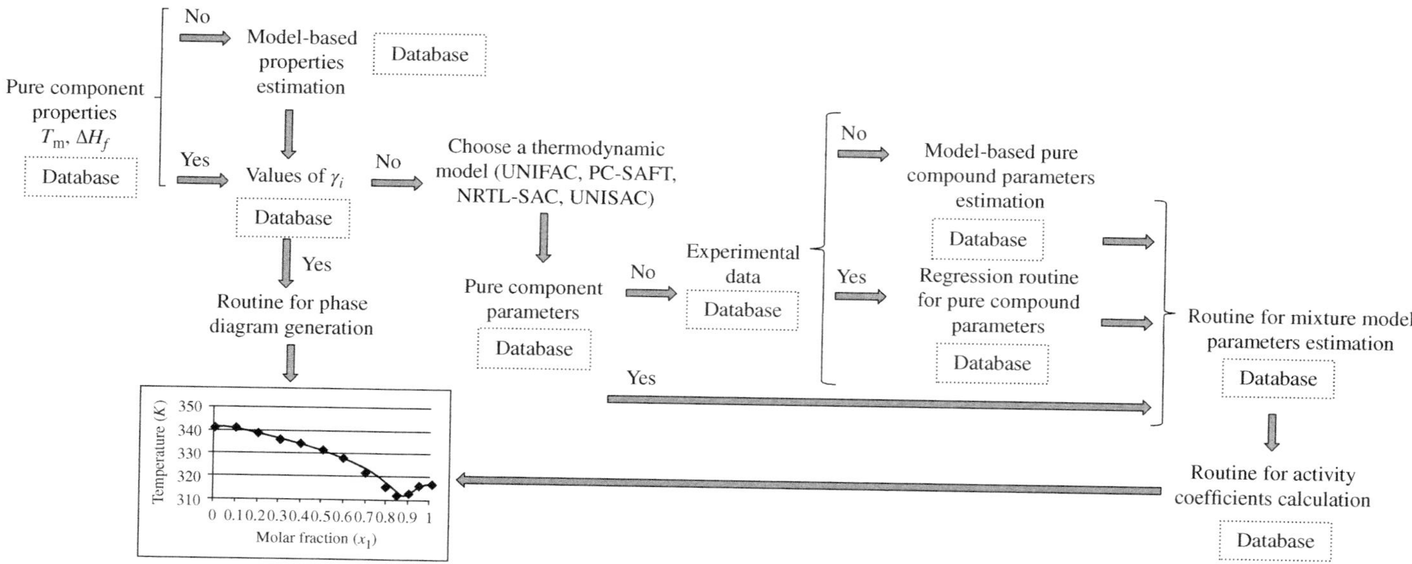

FIGURE 10.1 Work flow and dataflow for solid solubility calculations.

needed properties are not available, they are obtained from regressed model parameters by considering the model-based option. These estimated values, such as pure-component properties, regressed parameters (pure component and mixture), and the calculated activity coefficients from the GE-models are stored in the lipids database.

10.2.1 Thermodynamic Basis

For binary mixtures, the solute is usually the substance that is dissolved in a certain quantity of another substance—the solvent. Assuming that the solid (solute) is pure, the solubility of the solid (i.e., solute) in a liquid solution is obtained from the following relation [11]:

$$\ln \gamma_1 x_1 = -\frac{\Delta H_{\mathrm{fus}}}{RT}\left(1-\frac{T}{T_{\mathrm{m}}}\right)+\frac{\Delta C_{\mathrm{p}}}{R}\left(\frac{T_{\mathrm{m}}-T}{T}\right)-\frac{\Delta C_{\mathrm{p}}}{R}\ln\frac{T_{\mathrm{m}}}{T} \tag{10.1}$$

where ΔH_{fus} is the enthalpy of fusion, T_{m} is the melting temperature, x_1 is the molar fraction of component 1, γ_1 is the activity coefficient of component 1 for the pure component (Lewis–Randall) standard state, R is the gas constant, and T is the system temperature. Rigorous additional terms on the right-hand side of Equation 10.1 involving the difference in heat capacities of the solid and sub-cooled liquid (ΔC_{p}) have been ignored since they generally are small [11]. Further, it is assumed that no pure solid structure transitions occur between T and T_{m}.

10.2.2 The Necessary Property-Related Information for Solid Solubility Prediction and the Developed Databases

To apply Equation 10.1 for solid solubility calculations in a generic way, it is necessary to (i) develop a database containing experimentally measured values of properties of pure components (melting temperature, enthalpy of fusion, and specific heats) as well as mixture properties (activities and solubilities); (ii) analyze and assess the quality of the experimental data using thermodynamic consistency tests; and (iii) establish a systematic approach for performing parameter regression (including, the selection of the most appropriate objective function for the parameter regression). Also, there are numerous ways to obtain the needed data, for example, (i) by retrieving the property information available in databases/open literature, (ii) by performing laboratory measurements for the needed properties, and/or (iii) by employing suitable property prediction methods. A key limitation associated with the use of databases is the limited number of chemicals (and sometimes limited number of properties) stored in the database. Chemical and process industries that use computer-aided tools (e.g., process simulators such as PRO/II®, and ASPEN®) rely on the availability of data and models for properties listed in their built-in databases. Thus, lack of necessary physical and thermodynamic properties in the database restricts the use of computer-aided tools for synthesis-design, modeling-simulation of chemical processes.

For reliable regression of property model parameters, appropriately large data sets of experimentally measured property data for a wide range of chemicals (hydrocarbons, oxygenated chemicals, nitrogenated chemicals, polyfunctional chemicals, etc.) are necessary. The extended CAPEC database (Nielsen et al. [12]) provides data for melting temperatures (T_m) and enthalpies of fusion (ΔH_{fus}) for a wide range of chemicals (hydrocarbons, oxygenated components, nitrogenated components, polyfunctional components, etc.). In total, 5183 compounds provide the information for melting temperature and 761 for enthalpy of fusion. For lipids, a separate database (CAPEC_Lipids_Database) containing the most representative lipid compounds found in the lipid processing industry together with their physical property data has been developed. More than 13,500 data-points for different pure component properties covering nearly 290 lipids have been collected. In this database, the components with limited information are those commonly found in vegetable oils, more specifically acylglycerols and tocopherols. When experimental data is not available, property prediction models based on the Marrero and Gani (MG) method [13–15] are considered since this method is based exclusively on the molecular structure of the pure component and exhibits a good accuracy over a wide range of chemical-, biochemical-, and environment-related pure components. Joback and Reid [16] developed GC-based models for the prediction of different pure component properties including melting point. Recently, more sophisticated GC methods based on the multi-level property estimation approach have been developed by Constantinou and Gani [17], Marrero and Gani [13], and Hukkerikar et al. [14]. In many cases, the selected GC model may not have all the needed model parameters, that is, groups describing the molecular structure of a given pure component and/or their contributions. This issue is due to (i) the lack of necessary measured data of properties of pure components in the parameter regression step and (ii) the lack of necessary group definitions required to describe the complete molecular structure of pure components. To address these issues, the atom connectivity index (CI) method has been employed together with the GC method to create missing groups and/or to predict their contributions. This combined approach has led to the development of group contribution+ (GC+) method of a wider application range than before since the missing groups and their contributions can now be easily predicted through the regressed contributions of connectivity indices [18]. The model performance statistics for the pure component property models based on the model parameter values given by Hukkerikar et al. [14] is summarized in Cunico et al. [15]. One example for the Marrero and Gani [13] groups considered for a chosen API (paracetamol) including the multilevel property estimation approach is showed in Figure 10.2.

In the development of Capec_Lipids_Mixture_Database, a search of the literature was made to collect, within a limited time, as many data as possible. The criteria for data selection were details of measurement technique, measurement accuracy, different ranges of temperature, pressure, and molar fractions considered by the author, the use of a method of analysis or a model to describe the experimental data, and if the data source brought in results for different mixture systems for the class of components considered. The collected data are unlikely to be all those in the literature. However, the database is fully adequate to develop and test physical property models for the classes

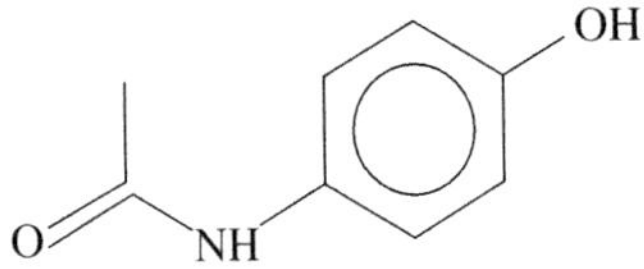

First-order groups

Times	Group
1	CH_3
4	aCH
1	aC–OH
1	aC–NHCO

Second-order groups

Times	Group
1	AROMRINGs1s4

Third-order groups

No group occurring

FIGURE 10.2 Marrero and Gani [13] group description for the compound paracetamol.

TABLE 10.1 Solid Solubility of Paracetamol (2) in Ethanol (1) [21]

T (K)	x^2
268.15	3.49E–02
273.15	3.80E–02
278.15	4.14E–02
283.15	4.54E–02
288.15	4.96E–02
293.15	5.49E–02

of lipids treated in this work. The database contains about 4500 measured data-points for 332 different phase equilibrium data sets, including binary and multicomponent systems (92 VLE, 91 LLE, 70 SLE, and 79 solubility data). Uncertainties of experimental measurements or quality estimates given by the authors of the experimental measured data were also considered. The published activity coefficients and parameter values from fitting different GE models (NRTL [4], UNIQUAC [5], and original UNIFAC [6]) for VLE and SLE binary systems are stored in the database. These models were used in parameter regressions for fine-tuning existing model parameters, improving VLE and SLE prediction, and obtaining model parameters not available in the literature. A total of 358 data sets from the DECHEMA® database for solid solubility systems, mainly for components important in the pharmaceutical industry were included in this work. More detailed description of the database (CAPEC_Lipids_Mixture_Database) can be found in Cunico et al. [15]. These data in addition to DECHEMA Chemistry Series Data Collection of Solid Solubility [19, 20] give a total of 11,351 different SLE data sets available for model parameter regression. One data set from DECHEMA Solid Solubility database [19, 20] containing an API was chosen to illustrate a solid solubility system and is given in Table 10.1. Paracetamol is the solute and the solid solubility of this chemical in different solvents, such as ethanol, methanol, 1-propanol, 1-butanol, 1-pentanol, acetronitrile, water, acetic acid, methyl ethyl ketone, and ethyl acetate, and can be found in the database [19, 20].

Lipids are often not tabulated in commonly used property databases, and their polyfunctional structure requires careful model analysis. Knowledge of the thermodynamic data and property models that consider the structure of the solid phase and consequently the polymorphism that may be present has been studied for lipids by others [22–31]. For example, triacylglycerols (TAGs), representing around 95% of the vegetable oils of interest, have been reported to have three polymorphs [22]. A thermodynamic model for fats and oils that consider the polymorphism of TAGs has been reported by Won [26]. Eutectic points and peritectic points can be observed in an SLE of lipid systems due to the similarity of the components involved, differing only in chain length and not in functional group representation. A characterization of peritectic point can be found in Slaughter and Doherty [32]. Costa et al. [33] report other mixtures where peritectic points occur, such as, binary systems of capric acid–myristic acid and lauric acid–myristic acid mixtures, mainly when the difference between the numbers of carbon atoms of the fatty acid chains in the mixture is less than 6. Costa et al. [33] demonstrated that the Slaughter and Doherty [32] approach for the prediction of the solid phases with an equilibrium constant for acid interactions provided good fits of the phase diagrams of systems with peritectic points. While the Slaughter and Doherty method [32] does not follow the Gibbs–Duhem equation, it has been used by many authors with good results, as in the work of Rocha and Guirardello [34].

10.2.3 SLE Thermodynamic Consistency Tests

"SLE data sets" are characterized here as those covering the entire composition range from the limits of pure-component melting points. The "solubility systems" are labeled as those data sets of limited composition range, where only one solid component precipitates. SLE systems can have similar types of errors as those found in VLE data sets. However, consistency tests based on the Gibbs–Duhem equation cannot be applied for solubility systems because there are no states where both component activities can be obtained simultaneously. In addition, there is normally a strong temperature dependence of the data, the pure component melting point limits are less well-identified than pure component vapor pressures, and the models typically used for describing nonideality in VLE may not be reliable for solid solubilities. Test-1 for SLE data is similar to the Qtest-5 of the ThermoData Engine (TDE) program developed by NIST [35–41] for VLE data, and the quality assessment algorithm is similar of those proposed by Kang et al. [42]. This test evaluates whether the mixture data asymptote to the pure-component melting points. The quality factor for Test-1 studied by Cunico et al. [3] presented some limitations and a more accurate proposal for quality factor calculation and Test-1 is proposed here:

$$Q_{\mathrm{SLE{\times}Test1}} = \left(\frac{2}{1+\left(\Delta t_1^0 + \Delta t_2^0\right)} - U \right) \tag{10.2}$$

where

$$\Delta t_1^0 = \left| \frac{T_{m1}^0 - t_1^0}{t_1^0} \right| \tag{10.3}$$

$$\Delta t_2^0 = \left| \frac{T_{m2}^0 - t_2^0}{t_2^0} \right| \tag{10.4}$$

and

$$U = \frac{1}{10}(\theta_1 + \theta_2) \tag{10.5}$$

In Equations 10.2–10.5, T_{mi}^0 is the measured or extrapolated melting point of the mixture in the limit $x_i \to 1$, x_i is the mole fraction of the component i, t_i^0 is the pure melting point temperature of component i, and θ_i is the absolute uncertainty in t_i^0. If the absolute uncertainty (θ_i) for the experimental data (t_i^0) is higher than the values found for the variable (Δt_i^0), then the quality factor can be considered equal to 1, once it is not possible to determine the quality of the mixture data end points.

Another method to analyze the consistency of the pure component data-points has been proposed by Kang et al. [43]. It analyzes the quality of the data-points close to the end points (molar fraction between $x_1 = 0$ and $x_1 = 0.2$, and between $x_1 = 0.8$ and $x_1 = 1$). In this methodology, the increase of the component 1 in the SLE can be calculated by considering the following equation:

$$\left(\frac{dT}{dx_1} \right)_{\mathrm{SLE}, x_1 = 1} = -\frac{RT^2}{\Delta H_f^0} \tag{10.6}$$

This requirement is based on the Gibbs energy and SLE condition:

$$\Delta G_{m,1}^0 = 0 \tag{10.7}$$

And its differential

$$d\Delta G_{m,1}^0 = \frac{\Delta H_{m,1}^0}{RT^2} dT + \left(\frac{\partial \ln(\gamma_1 x_1)}{\partial x_1} \right)_{T=\mathrm{const}} dx_1 \tag{10.8}$$

Here, $\Delta H_{m,1}^0$ is the enthalpy of transfer of component 1 from solid to liquid state (enthalpy of fusion if $x_1 = 0$). The activity coefficient can be considered the unit for x_1 ($x_1 = 1$), using the following approximation:

$$\left(\frac{\partial \gamma_1}{\partial x_1} \right)_{T=\mathrm{const}} = 0 \text{ at } x_1 = 1 \tag{10.9}$$

The quality factor is calculated using the following equation:

$$Q_{\text{SLE}\times\text{Test2}} = \left(\frac{0.2}{\Delta\text{Slope}_i}\right) \tag{10.10}$$

Here,

$$\Delta\text{Slope}_i = \left| \lim_{x_1 \to 0}\left[\frac{dx_2}{dT}\right]_{\text{model}} - \lim_{x_1 \to 0}\left[\frac{dx_2}{dT}\right]_{\text{expt}} \right| \tag{10.11}$$

Another consistency test, Test-3, is similar to that of Van Ness et al. [44] for VLE systems where the ability of a model to describe the data is assessed. The usefulness of this test depends on the reliability of the model for the described system. The earlier works [3, 15] used well-known GE model forms, such as the NRTL model. In order to evaluate whether any data might be assigned a lower Q_{SLE} because of model insufficiency instead of data error, an alternative activity coefficient model (Test-4, see Section 10.2.3.1) has been developed [3]. Both Test-3 and Test-4 can have the quality factor calculated by using the following equation:

$$Q_{\text{SLE}\times\text{Test3/4}} = \left(\frac{1}{1+\text{AAD}(\%)}\right) \tag{10.12}$$

Here average absolute deviation (AAD) in percent (%) is selected as the objective function for the regression:

$$\text{AAD}_{\%T} = \frac{100}{N}\sum_{i}^{N}\frac{\left|T_i^{\text{exp}} - T_i^{\text{calc}}\right|}{T_i^{\text{exp}}}, \quad \text{for } i = 1, N \tag{10.13}$$

$$\text{AAD}_{x_1} = \frac{100}{N}\sum_{i=1}^{N}\left|x_{1i}^{\text{exp}} - x_{1i}^{\text{calc}}\right|, \quad \text{for } i = 1, N \tag{10.14}$$

$$\text{AAD}_{\gamma_1} = \frac{100}{N}\sum_{i=1}^{N}\left|\gamma_{1i}^{\text{exp}} - \gamma_{1i}^{\text{calc}}\right|, \quad \text{for } i = 1, N \tag{10.15}$$

where T_i^{exp} is the measured temperature, x_{1i}^{exp} is the measured mole fraction, γ_{1i}^{exp} is the experimental activity coefficient, calculated by $\gamma_{1i}^{\text{exp}} = \exp\left[-\ln x_{1i}^{\text{exp}} + (\Delta H_{\text{fus}}/R)((1/T_{\text{m}}) - (1/T_i^{\text{exp}}))\right]$, and T_i^{calc}, x_{1i}^{calc}, and $\gamma_{1i}^{\text{calc}}$ are the temperature, mole fraction, and activity coefficient values calculated from the model at each of the N data-points, respectively. Note that values of $\text{AAD}_{\%T}$ are usually smaller than those of AAD_{x_1} since the former is a relative term while the latter is an absolute term; comparisons of the different AAD values should not be made. For consistency, calculated property values used only quantities from the correlations, not experimental values.

The use of the four tests provides the overall quality factor for the SLE data:

$$Q_{\mathrm{SLE}} = 0.25Q_{\mathrm{SLE\times Test1}} + 0.25Q_{\mathrm{SLE\times Test2}} + 0.25Q_{\mathrm{SLE\times Test3}} + 0.25Q_{\mathrm{SLE\times Test4}}, \quad Q_{\mathrm{SLE}} \leq 1 \tag{10.16}$$

10.2.3.1 Fluctuation Solution Theory (FST) At infinite dilution the solubility expression contains no hypothetical chemical potential of the solute [4, 45]. For dilute solutions, the Henry's law standard state can be more reliable than the pure component standard state since the unsymmetric convention activity coefficients, designated by γ_i^*, are often very close to unity. γ_i^* is related to γ_i by

$$\ln \gamma_i^* = \ln \gamma_i - \ln \gamma_i^\infty \tag{10.17}$$

where the infinite dilution activity coefficient is $\ln \gamma_i^\infty \equiv \lim_{x_i \to 0} \gamma_i$. This property is a function only of temperature or density and is often modeled with two parameters, a and b, simply as follows:

$$\ln \gamma_1^\infty = \frac{a+b}{T} \tag{10.18}$$

Fluctuation solution theory (FST) [46] shows that an expansion of the unsymmetric convention activity coefficient about infinite dilution has composition terms of the following form:

$$\ln \gamma_1^* = -f_2^0\left(2x_1 - x_1^2\right) - f_3^0\left(\frac{3}{2}x_1^2 - x_1^3\right) + \ldots \tag{10.19}$$

where the coefficients f_2^0 and f_3^0 are related to integrals of infinite-dilution molecular correlation functions and are functions only of temperature or density.

Combining Equations 10.1 and 10.19, and ignoring the ΔC_p term, yields an expression for solubility:

$$\ln x_1 = \frac{\Delta H_{\mathrm{fus}}}{R}\left(\frac{1}{T_\mathrm{m}} - \frac{1}{T}\right) + f_2^0\left(2x_1 - x_1^2\right) + f_3^0\left(\frac{3}{2}x_1^2 - x_1^3\right) + \left(\frac{a+b}{T}\right) \tag{10.20}$$

Sets of SLE data have been regressed with constant parameters, a and b, along with either constant f_2^0 or with $f_2^0 = c/T$. The temperature dependent $f_2^0 = c/T$ was found to be more accurate for the studied cases. For more complex systems where it is necessary to increase the temperature dependence of the thermodynamic model, the third-order term was also considered, with f_3^0 as constant or $f_3^0 = d/T$. Thus the FST model is

$$\ln x_1 = \frac{\Delta H_{\mathrm{fus}}}{R}\left(\frac{1}{T_\mathrm{m}} - \frac{1}{T}\right) + \frac{c}{T}\left(2x_1 - x_1^2\right) + \frac{d}{T}\left(\frac{3}{2}x_1^2 - x_1^3\right) + \left(\frac{a+b}{T}\right) \tag{10.21}$$

Here, the regression strategy was to choose a value of c and regress for a and b by modifying c and d until a minimum objective function value was found.

Once the parameter values are set, Equation 10.21 can be iteratively solved for temperature:

$$T = \frac{1}{\ln x_1}\left[\frac{\Delta H_{\text{fus}}}{R}\left(\frac{T}{T_{\text{m}}} - 1\right) + c\left(2x_1 - x_1^2\right) + d\left(\frac{3}{2}x_1^2 - x_1^3\right) + aT + b\right] \quad (10.22)$$

10.2.3.2 Parameter Regression For the regression of parameters for models, such as NRTL and UNIQUAC, the method of least squares is employed. In this method, the minimization of sum of the squares of the errors between the experimentally measured values and the calculated values using the selected model provides the estimated values of unknown model parameters and is given by

$$\Delta X = \sqrt{\sum_{i}^{\text{NC}} \frac{\left(X_i^{\text{lit}} - X_i^{\text{work}}\right)^2}{\text{NC}}}, \quad \text{for } i = 1, \text{NC} \quad (10.23)$$

where X_i^{lit} is the measured value, X_i^{work} is the calculated value (using the model) for each compound i, and NC is the total number of the compounds used in the parameter regression. In this work, X represents the pressure, the temperature, or the vapor molar fraction, depending on the problem formulation.

10.2.3.3 Case Studies for SLE Thermodynamic Consistency Tests and Model Performance The SLE consistency test and data evaluation is performed in a software containing options for data analysis, model analysis, and parameter regression. In this part of the chapter, data analysis for SLE data and the thermodynamic model performance (i.e., NRTL [4], UNIQUAC [5], and original UNIFAC [6]) are highlighted. Some examples of solid solubility and SLE data are selected to better exemplify the data evaluation. Figure 10.3 shows an example for the SLE thermodynamic consistency tests for a binary mixture of benzene and naphthalene. For this system, NRTL [4] and FST [3] give similar results in the representation of the experimental data, with final quality factor being 0.948. Pure component tests have also confirmed the quality of the end points ($x_1 = 0$ and $x_1 = 1$).

A solid solubility data set (salicylic acid and water) is selected to demonstrate the applicability of the thermodynamic consistency tests, and the results found for this system are highlighted in Figure 10.4. It is important to note that when the end points ($x_1 = 0$ and $x_1 = 1$) are not present in the data set, the quality factor for Test-1 and Test-2 is 0.5, that is, the quality of the data cannot be determined. The results shown in Figure 10.4 are quantified by the quality factors given in Table 10.2 (calculated using Eqs. 10.2, 10.10, 10.12, and 10.16).

The user has also the possibility to consider only the tests that are applicable. In the case of solid solubility data, for example, only Test-3 and Test-4, in the case that the end points are not given in many solid solubility data. Comparing regressions

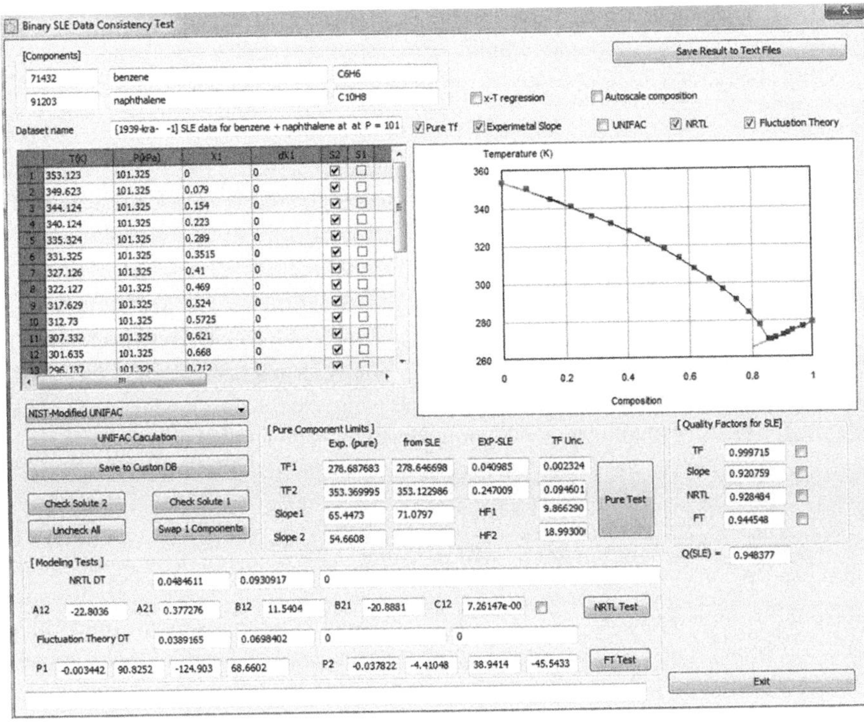

FIGURE 10.3 Screen shot from the software developed for thermodynamic consistency tests analysis. Experimental data for the binary mixture of benzene (1)+naphthlene (2) ■. Experimental data: Kravchenko [47] at pressure equal to 101.325 kPa using ■. Test-1 (Pure Test), — Test-2 (Slope), — Test-3 (NRTL model capability) and — Test-4 (FST).

from the NRTL and FST models point to some differences. For example, systems with noisy data are routinely better represented by the FST model [3].

A lipid data set containing peritectic point is also selected for analysis with the methodology for the SLE thermodynamic consistency tests, and the results are highlighted in Figure 10.5. The model performance observed here is confirmed by the results found in the uncertainty analysis of the parameter regression performed by Cunico et al. [3] for NRTL, UNIQUAC, UNIFAC, and FST models, where the regressed parameters play an important role in the intermediate points for NRTL, UNIQUAC, and original UNIFAC models, but for FST model, the parameters also influence the end points ($x_1=0$ and $x_1=1$), as stated in the examples shown by the authors [3]. It is possible to visualize in Figure 10.5 that the NRTL model tries to follow the tendency of the pure component data-points, which affects the model representation of experimental data.

Some examples of solid solubility and systems containing APIs were also selected to better illustrate the data analysis methodology for these kinds of compounds.

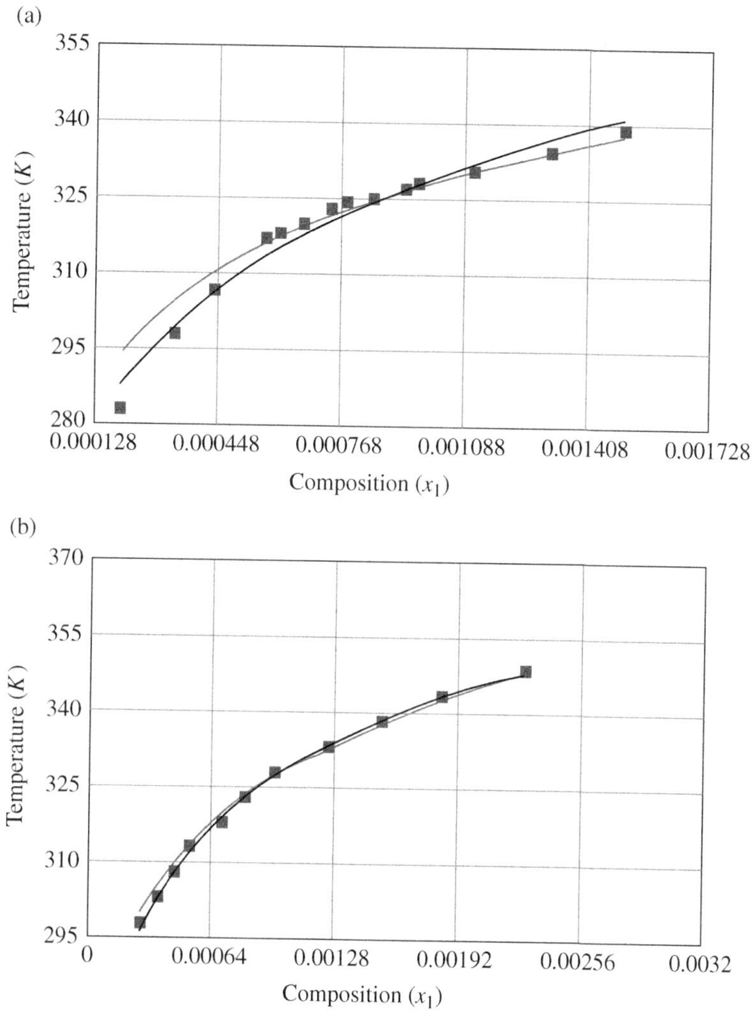

FIGURE 10.4 Binary mixture of salicylic acid (1) + water (2) ■. (a) Apelblat and Mazurola [48] and (b) Shalmashi and Eliassi [49] at pressure equal to 101.325 kPa using — Test-3 (NRTL model capability) and — Test-4 (FST).

TABLE 10.2 Quality Factors for the Binary Mixtures and the Proposed Methodology

\`t		Test-1	Test-2	Test-3	Test-4	Final Quality Factor
Salicylic acid + water	(a)	0.500	0.500	0.601	0.633	0.558
	(b)	0.500	0.500	0.739	0.850	0.648

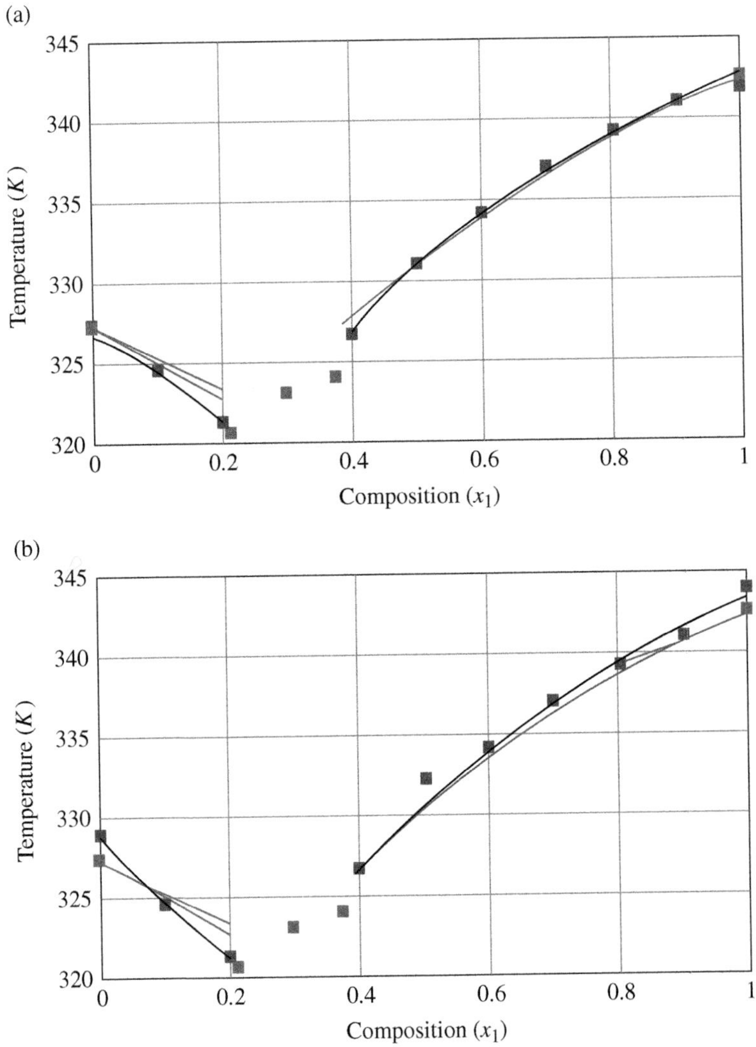

FIGURE 10.5 Binary mixture of myristic acid (1) + stearic acid (2) ■. Experimental data (a) Boros [50] and (b) Costa [51] at pressure equal to 101.325 kPa using ■. Data-points not used in the calculation (between eutectic and peritectic data-points) ■ Test-1 (Pure Test), — Test-2 (Slope), — Test-3 (NRTL model capability), and — Test-4 (FST).

Figure 10.6 shows the solid solubility of acetylsalicyclic acid (CAS-number 50-78-2) in two different solvents (water and ethanol). The application of the thermodynamic consistency tests showed high quality of the experimental data reported for this compound (acetylsalicylic, CAS-number 50-78-2), with a quality factor of 0.89 for both cases. Test-3 and Test-4 were considered in this analysis once the experimental data does not report the information for the end points ($x_1 = 0$ and $x_1 = 1$). The same was

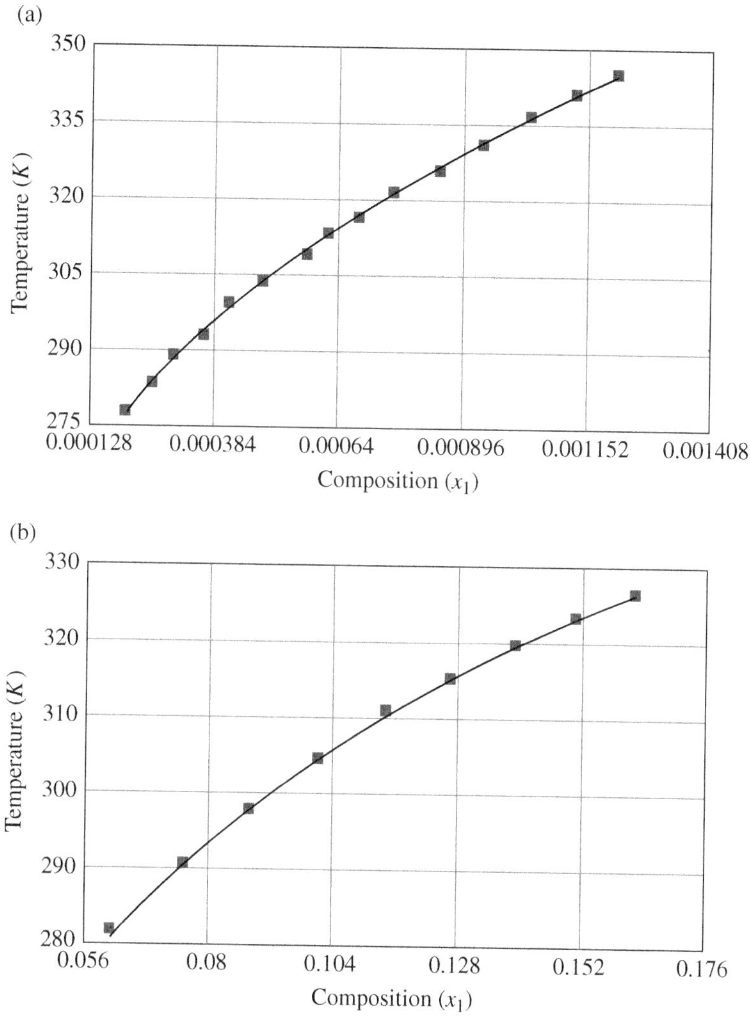

FIGURE 10.6 Binary mixture of: ■ (a) acetylsalicylic acid (1)+water (2), Apelblat and Mazurola [52] and (b) acetylsalicylic acid (1)+ethanol (2), Maia and Giulietti [53] at pressure equal to 101.325 kPa using — Test-3 (NRTL model capability) and — Test-4 (FST).

observed in the data sets containing anthracene+cyclohexane and paracetamol+*n*-propanol, showed in Figure 10.7. Here the quality factors calculated by the software are 0.918 and 0.887 for system (i) and system (ii), respectively.

10.2.3.4 UNIFAC Model Prediction It can be seen in Cunico et al. [3] that the well-known GE-models such as NRTL, UNIQUAC, and UNIFAC give only slightly different AAD values, with the FST model regression giving the lowest AAD and the original UNIFAC giving the highest, though the values are reasonably good. This is consistent with results shown in earlier work by the authors [15] on lipid VLE data.

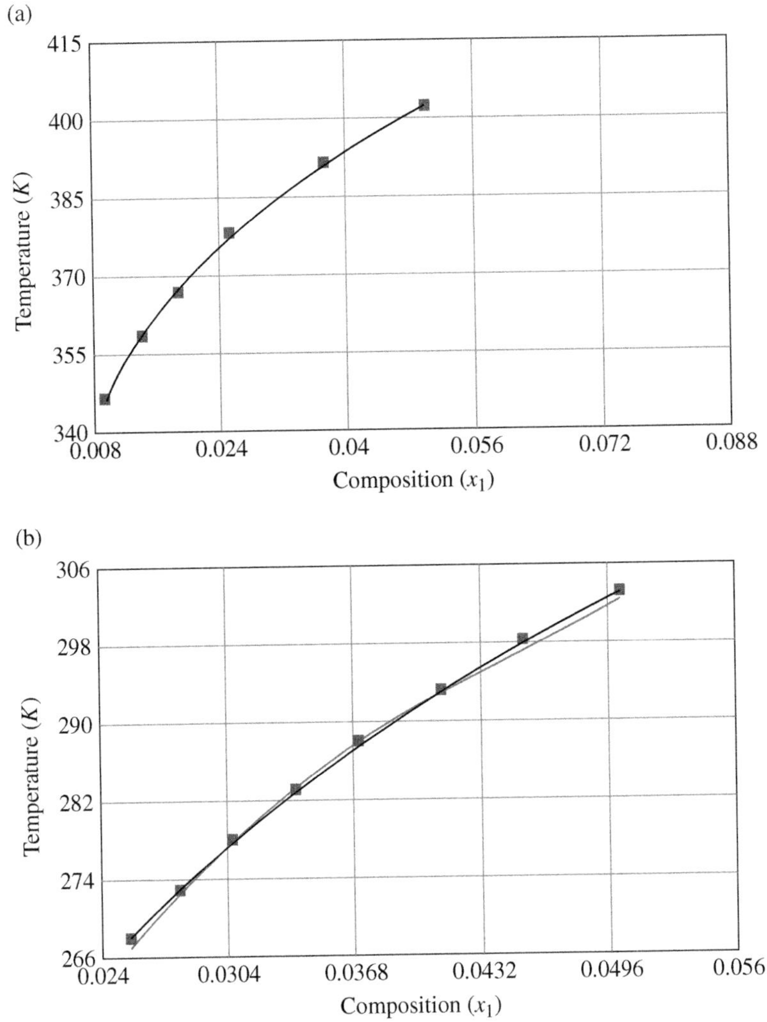

FIGURE 10.7 Binary mixture of: ■ (a) anthracene (1) + cyclohexane (2), Coon et al. [54] and (b) paracetamol (1) + *n*-propanol (2), Granberg and Rasmusson [21] at pressure equal to 101.325 kPa using Test-3 (NRTL model capability) and Test-4 (FST).

Similar observations have been also reported by Coelho et al. [55]. Some examples considering the NRTL model and the original UNIFAC model for lipid systems are selected and given in Table 10.3.

Since the original UNIFAC model parameters may not have been regressed with data from lipid systems, a possible way to improve the original UNIFAC performance is to fine-tune group interaction parameters using the lipid SLE data sets with their quality factors. This was done by regressing the interaction parameters for the functional group with the chain group, such as COOH with the CH3/CH2 group for

TABLE 10.3 SLE Model Performance Statistics for Lipid Systems

	Temperature	Parameters						
	AAD (%)	A_{12}/K[a]	A_{12}/K[b]	A_{21}/K[a]	A_{21}/K[b]	α_{12}[a]	α_{12}[b]	Reference
Lauric acid (1) + myristic acid (2) (316.94–327.48 K and 101,300 Pa)								
NRTL	0.102	−6719.60	−7476.75	35.98	448.67	0.97	0.31	[33]
Orig. UNIFAC	0.289	—		—		—		
Myristic acid (1) + palmitic acid (2) (327.07–335.02 K and 101,300 Pa)								
NRTL	0.062	574.25	755.89	−4570.86	4618.88	−0.49	−0.33	[56]
Orig. UNIFAC	0.098	—		—		—		
Methyl palmitate (1) + methyl stearate (2) (303.93–314.07 K and 101,300 Pa)								
NRTL	0.329	243.23	1096.17	−275.42	−1319.24	2.00	2.00	[57]
Orig. UNIFAC	0.337	—		—		—		

[a] *Aij*/K and α_{12} are the binary molecular parameters for the compounds *i* and *j* before the eutectic point.
[b] *Aij*/K and α_{12} are the binary molecular parameters for the compounds *i* and *j* after the eutectic point.

fatty acids. Note that groups used for original and modified UNIFAC parameter regression are presented in Cunico et al. [3] for lipid systems. This resulted in a lowering of the AAD, which was independent of the form of the objective function.

For pharmaceutical compounds, original UNIFAC shows poor model prediction for systems with large and complex molecules, mainly because some of the UNIFAC functional groups are missing or the functional group additivity rule is invalid [7]. An option for this problem is to regress the missing parameters considering only SLE solubility data, that is, the same procedure that has been successfully applied for lipid systems. Some examples of efforts to improve the original UNIFAC model performance have been attempted by Hahnenkamp et al. [58], who used the modified UNIFAC model to improve model representation of solubility of some pharmaceutical compounds, and, Abildskov and O'Connell [59], who have used minimal data and parameter fitting to obtain unknown UNIFAC parameters.

10.2.4 SolventPro

Although limited solubility data can be found in literature, the data task analysis can multiply rapidly when one considers the options of solvents and solvent–antisolvent mixtures, the effect of temperature dependence, the impacts of impurities, or the possibilities of multiple polymorphs, for example. Solvents are important as reaction mediums, reactants, or carriers in the chemical industry in general. This part of the chapter presents SolventPro, which is a computer-aided solvent selection and design framework, with a suite of models for solid solubility predictions and for solvent selection. The framework integrates different methods and tools needed to manage the complexity of pharmaceutical chemicals in an efficient, flexible, and robust way. In particular, the application of SolventPro with NRTL-SAC [7, 8] model for the solubility prediction, multilevel property estimation, and solution of some common problems encountered in the pharmaceutical industry are highlighted here. The NRTL-SAC model has been implemented and the obtained solid solubility results are compared with the combination of this model with the UNISAC model (this model predicts the segment parameters of the NRTL-SAC model) for some selected case studies. SolventPro provides also model regression features for the NRTL-SAC model with user-supplied experimental data. Alternatively, the UNISAC model may be used to predict the segment parameters, thereby avoiding the necessity of supplying experimental data.

10.2.4.1 NRTL-SAC The NRTL-SAC [7, 8] model provides a simple and practical thermodynamic framework for chemists and engineers to perform solubility modeling in pharmaceutical process design. While the original UNIFAC [6] model decomposes the molecules into a large set of predefined functional groups based on the chemical structure, NRTL-SAC [7, 8] maps molecules into a few predefined conceptual segments, or molecular descriptors, based on known characteristics of molecular interactions in solutions, what makes the use of this model more simple for the user. Specifically for each solute and solvent molecule, NRTL-SAC describes their effective surface interactions in terms of three types of conceptual segments: hydrophobic segment, polar segment, and hydrophilic segment. Equivalent numbers of the conceptual segments for each molecule are measures of the effective surface areas of the molecule that exhibit surface interaction characteristics of hydrophobicity (X), polarity (Y+ and Y−), and hydrophilicity (Z). According to Chen and Song [7] these conceptual segments are to be determined from experimental phase equilibrium data.

The NRTL-SAC model suggests that the activity coefficient for component i in solution is the sum of a combinatorial term and a residual term:

$$\ln \gamma_i = \ln \gamma_i^{\mathrm{C}} + \ln \gamma_i^{\mathrm{R}} \tag{10.24}$$

where, the combinatorial term is calculated from the Flory–Huggins equation for the combinatorial entropy of mixing, and the residual term is calculated from the local composition (lc) interaction contribution.

$$\ln \gamma_i^{\mathrm{R}} = \ln \gamma_i^{\mathrm{lc}} = \sum_m r_{m,I} \left[\ln \Gamma_m^{\mathrm{lc}} - \ln \Gamma_m^{\mathrm{lc},i} \right] \tag{10.25}$$

where the segment activity coefficient, Γ_m is calculated from the NRTL equation.

$$\ln \Gamma_m^{\mathrm{lc}} = \frac{\sum_j x_j G_{jm} \tau_{jm}}{\sum_k x_k G_{km}} + \sum_{m'} \frac{\sum_j x_{m'} G_{mm'}}{\sum_k x_k G_{km'}} \left(\tau_{mm'} - \frac{\sum_j x_j G_{jm'} \tau_{jm'}}{\sum_k x_k G_{km'}} \right) \tag{10.26}$$

$$\ln \Gamma_m^{\mathrm{lc},I} = \frac{\sum_j x_{j,I} G_{jm} \tau_{jm}}{\sum_k x_{k,I} G_{km}} + \sum_{m'} \frac{\sum_j x_{m',I} G_{mm'}}{\sum_k x_{k,I} G_{km'}} \left(\tau_{mm'} - \frac{\sum_j x_{j,I} G_{jm'} \tau_{jm'}}{\sum_k x_{k,I} G_{km'}} \right) \tag{10.27}$$

$$x_j = \frac{\sum_J x_J r_{j,J}}{\sum_I \sum_i x_J r_{i,I}} \tag{10.28}$$

$$x_{j,I} = \frac{r_{j,I}}{\sum_i r_{i,I}} \tag{10.29}$$

where i, j, k, m, and m' are the segment-based species indices, I and J are the component indices, x_j is the segment-based mole fraction of segment species j, x_J is the mole fraction of component J, $r_{m,I}$ are the number of segment species m contained in component I, Γ_m^{lc} is the activity coefficient of segment species m, and $\Gamma_m^{\mathrm{lc},I}$ is the activity coefficient of segment species m contained only in component I.

The segment-based parameters, that is, X, Y−, Y+, and Z are determined through regression of available experimental VLE or LLE data for binary systems of the reference compounds.

10.2.4.2 UNISAC The UNISAC model is based on a combination of the GC model [13], the CI model [18], and the NRTL-SAC [7, 8] itself. In other words, a GC model has been set up and developed to predict the NRTL-SAC [7, 8] segmental parameters that otherwise would require experimental data when group interaction parameters are missing. The CI method is used to supply the missing groups. GC together with CI is known as the GC+ approach. The result is a predictive model for estimation of NRTL-SAC model parameters.

GC model: The Marrero and Gani [13] GC method describes the properties of a pure compound using groups at three different levels: an initial approximation given by the contribution of first-order groups, an improvement provided by the second-order groups, that is further refined with the third-order groups. The detailed information of this model can be found in Marrero and Gani [13].

$$F(\theta)=\sum_i N_i C_i + w\sum_j M_j D_j + z\sum_k O_k E_k \tag{10.30}$$

CI model: The methodology proposed by Gani et al. [18] permits the creation of missing groups and the prediction of their contribution by using valence connectivity indices (${}^v\chi$) as described by Kier and Hall [60]. The property model equation is given by the following equation:

$$F(\theta)=\sum_i (a_i A_i)+b({}^v\chi^0)+2c({}^v\chi^1)+d \tag{10.31}$$

GC± model (GC ± CI): The combined GC+ model is given by the following equation:

$$F(\theta)=\sum_i N_i C_i + F(\theta^*)+ w\sum_j M_j D_j + z\sum_k O_k E_k \tag{10.32}$$

$$F(\theta^*)=\sum_k \left(n_k F(\theta)_k\right)+d, \quad \text{for } k=1\ldots K \tag{10.33}$$

where $F(\theta^*)$ is a function of the property θ for all missing groups/fragments and calculated by Equation 10.33. $F(\theta)_k$ is a function of the property θ for missing group/fragment k, which is calculated by Equation 10.30, n is the number of times a missing group/fragment appears in the molecule, and K is the total number of missing groups/fragments in a molecule.

10.2.4.3 Case Studies for SolventPro Paracetamol, also known as acetaminophen (CAS-number 103-90-32) is a widely used over-the-counter analgesic and antipyretic. In this case study the solubility curve of acetaminophen in methanol and acetone is generated through SolventPro using UNISAC model. NRTL-SAC parameters for acetaminophen were predicted by the GC+ model UNISAC using SolventPro as shown in Figure 10.8. SolventPro uses a created database of all the necessary groups and their contribution values to calculate the segment parameters. The description of the groups present in the solute molecule is easily obtained from the smiles string (e.g., for acetaminophen is CC(=O)NC1=CC=C(C=C1)O) of each compound.

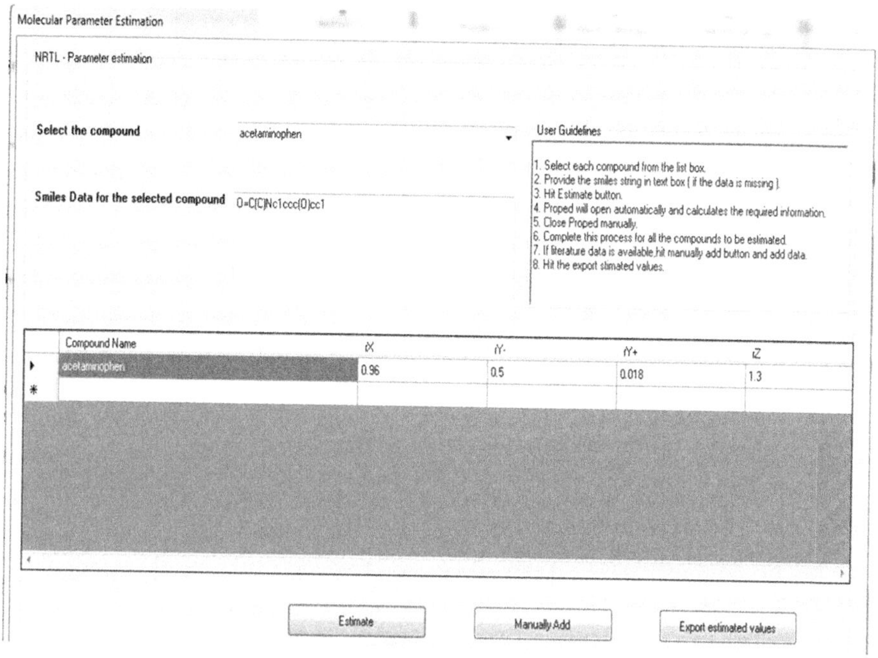

FIGURE 10.8 Screen shot from the SolventPro predicting SLE for the binary mixture of acetaminophen (1) + methanol (2) using UNISAC model.

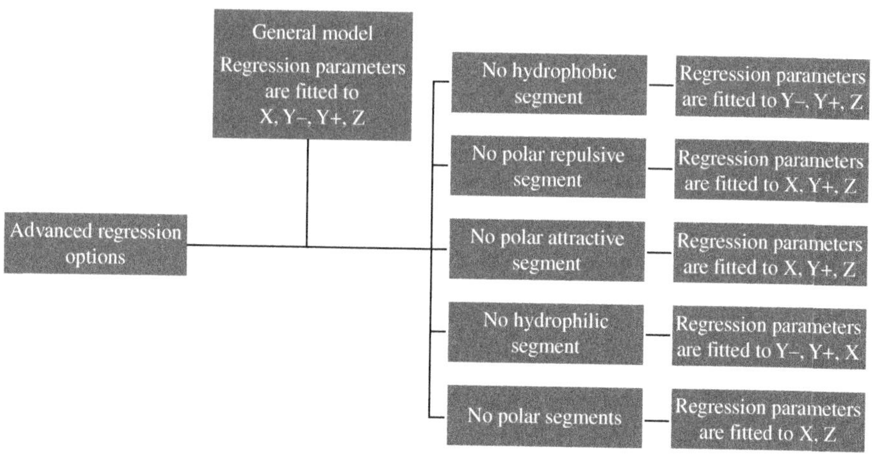

FIGURE 10.9 Screen shot of advanced regression options from the SolventPro software for fitting segmental parameters.

There are other options available for the parameter regression in the program, as showed in Figure 10.9. The program gives the option to the user to fit the segmental parameters based in the compound type. This is required since molecular parameters represent certain pairwise surface interaction characteristics, and often one or two

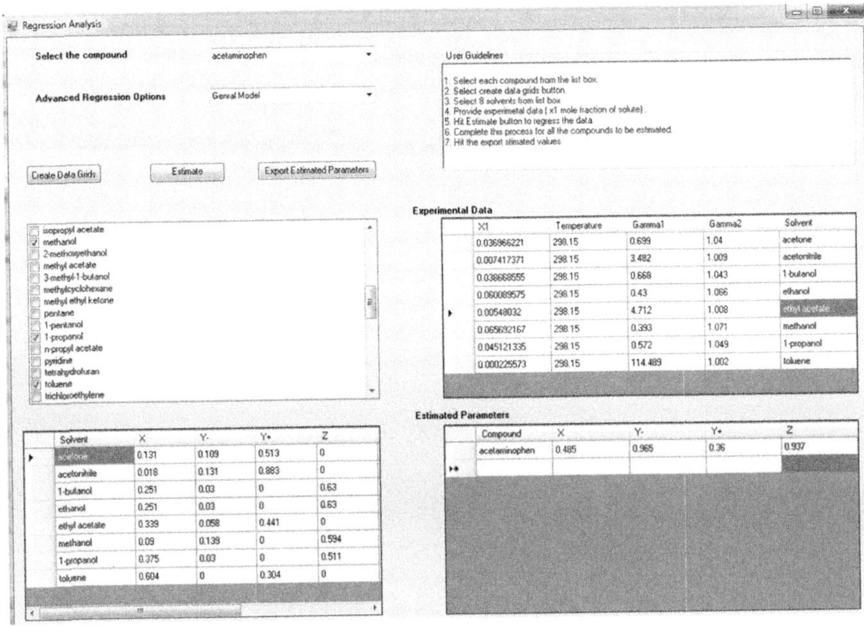

FIGURE 10.10 Screen shot from the SolventPro predicting NRTL-SAC parameters for acetaminophen using regression model.

molecular parameters are needed for most of the solvents. As examples, we have that alkanes are hydrophobic and well represented by only the hydrophobicity parameter, X. For compounds such as alcohols that are hybrids, hydrophobic and hydrophilic segments can be used and are primarily represented with X and Z, respectively. Figure 10.10 shows the SolventPro implemented regression analysis, which is based on the method proposed by Chen and Song [7], where solubility data of acetaminophen in eight different solvents are used to calculate the molecular parameters of acetaminophen.

The quality of NRTL-SAC model predictions depends on the quality of the experimental data used to identify the solute parameters, as observed previously in the SLE thermodynamic consistency tests subsection of this chapter. Figure 10.11 shows the comparison with experimental data, the predictions made with the parameters obtained through UNISAC model and through regression considering only NRTL-SAC. The prediction accuracy depends on the choice of the selected data. However, when the parameters or experimental data for the new compound is not available, UNISAC model becomes very useful.

Aspirin, also known as acetylsalicylic acid (CAS-number 50-78-2) is a salicylate drug often used as an analgesic to relieve minor aches and pains, as an antipyretic to reduce fever, and as an antiaspirin, also known as acetylsalicylic acid is a salicylate drug often used as an analgesic to relieve minor aches and pains, as an antipyretic to reduce fever, and as an inflammatory medication. In this case study, the solubility

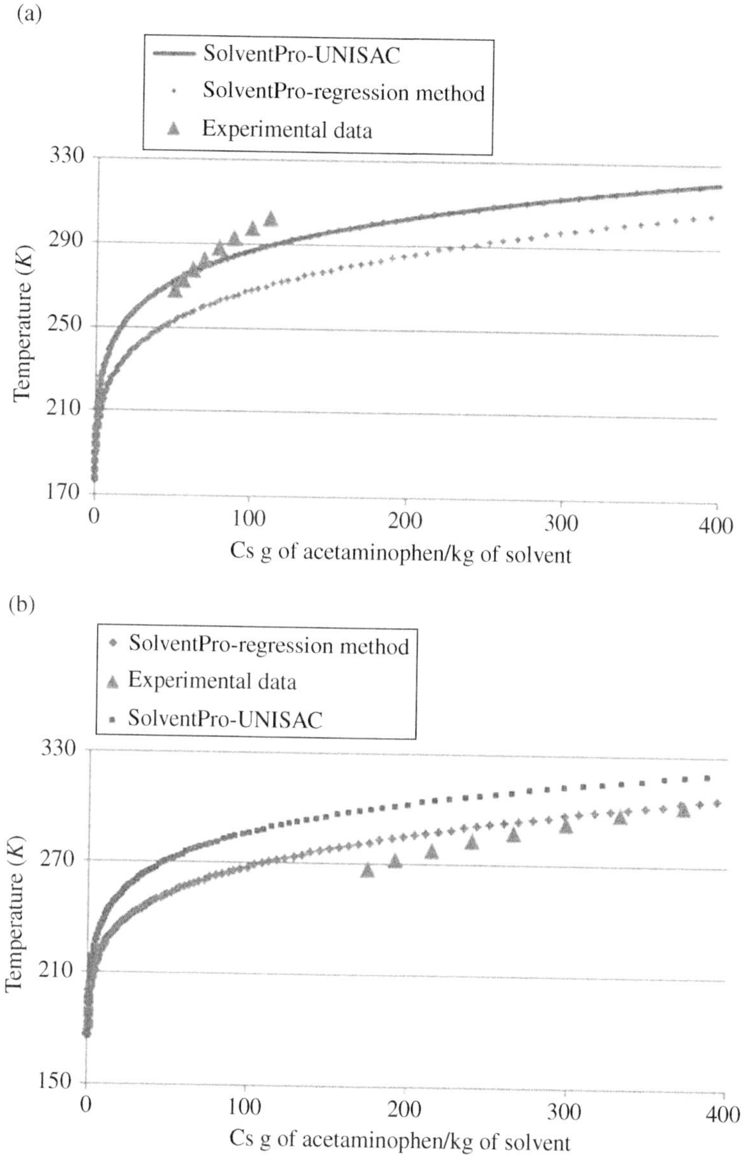

FIGURE 10.11 (a) SLE for binary mixture of acetaminophen (1) + methanol (2). (b) SLE for binary mixture of acetaminophen (1) + acetone (2). ▲ Experimental data [54].

curve of acetylsalicylic acid in 2-propanol and ethanol is generated through SolventPro using UNISAC model and regression model and the results are compared to available experimental data. NRTL-SAC parameters for acetylsalicylic acid were predicted by the UNISAC model. The results obtained are shown in Figure 10.12, where the solubility of aspirin in seven pure solvents was predicted. It is evident that

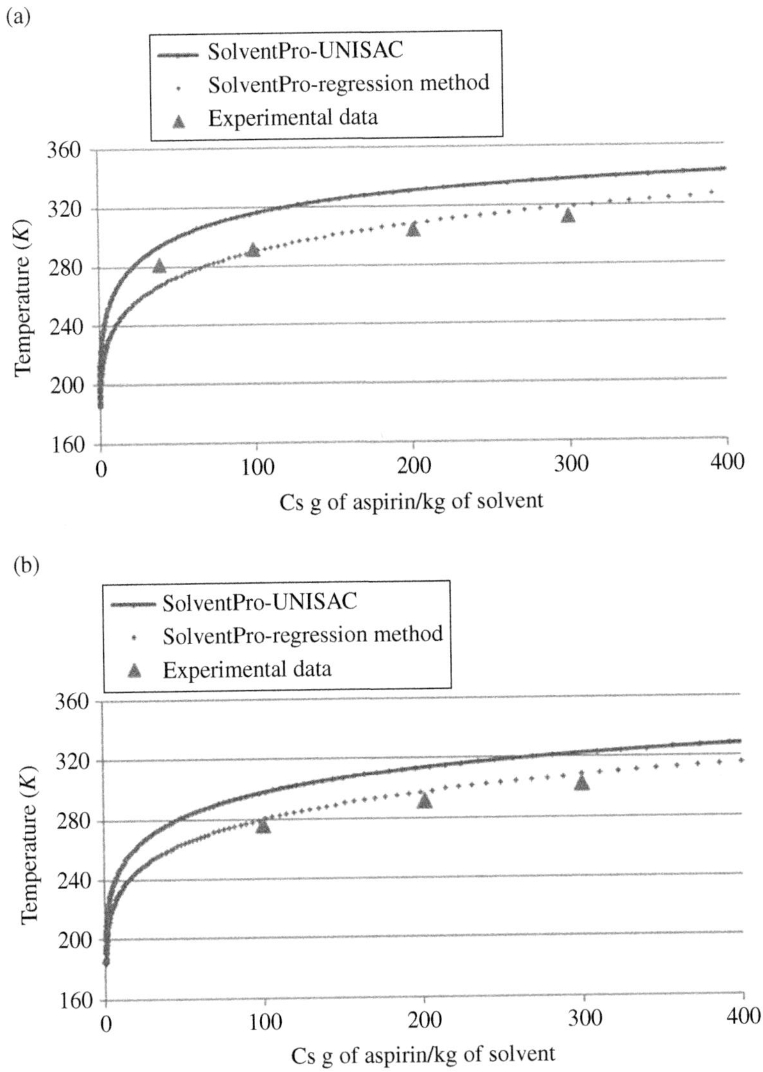

FIGURE 10.12 (a) SLE for binary mixture of aspirin (1) + 2-proponal (2). (b) SLE for binary mixture of aspirin (1) + ethanol (2). ▲ Experimental data [53].

the regression model closely matches with the solid solubility experimental data for all the solvents selected.

SolventPro also has the option to use the original UNIFAC model as one of the GE-model or the PC-SAFT equation of state [9, 10] to estimate the solid solubility as shown in Figure 10.13. Note that the PC-SAFT pure-component parameters are predicted through a GC method [9, 10].

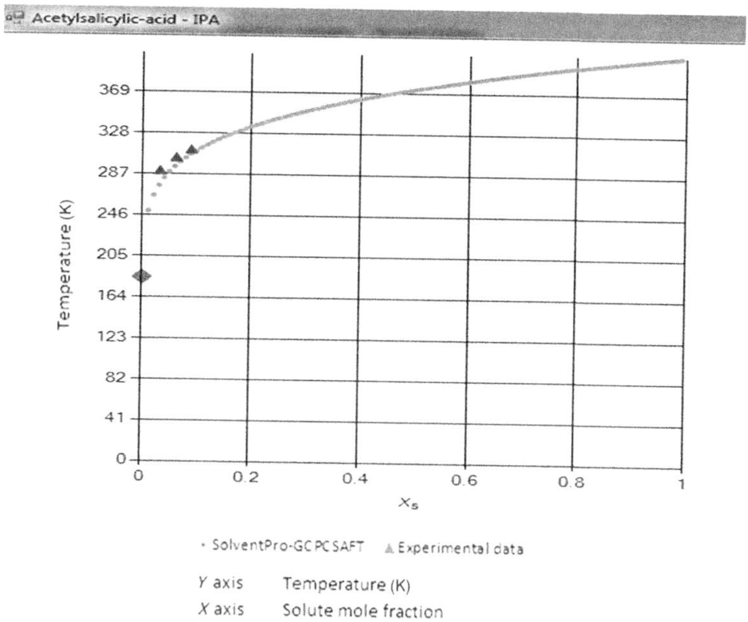

FIGURE 10.13 SLE for binary mixture of aspirin (1) + 2-proponal (2) using group contribution method for estimating PC-SAFT pure component parameters. ▲ Experimental data [53] ◆ Eutectic point.

10.3 CONCLUSION

The status of property and phase equilibria for solid solubility involving pharmaceutical and lipid systems has been reviewed and advanced by a more thorough investigation of SLE and solubility data, as well as by using an activity coefficient formulation based on fluctuation solution theory (FST). Though no rigorous consistency tests exist for such systems, using a reliable activity coefficient model along with comparing limits with independent pure component data allows quality factors to be established for complete composition range and limited range solubility SLE. It was found that the FST model is normally more accurate than either the NRTL or UNIQUAC models. The same approach adopted here for SLE quality factors including the FST model might also be applicable to liquid–liquid equilibrium systems.

The application of SolventPro with NRTL-SAC and the PC-SAFT models for solubility predictions, multi-level property estimation and solution of some common solvent-based calculations needed in pharmaceutical industry have been demonstrated using selected case studies. The NRTL-SAC model in combination with the UNISAC model and a GC version of the PC-SAFT have shown great potential in the calculation of solid solubility data. A comprehensive database of pure component data (or their reliable predicted values), however, is necessary for predictions of solid solubility for a wide range of chemical systems needed by the pharmaceutical industry. The option available in SolventPro is a very good start.

REFERENCES

[1] Petereit, A.; Saal, C., *Am. Pharm. Rev.* 2011, *14*, 68–73.

[2] Maximo, G. J.; Carareto, N. D. D.; Costa, M. C.; Santos, A. O.; Cardoso, L. P.; Krähenbühl, M. A.; Meirelles, A. J. A., *Fluid Phase Equilib.* 2014, *366*, 88–98.

[3] Cunico, L. P.; Ceriani, R.; Sarup, B.; O'Connell, J. P.; Gani, R., *Fluid Phase Equilib.* 2014, *362*, 318–327.

[4] Renon, H.; Prausnitz, J. M., *AIChE J.* 1968, *4*, 135–144.

[5] Abrams, D. S.; Prausnitz, J. M., *AIChE J.* 1975, *21*, 116–128.

[6] Fredenslund, A.; Jones, R. L.; Prausnitz, J. M., *AIChE J.* 1975, *21*, 1086–1099.

[7] Chen, C.-C.; Song, Y., *Ind. Eng. Chem. Res.* 2004, *43*, 8354–8362.

[8] Chen, C.-C.; Crafts, P., *Ind. Eng. Chem. Res.* 2006, *45*, 4816–4824.

[9] Gross, J.; Sadowski, G., *Ind. Eng. Chem. Res.* 2001, *40*, 1244–1260.

[10] Privat, R.; Gani, R.; Jaubert, J.-N., *Fluid Phase Equilib.* 2010, *295*, 76–92.

[11] Prausnitz, J. M., *Molecular Thermodynamics of Fluid-Phase Equilibria*, Prentice-Hall, Inc.: Englewood Cliffs, NJ, 1969.

[12] Nielsen, T.; Abildskov, J.; Harper, P.; Papaeconomou, I.; Gani, R., *J. Chem. Eng. Data.* 2001, *46*, 1041–1044.

[13] Marrero, J.; Gani, R., *Fluid Phase Equilib.* 2001, *183–184*, 183–208.

[14] Hukkerikar, A. S.; Sarup, B.; Ten Kate, A.; Abildskov, J.; Sin, G.; Gani, R., *Fluid Phase Equilib.* 2012, *321*, 25–43.

[15] Cunico, L. P.; Hukkerikar, A. S.; Ceriani, R.; Sarup, B.; Gani, R., *Fluid Phase Equilib.* 2013, *357*, 2–18.

[16] Joback, K. G.; Reid, R. C., *Chem. Eng. Commun.* 1987, *57*, 233–243.

[17] Constantinou, L.; Gani, R., *AIChE J.* 1994, *40*, 1697–1710.

[18] Gani, R.; Harper, P. M.; Hostrup, M., *Ind. Eng. Chem. Res.* 2005, *44*, 7262–7269.

[19] Marrero, J.; Abildskov, J., *Solubility and Related Properties of Large Complex Chemicals, Part 1: Organic Solutes Ranging from C4 to C40*, Volume *XV*, Chemistry Data Series. DECHEMA: Frankfurt, 2003.

[20] Abildskov, J., *Solubility and Related Properties of Large Complex Chemicals, Part 2: Organic Solutes Ranging from C2 to C41*, Volume *XV*, Chemistry Data Series. DECHEMA: Frankfurt, 2005.

[21] Granberg, R. A.; Rasmusson, A. C., *J. Chem. Eng. Data* 1999, *44*, 1391–1395.

[22] Sato, K., *Chem. Eng. Sci.* 2001, *56*, 2255–2265.

[23] De Loos, Th.; Giessen, R. V. D; Overbosch, P. Van Meeteren, J. A.; De Jong, S.; Peters, C.; Gandasasmita, I.; Royers, E.; Struik, M.; Grootscholten, P. A. M.; Don, A.; Wesdrop, H.; 2004, Liquid-Multiple Solid Phase Equilibria: Theory and Experiments, in *Fat Crystal Networks*, Marangoni, A. G., Eds. CRC Press: New York, 481–709.

[24] Himawan, C.; Starov, V. M.; Stapley, A. G., *Adv. Colloid Interface Sci.* 2006, *122*, 3–33.

[25] Teles Dos Santos, M.; Le Roux, C.; Gerbaud, V., *J. Am. Oil Chem. Soc.* 2011, *88*, 223–233.

[26] Won, K., *Fluid Phase Equilib.* 1993, *82*, 261–273.

[27] Wille, R. L.; Lutton, E. S., *J. Am. Oil Chem. Soc.* 1966, *43*, 491–496.

[28] Bouzidi, L.; Narine, S. S., *Chem. Phys. Lipids* 2012, *165*, 105–119.

[29] Vereecken, J.; De Graef, V.; Smith, K. W.; Wouters, J.; Dewettinck, K., *Food Res. Int.* 2010, *43*, 2057–2067.

[30] Campbell, S. D.; Goff, H. D.; Rousseau, D., *J. Am. Oil Chem. Soc.* 2004, *81*, 213–220.

[31] Widlak, N.; Hartel, R.; Narine, S., *Cristallization and Solid Properties of Lipids*, AOCS Press: Urbana, IL, 2001.

[32] Slaughter, D. W.; Doherty, M. F., *Chem. Eng. Sci.* 1995, *50*, 1679–1694.

[33] Costa, M. C.; Rolemberg, M. P.; Boros, L. A. D.; Krähenbühl, M. A.; Oliveira, M. G.; Meirelles, A. J. A., *J. Chem. Eng. Data* 2007, *52*, 30–36.

[34] Rocha, S.; Guirardello, R., *Fluid Phase Equilib.* 2009, *281*, 12–21.

[35] Frenkel, M.; Chirico, R. D.; Diky, V.; Yan, X.; Dong, Q.; Muzny, C., *J. Chem. Eng. Data* 2005, *45*, 816–838.

[36] Diky, V.; Muzny, C. D.; Lemmon, E. W.; Chirico, R. D.; Frenkel, M., *J. Chem. Inf. Model.* 2007, *47*, 1713–1725.

[37] Diky, V.; Chirico, R. D.; Kazakov, A. F.; Muzny, C. D.; Frenkel, M., *J. Chem. Inf. Model.* 2009, *49*, 503–527.

[38] Diky, V.; Chirico, R. D.; Kazakov, A. F.; Muzny, C. D.; Frenkel, M., *J. Chem. Inf. Model.* 2009, *49*, 2883–2896.

[39] Diky, V.; Chirico, R. D.; Kazakov, A. F.; Muzny, C. D.; Magee, J. W.; Abdulagatov, I.; Kang, J. W.; Kroenlein, K.; Frenkel, M., *J. Chem. Inf. Model.* 2011, *51*, 181–194.

[40] Kroenlein, K.; Muzny, C. D.; Diky, V.; Chirico, E. D.; Kazakov, A. F.; Chirico, R. D.; Magee, J. W.; Abdulagatov, I.; Frenkel, M., *J. Chem. Inf. Model.* 2011, *51*, 1506–1512.

[41] Diky, V.; Chirico, R. D.; Muzny, C. D.; Kazakov, A. F.; Kroenlein, K.; Magee, J. W.; Abdulagatov, I.; Kang, J. W.; Frenkel, M., *J. Chem. Inf. Model.* 2011, *52*, 260–276.

[42] Kang, J. W.; Diky, V.; Chirico, R. D.; Magee, J. W.; Muzny, C. D.; Abdulagatov, I.; Kazakov, A. F.; Frenkel, M., *J. Chem. Eng. Data* 2010, *55*, 3631–3640.

[43] Kang, J. W.; Diky, V.; Chirico, R. D.; Magee, J. W.; Muzny, C. D.; Kazakov, A. F.; Kroenlein, K.; Frenkel, M., *Chem. Eng. Data* 2014, *59*, 2283–2293.

[44] Van Ness, H. C.; Byer, S. M.; Gibbs, R. E., *AIChE J.* 1973, *19*, 238–244.

[45] O'Connell, J. P.; Prausnitz, J. M., *Ind. Eng. Chem. Fundam.* 1964, *3*, 347–351.

[46] O'Connell, J. P., *Mol. Phys.* 1971, *20*, 27–33.

[47] Kravchenko, V. M., *Zh. Fiz. Khim.* 1939, *13*, 133–145 (cited in Marrero, J.; Abildskov, J., Solubility and Related Properties of Large Complex Chemicals, Part 1: Organic Solutes ranging from C4 to C40, Volume XV, Chemistry Data Series. DECHEMA: Frankfurt, 2003).

[48] Apelblat, A.; Manzurola, E., *J. Chem. Thermodyn.* 1989, *21*, 1005–1008.

[49] Shalmashi, A.; Eliassi, A., *J. Chem. Eng. Data* 2008, *53*, 199–200.

[50] Boros, L. A. D., *Mathematical thermodynamics modeling and of solid-liquid equilibrium of fatty systems*, M.Sc. Thesis, University of Campinas, Campinas, 2005.

[51] Costa, M. C., *Experimental determination of solid-liquid equilibrium for binary systems of saturated fatty acids: a study detailed of the solid phase*, D.Sc. Thesis, University of Campinas, Campinas, 2008.

[52] Apelblat, A.; Manzurola, E., *J. Chem. Thermodyn.* 1999, *31*, 85–91.

[53] Maia, G. D.; Giulietti, M., *J. Chem. Eng. Data* 2008, *53*, 256–258.

[54] Coon, J. E.; Sediawan, W. B.; Auwaerter, J. E.; MacLaughlin, E., *J. Solut. Chem.* 1988, *17*, 519–534.

[55] Coelho, R.; Santos, P. G.; Mafra, M. F.; Cardozo-Filho, L.; Corazza, M. L., *J. Chem. Thermodyn.* 2011, *43*, 1870–1876.

[56] Rolemberg, M. P., *Solid liquid equilibrium of fatty acids and triacylglycerols: experimental determination and modeling*, Ph.D. Thesis, University of Campinas, Campinas, 2002.

[57] Costa, M. C.; Boros, L. A. D.; Coutinho, J. A. P.; Krähenbühl, M. A.; Meirelles, A. J. A., *Energy Fuels* 2011, *25*, 3244–3250.

[58] Hahnenkamp, I.; Graubner, G.; Gmehling, J., *Int. J. Pharm.* 2010, *388*, 73–81.

[59] Abildskov, J.; O'Connell, J. P., *Ind. Eng. Chem. Res.* 2003, *42*, 5622–5634.

[60] Kier, L. B.; Hall, L. H., *Environ. Res.* 2001, *12*, 55–74.

11

MOLECULAR SIMULATION METHODS TO COMPUTE INTRINSIC AQUEOUS SOLUBILITY OF CRYSTALLINE DRUG-LIKE MOLECULES

David S. Palmer[1] and Maxim V. Fedorov[2]

[1] *Department of Pure and Applied Chemistry, University of Strathclyde, Glasgow, UK*
[2] *Scottish Universities Physics Alliance (SUPA), Department of Physics, University of Strathclyde, Glasgow, UK*

11.1 INTRODUCTION

Accurate computational methods to predict the solubility of crystalline organic molecules in aqueous solutions are highly sought after in many fields of the biomolecular sciences and industry. For example, predictions of solubility are required in the pharmaceutical and agrochemical industries to assess the bioavailability of de novo designed drugs and the environmental fate of potential pollutants, respectively [1–4]. Due in part to the requirements of industry, interest in the prediction of solubility has risen dramatically in recent years, with hundreds of articles published in the last decade alone.

The most widely used methods to predict aqueous solubility from molecular structure are quantitative structure–property relationships (QSPRs) [4–6], which are empirical models that use experimental data to learn a statistical relationship between the physical property of interest (i.e., solubility) and molecular descriptors calculable from a simple computational representation of the molecule (e.g., counts of atoms or functional groups, polar surface area, and molecular dipole moment) [1]. The current

Computational Pharmaceutical Solid State Chemistry, First Edition. Edited by Yuriy A. Abramov.

state of the art allows the prediction of aqueous solubility with root mean square errors (RMSEs) of approximately 0.3–0.4 log units for simple organic molecules and 0.7–1.0 log units for drug molecules [7–11]. Although QSPR models are widely used for high-throughput *in silico* screening [12–14], they have some drawbacks. It is often observed, for example, that QSPR models are unreliable for molecules dissimilar to those in the training set. Furthermore, since QSPRs are not based on any fundamental physical/chemical theory, they provide little information about the underlying physical chemistry and are difficult to systematically improve. In all but a few cases [15–17], QSPR models predict solubility from molecular rather than crystal structure, which means they are not able to rationalize or predict different solubilities for different polymorphs of a molecule. New methods to predict solubility from molecular structure would have enormous scientific and economic value.

One promising approach to predicting aqueous solubility is to calculate it directly from theory and/or molecular simulation. In this chapter, we will discuss the different molecular theory and simulation-based approaches that have been used to calculate the intrinsic aqueous solubility of drug-like molecules. We also devote some of this chapter to the computation of other thermodynamic parameters that are required for the prediction of solubility such as hydration and sublimation free energies. Moreover, since a lack of accurate experimental data is currently a limiting factor in developing and testing all classes of computational solvent models, we also discuss methods to measure solubility and the extent and reliability of available data.

11.2 DEFINITIONS OF SOLUBILITY

Experimental measurements of solubility are influenced by many different factors, including the purity of the solute and solvent, presence of cosolvents, presence of salts, temperature, physical form of the undissolved solute, ionization state, and solution pH [18]. Consequently many different definitions of solubility are in common use in the published literature. Here we discuss the intrinsic aqueous solubility, S_0, which is defined as the concentration of the neutral form of the molecule in saturated aqueous solution at thermodynamic equilibrium at a given temperature [18–20]. Intrinsic aqueous solubility is used to calculate dissolution rate and pH-dependent solubility in models such as the Noyes–Whitney equation [21] and the Henderson–Hasselbalch equation [22, 23], respectively. Prediction of the intrinsic aqueous solubility of bioactive molecules is of great importance in the biochemical sciences because it is a key determinant in the bioavailability of novel pharmaceuticals [1, 3, 24–26] and the environmental fate of potential pollutants [27, 28].

11.3 SOLUBILITY AND THERMODYNAMICS

11.3.1 Solubility and Free Energy of Solution

The intrinsic aqueous solubility of a crystalline solute is measured at thermodynamic equilibrium between the undissolved crystalline form of the molecule and the neutral form of the molecule in solution, which can be written $X_s \rightleftarrows X_{aq}$. If the activity

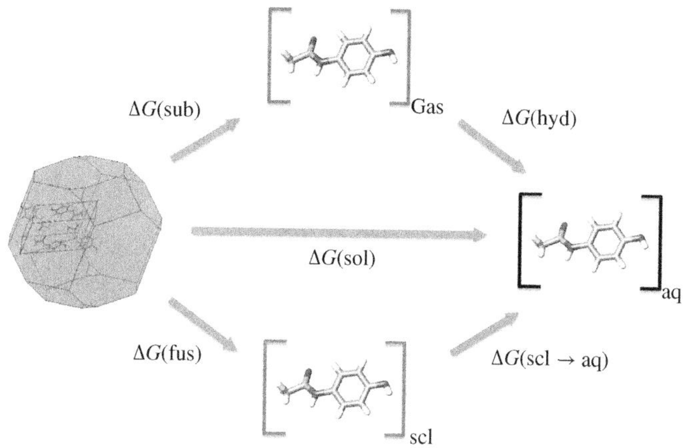

FIGURE 11.1 Illustration of two different thermodynamic cycles: (top) crystal to gas phase to solution phase and (bottom) crystal to supercooled liquid phase to solution phase.

coefficient for the solute in solution is assumed to be unity, then the relationship between intrinsic solubility (S_0) and the overall change in Gibbs free energy is [29]:

$$\Delta G^*_{\text{sol}} = -RT\ln\left(S_0 V_{\text{m}}\right) \quad (11.1)$$

where ΔG^*_{sol} is the Gibbs free energy for solution, R is the molar gas constant, T is the temperature (298 K), V_{m} is the molar volume of the crystal, S_0 is the intrinsic solubility in moles per liter, and the superscript * denotes that we are using the Ben-Naim terminology [30, 31], which refers to the Gibbs free energy for transfer of a molecule between two phases at a fixed center of mass in each phase.

The most tractable approach to calculating aqueous solubility from molecular simulation is via computation of the free energy of solution (ΔG_{sol}), which is the free energy change associated with transfer of the molecule from the crystalline phase to aqueous solution under standardized conditions. Since the solution free energy cannot easily be calculated from a single simulation, it is often decomposed into terms that can be computed in separate simulations (Fig. 11.1), as will be discussed later. The calculation of solubility by this approach is a significant challenge because it requires prediction of crystal structure and an accurate computation of solution free energy (at 298K an error of $RT\ln 10 \approx 5.7$ kJ/mol in ΔG_{sol} equates to a 10-fold error in solubility).

11.3.2 Computation of Solubility from the Thermodynamic Cycle of Solid to Supercooled Liquid to Aqueous Solution

The thermodynamic cycle of solid drug to supercooled liquid and then to aqueous solution has been the basis of several different methods to predict intrinsic aqueous solubility. A supercooled liquid is a hypothetical state in which a pure solute that would normally be solid at the experimental temperature is considered to behave as a liquid. Although the supercooled liquid state is not accessible by experiments, it is introduced into the thermodynamic cycle to allow the process of breaking down the crystal lattice (step 1: crystal to supercooled liquid) to be decoupled from the process of hydration (step 2: supercooled liquid to aqueous solution).

One of the popular methods for prediction of solubility from this thermodynamic cycle is the general solubility equation (GSE) [32, 33], which relates log$S$ to melting point (T_{m}) and the logarithm of the octanol–water partition coefficient (log$P$).

$$\log S = 0.5 - \log P - 0.01\left(T_{\mathrm{m}} - 25\right) \tag{11.2}$$

Although it is not strictly a molecular simulation method, we mention the GSE here since it can be derived from the thermodynamic cycle of crystal to supercooled liquid to solution provided that some assumptions are made about the entropy of melting, ΔS_{m}. The GSE provides useful estimates of solubility when experimental melting point and ΔS_{m} data are available [34]. However, the GSE is not usually applicable to unsynthesized molecules as the best empirical methods for predicting melting point give predictive errors of 40–50°C [35, 36]. The GSE has provided the basis of empirical methods to predict solubility such as the Solubility Forecast Index [37].

A thermodynamic cycle similar to that discussed in this paragraph has also been used as the basis of a systematic study of pure amorphous forms of drug-like molecules by molecular simulation methods [38–41]. The authors use Monte Carlo simulations and the free energy perturbation method to compute the free energies associated with transferring a drug molecule from gas phase into water (ΔG_{hyd}, hydration free energy) and from gas phase into pure amorphous drug phase (ΔG_{ga}) [38–41] at 25°C. An estimate of the solubility of the pure amorphous form of the drug-like molecule can then be obtained from the following equation [40]:

$$S_{\mathrm{a}} = \frac{\exp\left(-\Delta G_{\mathrm{aw}} / RT\right)}{V_{\mathrm{m,a}}} \tag{11.3}$$

Here, S_{a} is the solubility of the amorphous form, ΔG_{aw} is the free energy for transfer of the drug-like molecule from amorphous phase to aqueous solution ($\Delta G_{\mathrm{aw}} = -\Delta G_{\mathrm{ga}} + \Delta G_{\mathrm{hyd}}$), and $V_{\mathrm{m,a}}$ is the molar volume of the drug-like molecule in the amorphous phase. Since free energy perturbation simulations are too computationally expensive for routine use in computing solubility in the pharmaceutical industry, the authors of Refs. [38–41] also develop and test an approximate theory to generate the free energy differences required to determine solubility. In this "simplified response" (SR) theory, it is assumed that both ΔG_{hyd} and ΔG_{ga} can be computed from the same *ansatz*:

$$\Delta G_{\mathrm{aw}} = -\Delta G_{\mathrm{ga}} + \Delta G_{\mathrm{hyd}} = -\left(\Delta G_{\mathrm{cav,a}} + E_{\mathrm{LJ,a}} + \frac{E_{\mathrm{C,a}}}{2}\right) + \left(\Delta G_{\mathrm{cav,w}} + E_{\mathrm{LJ,w}} + \frac{E_{\mathrm{C,w}}}{2}\right) \tag{11.4}$$

Here, ΔG_{cav} is the free energy of cavity formation (obtained using a theory for hard oblate spheroids), E_{LJ} and E_C are the Lennard-Jones and Coulomb interaction energies, respectively, between the chosen molecule and the others in the fluid, and the subscripts a and w refer to amorphous and water phases, respectively. The SR theory assumes that the response of the system to electrostatic interactions can be described by linear response theory and that the response to dispersion or other induction and correlation interactions can be described by mean field theory [38–41].

Since there are relatively few measurements of amorphous solubility available in the published literature, the SR method was benchmarked on data obtained from experimental measurements of intrinsic aqueous solubility, entropy of melting (ΔS_m), and melting point (T_m), using the following relationship [40]:

$$S_a \approx S_0 \exp\left(\Delta S_m / R \ln\left(T_m / T\right)\right) \tag{11.5}$$

The results were found to be sensitive to the choice of force field and atomic partial charges used in the simulations, but for the best tested parameters a reasonable correlation between experimental and computed amorphous solubilities was observed [40, 41]. It was shown in Ref. [41] that after reparameterization of the model and introducing an empirical relationship through linear regression analysis of experimental and calculated data ($\Delta G_{aw,calc} = 2.04\Delta G_{aw,exp}$), amorphous solubility could be predicted with a correlation of $r^2 = 0.65$ and a root mean-square deviation (RMSD) of approximately 1 log$S$ unit [41]. At the current time, experimental ΔS_m and T_m data are required in order for this approach to be used to compute intrinsic aqueous solubility, but this may change in the future.

11.3.3 Computation of Solubility from the Thermodynamic Cycle of Solid to Gas Phase to Aqueous Solution

The thermodynamic cycle for transfer from crystal to vapor to solution (Fig. 11.2) provides a convenient method to compute solubility since both of the related thermodynamic parameters (sublimation and hydration free energy) are accessible by both experiment and computation. Perlovich et al. have published a series of papers that investigate the thermodynamic properties of a wide variety of structurally diverse drugs, by both experiment and computation, but they have not provided any molecular simulation methods for the prediction of solubility from structure alone without empirical parameterization [42, 43].

One of the successful computational methods to be based on the thermodynamic cycle of pure compound to gas phase to solution was the prediction of the solubility of liquids (and a small number of low-molecular-weight solids) from both experimental and calculated vaporization and hydration free energies by Thompson et al. [29]. The free energies were computed using the SM5.42R implicit continuum solvation model combined with density functional theory (DFT) (B3LYP functional and MIDI! basis set), where vaporization free energy was computed as a self-solvation free energy (i.e., where solute and solvent molecules are identical). Predictive mean unsigned errors in the range of 0.4–0.6 in log$S$ units were reported

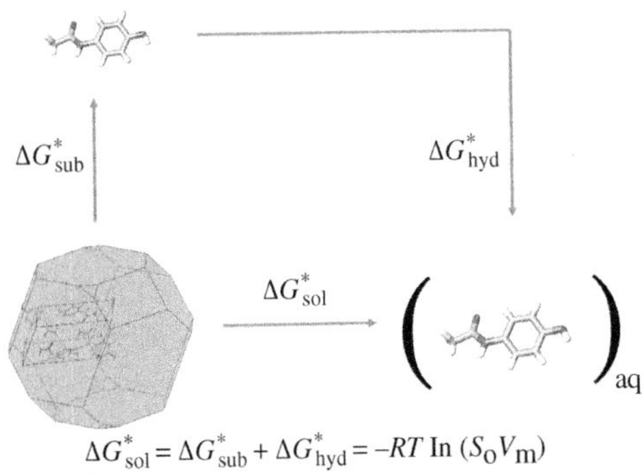

FIGURE 11.2 Thermodynamic cycle for transfer from crystal to gas and then to aqueous solution. Source: Palmer et al. [15]. Reprinted with permission of American Chemical Society.

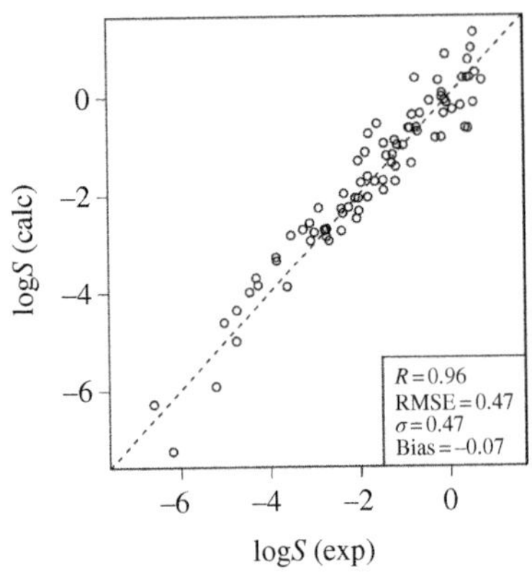

FIGURE 11.3 Plot of experimental versus calculated solubility data, where the latter were computed using the SM5.42R solvent model at the B3LYP level of theory combined with the MIDI! basis set. The dataset comprises low-molecular-weight organic molecules, most of which are liquids at the experimental temperature. This figure was prepared from the data in the supporting information provided in the original article by Thompson et al. [29].

for a dataset comprising simple (non-drug-like) low-molecular-weight organic compounds (Fig. 11.3). The problem with this approach is that it does not provide a theoretically rigorous method to model molecular crystals.

Schnieders et al. have recently proposed a method to predict the solubility of organic crystals from molecular dynamics simulations, using the polarizable force field AMOEBA (atomic multipole optimized energetics for biomolecular simulation) with an orthogonal space–based free energy algorithm [44]. For a small dataset comprising four low-molecular-weight n-alkylamides (acetamide, butanamide, pentanamide and hexanamide), solubility was predicted with a RMSE of 0.83 log*S* units (calculated from data presented in Table 8 of the original manuscript [44]). The molecular simulation results were used to rationalize the experimentally observed decrease in solubility as a function of *n*-alkylamide chain length, which for these molecules occurs due to an increasing stability of the crystalline state and to a lesser degree due to decreasing favorability of solvation (i.e., the hydrophobic effect). Reinwald et al. predicted the aqueous solubility of a more diverse selection of drugs from experimental enthalpies of sublimation, ΔH_{sub}, and calculated hydration energies [45]. Unfortunately, this method was only accurate to within 1 log$S$ unit for one molecule from a dataset of 12, and no computational procedure was suggested for the calculation of ΔH_{sub}. Palmer et al. have published two articles on predicting solubility from calculation of sublimation and hydration free energies [15, 46]. In the first work, where ΔG_{hyd} was calculated using an implicit solvent model based upon the Poisson–Boltzmann equation, *ab initio* results were not found to deliver the required accuracy, but after the introduction of a small number of empirical corrections, accurate predictions of a drug-like test set were obtained. The resulting linear regression model is for all intents and purposes a QSPR model, but one that includes a computed lattice energy as a molecular descriptor to account for the influence of crystal lattice interactions on solubility. The use of sublimation and free energies as molecular descriptors in machine learning QSPR models has also been investigated [10]. While direct theoretical calculation did not give accurate results in this approach, machine learning models gave predictions with a root mean square error (RMSE) of 1.1 log$S$ units for a dataset of 100 drug-like molecules. The hydration free energies used in that work were computed using implicit continuum solvation models. In the second article by Palmer et al., the intrinsic aqueous solubilities of 25 crystalline drug-like molecules were computed without direct empirical parameterization against experimental solubility data by combining sublimation free energies calculated using crystal lattice simulations with hydration free energies computed using the three-dimensional (3D) reference interaction site model (3D-RISM) of the integral equation theory (IET) of molecular liquids. The solubilities of 25 crystalline drug-like molecules taken from different chemical classes were computed with a correlation coefficient of $R=0.85$ and a RMSE equal to 1.45 log$S$ units. Although these results are not as accurate as those reported by purely empirical approaches, they are encouraging because there is great scope for systematic improvement. We note that this is the first "bottom-up" method in the area (i.e., the method is not directly parameterized against experimental solubility data), and it offers a full computational characterization of the thermodynamics of transfer of the drug molecule from crystal phase to gas phase to dilute aqueous solution.

11.4 CALCULATION OF ΔG_{hyd}

Generally, methods for calculating ΔG_{hyd} can be represented by two main categories: *implicit* or *explicit* solvent models [38, 47–58]. The main difference between these two categories is the representation of the solvent structure around the solute. *Implicit Continuum Solvent Models* (ICSMs) treat the solvent around the solvated molecule as a structureless polarizable medium characterized by a dielectric constant, ε [49, 59, 60]. In turn, in explicit solvent models (ESMs) both solute and solvent molecules in the solute–solvent systems are described at the atomistic level. There are two major classes of ESMs—(1) explicit solvent simulations methods based on direct simulation of a solute–solvent system using molecular mechanics (MM), quantum mechanics (QM), or combined QM/MM techniques [7, 38, 47, 48, 51, 61–66] and (2) Molecular Theories of Liquids (MTLs) that use an atomistic/molecular scale theory like, for example, molecular integral equations theory (MIET) [67, 68] or molecular DFT (MDFT) [69, 70] for describing the solute–solvent interactions. We will discuss advantages and disadvantages of all these methods later focusing on MTL-based methods because they provide a good compromise between accuracy and computational costs [67, 68, 71, 72].

11.4.1 Implicit Continuum Solvent Models

Typically, in ICSMs, ΔG_{hyd} is represented as a sum of several contributions:

$$\Delta G_{\text{hyd}} = \Delta G_{\text{elec}} + \Delta G_{\text{rep-disp}} + \Delta G_{\text{cav}}, \tag{11.6}$$

where ΔG_{elec} is the contribution due to polarization of the solute by the solvent environment, $\Delta G_{\text{rep-disp}}$ represents the repulsion–dispersion interactions between the solute and the solvent, and ΔG_{cav} is the contribution due to formation of a cavity in the solvent where the solute is embedded. For a given molecular geometry, the first term, ΔG_{elec}, can be calculated by solving either the generalized-Born or Poisson–Boltzmann equations; these equations can be solved for pure solvents as well as mixtures and electrolyte solutions [52]. The molecular 3D structure can be calculated by MM or QM calculations or in some cases extracted from crystallographic data.

A wide variety of methods have been proposed to calculate ΔG_{cav} and $\Delta G_{\text{rep-disp}}$. A common approach is to combine these terms and represent them by an equation of the form

$$\Delta G_{\text{cav}} + \Delta G_{\text{rep-disp}} = \gamma \text{SASA} + b, \tag{11.7}$$

where SASA is the solvent accessible surface area of the solute and γ and b are constants obtained by empirical fitting against experimental data for nonpolar molecules. The linear relationship between $\Delta G_{\text{cav}} + \Delta G_{\text{rep-disp}}$ and SASA only holds for simple organic molecules, which has led to the development of several more sophisticated models for ΔG_{cav} and $\Delta G_{\text{rep-disp}}$. However, despite significant progress during the last decade, low computational costs, and, as a consequence, their high popularity, implicit continuum models have one principal drawback—they ignore the microscopic solvent structure and, therefore, neglect many important effects such as solvent reorientation in the molecule solvation shell and specific solute–solvent interactions (e.g., hydrogen bonds).

11.4.2 Explicit Solvent Models: Atomistic Simulations

Within the scope of available ESMs, the *explicit solvent simulations* method is the most theoretically rigorous and most detailed approach to calculate ΔG_{hyd} [7, 38, 47, 48, 51, 61–66]. The method is based on *explicit* simulation of a solute–solvent system where both solute and solvent molecules are described at the atomistic level. In most of the studies to date, the inter- and intramolecular interactions are described at the MM level (i.e., classical mechanics); however, *ab initio* molecular dynamics (AIMD) and hybrid QM/MM methods are becoming increasingly popular in this area (we note though that these methods are more computationally expensive than the MM-based explicit solvent methods) [73–77].

Regardless of the level of theory used (MM, QM, or QM/MM), within the scope of the explicit solvent simulations approach the hydration (solvation) free energy can be calculated by four main groups of methods: [52] (1) thermodynamic integration, (2) free energy perturbation, (3) probability densities, and (4) nonequilibrium work methods.

Since the free energy (of solvation) is a state function, it is (formally) independent of the path that is used to calculate it. However, the numerical *convergence* of the calculations (and, consequently, the accuracy of the calculations) depends on the method as well as the roughness of the energy landscape of the system and the dimensionality of the system phase space. Overall, the explicit simulations methods are computationally much more expensive than ICSMs and MTL-based methods. Calculating ΔG_{hyd} for a drug-like molecule by direct simulations may take as long as several days on a modern desktop computer, which means that these methods are still too computationally expensive for routine use in computational screening of large databases of molecules. We refer the readers to the following references for more details of the methods and comparative analysis of their advantages and disadvantages [38, 47–58].

11.4.3 Explicit Solvent Models: Molecular Theories of Liquids

MTLs offer an alternative to expensive direct simulation methods; these methods use a statistical-mechanical molecular theory of solvation, for example, MIET-based RISM [78–82], molecular Ornstein–Zernike equation (MOZ) [82, 83], or MDFT [69, 70]. Similar to direct simulation methods, the MTLs also operate with an explicit description of the solute–solvent interactions via an all-atom force field. However, instead of operating with individual molecular trajectories they describe the solute–solvent system through correlation functions of solute and solvent species. This approach significantly reduces computational cost and allows one to efficiently compute the solvation structure and thermodynamics of molecular solvation from statistical mechanics [68–70]. Although several MTL-based methods have been developed to calculate ΔG_{hyd}, in the rest of this section we will focus only on RISM-based methods because these methods are the only ones (to the best of our knowledge) that have been implemented in popular molecular-modeling software like AMBER and Amber Tools [84], molecular operating environment (MOE) [85], CHARMM [86], and Amsterdam Density Functional (ADF) [87].

11.4.3.1 1D Reference Interaction Site Model The 1D RISM approach [78] and its extension to dipolar liquids (XRISM [79]) require a solution of a system of site–site Ornstein–Zernike (SSOZ) 1D integral equations that operate with the intermolecular *spherically symmetric* site–site correlation functions. To make the system of the integral equations solvable, it has to be combined with a closure relation [82, 83, 88–91]. More details on the theory and the algorithms for solving the equations can be found in Refs. [82, 92].

The main benefits of this approach are that (i) this is a *molecular* theory, and that (ii) due to the spherical symmetry of the correlation functions in the 1D RISM approach, the computational costs are significantly reduced compared to explicit solvent methods and high-dimensional molecular theories like 3D RISM [80–82, 93], MOZ [82, 83], and MDFT [69, 70]. For an average drug-like molecule a 1D RISM calculation of solvation free energy takes less than a minute on a desktop PC [67, 71, 72, 92]. This time scale is already comparable with the computational time scale for continuum methods (seconds). We note that an explicit solvent calculation for the same kind of molecules would take between hours and days [38, 47–58].

Due to the oversimplified representation of the solute–solute and solute–solvent correlations in the 1D RISM theory, it is generally less accurate than the more sophisticated 3D RISM [71].

11.4.3.2 3D Reference Interaction Site Model The 3D RISM [80–82, 93] is a theoretical method for modeling solution phase systems based on classical statistical mechanics. The 3D RISM equations relate 3D intermolecular *solvent site—solute* total correlation functions ($h_\alpha(\mathbf{r})$), and direct correlation functions ($c_\alpha(\mathbf{r})$) (index α corresponds to the solvent sites) [80, 82]:

$$h_\alpha(\mathbf{r}) = \sum_{\xi=1}^{N_{solvent}} \int_{R^3} c_\xi(\mathbf{r}-\mathbf{r}')\chi_{\xi\alpha}(|\mathbf{r}'|)d\mathbf{r}', \tag{11.8}$$

where $\chi_{\xi\alpha}(r)$ is the bulk solvent susceptibility function and N_{solvent} is the number of sites in a solvent molecule (Fig. 11.4). The solvent susceptibility function $\chi_{\xi\alpha}(r)$ describes the mutual correlations of sites ξ and α in solvent molecules in the bulk solvent. It can be obtained from the solvent intramolecular correlation function $\left(\omega_{\xi\alpha}^{\text{solv}}(r)\right)$, site–site radial total correlation functions $\left(h_{\xi\alpha}^{\text{solv}}(r)\right)$, and the solvent site number density (ρ_α): $\chi_{\xi\alpha}(r) = \omega_{\xi\alpha}^{\text{solv}}(r) + \rho_\alpha h_{\xi\alpha}^{\text{solv}}(r)$ (from here onward, we imply that each site is unique in the molecule so that $\rho_\alpha = \rho$ for all α) [82]. These solvent–solvent correlation functions can be obtained by solution of the 1D RISM equations of the pure solvent [82, 84].

In order to calculate $h_\alpha(\mathbf{r})$ and $c_\alpha(\mathbf{r})$, N_{solvent} *closure* relations are introduced:

$$\begin{gathered} h_\alpha(\mathbf{r}) = \exp\left(-\beta u_\alpha(\mathbf{r}) + h_\alpha(\mathbf{r}) - c_\alpha(\mathbf{r}) + B_\alpha(\mathbf{r})\right) - 1 \\ \alpha = 1,\ldots,N_{\text{solvent}} \end{gathered} \tag{11.9}$$

Here, $u_\alpha(\mathbf{r})$ is the 3D interaction potential between the solute molecule and α solvent site, $B_\alpha(\mathbf{r})$ are bridge functionals, $\beta = 1/k_{\text{B}}T$, k_{B} is the Boltzmann constant, and T is the temperature.

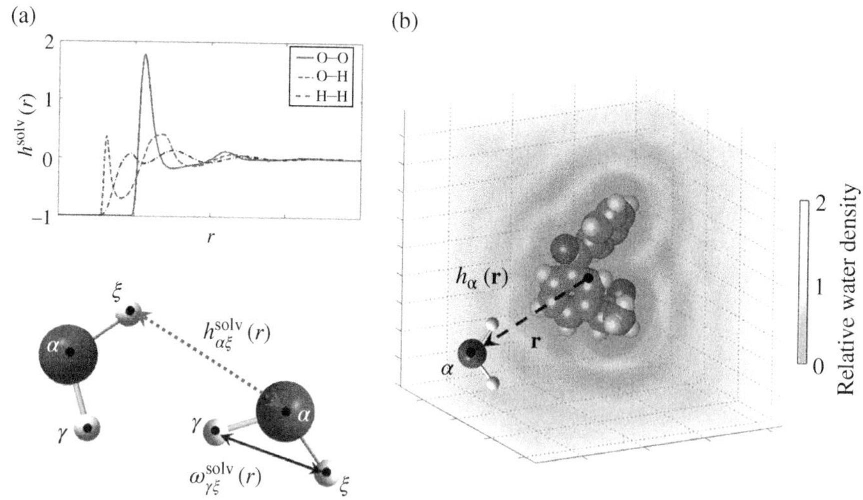

FIGURE 11.4 Correlation functions in the 3D RISM approach. (a) Site–site intramolecular ($\omega^{solv}_{\gamma\xi}(\mathbf{r})$) and intermolecular ($h^{solv}_{\alpha\xi}(\mathbf{r})$) correlation functions between sites of solvent molecules. The graph shows the radial projections of water solvent site–site density correlation functions: oxygen–oxygen (O–O, solid line), oxygen–hydrogen (O–H, dashed line), and hydrogen–hydrogen (H–H, dashed-dotted line). (b) Three-dimensional intermolecular solute–solvent correlation function $h_\alpha(\mathbf{r})$ around a model solute (diclofenac). Source: Palmer et al. [94]. Reprinted with permission of American Chemical Society. (*see insert for color representation of the figure.*)

In general, the exact bridge functions $B_\alpha(\mathbf{r})$ are represented as an infinite series of integrals over high-order correlation functions and are therefore practically incomputable, which makes it necessary to incorporate some approximations [82, 83, 89]. The most commonly used closure relationship is the KH closure proposed by Kovalenko and Hirata [91], which was designed to improve convergence rates and to prevent possible divergence of the numerical solution of the RISM equations [91]:

$$h_\alpha(\mathbf{r}) = \begin{cases} \exp(\Xi_\alpha(\mathbf{r})) - 1 & \text{when} \quad \Xi_\alpha(\mathbf{r}) \le 0 \\ \Xi_\alpha(\mathbf{r}) & \text{when} \quad \Xi_\alpha(\mathbf{r}) > 0 \end{cases} \tag{11.10}$$

Here, $\Xi_\alpha(\mathbf{r}) = -\beta u_\alpha(\mathbf{r}) + h_\alpha(\mathbf{r}) - c_\alpha(\mathbf{r})$. The 3D interaction potential between the solute molecule and α site of solvent ($u_\alpha(\mathbf{r})$) is estimated as a superposition of the site–site interaction potentials between solute sites and the particular solvent site, which depend only on the absolute distance between the two sites. The site–site interaction potential is commonly represented by a long-range electrostatic interaction term and a short-range term (Lennard-Jones potential) [95].

Within the framework of the RISM theory, there exist several approximate functionals that allow one to analytically obtain values of the hydration free energy (HFE) from the total $h_\alpha(\mathbf{r})$ and direct $c_\alpha(\mathbf{r})$ correlation functions [96–98]. Although these functionals have been extensively used to *qualitatively* model thermodynamics of different chemical systems [97, 99, 100], they generally give HFE values that are strongly biased from experimental data with a large standard deviation error [95–98, 101, 102].

The Kovalenko–Hirata free energy functional for 3D RISM is given by

$$\Delta G_{\text{hyd}}^{\text{KH}} = k_{\text{B}}T \sum_{\alpha=1}^{N_{\text{solvent}}} \rho_\alpha \int_{R^3} \left[\frac{1}{2} h_\alpha^2(\mathbf{r}) \Theta\left(-h_\alpha(\mathbf{r})\right) - \frac{1}{2} h_\alpha(\mathbf{r}) c_\alpha(\mathbf{r}) - c_\alpha(\mathbf{r}) \right] d\mathbf{r}, \quad (11.11)$$

where ρ_α is the number density of solvent sites α, and Θ is the Heaviside step function:

$$\Theta(x) = \begin{Bmatrix} 1 \text{ for } x > 0 \\ 0 \text{ for } x < 0 \end{Bmatrix} \quad (11.12)$$

The Gaussian fluctuation (GF) HFE functional was initially developed by Chandler, Singh, and Richardson, for 1D RISM and was adopted by Kovalenko and Hirata for the 3D RISM case [82, 103]:

$$\Delta G_{\text{hyd}}^{\text{GF}} = k_{\text{B}}T \sum_{\alpha=1}^{N_{\text{solvent}}} \rho_\alpha \int_{R^3} \left[-c_\alpha(\mathbf{r}) - \frac{1}{2} c_\alpha(\mathbf{r}) h_\alpha(\mathbf{r}) \right] d\mathbf{r}. \quad (11.13)$$

Here, ρ_α is the number density of solvent sites α. Unfortunately, HFEs calculated using the GF free energy functional have only a *qualitative* agreement with experiment. The error in hydration free energies calculated by the GF functional in 3D RISM is strongly correlated with the partial molar volume calculated by 3D RISM [67, 94, 104]. The 3D RISM/UC free energy functional developed from this observation is a linear combination of the $\Delta G_{\text{hyd}}^{\text{GF}}$, the dimensionless partial molar contribution, ρV, and a bias correction, b (intercept) [104]:

$$\Delta G_{\text{hyd}}^{\text{3D-RISM/UC}} = \Delta G_{\text{hyd}}^{\text{GF}} + a(\rho V) + b, \quad (11.14)$$

where the values of the scaling coefficient a and intercept b are obtained by linear regression against experimental data for simple organic molecules. For the AMBER-GAFF forcefield and SPC/E water model, the coefficients have the values $a = -3.2217$ kcal/mol and $b = 0.5783$ kcal/mol.

The solute partial molar volume V can be estimated via *solute–solvent site* correlation functions using the following 3D RISM theory expression [105, 106]:

$$V = k_{\text{B}}T\eta \left(1 - \rho_\alpha \sum_{\alpha=1}^{N_{\text{solvent}}} \int_{R^3} c_\alpha(\mathbf{r}) d\mathbf{r} \right). \quad (11.15)$$

Here, η is the pure solvent isothermal compressibility, and ρ_α is the number density of solute sites α.

The 3D RISM/UC method has been shown to give accurate hydration free energies for both simple organic molecules and bioactive (drug-like) molecules [67, 94, 104].

11.5 CALCULATION OF ΔG_{sub}

11.5.1 Crystal Polymorphism

The intrinsic aqueous solubility of a solute depends on the polymorphic form of the crystalline precipitate that is present at thermodynamic equilibrium during the solubility experiment. The repeated dissolution and re-precipitation of the solute that occurs during a single solubility measurement often promotes changes in crystal polymorph, from less to more thermodynamically stable forms in accordance with Ostwald's law of stages [107–109], which means that simply characterizing the precipitate at the beginning of a solubility experiment is not satisfactory. Since the majority of experimental data in the published literature are reported without characterization of the crystalline form of the precipitate, it is difficult to compile an accurate database of polymorph solubility. Nevertheless, the average difference in molar solubilities between polymorphs has been estimated to be approximately two-fold [110] (or even lower, 1.5-fold, according to the data on 153 polymorphic pairs published in Ref. [17]). As this value is smaller than the average error in models to predict solubility (5- to 10-fold molar solubility) [4–6, 111], it has previously been suggested that a simulated low-energy crystal structure may be still a good starting point from which to predict solubility, even if small errors exist in the calculated polymorphic landscape [15, 16].

11.5.2 Crystal Structure Prediction

The aim of crystal structure prediction is to predict stable polymorphic forms from molecular structure. Although a large number of different methods have been proposed for crystal structure prediction, most of them are based on the same approach, whereby stable polymorphs are identified by minimizing the lattice energy (U_{latt}) of a large number of candidate structures [112]. The use of lattice energy rather than free energy to assess polymorph stability is based on the assumption that the difference in entropy of organic polymorphs is negligible [113]. While this is a reasonable assumption, it is not always sufficiently accurate; significant entropy differences have been measured for model organic molecules (e.g., paracetamol) [42, 43, 114]. Where necessary, the ranking of polymorphs can sometimes be usefully refined by calculating lattice vibration entropies [115]. Without knowledge of the entropic difference, it is not possible to account for enantiotropic relationships between polymorphs [116].

The key challenges faced in crystal structure prediction are in sampling all possible configurational and conformational degrees of freedom in the crystal (as given by space group, unit cell dimension and position of atoms in the unit cell) and in computing the lattice energy accurately for each hypothetical crystal structure [117]. For small organic molecules with few degrees of conformational freedom, it is often sufficient to keep the conformational degrees of freedom of each molecule fixed (e.g., at a gas phase geometry) and optimize only the configurational degrees of freedom in the crystal phase. The size of the search space can be reduced by considering candidates from the most common space groups only [112]. Since crystal structure prediction can be viewed as a global optimization problem, stochastic search algorithms have also been used to improve sampling, for example, particle swarm algorithms, genetic algorithms, Monte Carlo parallel tempering algorithms, and simulated annealing [118, 119]. The success of crystal structure prediction depends to a large extent on the accuracy of the method for computing lattice energy [113, 120]. Crystal geometry optimization is normally performed using model potentials, but the use of QM to refine the calculated lattice energy for the energy minimized structure is becoming more common [114]. Methods to compute lattice energy will be discussed further in Section 11.5.4.

Over the past 15 years, the Cambridge Crystallographic Data Centre (CCDC) has hosted six blind challenges (one of which is still ongoing at the time of writing), in which the scientific community have been provided with molecular structures and invited to make predictions of the structures of previously unpublished organic crystals [121–126]. The current state of the art allows the polymorphic landscape of rigid and semiflexible organic molecules to be calculated with reasonable confidence, with some recent successes also reported for crystal structure prediction of molecules with multiple rotatable bonds [127].

11.5.3 Calculation of ΔG_{sub}

The existing methods to compute ΔG_{sub} fall in one of two categories: (1) those that compute ΔG_{sub} from separate estimates of ΔH_{sub} and ΔS_{sub} by $\Delta G_{\text{sub}} = \Delta H_{\text{sub}} - T\Delta S_{\text{sub}}$; and (2) those that compute ΔG_{sub} directly from molecular simulation.

The sublimation free energy data that is reported in the published literature for organic crystals is often given relative to a 1 atm standard state in the gas phase ($\Delta G^{o}_{\text{sub}}$), while hydration free energies are more commonly given relative to a molar standard state in the gas phase ($\Delta G^{*}_{\text{hyd}}$). Sublimation free energy data reported as $\Delta G^{o}_{\text{sub}}$ can be converted to $\Delta G^{*}_{\text{sub}}$ using the following equation, which may be derived considering the work for isothermal expansion of an ideal gas [15, 128]:

$$\Delta G^{*}_{\text{sub}} = \Delta G^{o}_{\text{sub}} - RT \ln\left(\frac{V_{\text{m}} p_o}{RT}\right) \tag{11.16}$$

Here, V_{m} is the molar volume of the crystal, and p_o is the standard atmospheric pressure (1 atm = 101.325 kPa).

11.5.4 Calculation of ΔH_{sub}

The enthalpy of sublimation ΔH_{sub} can be approximated from the crystal lattice energy, U_{latt}, by $\Delta H^{o}_{\text{sub}} = -U_{\text{latt}} - 2RT$. The $-2RT$ term arises because the lattice energy does not include lattice vibrational energies (which can be approximated by $6RT$ for crystals of rigid molecules oscillating in a harmonic potential), the energy of the vapor is $3RT$, and a $PV = RT$ correction is necessary to change energies into enthalpies, thus yielding $-6RT + 3RT + RT = -2RT$ [129].

The lattice energy of a molecular crystal is defined as the energy for formation of the crystal lattice from molecules that are infinitely separated in the gas phase. U_{latt} can be decomposed into two components:

$$U_{latt} = U_{inter} + \Delta E_{intra} \quad (11.17)$$

Here, U_{inter} is the intermolecular interactions between molecules in the crystal, and ΔE_{intra} is the change in the molecular energy between the crystal and gas. Since molecules generally adopt low-energy conformations within their observed crystal structures, ΔE_{intra} is often a small contribution to the overall lattice energy (and is negligible for rigid molecules). However, for large conformationally flexible molecules it is important to assess the influence of ΔE_{intra} on U_{latt}. This is especially true when molecules with an intramolecular hydrogen bond change their conformation on crystallization to adopt a more stable intermolecular hydrogen bond.

U_{inter} may be understood through the theory of intermolecular interactions. As a rough approximation, directional interactions (e.g., hydrogen bonding, pi–pi interactions, and halogen bonding) are important in determining the relative orientations of molecules in the crystal lattice, while repulsion interactions are critical in determining close contact distances, and dispersion interactions provide the universal attractive contributions that favor denser structures. U_{inter} can be computed using either QM or model potentials. While QM provides a theoretically rigorous method for computing lattice energies, it is computationally expensive, especially for use in lattice energy minimization of pharmaceutical crystal structures. Nevertheless, there have been some recent successes in modeling organic crystals using *ab initio* or density functional–based methods, typically including some form of dispersion correction. Alternatively, U_{inter} may be computed using either isotropic or anisotropic model potentials. It has been shown that representing the molecular electron distribution using distributed multipoles (anisotropic) rather than atomic partial charges (isotropic) leads to more accurate crystal structure modeling. A key concern in developing model potentials is that the contributions to U_{inter} (including dispersion and polarization terms) are both accurate and well balanced with the calculated intramolecular forces. The evaluation of lattice energies of small organic molecules is currently a very active research area for computational chemistry. As shown in a series of blind tests of crystal structure prediction algorithms [122–125], there is currently no method of evaluating the lattice energy of organic molecules that is ideal for all molecular systems.

11.5.5 Calculation of ΔS_{sub}

ΔS_{sub} can be computed as the difference between the entropy of an ideal gas and the entropy of the crystal at a given temperature and pressure. If the intra- and intermolecular contributions to the entropy of the crystal are considered to be decoupled, such that the change in intramolecular vibrational entropy for transfer from crystal to gas can be taken to be zero, then the sublimation entropy of rigid molecules can be approximated by $\Delta S^{o}_{sub} = S_{rot,gas} + S_{trans,gas} - S_{ext,cryst}$, where $S_{rot,gas}$ and $S_{trans,gas}$ are the rotational and translational contributions to the entropy of the gas at temperature T, respectively, and $S_{ext,cryst}$ is the intermolecular vibrational contribution to the entropy of the crystal at T. Here it is assumed that the crystalline phase is infinite and perfect, that translation and rotational entropies of the crystal are negligible and that the molecule is in its electronic ground state in both crystal and gas phase. S_{rot} and S_{trans} can be calculated from statistical thermodynamics as described in physical chemistry textbooks (see e.g., Refs. [128, 129]) and as implemented in many popular molecular modeling programs (e.g., Gaussian09 [130]). $S_{ext,cryst}$ is determined by the intermolecular vibrational (phonon) modes of the crystal, which may be computed by rigid-molecule lattice dynamics employing the same model potential used in lattice energy minimization. Only the $6N–3$ (where N is the number of molecules in the unit cell) optical zone-center ($k = 0$) phonons need to be calculated; the remaining three acoustic modes have zero frequency at $k = 0$. The density of states can be calculated using a hybrid Debye–Einstein approximation for $k \neq 0$, where the frequencies of the optical phonons are assumed to be independent of k and the acoustic contribution is modeled by the Debye approximation, with the Debye cutoff frequency estimated by extrapolating the acoustic modes to the zone boundary, using sound velocities calculated from the elastic stiffness tensor. The resulting free energy expression is given in Anghel et al. [131]. In this approach, it is assumed that vibrations are harmonic, and coupling between inter and intramolecular vibrations is ignored.

11.5.6 Other Methods to Compute ΔG_{sub}

Over the previous 50 years a large number of methods have been developed to compute HFE using molecular dynamics or Monte Carlo simulations coupled with statistical mechanics, whereas relatively few such methods have been developed and applied to computing of the sublimation free energy of drug-like molecules. One such approach is the GAUCHE methodology [132], which computes the sublimation free energy as follows:

$$-\Delta G_{sub} = -k_B T \ln\left(\frac{V_c}{V_g}\right) + \Delta G_{AU} + \Delta G_{AU \to UC}. \tag{11.18}$$

Here, V_g is the volume of a gas at 1 molar concentration, V_c is the molar volume of the crystal, ΔG_{AU} is the deposition free energy for a system composed of only N_{AU} asymmetric unit molecule(s), and $\Delta G_{AU \to UC}$ is the change in free energy to expand the asymmetric unit into a unit cell composed of N_{UC} independent molecules. The first term corresponds to the change of standard state discussed in Section 11.5.3, the latter two terms are computed from separate simulation steps. The current implementations of the GAUCHE method require knowledge of the crystal space group and unit cell parameters from experiment, but not *a priori* knowledge of crystalline atomic coordinates. Using the GAUCHE methodology and the polarizable AMOEBA force field, a RMSE of 1.7 kcal/mol was reported for a dataset of five organic molecules, which is encouraging, but probably not yet accurate for routine use in solubility prediction [132].

11.6 EXPERIMENTAL DATA

A critical understanding of the availability and accuracy of experimental data in the published literature is an important step in developing or testing computational methods. This is especially true since there is currently a lack of accurate and well-documented experimental data for polyfunctional organic molecules [53, 133].

The traditional "gold-standard" method for measuring intrinsic aqueous (and pH-dependent) solubility is the shake-flask method [134]. The sample drug is added to a buffered solution until precipitation occurs. The solution is then agitated until thermodynamic equilibrium has been established, at which point the excess precipitate is removed, by filtration or centrifugation, and the concentration of drug in solution is measured often by HPLC with UV detection. The main problem with the shake-flask method is that the solution must be left to equilibrate for long periods of time, typically between 24 and 72 h. Additional errors may also be incurred by the filtration or centrifugation step and the use of buffers [134]. Care must be taken in characterizing the physical form of the undissolved precipitate as different crystal forms will have different apparent solubilities [135]. Recently, potentiometric methods such as the dissolution template titration (DTT) method [136] and the CheqSol method [137] have found more widespread use. The potentiometric methods use accurate acid–base titrations to control the precipitation of the solute by altering the concentration of the nonionized (and ionized) form of the solute in solution. Such experiments are commonly visualized by a Bjerrum plot, which is a graph of the average number of bound protons as a function of pH [138]. Often precipitation can be identified as a deviation between the theoretical and observed titration curves at a specific pH, which occurs as the nonionized solute precipitates, thus affecting the equilibrium hydronium ion concentration. The principal benefit of the potentiometric methods is that they increase the rate at which thermodynamic equilibrium is obtained, but they can typically only be used for solutes with a pK_a in the range 2–12 (we note that the majority of drugs do have such a pK_a).

Solubility data for small organic molecules and several hundred pharmaceuticals are available in either the primary literature or from sources such as the AquaSol [139] and PhysProp [140] databases or the Merck Index [141]. However, data from these sources typically contain large uncertainties. A good example of this problem is provided by caffeine, for which the solubility is reported by different authors to be as low as 2.132 g/l [142] and as high as 896.2 g/l [143]. To address this problem, several groups have recently reported new experimental measurements of solubilities for drug-like molecules. For example, Llinas et al. published a dataset of 132 drug-like molecules for which intrinsic aqueous solubilities (and pK_a's) were measured by standard methods [6, 144]. Since accurate melting points and octanol–water partition coefficients are also available for these molecules, the dataset provides a useful starting point to study the physical chemistry of drug-like molecules and their solvation behavior. Unfortunately, pure compound vapor pressures have not been reported for the majority of these molecules, which makes it impossible to evaluate their hydration free energies. We note that measurement of these data for these molecules would significantly benefit the development of new computational solvent models.

11.7 CONCLUSION AND FUTURE OUTLOOK

One of the benefits of the methods discussed in this chapter is that they provide a complete characterization of the thermodynamics of transfer of solute from crystal to aqueous solution. Since the solubility of a crystalline solute depends upon the properties of the undissolved crystalline precipitate as well as the properties of the solution, the thermodynamic data provides valuable information in understanding not only which of the two molecules is more soluble but also why the selected molecule is more soluble. By contrast, QSPR models, which are statistical rather than first-principles approaches, provide only limited statistical information about the underlying physicochemical processes. Moreover, since most QSPR models predict solubility from molecular rather than crystal structure, they are not able to rationalize or predict different solubilities for different polymorphs of a molecule. Therefore, we believe that the *bottom-up* methods that utilize efficiently molecular-scale information about the solute and solvent structure will attract more attention in the future in terms of both practical applications and fundamental studies of solubility of drug-like molecules.

ACKNOWLEDGMENTS

DSP and MVF thank the Scottish Universities Physics Alliance (SUPA) for funding through MVFs SUPA2 start-up funds. DSP and MVF are also grateful for partial financial support from the Deutsche Forschungsgemeinschaft (DFG)—German Research Foundation, Research Grant FE 1156/2-1, and for the use of the computing facilities at the John von Neumann-Institut für Computing (NIC), Jülich Supercomputing Centre (JSC), Forschungszentrum Jülich GmbH, Germany—Project IDs: HLZ18 and HLZ16. DSP is grateful for funding from the European Commission through a Marie Curie Intra-European Fellowship within the 7th European Community Framework Program (FP7-PEOPLE-2010-IEF).

REFERENCES

[1] van de Waterbeemd, H.; Gifford, E. *Nat. Rev. Drug Discov.* 2003, *2*, 192–204.

[2] Kubinyi, H. *Nat. Rev. Drug Discov.* 2003, *2*, 665–668.

[3] Jorgensen, W. L. *Science* 2004, *303*, 1813–1818.

[4] Balakin, K. V.; Savchuk, N. P.; Tetko, I. V. *Curr. Med. Chem.* 2006, *13*, 223–241.

[5] Dearden, J. C. *Expert Opin. Drug Discov.* 2006, *1*, 31–52.

[6] Hopfinger, A. J.; Esposito, E. X.; Llinas, A.; Glen, R. C.; Goodman, J. M. *J. Chem. Inf. Model.* 2009, *49*, 1–5.

[7] Jorgensen, W. L.; Duffy, E. M. *Adv. Drug Deliv. Rev.* 2002, *54*, 355–366.

[8] Ali, J.; Camilleri, P.; Brown, M. B.; Hutt, A. J.; Kirton, S. B. *J. Chem. Inf. Model.* 2012, *52*, 420–428.

[9] Lian, B.; Yalkowsky, S. H. *J. Pharm. Sci.* 2014, *103*, 2710–2723.

[10] McDonagh, J. L.; Nath, N.; De Ferrari, L.; van Mourik, T.; Mitchell, J. B. O. *J. Chem. Inf. Model.* 2014, *54*, 844–856.

[11] Palmer, D. S.; Mitchell, J. B. O. *Mol. Pharm.* 2014, *11*, 2962–2972.

[12] Bajorath, F. *Nat. Rev. Drug Discov.* 2002, *1*, 882–894.

[13] Wang, J.; Hou, T. *Comb. Chem. High. T. Scr.* 2011, *14*, 328–338.

[14] Talevi, A.; Goodarzi, M.; Ortiz, E. V.; Duchowicz, P. R.; Bellera, C. L.; Pesce, G.; Castro, E. A.; Bruno-Blanch, L. E. *Eur. J. Med. Chem.* 2011, *46*, 218–228.

[15] Palmer, D. S.; Llinas, A.; Morao, I.; Day, G. M.; Goodman, J. M.; Glen, R. C.; Mitchell, J. B. O. *Mol. Pharm.* 2008, *5*, 266–279.

[16] Johnson, S. R.; Chen, X. Q.; Murphy, D.; Gudmundsson, O. *Mol. Pharm.* 2007, *4*, 513–523.

[17] Abramov, Y.; Pencheva, K. Thermodynamics and Relative Solubility Prediction of Polymorphic Systems. In *Chemical Engineering in the Pharmaceutical Industry: from R&D to Manufacturing*; Ende, D. J., Ed.; Wiley: New York, 2011, pp. 477–490.

[18] Yalkowsky, S. H. *Solubility and Solubilization in Aqueous Media*; Oxford University Press: New York, 1999.

[19] Horter, D.; Dressman, J. B. *Adv. Drug Deliv. Rev.* 2001, *46*, 75–87.

[20] Avdeef, A. *Absorption and Drug Development: Solubility, Permeability, and Charge State*; Wiley-Interscience: Hoboken, NJ, 2003.

[21] Noyes, A. A.; Whitney, W. R. *J. Am. Chem. Soc.* 1897, *19*, 930–934.

[22] Hendersen, L. J. *Am. J. Physiol.* 1908, *21*, 173–179.

[23] Hasselbalch, K. A. *Biochem. Z.* 1917, *78*, 112–144.

[24] Paolini, G. V.; Shapland, R. H. B.; Hoorn, W. P. van; Mason, J. S.; Hopkins, A. L. *Nat. Biotechnol.* 2006, *24*, 805–815.

[25] Lipinski, C. A.; Lombardo, F.; Dominy, B. W.; Feeney, P. J. *Adv. Drug Deliv. Rev.* 1997, *23*, 3–25.

[26] Walters, W. P.; Namchuk, M. *Nat. Rev. Drug Discov.* 2003, *2*, 259–266.

[27] Chiou, C. T.; Peters, L. J.; Freed, V. H. *Science* 1979, *206*, 831–832.

[28] Mackay, D.; Shiu, W. Y.; Ma, K. C. *Illustrated Handbook of Physical-Chemical Properties and Environmental Fate for Organic Chemicals, Volume 2, Polynuclear Aromatic Hydrocarbons, Polychlorinated Dioxins, and Dibenzofurans*; Lewis Publishers: Boca Raton, FL, 1992.

[29] Thompson, J. D.; Cramer, C. J.; Truhlar, D. G. *J. Chem. Phys.* 2003, *119*, 1661–1670.

[30] Ben-Naim, A. *J. Phys. Chem.* 1978, *82*, 792–803.

[31] Ben-Naim, A.; Marcus, Y. *J. Chem. Phys.* 1984, *81*, 2016–2027.

[32] Yalkowsky, S. H.; Valvani, S. C. *J. Pharm. Sci.* 1980, *69*, 912–922.

[33] Jain, N.; Yalkowsky, S. H. *J. Pharm. Sci.* 2001, *90*, 234–252.

[34] Wassvik, C. M.; Holmen, A. G.; Bergstrom, C. A. S.; Zamora, I.; Artursson, P. *Eur. J. Pharm. Sci.* 2006, *29*, 294–305.

[35] Hughes, L. D.; Palmer, D. S.; Nigsch, F.; Mitchell, J. B. O. *J. Chem. Inf. Model.* 2008, *48*, 220–232.

[36] O'Boyle, N. M.; Palmer, D. S.; Nigsch, F.; Mitchell, J. B. O. *Chem. Cent. J.* 2008, *2*, 21.

[37] Hill, A. P.; Young, R. J. *Drug Discov. Today* 2010, *15*, 648–655.

[38] Westergren, J.; Lindfors, L.; Hoglund, T.; Luder, K.; Nordholm, S.; Kjellander, R. *J. Phys. Chem. B* 2007, *111*, 1872–1882.

[39] Luder, K.; Lindfors, L.; Westergren, J.; Nordholm, S.; Kjellander, R. *J. Phys. Chem. B* 2007, *111*, 1883–1892.

[40] Luder, K.; Lindfors, L.; Westergren, J.; Nordholm, S.; Kjellander, R. *J. Phys. Chem. B* 2007, *111*, 7303–7311.

[41] Luder, K.; Lindfors, L.; Westergren, J.; Nordholm, S.; Persson, R.; Pedersen, M. *J. Comput. Chem.* 2009, *30*, 1859–1871.

[42] Perlovich, G. L.; Volkova, T. V.; Manin, A. N.; Bauer-Brandl, A. *AAPS PharmSciTech* 2008, *9*, 205–216.

[43] Perlovich, G. L.; Volkova, T. V.; Manin, A. N.; Bauer-Brandl, A. *J. Pharm. Sci.* 2008, *97*, 3883–3896.

[44] Schnieders, M. J.; Baltrusaitis, J.; Shi, Y.; Chattree, G.; Zheng, L.; Yang, W.; Ren, P. *J. Chem. Theory Comput.* 2012, *8*, 1721–1736.

[45] Reinwald, G.; Zimmermann, I. *J. Pharm. Sci.* 1998, *87*, 745–750.

[46] Palmer, D. S.; McDonagh, J. L.; Mitchell, J. B. O.; van Mourik, T.; Fedorov, M. V. *J. Chem. Theory Comput.* 2012, *8*, 3322–3337.

[47] Kollman, P. *Chem. Rev.* 1993, *93*, 2395–2417.

[48] Jorgensen, W. L.; TiradoRives, J. *Perspect. Drug Discovery Des.* 1995, *3*, 123–138.

[49] Roux, B.; Simonson, T. *Biophys. Chem.* 1999, *78*, 1–20.

[50] Kollman, P. A.; Massova, I.; Reyes, C.; Kuhn, B.; Huo, S. H.; Chong, L.; Lee, M.; Lee, T.; Duan, Y.; Wang, W.; Donini, O.; Cieplak, P.; Srinivasan, J.; Case, D. A.; Cheatham, T. E. *Acc. Chem. Res.* 2000, *33*, 889–897.

[51] Duffy, E. M.; Jorgensen, W. L. *J. Am. Chem. Soc.* 2000, *122*, 2878–2888.

[52] Orozco, M.; Luque, F. J. *Chem. Rev.* 2000, *100*, 4187–4225.

[53] Nicholls, A.; Mobley, D. L.; Guthrie, J. P.; Chodera, J. D.; Bayly, C. I.; Cooper, M. D.; Pande, V. S. *J. Med. Chem.* 2008, *51*, 769–779.

[54] Guthrie, J. P. *J. Phys. Chem. B* 2009, *113*, 4501–4507.

[55] Klamt, A.; Eckert, F.; Diedenhofen, M. *J. Phys. Chem. B* 2009, *113*, 4508–4510.

[56] Marenich, A. V.; Cramer, C. J.; Truhlar, D. G. *J. Phys. Chem. B* 2009, *113*, 4538–4543.

[57] Mobley, D. L.; Bayly, C. I.; Cooper, M. D.; Dill, K. A. *J. Phys. Chem. B* 2009, *113*, 4533–4537.

[58] Sulea, T.; Wanapun, D.; Dennis, S.; Purisima, E. O. *J. Phys. Chem. B* 2009, *113*, 4511–4520.

[59] Tomasi, J.; Persico, M. *Chem. Rev.* 1994, *94*, 2027–2094.

[60] Tomasi, J.; Mennucci, B.; Cammi, R. *Chem. Rev.* 2005, *105*, 2999–3094.

[61] Reynolds, C. A.; King, P. M.; Richards, W. G. *Mol. Phys.* 1992, *76*, 251–275.

[62] Matubayasi, N.; Nakahara, M. *J. Chem. Phys.* 2000, *113*, 6070–6081.

[63] Matubayasi, N.; Nakahara, M. *J. Mol. Liq.* 2005, *119*, 23–29.

[64] Shirts, M. R.; Pande, V. S. *J. Chem. Phys.* 2005, *122*, 134508.

[65] Matubayasi, N. *Front. Biosci.* 2009, *14*, 3536–3549.

[66] Knight, J. L.; Brooks, C. L. *J. Comput. Chem.* 2009, *30*, 1692–1700.

[67] Palmer, D. S.; Chuev, G. N.; Ratkova, E. L.; Fedorov, M. V. *Curr. Pharm. Des.* 2011, *17*, 1695–1708.

[68] Kovalenko, A. *Pure Appl. Chem.* 2013, *85*, 159–199.

[69] Borgis, D.; Gendre, L.; Ramirez, R. *J. Phys. Chem. B* 2012, *116*, 2504–2512.

[70] Jeanmairet, G.; Levesque, M.; Vuilleumier, R.; Borgis, D. *J. Phys. Chem. Lett.* 2013, *4*, 619–624.

[71] Frolov, A. I.; Ratkova, E. L.; Palmer, D. S.; Fedorov, M. V. *J. Phys. Chem. B* 2011, *115*, 6011–6022.

[72] Sergiievskyi, V. P.; Fedorov, M. V. *J. Chem. Theory Comput.*2012, *8*, 2062–2070.

[73] Leung, K.; Rempe, S. B.; Lilienfeld, O. A. *J. Chem. Phys.* 2009**,** *130*, 204507.

[74] König, G.; Hudson, P. S.; Boresch, S.; Woodcock, H. L. *J. Chem. Theory Comput.* 2014, *10*, 1406–1419.

[75] Hassanali, A. A.; Cuny, J.; Verdolino, V.; Parrinello, M. *Phil. Trans. R. Soc. A* 2014, *372*, 20120482.

[76] Acevedo, O., Jorgensen, W. L., *WIREs Comput. Mol. Sci.* 2014, *4*, 422–435.

[77] Holmberg, N.; Chen, J.-C.; Foster, A.S.; Laasonen, K. *Phys. Chem. Chem. Phys.* 2014, *16*, 17437–17446.

[78] Chandler, D.; Andersen, H. C. *J. Chem. Phys.* 1972, *57*, 1930–1937.

[79] Hirata, F.; Rossky, P. J. *Chem. Phys. Lett.* 1981, *83*, 329–334.

[80] Beglov, D.; Roux, B. *J. Phys. Chem.* 1997, *101*, 7821–7826.

[81] Du, Q. H.; Beglov, D.; Roux, B. *J. Phys. Chem. B* 2000, *104*, 796–805.

[82] Hirata, F., Ed. *Molecular Theory of Solvation*; Kluwer Academic Publishers: Dordrecht, Netherlands, 2003.

[83] Hansen, J.-P.; McDonald, I. R. *Theory of Simple Liquids*, 4th ed.; Elsevier Academic Press: Amsterdam, the Netherlands, 2000.

[84] Luchko, T.; Gusarov, S.; Roe, D. R.; Simmerling, C.; Case, D. A.; Tuszynski, J.; Kovalenko, A. *J. Chem. Theory Comput.* 2010, *6*, 607–624.

[85] Molecular Operating Environment (MOE), 2012; Chemical Computing Group Inc., 1010 Sherbooke St. West, Suite #910, Montreal, QC, Canada, H3A 2R7, 2012.

[86] Brooks, B. R.; Brooks, C. L.; Mackerell, A. D.; Nilsson, L.; Petrella, R. J.; Roux, B.; Won, Y.; Archontis, G.; Bartels, C.; Boresch, S.; Caflisch, A.; Caves, L.; Cui, Q.; Dinner, A. R.; Feig, M.; Fischer, S.; Gao, J.; Hodoscek, M.; Im, W.; Kuczera, K.; Lazaridis, T.; Ma, J.; Ovchinnikov, V.; Paci, E.; Pastor, R. W.; Post, C. B.; Pu, J. Z.; Schaefer, M.; Tidor, B.; Venable, R. M.; Woodcock, H. L.; Wu, X.; Yang, W.; York, D. M.; Karplus, M. *J. Comput. Chem.* 2009, *30*, 1545–1614.

[87] Gusarov, S.; Ziegler, T.; Kovalenko, A. *J. Phys. Chem. A* 2006, *110*, 6083–6090.

[88] Monson, P. A.; Morriss, G. P. *Adv. Chem. Phys.* 1990, *77*, 451–550.

[89] Duh, D. M.; Haymet, A. D. J. *J. Chem. Phys.* 1995, *103*, 2625–2633.

[90] Singer, S. J.; Chandler, D. *Mol. Phys.* 1985, *55*, 621–625.

[91] Kovalenko, A.; Hirata, F. *J. Phys. Chem. B* 1999, *103*, 7942–7957.

[92] Sergiievskyi, V. P.; Hackbusch, W.; Fedorov, M. V. *J. Comput. Chem.* 2011, *32*, 1982–1992.

[93] Imai, T.; Oda, K.; Kovalenko, A.; Hirata, F.; Kidera, A. *J. Am. Chem. Soc.* 2009, *131*, 12430–12440.

[94] Palmer, D. S.; Frolov, A. I.; Ratkova, E. L.; Fedorov, M. V. *Mol. Pharm.* 2011, *8*, 1423–1429.

[95] Palmer, D. S.; Sergiievskyi, V. P.; Jensen, F.; Fedorov, M. V. *J. Chem. Phys.* 2010, *133*, 044104.

[96] Ratkova, E. L.; Chuev, G. N.; Sergiievskyi, V. P.; Fedorov, M. V. *J. Phys. Chem. B* 2010, *114*, 12068–12079.

[97] Genheden, S.; Luchko, T.; Gusarov, S.; Kovalenko, A.; Ryde, U. *J. Phys. Chem. B* 2010, *114*, 8505–8516.

[98] Ten-no, S.; Jung, J.; Chuman, H.; Kawashima, Y. *Mol. Phys.* 2010, *108*, 327–332.

[99] Drabik, P.; Gusarov, S.; Kovalenko, A. *Biophys. J.* 2007, *92*, 394–403.

[100] Blinov, N.; Dorosh, L.; Wishart, D.; Kovalenko, A. *Biophys. J.* 2010, *98*, 282–296.

[101] Ten-no, S. *J. Chem. Phys.* 2001, *115*, 3724–3731.

[102] Ratkova, E. L.; Fedorov, M. V. *J. Chem. Theory Comput.* 2011, *7*, 1450–1457.

[103] Chandler, D.; Singh, Y.; Richardson, D. M. *J. Chem. Phys.* 1984, *81*, 1975–1982.

[104] Palmer, D. S.; Frolov, A. I.; Ratkova, E. L.; Fedorov, M. V. *J. Phys. Condens. Matter* 2010, *22*, 492101.

[105] Harano, Y.; Imai, T.; Kovalenko, A.; Kinoshita, M.; Hirata, F. *J. Chem. Phys.* 2001, *114*, 9506–9511.

[106] Imai, T.; Harano, Y.; Kovalenko, A.; Hirata, F. *Biopolymers* 2001, *59*, 512–519.

[107] Llinas, A.; Burley, J. C.; Box, K. J.; Glen, R. C.; Goodman, J. M. *J. Med. Chem.* 2007, *50*, 979–983.

[108] Llinas, A.; Box, K. J.; Burley, J. C.; Glen, R. C.; Goodman, J. M. *J. Appl. Crystallogr.* 2007, *40*, 379–381.

[109] Llinas, A.; Burley, J. C.; Prior, T. J.; Glen, R. C.; Goodman, J. M. *Cryst. Growth Des.* 2008, *8*, 114–118.

[110] Pudipeddi, M.; Serajuddin, A. T. M. *J. Pharm. Sci.* 2005, *94*, 929–939.

[111] Palmer, D. S.; O'Boyle, N. M.; Glen, R. C.; Mitchell, J. B. O. *J. Chem. Inf. Model.* 2007, *47*, 150–158.

[112] Price, S. L. *Chem. Soc. Rev.* 2014, *43*, 2098–2111.

[113] Price, S. L.; Leslie, M.; Welch, G. W. A.; Habgood, M.; Price, L. S.; Karamertzanis, P. G.; Day, G. M. *Phys. Chem. Chem. Phys.* 2010, *12*, 8478–8490.

[114] Perlovich, G. L., Volkova, T. V., Bauer-Brandl, A.; *J. Therm. Anal. Calorim.* 2007, *89*, 767–774.

[115] Day, G. M.; Price, S. L.; Leslie, M. *J. Phys. Chem. B* 2003, *107*, 10919–10933.

[116] Abramov, Y. A.; *Org. Process Res. Dev.* 2013, *17*, 472–485.

[117] Day, G. M. *Crystallogr. Rev.* 2011, *17*, 3–52.

[118] Wang, Y.; Lv, J.; Zhu, L.; Ma, Y. *Phys. Rev. B* 2010, *82*, 094116.

[119] The generation, ranking, and characterization engine software package [GRACE], 2014. http://www.avmatsim.de/services/software. Accessed July 21, 2014.

[120] Day, G. M.; Price, S. L. *J. Am. Chem. Soc.* 2003, *125*, 16434–16443.

[121] Lommerse, J. P. M.; Motherwell, W. D. S.; Ammon, H. L.; Dunitz, J. D.; Gavezzotti, A.; Hofmann, D. W. M.; Leusen, F. J. J.; Mooij, W. T. M.; Price, S. L.; Schweizer, B.; Schmidt, M. U.; van Eijck, B. P.; Verwer, P.; Williams, D. E. *Acta Crystallogr. Sect. B: Struct. Sci.* 2000, *56*, 697–714.

[122] Motherwell, W. D. S.; Ammon, H.; Dunitz, J.; Dzyabchenko, A.; Erk, P.; Gavezzotti, A.; Hofmann, D. W. M.; Leusen, F. J. J.; Lommerse, J. P. M.; Mooij, W. T. M.; Price, S. L.; Scheraga, H.; Schweizer, B.; Schmidt, M. U.; van Eijck, B. P.; Verwer, P.; Williams; D.E. *Acta Crystallogr. Sect. B: Struct. Sci.* 2002, *58*, 647–661.

[123] Day, G. M.; Motherwell, W. D. S.; Ammon, H. L.; Boerrigter, S. X. M.; Della Valle, R. G.; Venuti, E.; Dzyabchenko, A.; Dunitz, J. D.; Schweizer, B.; van Eijck, B. P.; Erk, P.; Facelli, J. C.; Bazterra, V. E.; Ferraro, M. B.; Hofmann, D. W. M.; Leusen, F. J. J.; Liang, C.; Pantelides, C. C.; Karamertzanis, P. G.; Price, S. L.; Lewis, T. C.; Nowell, H.; Torrisi, A.; Scheraga, H. A.; Arnautova, Y. A.; Schmidt, M. U.; Verwer, P. *Acta Crystallogr. Sect. B: Struct. Sci.* 2005, *61*, 511–527.

[124] Day, G. M.; Cooper, T. G.; Cruz-Cabeza, A. J.; Hejczyk, K. E.; Ammon, H. L.; Boerrigter, S. X. M.; Tan, J. S.; Della Valle, R. G.; Venuti, E.; Jose, J.; Gadre, S. R.; Desiraju, G. R.; Thakur, T. S.; van Eijck, B. P.; Facelli, J. C.; Bazterra, V. E.; Ferraro, M. B.; Hofmann, D. W. M.; Neumann, M. A.; Leusen, F. J. J.; Kendrick, J.; Price, S. L.; Misquitta, A. J.; Karamertzanis, P. G.; Welch, G. W. A.; Scheraga, H. A.; Arnautova, Y. A.; Schmidt, M. U.; van de Streek, J.; Wolf, A. K.; Schweitzer, B. *Acta Crystallogr. Sect. B: Struct. Sci.* 2009, *65*, 107–125.

[125] Bardwell, D. A.; Adjiman, C. S.; Arnautova, Y. A.; Bartashevich, E.; Boerrigter, S. X. M.; Braun, D. E.; Cruz-Cabeza, A.; Day, G. M.; Della Valle, R. G.; Desiraju, G. R.; van Eijck, B. P.; Facelli, J. C.; Ferraro, M. B.; Grillo, D.; Habgood, M.; Hofmann, D. W. M.; Hofmann, F.; Jose, K. V. J.; Karamertzanis, P. G.; Kazantsev, A. V.; Kendrick, J.; Kuleshova, L. N.; Leusen, F. J. J.; Maleev, A. V.; Misquitta, A. J.; Mohamed, S.; Needs, R. J.; Neumann, M.; Nikylov, D.; Orendt, A. M.; Pal, R.; Pantelides, C. C.; Pickard, C. J.; Price, L. S.; Price, S. L.; Scheraga, H. A.; van de Streek, J.; Thakur, T. S.; Tiwari, S.; Venuti, E.; Zhitkov, I. K. *Acta Crystallogr. Sect. B: Struct. Sci.*2011, *67*, 535–551.

[126] Groom, C. R.; Reilly, A. M. *Acta Crystallogr. Sect. B: Struct. Sci.* 2014, *70*, 776–777.

[127] Kazantsev, A. V.; Karamertzanis, P. G.; Adjiman, C. S.; Pantelides, C. C.; Price, S. L.; Galek, P. T. A.; Day, G. M.; Cruz-Cabeza, A. J. *Int. J. Pharm.* 2011, *418*, 168–178.

[128] Atkins, P.; de Paula, J. *Physical Chemistry*, 9th ed.; Oxford University Press: Oxford, UK, 2009.

[129] Gavezzotti, A.; Filippini, G. Energetic Aspects of Crystal Packing: Experiment and Computer Simulations. In *Theoretical Aspects and Computer Modeling of the Molecular Solid State*; Gavezzotti, A, Ed.; John Wiley & Sons: Chichester, 1997; pp. 61–99.

[130] Frisch, M. J.; Trucks, G. W.; Schlegel, H. B.; Scuseria, G. E.; Robb, M. A.; Cheeseman, J. R.; Scalmani, G.; Barone, V.; Mennucci, B.; Petersson, G. A.; Nakatsuji, H.; Caricato, M.; Li, X.; Hratchian, H. P.; Izmaylov, A. F.; Bloino, J.; Zheng, G.; Sonnenberg, J. L.; Hada, M.; Ehara, M.; Toyota, K.; Fukuda, R.; Hasegawa, J.; Ishida, M.; Nakajima, T.; Honda, Y.; Kitao, O.; Nakai, H.; Vreven, T.; Montgomery, J. A., Jr.; Peralta, J. E.; Ogliaro, F.; Bearpark, M.; Heyd, J. J.; Brothers, E.; Kudin, K. N.; Staroverov, V. N.; Kobayashi, R.; Normand, J.; Raghavachari, K.; Rendell, A.; Burant, J. C.; Iyengar, S. S.; Tomasi, J.; Cossi, M.; Rega, N.; Millam, J. M.; Klene, M.; Knox, J. E.; Cross, J. B.; Bakken, V.; Adamo, C.; Jaramillo, J.; Gomperts, R.; Stratmann, R. E.; Yazyev, O.; Austin, A. J.; Cammi, R.; Pomelli, C.; Ochterski, J. W.; Martin, R. L.; Morokuma, K.; Zakrzewski, V. G.; Voth, G. A.; Salvador, P.; Dannenberg, J. J.; Dapprich, S.; Daniels, A. D.; Farkas; Foresman, J. B.; Ortiz, J. V.; Cioslowski, J.; Fox, D. J. Gaussian 09, Revision D.01, Gaussian, Inc.: Wallingford, CT, 2009.

[131] Anghel, A. T.; Day, G. M.; Price, S. L. *CrystEngComm* 2002, *4*, 348–355.

[132] Park, J.; Nessler, I.; McClain, B.; Macikenas, D.; Baltrusaitis, J.; Schnieders, M. J. *J. Chem. Theory Comput.* 2014, *10*, 2781–2791.

[133] Geballe, M. T.; Skillman, A. G.; Nicholls, A.; Guthrie, J. P.; Taylor, P. J. *J. Comput. Aided Mol. Des.* 2010, *24*, 259–279.

[134] Avdeef, A. *Adv. Drug Deliv. Rev.* 2007, *59*, 568–590.

[135] Llinas, A.; Goodman, J. M. *Drug Discov. Today* 2008, *13*, 198–210.

[136] Avdeef, A.; Berger, C. M. *Eur. J. Pharm. Sci.* 2001, *14*, 281–291.

[137] Stuart, M.; Box, K. *Anal. Chem.* 2005, *77*, 983–990.

[138] Avdeef, A.; Kearney, D. L.; Brown, J. A.; Chemotti, A. R. *Anal. Chem.* 1982, *54*, 2322–2326.

[139] Yalkowsky, S. H.; He, Y. *The Handbook of Aqueous Solubility Data*; CRC Press LLC: Boca Raton, FL, 2003.

[140] PhysProp database: 2005; see: http://www.syrres.com. Accessed July 21, 2010.

[141] O'Neil, M.J., ed. (2006) *The Merck Index: An Encyclopedia of Chemicals, Drugs, and Biologicals*, 14th ed. Merck & Co. Inc.: Whitehouse Station, NJ.

[142] Mandala, O. V. E. *Gazz. Chim. Ital.* 1926, *56*, 896.

[143] Ochsner, A. B.; Belloto, R. J.; Sokoloski, T. D. *J. Pharm. Sci.* 1985, *74*, 132.

[144] Llinas, A.; Glen, R. C.; Goodman, J. M. *J. Chem. Inf. Model.* 2008, *48*, 1289–1303.

12

CALCULATION OF NMR TENSORS: APPLICATION TO SMALL-MOLECULE PHARMACEUTICAL SOLIDS

Luís Mafra[1], Sérgio Santos[1], Mariana Sardo[1], and Heather Frericks Schmidt[2]

[1] *CICECO - Aveiro Institute of Materials, Department of Chemistry, University of Aveiro, Aveiro, Portugal*

[2] *Pfizer Worldwide Research & Development, Groton, CT, USA*

12.1 SSNMR SPECTROSCOPY: A SHORT INTRODUCTION

Pharmaceutical solids exhibit a range of specific mechanical, thermodynamic and pharmacokinetic properties that impact their therapeutic performance. Insight into the structure and dynamics of these solids over a range of length- and time scales is key for rationalizing, engineering and improving these properties. *Solid-state nuclear magnetic resonance* (SSNMR) spectroscopy is a well-established method for studying inter- and intramolecular interactions based on the magnetic and electric properties of the molecule with extreme sensitivity and nuclear specificity, and it is also commonly used to study molecular dynamics over a wide range of time scales. Atomic-scale information such as nuclear proximities, torsion angles, coordination numbers, and molecular packings may also be obtained via local nuclear interactions that may be probed by a variety of SSNMR methodologies. Small molecules such as drugs are the building blocks of supramolecular entities arranged in space through non-covalent interactions that ultimately determine their physicochemical properties. Although SSNMR is capable to sense even very weak molecular packing interactions, *ab initio* calculations are very often required to support NMR studies in complex crystal structures.

Computational Pharmaceutical Solid State Chemistry, First Edition. Edited by Yuriy A. Abramov.

Many reviews are available on the SSNMR technique and its application to study materials and polymers [1–3]; however, in this chapter, we focus on the use of SSNMR for studying small pharmaceutical molecules in synergy with theoretical calculations of NMR parameters. This chapter emphasizes computer modeling of ***chemical shift tensors*** (CSTs) to overcome the difficulty in using experimental ***chemical shifts*** (CSs) alone to determine specific structural features. The number of experimental NMR techniques available greatly outnumbers the existent theoretical approaches. However, over the past decade several scientific advancements have lead to faster computers, more efficient codes/algorithms and new levels of theory used to compute NMR parameters. Calculations then became more accurate and capable to model NMR chemical shieldings of large systems (hundreds of atoms). Consequently, this allowed the use of NMR modeling to support experimental evidence on a more routine basis.

The remainder of this chapter is organized in five sections. Sections 12.2–12.4 provide a brief review of the most relevant computational methods employed to calculate NMR chemical shieldings, and Section 12.5 reports a number of case studies where computer modeling is used in tandem with SSNMR spectroscopy and ***X-ray diffraction*** (XRD) to provide structure details in solid ***active pharmaceutical ingredients*** (APIs). Section 12.5 focuses, essentially, in correlating calculated NMR parameters with relevant structure features and is further divided in four subsections: NMR assignment of drug polymorphs (Sections 12.5.1 and 12.5.2), crystal packing interactions (Section 12.5.3), and the use of CSs and ***crystal structure prediction*** (CSP) methods for crystal structure elucidation/determination (Section 12.5.4).

12.2 THE CHEMICAL SHIELDING TENSORS: FUNDAMENTALS

The application of a magnetic field to a medium induces in it an electronic current as the result of the modification of the electronic ground state, which itself induces back an additional inhomogenous magnetic field. The "effective" local magnetic field $\mathbf{B}_{\text{local}}$ sensed by a probe nucleus at position $\mathbf{r}$ is then the sum of the strong external homogeneous magnetic field $\mathbf{B}_{\text{ext}}$ and the magnetic field $\mathbf{B}_{\text{ind}}$ induced by internal currents in the system [4]:

$$\mathbf{B}_{\text{local}}(\mathbf{r}) = \mathbf{B}_{\text{ext}} + \mathbf{B}_{\text{ind}}(\mathbf{r}) \tag{12.1}$$

The induced field is determined by the total electronic current density $\mathbf{j}(\mathbf{r})$ by means of the Biot–Savart law as follows:

$$\mathbf{B}_{\text{ind}}(\mathbf{r}) = \frac{\mu_0}{4\pi} \int \frac{\mathbf{r}' - \mathbf{r}}{\left|\mathbf{r}' - \mathbf{r}\right|^3} \times \mathbf{j}(\mathbf{r}')\,d\mathbf{r}' \tag{12.2}$$

where μ_0 is the permeability of the vacuum (SI units).

In a crystalline solid, in which the Born–von Karman periodic boundary conditions are applicable, the current density **j**(**r**) will be periodic, and $\mathbf{B}_{\text{ind}}$ can be conveniently calculated in reciprocal space using the Fourier transform of Equation 12.2 [5]:

$$\mathbf{B}_{\text{ind}}(\mathbf{G} \neq 0) = -\mu_0 i \frac{G}{|G|} \times \mathbf{j}(G) \tag{12.3}$$

where **G** is a vector in the reciprocal space. The proportionality factor between the induced and the externally applied magnetic fields is a tensorial quantity called the chemical shielding tensor ($\ddot{\sigma}$):

$$\mathbf{B}_{\text{ind}}(\mathbf{r}_s) = -\ddot{\sigma}(\mathbf{r}_s) \cdot \mathbf{B}_{\text{ext}}, \quad \sigma_{\alpha\beta}(\mathbf{r}_s) = -\frac{\partial \mathbf{B}_{\text{ind},\alpha}(\mathbf{r}_s)}{\partial \mathbf{B}_{\text{ext},\beta}} \tag{12.4}$$

where the subscript *s* indicates that the corresponding quantity is to be taken at the position of nucleus *s* (i.e., $\mathbf{r}_s$). It is common to report the ***isotropic chemical shift*** (ICS) $\delta_{\text{iso,s}}$ rather than the chemical shielding tensor, which is conventionally defined as

$$\delta_{\text{iso,s}} = \sigma_{\text{iso,ref}} - \sigma_{\text{iso,s}} \tag{12.5}$$

where $\sigma_{\text{iso,ref}}$ is the isotropic chemical shielding ($\sigma_{\text{iso}} = (1/3)Tr(\ddot{\sigma})$) of a reference compound. From Equations 12.2 to 12.4, it is quite clear that the question of calculating the chemical shielding essentially resorts to the calculation of the induced current. However, this is, by no means, a trivial task.

In the framework of quantum mechanics, the all-electron Hamiltonian $\hat{H}$ is, in atomic units,

$$\hat{H} = \frac{\mathbf{p}^2}{2} + V(\mathbf{r}) \tag{12.6}$$

where *V*(***r***) is the Hartree/Kohn–Sham potential and **p** the linear momentum operator. In a crystalline solid, that is, an extended system with translational symmetry, the potential *V*(**r**) has the same periodicity as the underlying *Bravais* lattice:

$$V(\mathbf{r}) = V(\mathbf{r} + \mathbf{R}) \tag{12.7}$$

for all **R** belonging to the appropriate *Bravais* lattice vector.

According to Bloch's theorem [6], the solutions to Schrödinger's equation must take the form:

$$\psi_{nk} = e^{i\mathbf{k}\cdot\mathbf{r}} u_{nk}(\mathbf{r}) \tag{12.8}$$

where $u_{nk}(\mathbf{r})$ is a function with the same periodicity as the lattice itself:

$$u_{nk}(\mathbf{r}) = u_{nk}(\mathbf{r} + \mathbf{R}) \tag{12.9}$$

k is a vector within the reciprocal unit cell (first Brillouin zone) and n and k label the eigenstates and reciprocal vectors, respectively [7].

The explicit periodicity of the Bloch wave function poses a major challenge for the calculation of aforementioned chemical shielding tensor in the solid state. Under these circumstances, the calculation of NMR parameters requires the inclusion of a macroscopic magnetic field ($\mathbf{B}_{ext}$), described by a nonperiodic vector potential, yielding a new Hamiltonian:

$$\hat{H} = \hat{H}^{(0)} + \mathbf{A}_{ext}\mathbf{p} + \frac{1}{2}\mathbf{A}_{ext}^{2} \tag{12.10}$$

where $\mathbf{A}_{ext}$ is the vector potential describing the external magnetic field. Unfortunately, there are multiple choices for $\mathbf{A}_{ext}$ for a given magnetic field such as, for instance,

$$\mathbf{A}_{ext} = \frac{1}{2}\mathbf{B}_{ext} \times (\mathbf{r} - \mathbf{R}) \tag{12.11}$$

where **R** is an arbitrary vector. The arbitrariness of **R** leads to the so-called *gauge origin problem*, in that the inclusion of the vector potential into the Hamiltonian is simply not compatible with the aforementioned Bloch symmetry. This drawback has rendered the calculation of NMR parameters in extended crystalline systems impossible until 1996, when Mauri and co-workers [8] developed a linear-response approach for the calculation of such parameters. Until then, calculation of NMR parameters was limited to the cluster approach, with various approximations developed to circumvent the gauge origin problem. The particularities of both the cluster and periodic approaches will be addressed in the following sections.

12.3 COMPUTATIONAL APPROACHES TO THE CALCULATION OF CHEMICAL SHIFT TENSORS IN SOLIDS

12.3.1 Cluster Approach

The cluster approach was, for a long time, the only available mechanism for calculating NMR parameters. A crystalline system can be, up to some extent, approximated as a cluster of units (molecules/ions/atoms) whose configuration resembles the crystal. The local environment in the center of this cluster is then an approximation to the bulk solid. The resemblance of the cluster with the true crystal tends to increase as the size of the cluster grows. Therefore, the size of the cluster is critical for the accurate calculation of the desired parameters, as small clusters do not really mimic the crystal environment, whereas big clusters imply heavy computational burdens. When using this approach, the optimal choice resides in a tradeoff between accurate crystal description and acceptable computational cost.

Generally speaking, the calculation of the induced current density, required for the determination of the NMR CSs (Eqs.12.2–12.4 above), is performed based on the static linear-response theory (perturbation theory) of the one-particle Hamiltonian and Hartree–Fock (or Kohn–Sham) orbitals [4]:

$$\begin{aligned}\hat{h} &= \hat{h}^{(0)} + i\hat{h}^{(1)} + O\left(\mathbf{B}_{\text{ext}}^2\right)\\ \psi &= \psi_k^{(0)} + i\psi_k^{(1)} + O\left(\mathbf{B}_{\text{ext}}^2\right)\end{aligned} \tag{12.12}$$

where $\psi_k^{(0)}$ are the Hartree–Fock (or Kohn–Sham) orbitals in the absence of the external magnetic field and $i\psi_k^{(1)}$ the first-order response of those orbitals to the magnetic field; $\hat{h}^{(0)}$ and $\hat{h}^{(1)}$ are the zeroth and first -order expansions of the one-particle (Hamiltonian $O\left(\mathbf{B}_{\text{ext}}^2\right)$ are higher order expansion terms). In the absence of $\mathbf{B}_{\text{ext}}$, the current density vanishes leading to

$$\mathbf{j} = \mathbf{j}^{(1)} + O\left(\mathbf{B}_{ext}^2\right) \tag{12.13}$$

where $\mathbf{j}^{(1)}$ is the sum of two terms: the paramagnetic ($\mathbf{j}_{\text{para}}^{(1)}$) and the diamagnetic ($\mathbf{j}_{\text{dia}}^{(1)}$) part of the current density; $\mathbf{j}_{\text{para}}^{(1)}$ involves the linear response of the orbitals, while $\mathbf{j}_{\text{dia}}^{(1)}$ can be evaluated without knowing the perturbed orbitals. However, under a gauge transformation (i.e., a change in $\mathbf{R}$ in Eq. 12.11), the paramagnetic and diamagnetic parts of the first-order current density transform accordingly, leaving the sum of the two unchanged. When the orbitals are expanded in a finite basis set, the representation of the perturbed orbitals $\psi_k^{(1)}$ largely depends on the choice of the gauge origin $\mathbf{R}$, thus leading to the well-known *gauge origin problem* [5]. To address this problem, several strategies have been developed that, somehow, distribute the gauge origin over the system. The two most successful and widely employed are the ***i****ndividual* ***g****auge for* ***l****ocalized* ***o****rbital* (IGLO) [9, 10]) and ***g****auge* ***i****ncluding* ***a****tomic* ***o****rbital* (GIAO)[11, 12]) approaches. In IGLO, individual gauge origins are assigned to the charge centers; whereas in GIAO, they are directly incorporated into the basis functions. The main requirement of GIAO is that it needs the use of atom-centered basis functions, although it can be used with wave functions of any type such as multi-configuration Hartree–Fock wave functions. Other methods for computing $\mathbf{j}^{(1)}$ have been devised such as the ***c****ontinuous* ***s****et of* ***g****auge* ***t****ransformations* (CSGTs) [13–15], in which the gauge depends on the position where the induced current is to be calculated, and the ***i****ndividual* ***g****auge for* ***a****toms in* ***m****olecules* (IGAIMs) [13, 14]. Although the results obtained using these methods are satisfactory, they generally do not outperform those obtained with GIAO and IGLO.

12.3.2 Periodic Approach

The strategy of approximating a crystal as a subset of the crystal components, as is done in the cluster approach, can introduce severe limitations in the accuracy of the description of the crystal itself. From a practical point of view, it is impossible to make a cluster big enough such that it properly describes the quantum mechanical and electrostatic effects governing the crystal. In order to eliminate these limitations in calculating NMR parameters in truly solid-state systems, some methodologies were developed based on ***d**ensity **f**unctional **t**heory* (DFT) under periodic boundary conditions.

In 1996, Mauri and co-workers [8] developed a framework for the first-principles calculation of NMR parameters in the solid state, by developing a way for eliminating the need for explicitly calculating the expectation values of the position operator over an extended system (the position operator cannot be defined under ***p**eriodic **b**oundary **c**onditions* or PBC). They circumvented the problem by obtaining the magnetic response to a uniform magnetic field as the long wavelength of a periodically modulated field. In addition, Pickard and Mauri introduced the ***g**auge-**i**ncluding **p**rojector-**a**ugmented-**w**ave* (GIPAW) formalism [16], as an extension of Blöchl's **p**rojector **a**ugmented **w**ave (PAW) method [17], leading to the possibility of calculating NMR chemical shielding using plane waves and pseudopotentials with an all-electron accuracy. In fact, similar to GIAO in the cluster approach, GIPAW has become the standard method in the calculation of these parameters in solids. An alternative to the GIPAW method has been developed by Sebastiani and Parrinelo [5], also within the density functional perturbation theory approach, in which the magnetic perturbations are evaluated in terms of localized Wannier orbitals, thus taking advantage of their exponentially decaying nature, coupled to a jig-saw representation of the position operator. Compared to GIPAW, this formulation has lower computational requirements and for very small unit cells, but requires the use of the supercell technique to enable the determination of the localized Wannier orbitals. In such case, the GIPAW method of Mauri et al. is generally more efficient.

Recently, an alternative to the aforementioned *direct* formalisms has been devised by Thonhauser et al. [18] and adequately coined the *converse* approach, which eliminates the need for a linear-response framework; instead of determining the induced current response to an external magnetic field, the chemical shielding tensor of a probe nuclei is determined as the derivative of the orbital magnetization with respect to a magnetic point dipole placed at the site of the probe. Rather than applying a constant magnetic field to an infinite periodic system and calculating the induced current, an infinite array of magnetic dipoles is applied to all equivalent sites and the change in the orbital magnetization is calculated. This change in magnetization is calculated through finite differences of ground-state calculations, thus allowing the use of more complex exchange-correlation functionals such as DFT + U and hybrid functionals or even implementation within the framework of quantum Monte Carlo or any wave function–based method.

12.3.3 Pitfalls and Practical Considerations

When performing calculation of CSs, several factors may directly influence the accuracy of the final results. When using the cluster approach, its size is often the main source of inaccuracy. Also, the level of theory and size/quality of the basis set plays an important role in the quality of the obtained results. In contrast, although the use of a fully periodic approach eliminates the problems associated with the cluster size, the type of exchange-correlation functionals and the nature of the pseudopotentials employed have an impact on the final results. In addition, the size (controlled through the cutoff energy) of the plane wave basis set deeply influences the results. To mitigate the influence of the finite basis set size in both cases, in the final results, it is required that the variation of the calculated properties systematically converge with respect to the number of basis functions. In the periodic case, the *k-point* grid used to average properties over the Brillouin zone should be dense enough such that it is adequately sampled, and, hence, convergence tests should also be performed. In DFT, the better the functional, the closer the result will be to the exact solution of Schrödinger's equation. From the practical point of view, the gradient-corrected approximation yields better results than the local density approximation.

Other situations exist that may potentially lead to discrepancies between calculated and measured quantities. For example, a typical NMR calculation first starts with a structure relaxation, in which atomic positions and, eventually, cell size and shape (in the periodic case) are adjusted as to minimize the forces acting on them, yielding an energy-minimized configuration, over which the NMR parameters are then calculated. This is equivalent to a single calculation over an equilibrium structure at 0 K. In real systems, thermal motion drives configurations away from the equilibrium ones, such that the measured quantities reflect the motional averaging of all spanned configurations. This limitation can be overcome by averaging NMR parameters over snapshots taken from molecular dynamics simulations, or by averaging over vibrational modes [18–20]. When using crystal structures obtained from diffraction-based techniques as input geometries for the calculation of NMR parameters, special attention should be paid to the quality of the structure and, in particular, to the positioning of lighter nuclei such as hydrogens. Most of X-ray diffraction-derived structures have hydrogens placed according to empirical criteria and/or differences in electron density maps and are, thus, mispositioned. It is, therefore, critical to perform a geometry optimization of the hydrogen and/or all atom positions prior to the calculation of the NMR parameters to ensure reliable results. If full structure relaxation is required, especially when using the cluster approach, one should ensure that the overall crystal configuration is not lost.

Another important topic is the calibration of the calculated chemical shieldings. From Equation 12.5, the ICS is calibrated relatively to the isotropic shielding of a reference compound. It is, therefore, desirable that the shielding tensor(s) of the reference compound are calculated under the same conditions (basis set, level of theory/functional, *k*-points grid (when applicable)) as those of the sample under study to ensure consistency between the two sets of results. An alternative way of calibrating chemical shieldings is to directly compare them to the measured CSs by means of a linear regression. In such case, the calculated chemical shieldings are plotted against the measured CSs such that the y-intercept value corresponds to $\sigma_{iso,ref}$. This way, errors in the calibration of the experimental spectra are absorbed into the intercept while simultaneously eliminating the need for additional calculations of the reference shieldings. In an ideal situation, the obtained slope should be equal to −1. This is generally, however, not the case. Deviation from unity encapsulates some of the approximations used in the calculation, the most important of which is the absence of temperature/dynamics effects that lead to a distribution of measured CSs rather than a single value per site. This can, however, be tackled experimentally by measuring two sets of CSs at two very distinct temperatures and then extrapolating them to 0 K prior to the calibration of the calculated chemical shieldings (assuming a linear dependency on temperature, which might need checking). Some authors have taken the correlation approach a step further by partitioning the CSs into functional groups and performing the linear fit for each set individually [21, 22], while others have calibrated the computed shieldings by equating the averaged computed shift to the average observed one [23].

12.4 NICS

In NMR experiments one can only measure the CSs in a limited set of points in space (the nuclei positions $\mathbf{r}_s$). However, the induced field and, consequently, the nuclear shielding tensor, are defined and can be computed for all points in space such that one can drop the index s from Equation 12.4 and calculate the shielding tensor at any arbitrary position $\mathbf{r}$, yielding the so-called ***nucleus-independent chemical shifts*** (NICS) [24]. NICS values are, thus, not referenced with respect to any nucleus and that is why they are nucleus independent. Also, and similarly to site-specific shieldings, rather than dealing with the shielding itself, one frequently reports the isotropic NICS shift (δ_{NICS}) defined as follows:

$$\delta_{\mathrm{NICS}}(\mathbf{r}) = \frac{1}{3}\mathrm{Tr}\left[\vec{\sigma}(\mathbf{r})\right]$$

This generalization of the CS into a scalar field was first proposed by Johnson and Bovey [25], while the implementation and application was pioneered by Schleyer et al. [26]. The applications of NICS values and maps are quite diverse and spread across multiple areas of research. These are predominantly used in the characterization of molecular aromaticity and the quantification of induced ring currents, sometimes focusing on their topological particularities. NICS have also been used for decomposing the magnetic-response properties into contributions of individual orbitals and/or functional groups. More recently, NICS maps have been used for studying the packing contributions to the solid-state CSs of organic molecular crystals [20, 27].

12.5 CASE STUDIES COMBINING EXPERIMENTAL AND COMPUTATIONAL NMR METHODS

This section will briefly describe selected examples encompassing the importance of a multidisciplinary approach involving experimental and computational NMR methods as well as the correlation with information taken from single-crystal and ***powder XRD*** (PXRD), to better understand the supramolecular structure and function of small-molecule pharmaceuticals. The use of *ab initio* calculations of NMR parameters plays a key role in topics such as using calculated NMR tensors for resonance assignment in drugs and polymorphism studies (see Sections 12.5.1 and 12.5.2), NMR sensitivity to crystal packing interactions (Section 12.5.3), and crystal structure determination (Section 12.5.4).

12.5.1 NMR Assignment of Polymorphs Aided by Computing NMR Parameters

SSNMR spectroscopy contributes significantly to polymorph detection and characterization since it is complementary to diffraction techniques, providing information on molecular symmetry, intermolecular interactions, molecular mobility, disorder, and interatomic distances. SSNMR presents unique capabilities in distinguishing between the different polymorphic forms, anhydrate, hydrate, and other solvate forms of APIs, monitoring transitions as well as detecting tautomeric forms [28–31]. Although NMR spectroscopy is sensitive to the local structural environment, the interpretation of the obtained high-resolution spectra is very often not straightforward, presenting a considerable challenge in organic systems with a variety of similar functional groups. In addition, complications arise when more than one molecule is part of the crystallographic asymmetric unit ($Z' > 1$) that is, when multiple molecules are not crystallographically equivalent. This subsection shows the use of both ICSs and NMR techniques to exploit the *J*-coupling (Section 12.5.1.1) and quadrupolar (Section 12.5.1.2) interactions to provide structural information.

12.5.1.1 Combining Calculated Chemical Shifts and J-mediated NMR Experiments

In the calculation of NMR parameters for solids, the ICS has been the most assessed parameter as it is directly obtained from the simplest NMR experiments. This property is highly sensitive to the chemical surroundings and a vast number of theoretical approaches to describe ICSs attained a level of accuracy comparable to those obtained experimentally. This is mainly true for ^{1}H and ^{13}C, the main nuclei present in pharmaceuticals, where Harris et al. have shown a number of pioneering work on calculated CS applied to pharmaceutical systems [30, 32]. The through-bond dipolar coupling, so-called *J*-coupling interaction, provides an additional sensitive probe of chemical bonding and molecular geometry. The experimental determination of *J*-couplings in solids cannot, in most cases, be obtained directly from ordinary 1D NMR spectra as *J*-couplings have small magnitudes of 1–100 Hz. The ***m**agic **a**ngle **s**pinning* (MAS) line widths of resonances in rigid organic molecules are typically dominated by anisotropic interactions (e.g., through-space dipolar couplings and ***c**hemical **s**hift **a**nisotropy* or CSA) and other broadening mechanisms having magnitudes in the kilohertz range thus masking the direct observation of *J*-coupling splittings. However, a simple Hahn echo can be incorporated in 2D MAS experiments (e.g., *J*-resolved, *J*-INEPT (***in**sensitive **n**uclei **e**nhanced by **p**olarization*), and *J*-HMQC (***h**eteronuclear **m**ultiple **q**uantum **c**oherence*)) to refocus anisotropies and provide transverse dephasing times (T_2'), which are typically much longer than the non-refocused NMR signal decay rate (T_2^*) that govern the free induction decay.

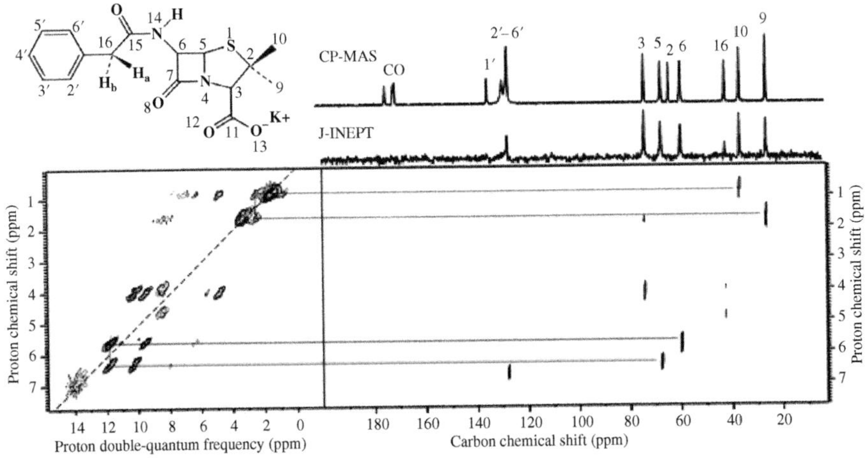

FIGURE 12.1 2D through-bond INEPT-HSQC spectrum of K salt of penicillin G (left) and 2D ^{1}H–^{1}H DQ spectrum (right). The lines illustrate the type of sequential connectivity used to finalize the assignments, with the example of the assignment of C9 and C10 through the near-neighbor proton contact, and C5 and C6 similarly. Reprinted with permission from Ref. [35]. Copyright 2006, RSC Publishing.

In such conditions, a refocused line width is achieved and subsequently *J*-coupling splittings may be assessed. Over the past decade, methodological developments, related to more efficient ^{1}H–^{1}H decoupling schemes, have enabled the experimental determination of the small *J*-couplings (e.g., $^1J_{CH}$ < 10 Hz) common for small organic molecules including drugs [33]. Very recently, methods for the calculation of *J* tensors for solids have been developed and applied to various systems showing good agreement with experiment [34]. However, we are not aware of such calculations in organic pharmaceuticals, and therefore this section is restricted to the use of *J*-coupling-based NMR experiments for spectral editing combined with the calculation of ICSs.

Mifsud et al. [35] presented one of the earliest examples showing correlation between ^{13}C and ^{1}H CS predictions from X-ray crystal structures, using the GIPAW-DFT approach, with experimental values, using K salt of penicillin G as a model compound (Fig. 12.1). A vital step toward the identification of different crystalline forms of a drug is the unambiguous assignment of both ^{13}C and ^{1}H resonances in the NMR spectra. The authors used a 2D ^{1}H–^{1}H ***double quantum*** (DQ) ***combined rotation and multiple pulse spectroscopy*** (CRAMPS) in combination with through-bond 2D ^{1}H–^{13}C ***heteronuclear single-quantum correlation*** (HSQC) (Fig. 12.1) experiments for a tentative full assignment. The INEPT-HSQC experiment had the advantage of showing only correlations between bonded C–H pairs and enabled the unambiguous assignment of the non-protonated quaternary carbons C_2 and C_1' as they vanish from the spectrum. The remaining unresolved ^{13}C peaks, aromatics and three carbonyls, were assigned based on the calculated ICS values determined from the geometry optimized crystal structure.

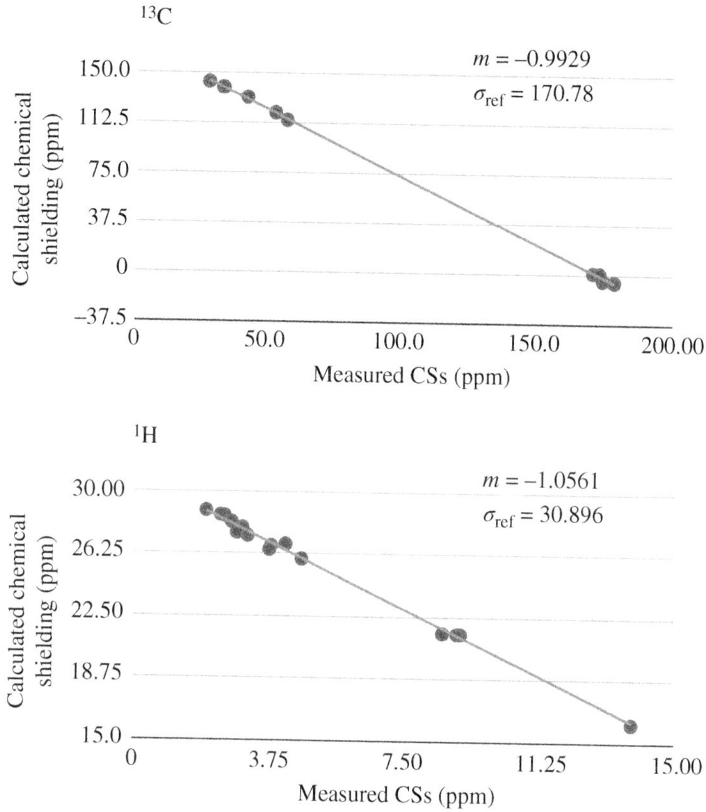

FIGURE 12.2 Calculated chemical shieldings versus measured CSs for ^{13}C and ^{1}H using CASTEP. Reprinted with permission from Ref. [36]. Copyright 2012, American Chemical Society.

This work presents the first comparison between calculated and experimental ^{1}H CS in a drug molecule because at that time only recently had the ^{1}H NMR spectra of solids reached the required resolution to separate proton resonances with the help of CRAMPS decoupling techniques. The ^{1}H–^{1}H DQ CRAMPS spectrum was able to separate 16 ^{1}H CSs, and a very good correlation between calculated and experimental CSs was reported for both ^{1}H and ^{13}C showing an average pairwise ***root-mean-square deviation*** (RMSD) between both of 2.5 and 0.3 ppm, respectively. The calculated ^{1}H CSs were particularly useful to unambiguously assign the two diastereotopic protons (H_a and H_b) of the CH_2 group (Fig. 12.1). A previous contribution from our group has also compared calculated versus experimental ^{1}H and ^{13}C CS on glutathione tripeptide showing an average RMSD of 1.32/1.50 and 0.04/0.04 for ^{13}C and ^{1}H resonances, respectively, using the two most popular codes (CASTEP/Quantum ESPRESSO) having the GIPAW approach implemented (Fig. 12.2) [36]. Very good correlations were also obtained.

In another example reported by Harris and co-workers [23], SSNMR was used, together with single-crystal and PXRD, to study the ***hydrogen bonding*** (H-bonding) network differences of several crystal forms of terbutaline sulfate (TBS), a drug used for asthma therapy. Full assignment of the 13C spectrum of the most stable TBS modification—anhydrate B (two terbutaline cations and one sulfate anion present in the asymmetric unit), was achieved by comparison with solution-state NMR spectra. In addition, dipolar dephasing experiments and the observation of the resonance broadening at low magnetic fields of 200 and 300 MHz caused by 14N second-order quadrupolar interaction in adjacent carbons forming C–N bonds were also performed. The ICSs for TBS anhydrate B form were calculated using the GIPAW-DFT method and the PBE functional. A moderately good agreement between the observed and calculated ^{13}C CSs values was obtained. The data obtained from GIPAW CS calculations were relevant to assign the two lowest-frequency peaks in the range 100–110 ppm to aromatic carbons.

12.5.1.2 Computing Electric Field Gradient Tensors to Study ^{35}Cl, ^{14}N and ^{17}O Quadrupolar Spins Recent advances in theory and computing power have made the computation of the ***electric field gradient*** (EFG) tensor parameters possible by first-principles methods and made quadrupolar spins a more valuable source of chemical information in solids. This section focuses on recent advances on the computational and experimental studies of APIs exploiting ^{35}Cl, ^{17}O and ^{14}N quadrupolar nuclei.

Approximately 50% of all pharmaceutical salts are crystallized in the HCl form, and chlorine is present in final formulations of approximately 25% of drugs. So ^{35}Cl NMR seems to be a good choice to probe the chlorine ion-binding environment in HCl amino acids (good sensitivity demonstrated by the wide range of chlorine C_Q values) [37] and was used to distinguish between pseudopolymorphs in chlorine-containing coordination compounds [38]. Hamaed et al. [39] presented the first application of ^{35}Cl for the identification and structural characterization of hydrochloride (HCl) anesthetics, namely Procaine HCl (PH), Tetracaine HCl (TH), Lidocaine HCl Monohydrate (LH), and Bupivacaine HCl Monohydrate (BH). The authors showed that the sensitivity of the ^{35}Cl EFG and CST parameters to the chlorine chemical environment allows for the prediction of the number of short H-bonds involving the chlorine ion [39]. For this purpose, *ab initio* calculations of the ^{35}Cl EFG and CST (using the Restricted Hartree–Fock and B3LYP method, respectively) parameters were conducted and compared with the experimental ones taken from simulations of the MAS spectra of the ^{35}Cl central transition recorded at 21.1 T. The calculated EFG tensor principal values, V_{11}, V_{22}, and V_{33}, allowed to obtain the quadrupolar coupling constants—C_Q and the quadrupolar asymmetry parameters—η_Q, while the calculated principal values of the CST, δ_{11}, δ_{22}, and δ_{33}, allowed the determination of certain anisotropic CS parameters such as the span, Ω and the skew, κ (following the Maryland notation) [40, 41].

Static ^{35}Cl NMR spectra allowed to disentangle the line shape contribution due to the EFG and CSTs and to extract the CSA parameters. As an example, ^{35}Cl static and MAS NMR spectra for two of the four pharmaceuticals are shown in Figure 12.3. The principal components and tensor orientations with respect to the molecular

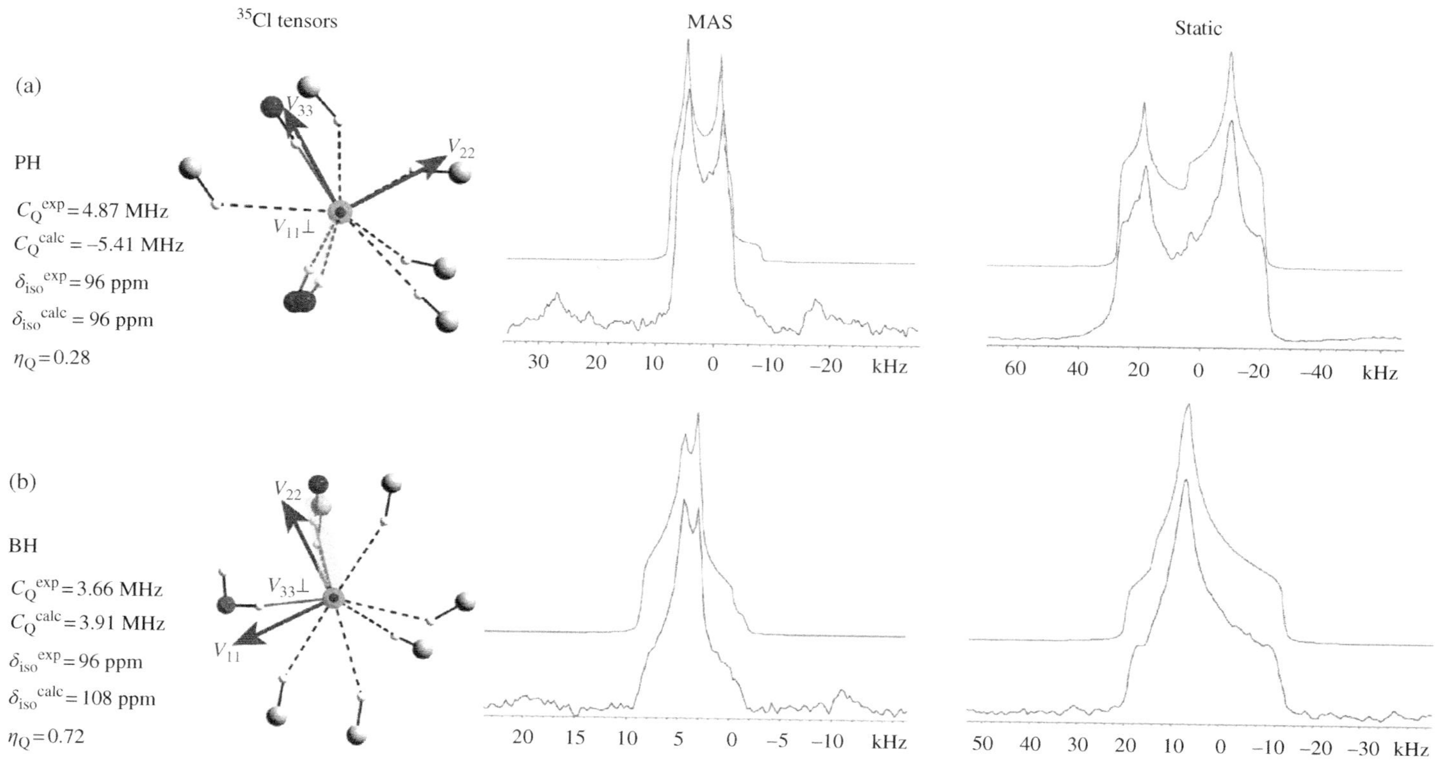

FIGURE 12.3 ^{35}Cl EFG tensor orientations and ^{35}Cl SSNMR spectra at 21.1 T for (a) Procaine HCl and (b) Bupivacaine HCl Monohydrate. Simulations for the static and MAS spectra are shown on top (red line) of the experimental lines (in blue). The yellow-shaded region represented on the tensor orientation diagrams highlights the N–H···Cl interactions. Experimental and calculated C_Q and δ_{iso} values are also shown for each compound. Adapted with permission from Ref. [39]. Copyright 2008, American Chemical Society. (*see insert for color representation of the figure.*)

coordinates of the four HCl pharmaceuticals were carefully examined. Short Cl···HN contacts have their bond axis oriented close to V_{33} (e.g., for PH in Figure 12.3a and for TH—figure not shown). Furthermore, V_{11} and V_{22} are almost similar in the two compounds because there are no other short contacts apart from Cl···HN. Also η_Q is closer to 0 than to 1. V_{11} and V_{22} are only differentiated in the presence of an additional short Cl···HO contact with H_2O molecules ((LH: not shown and BH: Figure 12.3b)). The Cl···HN contact in BH is the longest of the four pharmaceuticals, and the Cl···HO contact is the shortest. As a result, the ^{35}Cl quadrupolar parameters for BH are very different from those of the other complexes discussed in the paper, with the V_{33} principal value of the EFG tensor, no longer dominated by a short Cl···HN H-bond distance. Also, V_{22} is now nearly oriented along the Cl···HN bond and the V_{33} component has opposite sign compared to the other four compounds.

NMR spectroscopy of nitrogen is relevant for pharmaceutical studies since nitrogen often directly participates in H-bonding interactions. The NMR observation of the spin-1/2 ^{15}N isotope is difficult due to its very low natural abundance (0.364%), thus its observation is usually performed in ^{15}N-labeled systems. 2D ^{14}N-based NMR methods, on the other hand, offer a more sensitive alternative to explore nitrogen nuclides thus resulting in much faster experiments due to its much higher natural abundance (99.6%). ^{14}N is a quadrupolar nucleus with integer spin (I = 1), and despite of the recent advances to observe ^{14}N spins [42, 43], it is still a challenge to excite ^{14}N spins and obtain reliable structural information (due to spectral distortions related with the absence of a central transition). ^{14}N nuclei may provide access to additional structural information through the ^{14}N ***quadrupolar-induced shift***—QIS (or isotropic second-order quadrupolar shift), which depends on the magnitude of the quadrupolar interaction and on the magnetic field strength. Analysis of the QIS complements the ^{15}N ICSs that can be obtained in 1D experiments regarding their sensitivity to changes in H-bonding configurations.

2D ^{14}N–^{1}H HMQC experiments recorded at a 60 kHz MAS rate were recently performed, in drugs (acetaminophen), polymer (PVP), and polymer–drug dispersions, to assess proximities between nitrogen and proton nuclei involved in H-bonds [44]. This 2D ^{14}N–^{1}H HETCOR NMR experiment considerably shortens the acquisition time required compared to a ^{15}N–^{1}H HETCOR spectrum (6 h vs 48 h) [45]. The GIPAW-DFT approach was employed to calculate ^{15}N ICSs and ^{14}N EFG parameters (C_Q and η_Q) in order to assist in the resonance assignment process. The observed changes in the isotropic shifts for both ^{14}N and ^{15}N were evaluated because while ^{15}N shifts are only affected by the ICS, ^{14}N shifts combine the effect of the ICS (the same for both nitrogen isotopes) and the additional QIS contribution. The ^{14}N shift of the acetaminophen amide resonance changes downfield by 65 ppm when incorporated in a solid amorphous dispersion. Extraction of ^{15}N ICS values from ^{15}N CP MAS experiments showed that there is no significant change between the ^{15}N isotropic CS values observed for the amide nitrogen in acetaminophen and the dispersion; therefore, the change observed in the ^{14}N isotropic shift was attributed to different contributions from the QIS owing to different H-bonding environment in the amorphous dispersion.

In another contribution, a 2D ^{14}N–^{1}H HMQC experiment and $^{1}J_{15N-1H}$ spectral editing techniques were used in combination to identify proton transfer and H-bonding of

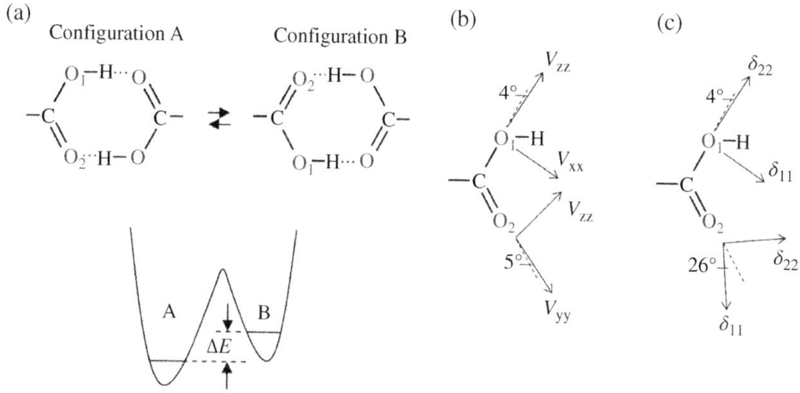

FIGURE 12.4 (a) Diagram illustrating the concerted double proton transfer in a carboxylic acid dimer and the corresponding double-well potential curve. Computed ^{17}O (b) *quadrupolar coupling* (QC) and (c) CS tensor orientations in salicylic acid and aspirin. V_{yy} for O_1 and V_{xx} for O_2 are perpendicular to the O_1–C–O_2 plane. For both O_1 and O_2, δ_{33} is perpendicular to the O_1–C–O_2 plane. Adapted with permission from Ref. [46]. Copyright 2013, American Chemical Society.

nitrogens in two drugs at natural isotopic abundance [45]. Protons directly attached to a nitrogen produce a signal modulation induced by the $^1J_{15N-1H}$ coupling in the corresponding ^{15}N resonance, which does not happen in non-protonated nitrogens. This strategy was used to assign N–H nitrogen environments. The ^1H and $^{15}N/^{14}N$ resonance assignment of the experimental spectra was also aided by GIPAW-DFT calculations.

Applying ^{17}O NMR techniques to drug molecules can be challenging due to its extremely low natural abundance (0.037%) and strong broadening effects associated to the quadrupolar coupling typical of quadrupolar nuclei such as the ^{17}O isotope. However, such problem can be overcome using ^{17}O labeling and high field magnets. In a recent contribution [46], static, MAS, and variable temperature ^{17}O SSNMR spectra have been recorded for salicylic acid enriched in positions [1,2-$^{17}O_2$] and [3-^{17}O] and aspirin enriched in positions [1,2-$^{17}O_2$], [3-^{17}O], and [4-^{17}O]. The ^{17}O EFG and CST parameters for O_1 (carboxylic acid), O_2 (carboxylic acid), O_3 (ester), and O_4 (ester) were determined experimentally and by means of GIPAW-DFT calculations, and a comparison between their ^{17}O CST values was made in light of the structural features. ^{17}O EFG tensor parameters (C_Q, η_Q) and ICSs for all oxygen sites in both compounds were obtained by simulation of the ^{17}O MAS NMR spectra recorded at two different magnetic fields. The parameters were then used to simulate the ^{17}O static NMR spectra that allowed to measure the CST parameters and relative orientation between EFG and CSTs. Interestingly, it was noticed that despite the aspirin and salicylic acid drugs being involved in a very similar cyclic carboxylic dimer, their ^{17}O EFG and CST parameters are significantly different (δ_{iso} (O_1)=168 ppm and δ_{iso} (O_2)=284 ppm for salicylic acid; δ_{iso} (O_1)=215 and δ_{iso} (O_2)=273 ppm for aspirin). An interesting observation was related with the averaging effect due to concerted double proton transfer in carboxylic acid dimers involving O_1 and O_2 hydroxyls in both (1,2-$^{17}O_2$) labeled compounds (Fig. 12.4). One has to take this dynamic effect into account in order to compare experimental and computed ^{17}O CS and EFG tensors for carboxylic acid functional groups. Salicylic acid and aspirin form cyclic H-bonded dimers in the crystal lattice, adopting configurations A and B (Fig. 12.4a). This means that all experimental NMR tensors in a carboxylic acid dimer are averaged between the corresponding "rigid" tensors found in configurations A and B. The observed ^{17}O CS and EFG tensors for O_1 and O_2 (σ_1^{obs} and σ_2^{obs} for CST) were thus considered as a weighted average between the two "rigid" tensors, that is, $\sigma_1^{obs} = P_A\sigma^A{}_{(C=O)} + P_B\sigma^B{}_{(C-OH)}$ and $\sigma_2^{obs} = P_A\sigma^A{}_{(C-OH)} + P_B\sigma^B{}_{(C=O)}$ for the case of CSTs. It is clear that both computed ^{17}O EFG and CSTs change their orientations (Fig. 12.4b and 12.4c) as a result of this double proton transfer. The populations of conformers A $\left(P_A = e^{(\Delta E/RT)} / \left(1 + e^{(\Delta E/RT)}\right)\right)$ could be assessed by fitting P_A employing the equations σ_1^{obs} and σ_2^{obs} (for the EFG and CS tensor components). The mean value of P_A obtained for aspirin and salicylic acid was 0.78 ± 0.04 and 0.98 ± 0.02, respectively (Fig. 12.5a and b). To complement this theoretical analysis, variable temperature ^{17}O experiments were recorded, and the energy asymmetry (ΔE) was found to be $\Delta E = 3.0 \pm 0.5$ kJ/mol and $\Delta E > 10$ kJ/mol for aspirin and salicylic acid, respectively. These results show that the potential curves for the concerted double proton transfer are significantly different, being nearly symmetrical for aspirin and highly asymmetrical in salicylic acid. These asymmetrical features in potential energy curves were confirmed by plane-wave DFT computations.

With the description of a proper averaging model for data analysis, correlation coefficients of ~0.99 were found between measured and computed ^{17}O NMR tensors in salicylic acid and aspirin (Fig. 12.5c and d).

Other studies are also worth mentioning in this section due to their potential application using distinct quadrupolar nuclei. A recent work by Bonhomme et al. [47] uses ^{87}Sr NMR techniques to provide information on the local structure around the Sr atom and obtain insight into the structure of antiosteoporotic pharmaceuticals and bioactive glasses. The authors calculated CS and EFG tensor parameters and found that GIPAW-DFT calculations can accurately compute ^{87}Sr NMR parameters. Additionally, Facey et al. [48] performed ^{2}H static NMR experiments at variable temperature to study the molecular dynamics of pyridine as a guest molecule in a clathrate. Despite not being a drug, this study is of great value to pharmaceuticals since hydrates and solvates are very common pharmaceutical solids.

12.5.2 Calculated vs Experimental Chemical Shift Tensors Using Different NMR Methods

The vast majority of applications of GIPAW so far have focused on the ICS and the quadrupolar coupling (for nuclei with spin quantum number I > 1/2), and there have been far fewer studies that have exploited interactions such as the CSA. This is mainly a result related to the difficulty in accurately measuring such interactions for all but the simplest systems, although developments in experimental methodologies will enable the determination of parameters across a wider range of increasingly complex systems.

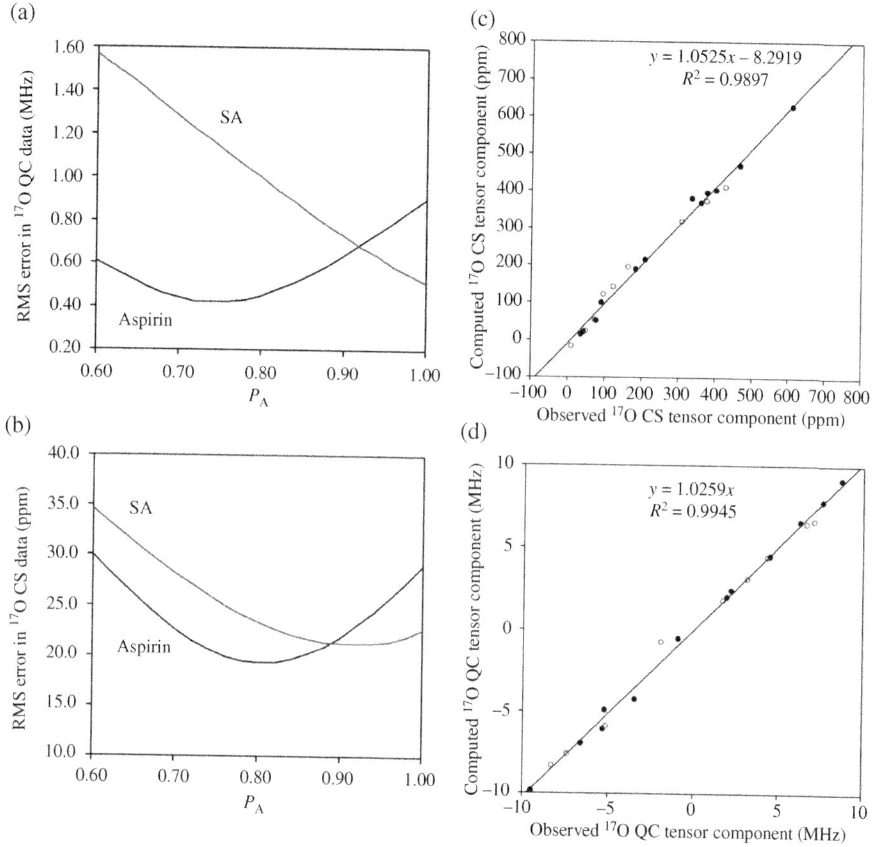

FIGURE 12.5 Quality of the data fitting as a function of P_A for (a) ^{17}O QC tensor components (e^2Qq_{ii}/h, $i=x, y, z$) and (b) ^{17}O CST components obtained at 298 K. Correlations between experimental and CASTEP-computed (c) ^{17}O CST components and (d) ^{17}O QC tensor components for salicylic acid (open circles) and aspirin (closed circles). Adapted with permission from Ref. [46]. Copyright 2013, American Chemical Society.

The 2D ***phase-adjusted spinning sideband*** (PASS) experiment, together with DFT-based calculations of the local fields, has been used by Smith and co-workers [49] to analyze the structural information available in the SSNMR spectra of 5-methyl-2-((2-nitrophenyl)amino)-3-thiophenecarbonitrile, producing red, orange, and yellow (ROY) polymorphs. It was shown that CS and dipolar coupling information obtained with such experiments are different for the distinct polymorphic forms and were thus used to quantitatively determine aspects of the molecular structure, including the coplanar angle between the phenyl and thiophene rings. ^{13}C–^{14}N residual dipolar couplings have been exploited to measure C13–N14–C16 (atom numbering shown in Figure 12.6a) bond angles in the three ROY compounds, and a good agreement was obtained with respect to the X-ray angle values.

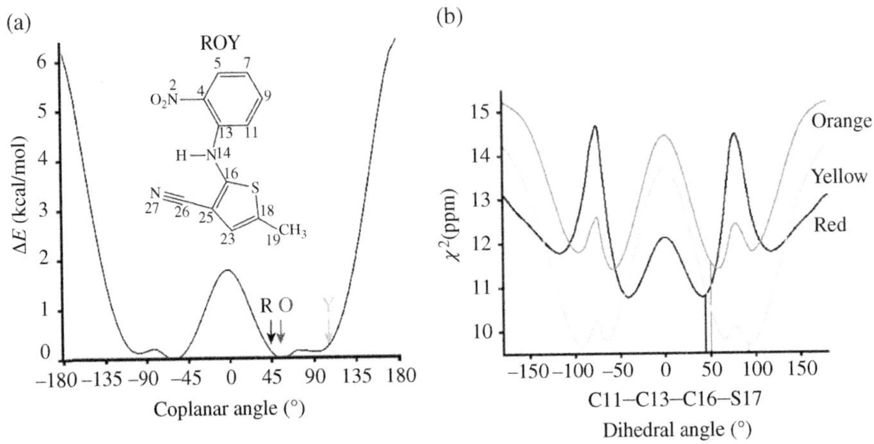

FIGURE 12.6 (a) Potential energy surface for thiophene ring rotation in ROY. The angles corresponding to X-ray values are labeled by form. (b) Least-squares comparison of all ^{13}C and ^{15}N tensor values between experiment and molecular modeling. The X-ray values of the C11–C13–C16–S17 dihedral are marked with vertical lines. Adapted with permission from Ref. [49]. Copyright 2006, American Chemical Society.

Results from DFT chemical shielding calculations for each carbon site were used to solve assignment ambiguities. In some situations, the agreement between theory and experiment was sufficient to allow for the unambiguous assignment of the spectrum. However, closely spaced or overlapping resonances between the distinct ROY polymorphs are difficult to assign reliably using only the ICS. To overcome this, the authors performed molecular modeling of the ROY compounds and an extensive comparison between calculated and experimental principal values of the ^{13}C, ^{15}N CST. The three ROY forms have a structural relationship that depends on the phenyl-thiophene coplanar angle (Fig. 12.6a). A particularly interesting result was the attempt to correlate this coplanar angle with the CST values. For this purpose, DFT calculations were performed to partially optimize each ROY conformer geometry after rotating the C11-C13-C16-S17 dihedral angle (through rotation of the thiophene ring) by steps of 5° until completing 360°. The CST principal values were then computed for every conformer and evaluated. Figure 12.6a shows the potential energy surface due to the different coplanar angles showing a total of four wells. The ROY conformer energies are also indicated. The dependence of the calculated CST values of selected carbons (nitrogens were also used but are not shown here) for each phenyl-thiophene coplanar angles was also studied. For example, only the C13 (phenyl ring) showed a significant change in its principal tensor values compared to other phenyl carbons (e.g., C7) as it is involved in the coplanar angle (Figure 12.7a and b). Despite of the constant values of δ_{11}, δ_{22}, and δ_{33} for C7, its ICS and asymmetry parameter changed significantly. On the other hand, carbons from the thiophene ring such as C18 and C25 were the most sensitive to the change of coplanar angles, with net changes in the breadth of the CSA as large as 50 ppm.

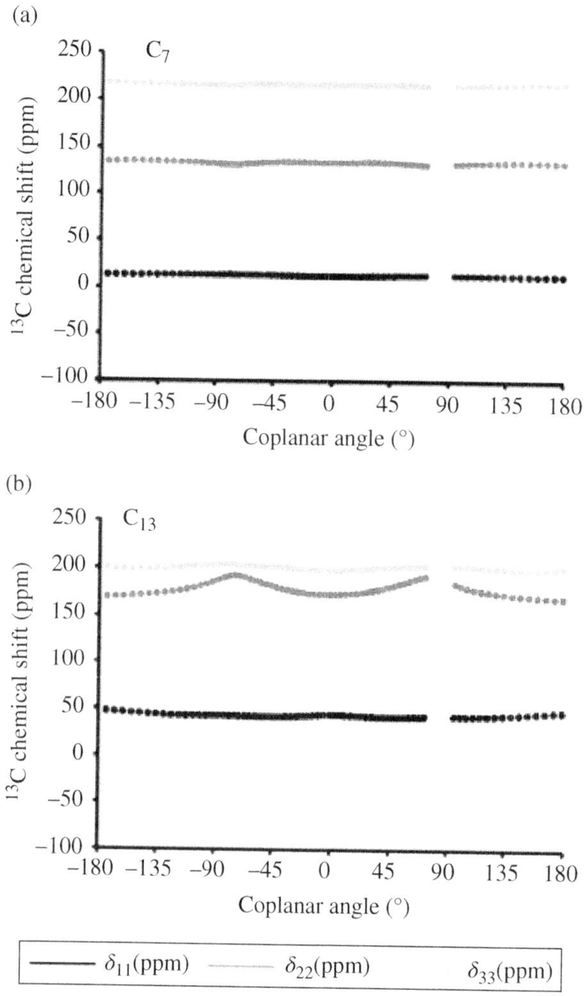

FIGURE 12.7 The calculated CSs for the phenyl ring selected carbons (a) C7 and (b) C13 as a function of coplanar angle. Adapted with permission from Ref. [49]. Copyright2006, American Chemical Society.

In addition, the CST values were used as molecular constraints in determining the model structure that provides the best correspondence between theory and experiment using a least-squares minimization function for all three ROY polymorphs:

$$\chi^2 \equiv \sum_n \left[\sum_{i,j} \frac{A^{(n)}_{\mathrm{exp}_{i,j}} - A^{(n)}_{\mathrm{model}_{i,j}}}{\varepsilon^{(n)}_{\mathrm{exp}_{i,j}}} \right] \rightarrow 0$$

where $A^{(n)}_{\mathrm{exp}}$ and $A^{(n)}_{\mathrm{model}}$ are the experimental and calculated tensors for the n sites in ROY molecules. $\varepsilon^{(n)}_{\mathrm{exp}}$ is the error associated with the experimental tensor values.

A plot of the χ^2 values derived from the different models versus the coplanar angle gave their minima at angles very close to the ones obtained by XRD (i.e., 41.0° (modeling) versus 43.6° (X-ray) for ROY-R; 58.0° versus 50.1° for ROY-O and 91.9° versus 91.8° for ROY-Y); see Figure 12.6b.

With better computing power and speed, these methods might be important for structural elucidation of other pharmaceutical systems.

The 3D structure of the anhydrous polymorph of anticancer drug paclitaxel was established by SSNMR ^{13}C and ^{15}N CST analysis and ^{1}H–^{13}C HETCOR correlation data by Heider et al. without previous X-ray data. This is the first SSNMR-based characterization of a drug with $Z' > 1$ ($Z' = 2$; two molecules in the asymmetric unit) [50]. All the experimental data are correlated with structure through a total of 650 computational models that extensively sample all possible conformations. For each model, the CST values are calculated (at the B3PW91/D95** level of theory) to allow comparison with the experiment and the model is either retained or eliminated at a statistical probability. Only 13 models were found to match experiment. These models required only fragments of the paclitaxel structure, since CS principal values are dominated by the local environment. In addition, the rigidity of paclitaxel's baccatin moiety minimizes the global search of conformation. This approach is computationally demanding; however, the comparison of CS principal values reduces the chance of becoming trapped in a local minima and provides structures that are difficult to determine by other methods. For the success of this approach, the spectra have to be correctly assigned and the principal values need to be measured. The structure models were compared with CS principal values (obtained experimentally using the FIREMAT method [51]) for the relevant nuclei using an F-test [52]. Torsion angles were examined by individually rotating each flexible dihedral in 30.0° increments over the 360° range. At each point, the chosen angle was fixed and the remaining structure allowed to relax (via partial geometry optimization) before chemical shielding values were computed. The best-fit dihedral angle was retained and the process continued for the next flexible bond. This modeling was continued until all conformational combinations had been explored and the overall best fit determined.

The structural information contained in the isotropic and anisotropic CSs was explored to model structure conformations of three flavonoids [53]. The principal values of the ^{13}C CST were determined under slow sample spinning (2 kHz) using the 2D-PASS NMR technique, and verified by DFT-GIAO calculations of ^{13}C CSs. Analysis of the ^{13}C CST components and comparison with shielding parameters calculated for different conformers of the three compounds enabled the selection of the most reliable geometry in the solid phase. Conformational analysis was performed by modeling the structures using the semiempirical PM3 method and the results obtained for baicalein were in agreement with XRD data.

SSNMR methods have been used to investigate several molecular cocrystals and complexes, even in the presence of impurities. A protocol for the application of methods based on dipolar connectivity, CS information, and relaxation measurements using several cocrystals was presented [54]. A plethora of NMR techniques were explored in order to study H-bonding and other intermolecular contacts. In addition, chemical shielding calculations were performed to assign ^{1}H and ^{13}C resonances.

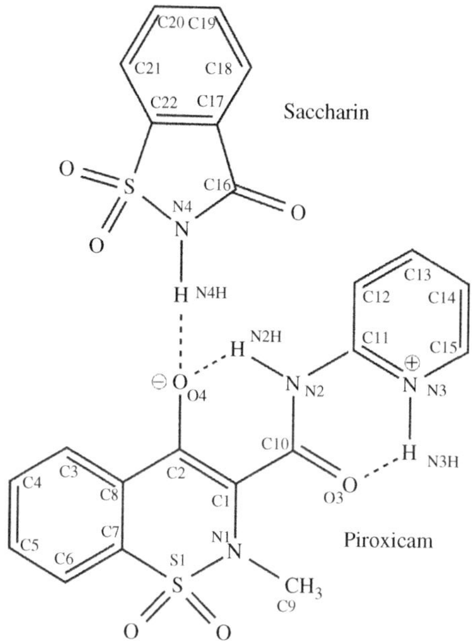

FIGURE 12.8 Scheme including atom numbering of the piroxicam saccharin cocrystal.

CS calculations were performed in gas-phase clusters extracted from the optimized solid-state structure and the ***embedded ion method*** (EIM) was used to help replicate the electrostatic potential of the crystal structure. The authors also found that the ^{1}H and ^{13}C CS calculated values were slightly more accurate using the EIM. As an example, experimental and calculated ^{13}C CST principal values were explored for the zwitterionic piroxicam saccharin cocrystal (Fig. 12.8) using the EIM, to provide information about the ionization state and assist with spectral assignment. For that purpose, the CP-FIREMAT technique was employed to extract the experimental ^{13}C CST principal values (Fig. 12.9a). This experiment helped assign overlapped ^{13}C positions. For instance, the sideband pattern for the signal at 167.82 ppm was fitted to a two-site model, allowing for a reliable fit and enabling the assignment of this position to carbons C2 and C10.

In Figure 12.9b, the experimental CST principal values, extracted from the CP-FIREMAT experiment (Fig. 12.9a), and the ICS values are compared for selected carbon sites in piroxicam saccharin and β-piroxicam. These data help uncover some useful CS trends; for example, the cancelation of the δ_{11} and δ_{22} components for C11 leads to little change in the ICS between piroxicam saccharin and β-piroxicam, but large (~25 ppm) differences in the δ_{11} and δ_{22} principal values are easily observed, which may be related to specific structural features.

For cocrystal systems containing phosphorous (triphenylphosphine oxide 6-chloro-2-pyridone), ^{31}P MAS NMR at slow MAS rates offers a more rapid way to confirm H-bonding and cocrystal formation fitting the CSA profile of the ^{31}P spectrum. The anisotropy of the CST of a phosphine oxide group is strongly related to the

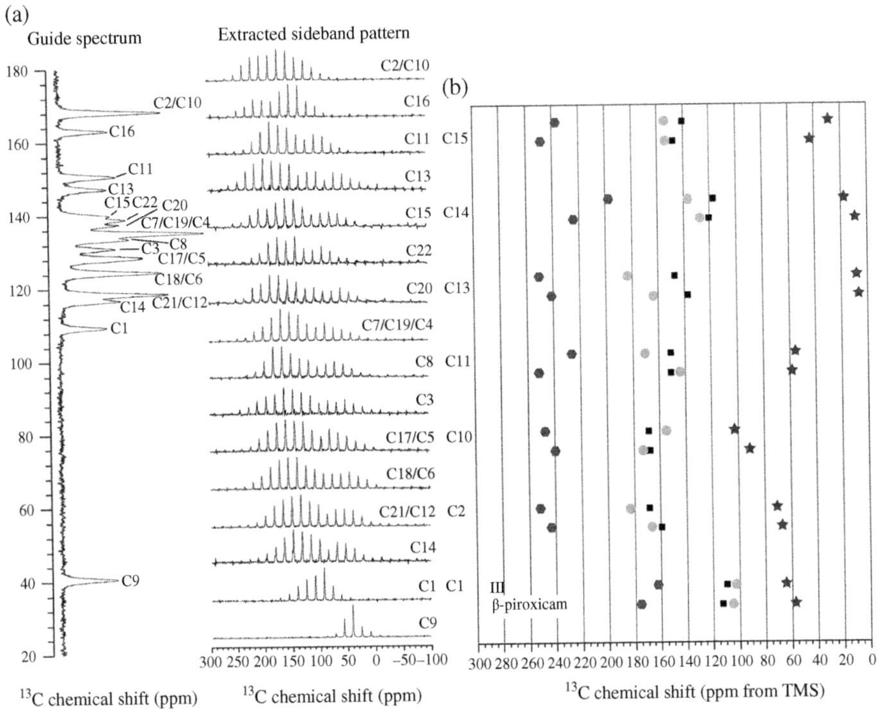

FIGURE 12.9 (a) 13C CP-FIREMAT data for piroxicam saccharin showing overlaid extracted experimental sideband patterns and fitted sideband patterns (middle), as well as the overlaid experimental and fitted guide spectra (left) (b) selected ^{13}C CSs for β-piroxicam and III (piroxicam saccharin) from CP-FIREMAT experiments are shown, comparing the principal components of the CST and the ICS. The δ_{11} (●), δ_{22} (●), δ_{33} (★), and δ_{iso} (□) values for III are at the top of each pair while the values for β-piroxicam are the bottom. Reprinted with permission from Ref. [54]. Copyright 2008, American Chemical Society.

number and length of H-bonds that the P=O group accepts. ^{31}P CSA tensor values of pure TPPO (triphenylphosphine), used as a cocrystal former in this study, were calculated using B3LYP/6-311+G(3df,3pd) level of theory and compared to the calculated values from the cocystal showing interesting differences.

In a recent contribution, Wang et al. [55] performed structural modeling of a hydrate salt verified by SSNMR correlation and ^{13}C CST calculations. Atorvastatin calcium Form I (ATC-I) is the most stable crystalline form known for this drug and no single-crystal X-ray structure was solved for this form. A complete ^{13}C, ^{19}F, and ^{15}N NMR spectral assignment of ATC-I, containing two molecules in the asymmetric unit, was achieved. In addition, a possible local structure for ATC-I based on the simultaneous interpretation of the experimental SSNMR data and DFT calculations of ^{13}C CSs (B3LYP/6-31G(d)) was proposed. ^{13}C CST principal values of all resolvable carbon resonances were measured through a 2D ^{13}C ***separation of undistorted power***

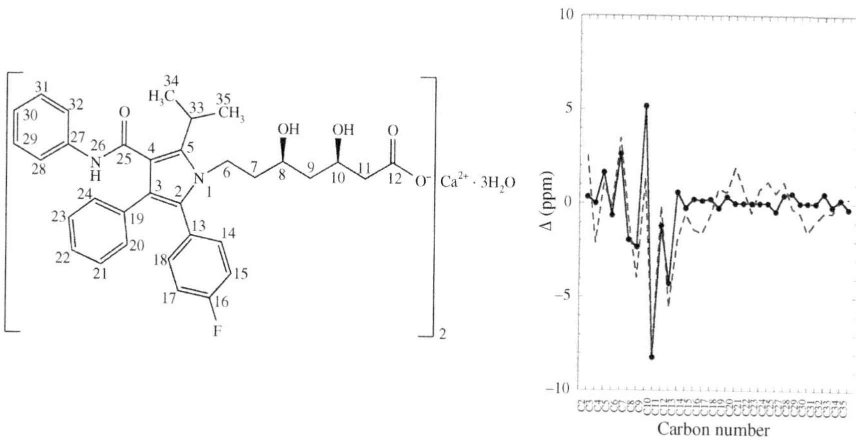

FIGURE 12.10 Molecular structure of atorvastatin, with atom numbering (left). CS difference between atorvastatin "a" and "b" (right). The calculated CS differences based on the proposed model (dashed line) are compared with experimental values (●). Reprinted with permission from Ref. [55]. Copyright 2012, American Chemical Society.

*patterns by **e**ffortless **r**ecoupling* (SUPER) experiment [56], another variant to recover the CSA profile under MAS rotation. Figure 12.10 compares the ^{13}C ICSs of each carbon site between the two independent molecules "a" and "b" in the asymmetric unit ($Z'=2$), defined as $\Delta\delta=\delta_a-\delta_b$. The variation in $\Delta\delta$ with carbon positions served as a benchmark in evaluating structural models based on predicted CSs from the DFT calculations. The benchmark used 66 ^{13}C resonances (33 carbons for "a" and "b" molecules), which is statistically significant. More importantly, the comparison of ICS differences helps eliminate inaccuracies that might arise from offsets in the DFT calculations or the choice of basis sets. Differences observed between the crystal model (dashed line) and the experimental (solid line) values in Figure 12.10 are similar; however, the π···π ring currents might be the explanation for some discrepancies as the calculation method used (DFT) does not account for ring current effects.

A new approach to interpret the relationships between intermolecular H-bonding, redistribution of electron density, and NMR CSTs has been presented by Marek and co-workers [57]. In this work, the authors analyze how the cluster model can be used for identification of intermolecular contacts that govern the packing motifs in crystals. The clusters were based on the XRD data and were carefully chosen to study the "isolated" effect of a particular intermolecular interaction (H-bond and π···π stacking). Differences in the electron density distribution between the isolated molecule and selected clusters were calculated and visualized as ***e**lectron **d**eformation **d**ensity* (EDD) patterns. The EDD patterns can be correlated with the changes in the CSTs of the nuclei involved in the intermolecular contact. For a better understanding of this relationship, the changes in the isotropic magnetic shielding (***s**hielding **d**eformation **d**ensity*, SDD) in the proximity of the nucleus in question can also be calculated and visualized in the real space. Quantum chemical calculation of EDD, SDD, and CSTs employed the B3LYP functional and the 6-311G** basis set. EDD ($\Delta\rho$) is calculated

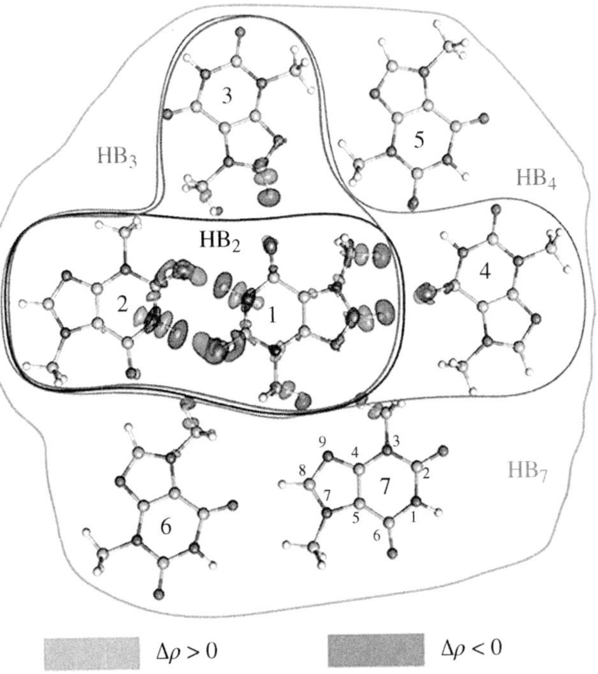

FIGURE 12.11 Scheme illustrating the composition of clusters HB2, HB3, HB4, and HB7 of theobromine emphasizing the EDD for the H-bonded cluster HB7. Atom labeling is shown in molecule 7. Adapted with permission from Ref. [58]. Copyright 2013, American Chemical Society.

and visualized as the difference between the electron density of a molecule placed in its crystal environment (ρ_1) and the sum of the electron densities of the isolated molecule (ρ_2) and its crystal environment (ρ_3): $\Delta\rho=\rho_1-(\rho_2+\rho_3)$. SDD is defined similarly to EDD, but instead it accounts with the changes in the isotropic values of the magnetic shielding ($\Delta\sigma$) in real space: $\Delta\sigma=\sigma_1-(\sigma_2+\sigma_3)$. An interesting feature in this work is that not only the presence but also the strength of an H-bond is encoded in the EDD pattern.

This approach has been applied to theobromine clusters [58], based on the crystal structure available at the CSD database, to identify and classify the intermolecular contacts that govern the packing motifs and understand how the crystal packing effect translates into the ^{13}C NMR CST changes. Comparison of the calculated ^{13}C CSTs involving a total of six supramolecular clusters and for an isolated molecule were performed and related to the experimental CST values. An example of four different types of in-plane clusters (HB2, HB3, HB4, and HB7) are shown in Figure 12.11, where the intermolecular contacts involving the two H-bonded theobromine molecules 1 and 2 in the asymmetric unit ($Z'=2$) are emphasized in the form of EDD patterns [58].

Figure 12.12 is a zoom in of the C_2=O···H–NH-bond involved in the formation of the theobromine dimer observed in Figure 12.11 in order to better visualize the EDD

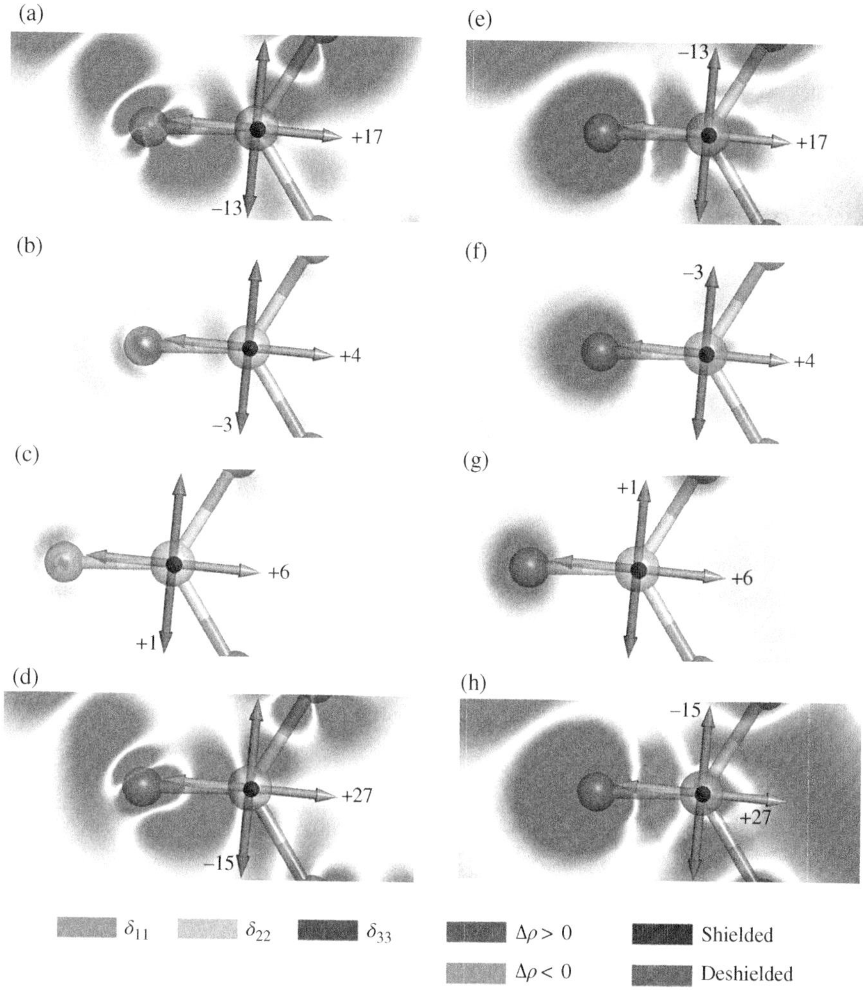

FIGURE 12.12 Visualization of the in-plane ($\delta_{11}|\delta_{22}$) slices of the EDD (left) and SDD (right) for atom C2 in theobromine. (a and e) HB2; (b and f) cluster composed of molecules 1 and 6 (weak H-bonding, see Fig. 12.11); (c and g) S7; and (d and h) HBS13. The arrows schematically represent the calculated relative orientations of the in-plane principal components of the ^{13}C CST, and the numbers denote the calculated changes in the magnitude of these components upon formation of the clusters. Adapted with permission from Ref. [58]. Copyright 2013, American Chemical Society. (*see insert for color representation of the figure.*)

and SDD maps determined based on four distinct clusters as described in the caption. The polarization of the C=OH-bond acceptor is clearly observed as the electron density is accumulated at the oxygen atom (blue area, Fig. 12.12, left column a–d) and decreases at the C_2 atom (red area). A decrease in electron density around C_2 results in an increase in CS; if the calculated $\delta_{iso}(C_2)$ is compared between an isolated molecule and the HB_2 cluster, it increases from 148.5 ppm to 150.2 ppm as expected. According to Figure 12.12, right column e–h, the magnetic shielding decreases at C_2

in the δ_{22} direction parallel to the C═O bond (red area) and increases along the approximately perpendicular δ_{11} direction (blue area). This is in good agreement with the calculated NMR CST for the HB2 cluster. From Figure 12.11, molecules 6 and 3 are also involved in intermolecular contacts with the central molecule 1 and are involved in weak H-bonds of the type C2═O···H–CH3 and C6═O···H–C8, respectively.

To take into account stacking interactions, another cluster, S7 cluster (not shown) consisting of seven molecules, was considered where three molecules below and another three above the central molecule 1 were assembled. The total effect of the packing interactions in the ^{13}C CST can be obtained considering the largest cluster used, that is, HBS13 which is a sum of clusters HB7 (Fig. 12.11) and S7. The differences in chemical shieldings ($\Delta\delta_{ii}$) calculated for HB13 for δ_{22} and δ_{11} are illustrated in Figure 12.12 and agree with the sum of the ones obtained for each subcluster HB7 and S7.

The analysis of the EDD and SDD maps for the identification of the multiple packing effects based on the quantification of CST components in each cluster was important as sometimes effects of intermolecular interactions compensate; such effects are invisible by just considering the largest cluster (HBS13) or a single molecule. This study uses a similar philosophy as done by Uldry et al. [59] and Mafra et al. [20] (see next section) by evaluating the CS parameters in the isolated molecule and for the molecule embedded in the crystal packing. This is an interesting approach that attempts to rationalize how the effect of crystal packing is encoded in the CSTs in terms of the redistribution of the ground-state electron density induced by intermolecular interactions.

12.5.3 Studying Crystal Packing Interactions

Changes in a drug's crystal packing interactions will impact its solubility, stability, manufacturing ease, and bioavailability. These changes can be due to polymorph transformations or salt/cocrystal formation. Comparison between SSNMR experimental and calculated CST have been used to prove molecular association, identify H-bond donors and acceptors, show aromatic ring-currents, and report on the strength of these interactions.

12.5.3.1 Characterization of H-Bondings in Cocrystal Systems Using ^{1}H and ^{15}N Chemical Shifts In a recent U. S. Food and Drug Administration guidance, salts and cocrystals are given different regulatory classifications. This has put further pressure on the industry to identify a complex as a salt or cocrystal [60], adding further value to NMR characterization of H-bonding in cocrystal complexes. One of the earliest studies involving CST calculations on cocrystals provides guidance for using ^{15}N NMR tensors to classify a complex as a salt or cocrystal, and to rank bond strength. In Li et al., the salt–cocrystal continuum was examined for a drug with two weakly basic heterocyclic nitrogens [61]. The drug was crystallized with three dicarboxylic acids of increasing strength. The ^{15}N isotropic shifts for these complexes differed up to −120 ppm from the crystalline free base, demonstrating the high sensitivity of ^{15}N to intermolecular interactions.

Each drug complexes examined, contained five nitrogens thus complicated CS assignments. A combination of solution NMR and short contact time CPMAS experiments were used to assign the CSs. Despite these experiments, there were still ambiguous assignments and NMR tensor calculations were necessary. These calculations were completed using a cluster approach and GIAO methods using a B3LYP/6-31G* basis set. Clusters were built from the single-crystal structures, comprising of a single API molecule and any ionic or H-bonded acid partners. This approach reduced the computational demands, while capturing the key interaction with the basic nitrogens. Of course, ring stacking and other weak contacts would not be accounted for by this method, which can increase error. For complexes with proton transfer the ^{15}N ICSs moved upfield by 80–120 ppm from the crystalline free base. The cocrystal complexes produced upfield shifts of 20–40 ppm due to strong H-bonding between the nitrogen acceptor and the carboxylic acid donor. The observed ICS differences (Δppm) correlate well with the ΔpK_a between the drug's two basic nitrogens and the acids and with the rank ordering of the bond strengths based on donor acceptor (D···A) bond distances extracted from the single crystal structures. Based on these trends, guidance for discerning between salts and cocrystals, and rank ordering bond strength based on ^{15}N Δppm of heterocyclic aromatic nitrogen was provided.

Similar methods were applied by Kahn et al. to the quinidine methyl paraben cocrystal and quinidine 4-hydroxybenzate salt [62, 63]. Again DFT tensor calculations using a GIAO cluster approach was performed to assign the ^{15}N resonances and confirm bonding interactions. The ^{15}N Δppm values for the cocrystal and salt from the free quinidine were much smaller than seen in the previous study. These differences can be explained by the formation of moderate to weaker bonds in these complexes compared to the free quinidine. Overall the ^{15}N ICS still follow the ranking of the bond strengths, but identifying proton transfer from experimental values alone may be more challenging. In this case, CSTs were necessary to identify the protonation state of these complexes.

Improvements in high-resolution ^{1}H SSNMR techniques derived from probe technology and pulse sequence developments allowed to take advantage of the high sensitivity of ^{1}H nuclei. More recent studies have focused on using ^{1}H ICS and ^{1}H homo and heteronuclear correlation experiments to examine cocrystals [54, 62]. In these studies, commercially available fast MAS probes and CRAMPS were used to resolve ^{1}H ICSs. For complexes with moderate to strong H-bonding (where the D···A distance is less than the sum of their van der Waals radii), the ^{1}H ICSs are greatly deshielded and clearly resolved using MAS rates ≥35 kHz. For the methyl paraben quinidine cocrystal a moderate H-bond (H48) is resolved at 9.39 ppm using 50 kHz spinning in the ^{1}H MAS, ^{1}H–^{1}H DQ BABA and ^{1}H-^{13}C CP-HETCOR spectra (Fig. 12.13) [62].

Once ^{1}H CSs are resolved, correlation experiments and CST calculation can be used to prove cocrystal formation and examine the bond properties. The crystallization of methyl paraben and quinidine produced two new peaks at 9.39 and 13.45 ppm, suggesting cocrystal formation [62]. These two new peaks were assigned using ^{2}H labeling to the hydroxyl protons of quinidine (H48) and methyl paraben (H67),

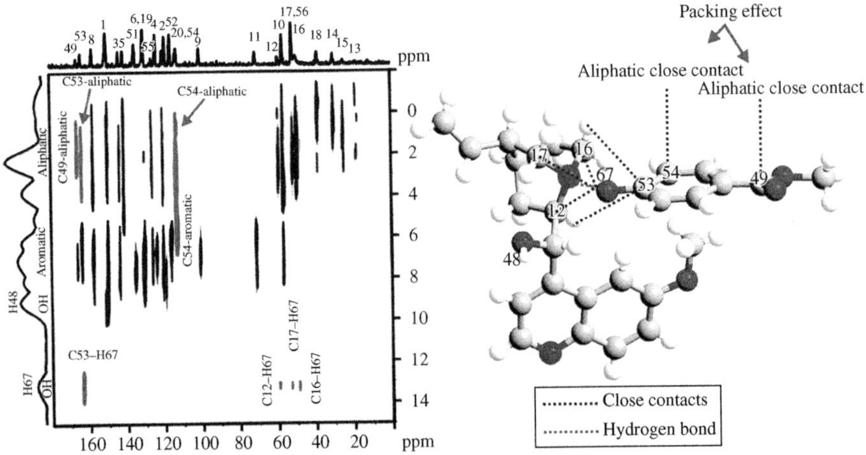

FIGURE 12.13 ^1H–^{13}C CP-HETCOR spectrum of the methyl paraben quinidine cocrystal acquired at 850 MHz using a commercially available 1.3 mm MAS probe at a MAS rate = 50 kHz (left). X-ray crystal structure of methyl paraben quinidine cocrystal showing intermolecular contacts (right). Reprinted with permission from Ref. [62]. Copyright 2010, American Chemical Society.

respectively. Correlations were observed in ^1H-^1H DQ and ^1H-^{13}C HETCOR experiments showing the formation of H-bonds between H67 and quinuclidinic ring nitrogen (N24), and H48 and quinoline ring nitrogen (N23) (highlighted in Fig. 12.13). This demonstrates the successful formation of O–H···N heterosynthon. DFT CST calculations using a GIAO cluster approach (B3LYP functional and the 6-311G** basis set) verified this interaction, predicting a ^1H CS within 0.3 ppm of the observed value.

Other reports show that SSNMR methods were explored to prove cocrystal formation of the solids produced during solid form screening activities [54]. ^1H ICS and ^1H correlation experiments were found to be the fastest and most definite methods to prove molecular association. The authors first recommend screening solids by ^{13}C and ^1H spectra to assess phase purity, confirm the presence of new crystalline species, and identify strong H-bonding between conformers from the downfield ^1H ICS. ^1H–^{13}C CP-HETCOR and ^1H–^1H DQ recoupling experiments can then provide dipolar connectivity to demonstrate molecular association between components. CST calculations can further confirm molecular association. This protocol was vetted with nine cocrystal systems to prove complexation.

12.5.3.2 Characterization of Weak H-Bondings in Hydrates Using NICS Much of the SSNMR literature has focused on strong intermolecular interactions, such as ionic and H-bonding. Weak interactions are much more difficult to examine both experimentally and computationally. The true impact of weak H-bonds or van der Waals contacts may be obscured by other packing constraints. Although these interactions have lower energies, –1 to –5 kcal/mol for CH···π contacts versus –5 to –40 kcal/mol for moderate/strong H-bonds, they can drive polymorph transformations. In work by

Uldry et al. on uracil and 4-cyano-4′-ethynylbiphenyl, and Mafra et al. on ciprofloxacin polymorphs, GIPAW CST and NICS calculations on the piece-wise deconstructed crystalline lattice are used to deconvolute these effects [20, 59].

For the antibiotic ciprofloxacin, the packing interactions (CH···π, CH···$O^{(-)}$, π···π and $^{(+)}$NH···$O^{(-)}$) and the impact of hydration on the ^{1}H and ^{13}C ICSs are examined for the anhydrate, Form I, and channel hydrate, Form II. NMR tensor calculations were completed using the GIPAW approach. To evaluate hydration effects, the water molecules were removed *in silico* from the optimized Form II structure and the CST recalculated. To examine packing effects GIPAW CST calculations were performed on a 1D chain of zwitterionic ciprofloxacin molecules connected through $^{(+)}$NH···$O^{(-)}$ H-bonds. The supercell dimension was expanded to minimize interactions between neighboring chains. For neutral molecules, calculations could be performed on a single molecule in an enlarged supercell instead of a 1D chain.

The crystal packing and hydration effects on the calculated ^{1}H ICS of forms I and II are plotted as blue and red stems in Figure 12.14. For instance, the CS difference ($\Delta\delta^{hydration}$) values (red stems) were determined from the GIPAW calculated values for 'dehydrated' Form II subtracted from the water occupied crystal. The presence of water in Form II results in a maximum $\Delta\delta^{hydration} = 2$ ppm and 5 ppm for ^{1}H and ^{13}C, respectively. Hydration has the greatest impact on the atoms in cyclopropane and piperazine rings that interface with the water channel, as the polarizing effect of the water oxygens considerably deshields these protons.

The GIPAW calculated ^{1}H CS differences ($\Delta\delta^{packing}$) between the complete crystal and 1D chains (blue) stems in Fig. 12.14) show the packing contributions of both forms. It is not surprising to see the largest $\Delta\delta^{packing}$ is for the HN proton involved in the zwitterionic $^{(+)}$NH···$O^{(-)}$ H-bond. Large packing effects were observed for CH···π interactions, shielding ^{1}H CS values by up to 3.6 ppm (H3′ in Fig. 12.14). These and π···π interactions are also impacted by intermolecular ring currents, which influence the ^{1}H and ^{13}C shieldings. NICS were determined to estimate the contribution from aromatic ring currents alone. The $\Delta\delta^{packing}$ values were corrected by NICS for selected crystal packing interactions. These corrected values show that weak H-bonds, such as the CH···π intermolecular interactions in ciprofloxacin, can contribute up to −4 ppm to the ^{1}H ICS. Smaller packing contributions, up to 2.3 ppm, were calculated for the weak CH···$O^{(-)}$ H-bonds. Similar values were obtained for the weak CH···X H-bonding in maltose [64], uracil and 4-cyano-4′-ethynylbiphenyl [59] by comparable CST calculation methods.

12.5.4 Employing Chemical Shifts for Crystal Structure Elucidation/Determination

12.5.4.1 Combining XRD, DFT, and NMR for Structure Validation The most relevant point in combining PXRD, NMR, and DFT calculations is that the quality of the structure is assessed against both the experimental PXRD data and the experimental SSNMR data, validated by NMR CST calculations, providing a very robust confirmation of the reliability of the structure. A few key examples of crystal structure determination are given later where such interdisciplinary approach is exploited.

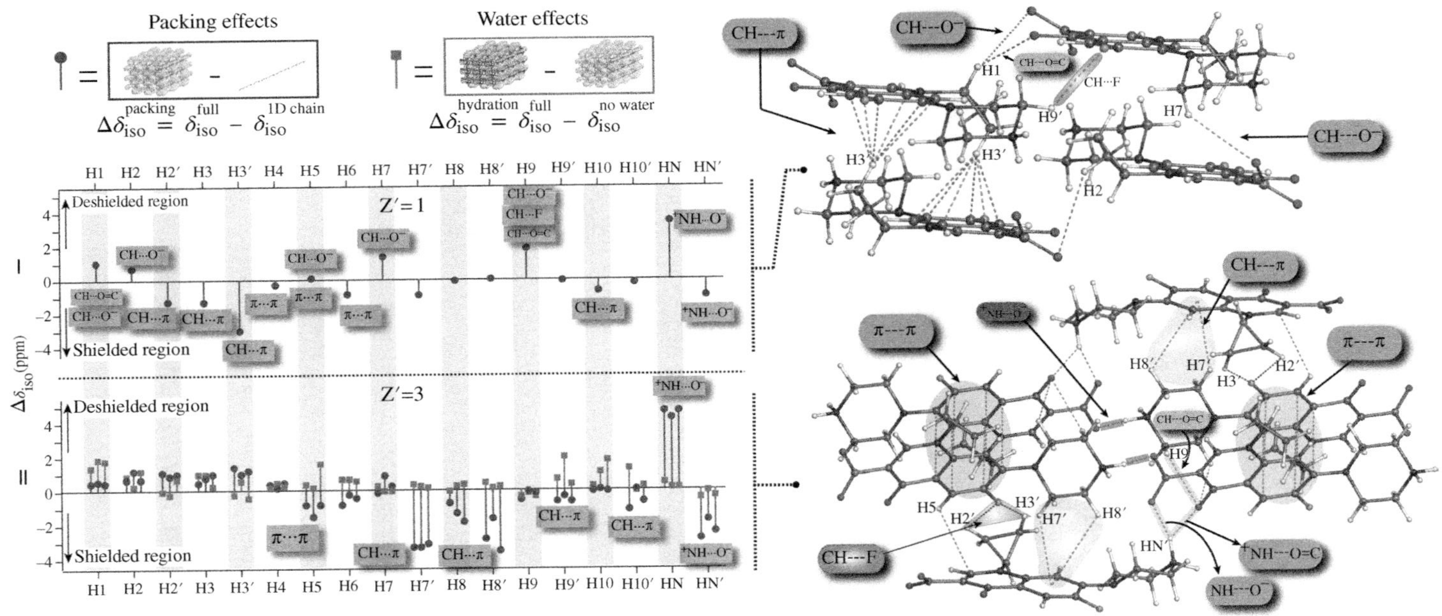

FIGURE 12.14 (Left) $\Delta\delta$(ppm) plot showing crystal packing contributions (blue stems) and hydration contributions (red stems) to the GIPAW calculated ^{1}H ICS of ciprofloxacin Form I and Form II. In II, each of the three stems per proton corresponds to a crystallographic unique molecule ($Z'=3$). Right: Detailed view of intermolecular interactions in the ciprofloxacin Forms I and II structures. Reprinted with permission from Ref. [20]. Copyright 2012, American Chemical Society. (*see insert for color representation of the figure.*)

Dudenko et al. have very recently applied this combined XRD, DFT, and NMR approach to the structure validation of the 1:1 cocrystal of indomethacin and nicotinamide (IND-NIC) [65]. Subsequent to the structure solution process by PXRD, first-principles DFT-based techniques within the GIPAW approach were exploited to calculate the SSNMR data for the structure. The calculated ^{13}C and ^{1}H CSs were found to be in excellent agreement with the corresponding measured CSs. In particular, the mean and highest differences between experimental and calculated CSs were for ^{1}H, 0.4 ppm (mean), 2.2 ppm (highest) and for ^{13}C, 1.6 ppm (mean), 4.3 ppm (highest). This approach was found to provide a robust assessment of the validity and quality of a refined crystal structure.

PXRD methods, lattice energy calculations, ^{13}C and ^{15}N SSNMR, and a systematic search algorithm have been applied to anhydrous theophylline [66]. The trial crystal structure generation unraveled two different H-bonding motifs involving N–H···N and N–H···O intermolecular interactions in a fixed unit cell. These structures were "indistinguishable" in terms of calculated lattice energy and XRD profile fitting to experimental data. SSNMR spectroscopy was applied successfully in this example to distinguish subtle conformational and packing differences by comparison of experimental and calculated isotropic ^{15}N and ^{13}C CSs. BPW91/TZVP level of theory and the LORG47 method were used to calculate the NMR CSs. ICSs were calculated for trimer arrangements, optimized in vacuo, representative of the N–H···N and N–H···O trial crystal structures as performing high-level DFT calculations on full crystal structures was not feasible. Good agreement between the experimental and theoretical ICSs revealed that the structure showing the N–H···N networks is indeed the correct one. Four trial crystal structures were progressed for Rietveld refinement. Two were representative of the N–H···N motif and other two of the N–H···O motif, that is, the two highest ranked trial structures from the systematic searches in terms of lattice energy and powder X-ray profile fit, respectively. Structural differences between the profile difference plots obtained after refinement of the four structures were imperceptible. Although the resulting structural model was found to be extremely similar to the one refined from the best N–H···N trial structure, the results of the Rietveld refinements were not sufficiently convincing to justify disregarding the N–H···O motif as the true crystal structure. However, analysis of the NMR data showed that the N–H···N model accounts for the experimental ^{15}N CSs more adequately than does the N–H···O structural model.

A high-value approach for the structure determination of an enantiotropically-related dimorphic system having low solid–solid conversion temperatures has been presented [67]. The crystal structure of the thermodynamically more stable form at room temperature was determined by single-crystal XRD (polymorph 1, $Z'=4$, $Z=16$). The crystal structure of the other form (polymorph 2, $Z'=1$, $Z=4$) was determined using iterative PXRD structure solution methods, assisted by SSNMR experiments (dipolar connectivity and CS measurements). DFT geometry optimizations were used in tandem with Rietveld refinement and NMR CS calculations to improve and verify the structure for polymorph 2.

Gas-phase DFT NMR calculations using the GIAO cluster approach for determining NMR shielding tensors, the hybrid B3LYP DFT functional and the

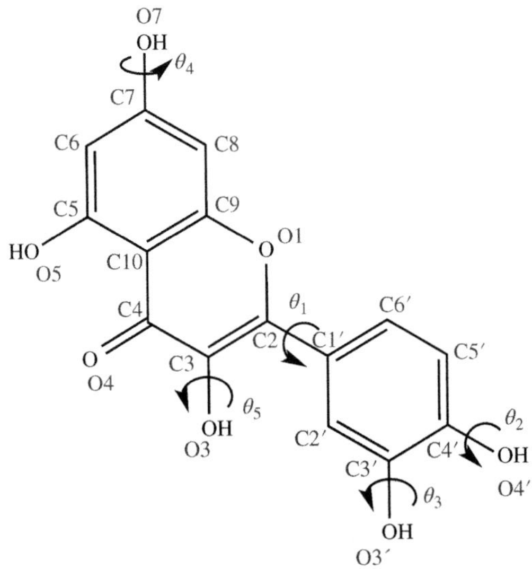

FIGURE 12.15 Molecular structure of quercetin, with atom and angle labeling included.

6-311+G(2d,p) basis set were applied [68, 69]. Several experimental techniques were applied to the characterization of both polymorphs, namely ^{1}H–^{19}F–^{13}C double-CP and ^{1}H-^{13}C CP-HETCOR with ***frequency switched Lee-Goldburg*** (FSLG) decoupling, ^{1}H–^{1}H DQ-SQ spectroscopy, and relaxation measurements. Furthermore, the ^{13}C CST values were calculated and experimentally determined using a 2D ^{13}C CP-PASS NMR experiment.

The comparison between experimental and calculated ^{13}C CSs highlighted some discrepancies, primarily for two carbons in the thiazole ring and the directly bonded benzenyl carbon. Although the calculated shifts for all the carbon sites improved significantly throughout the refinement stages, the errors remained for three of the mentioned carbons in the final structure. Such errors were claimed to be caused by several sources, including a minor structural deficiency in the thiazole ring, the accuracy of the DFT calculations themselves, or experimental error in the CP-PASS data.

NMR was also combined with PXRD for the structural characterization of microcrystalline powder anhydrous quercetin, previously unknown [70]. PXRD, $^{13}C/^{1}H$ NMR, and molecular modeling by both *ab initio* and ***molecular mechanics*** (MM) methods were used and three distinct structure solutions were obtained depending on the accuracy level with which the input molecular structures were generated. From the analysis of the PXRD data, multiple structure solutions emerged. In order to reduce ambiguities and to further refine the PXRD structure solution, $^{13}C/^{1}H$ SSNMR data and molecular modeling (by *ab initio* and MM methods) were incorporated in the process. Three distinct models were discussed in this work: model 1 was drawn from scratch and its geometry roughly optimized at the MM level of theory; model 2 was built using an input molecular structure resulted from a systematic grid search over flexible torsion angles θ_1–θ_5 (Fig. 12.15) with all of the generated approximately 250,000 molecular conformations being optimized to the closest local minimum in the total energy surface; finally model 3 was obtained using an isolated molecule taken from the reported X-ray single-crystal structure of quercetin dihydrate, which help assess the robustness of the other two models. Each model underwent first-principle geometry optimization followed by ^{1}H and ^{13}C CS calculations using GIPAW-DFT approach for routine structure validation.

The analysis of calculated CSs through the RMSDs confirmed that model 1 was the worst since it presented the highest RMSD, that is, a particularly poor agreement between the computed and experimental C5 and C8 CSs, related with the particular conformation of the O7–H hydroxyl, preventing the formation of the O7–H···O5 H-bonds observed in models 2 and 3. Furthermore, the authors compared the calculated CSs from models 2 and 3 with previously calculated values for an isolated molecule and reported the differences in ^{13}C CSs, arising from the fact that the full crystal incorporated H-bonding effects.

The supramolecular architectures and consequent H-bonding network identified in model 2 were in perfect agreement with the observed and predicted $^{13}C/^{1}H$ CSs and the peak correlations present in the 2D ^{1}H–^{1}H DQ-SQ spectrum. This work is a rather interesting case study for improving NMR crystallography methods since it shows examples (model 1) where small errors in the position of hydrogen atoms lead to wrong structure solutions that are impossible to be further adjusted, most likely because structures are trapped around a local energy minimum.

A complete NMR crystallography protocol for the structure determination of molecular crystals has also been presented by Salager and co-workers [71, 72]. The protocol combines ***proton spin-diffusion*** (PSD) restraints with molecular modeling to determine the crystalline structure of thymol. The best structures then undergo structure refinement using DFT geometry optimization and are selected based on their agreement with experimental ^{1}H and ^{13}C CS.

The build-up curves extracted from the PSD spectra are used as constraints (with input from PXRD determined cell parameters) to evaluate the several trial structures generated by molecular modeling using the Xplor-NIH package. This molecular modeling approach includes a standard force field to ensure reasonable geometries are generated, which is added by a PSD force field term (E_{PSD}), defined as the weighted goodness of fit (χ^2_{PSD}) between calculated and experimental PSD build-up curves. The weight ensures that the agreement with the PSD experimental data will be the dominant term in the modified initial force field.

The build-up curves do not provide a direct measure of distance information, as each individual cross-peak (Fig. 12.16a) contains contribution from tens of equivalent pairs of exchanging protons. To overcome this problem, the authors have proposed using a phenomenological rate matrix description of the build-up curves ($\mathbf{P}_{\mathrm{ij}}(\tau_{\mathrm{SD}})$), inspired in previous contributions [73]. With this approach intra- and intermolecular distances r_{ij} between protons i and j in the generated candidate structures are computed to simulate $\mathbf{P}_{ij}(\tau_{SD})$ defined as follows:

$$\mathbf{P}_{ij}\left(\tau_{SD}\right) = \exp\left(-K\tau_{SD}\right)_{ij} M_z^{j,0}$$

$M_z^{j,0}$ accounts for the volume of the diagonal peaks at $\tau_{\mathrm{SD}} = 0\,\mathrm{s}$ and the elements of the $\mathbf{K}$ matrix ($\mathbf{K} \propto 1/r_{ij}$) may be found in Ref. [71]. Each cross-peak build-up curve

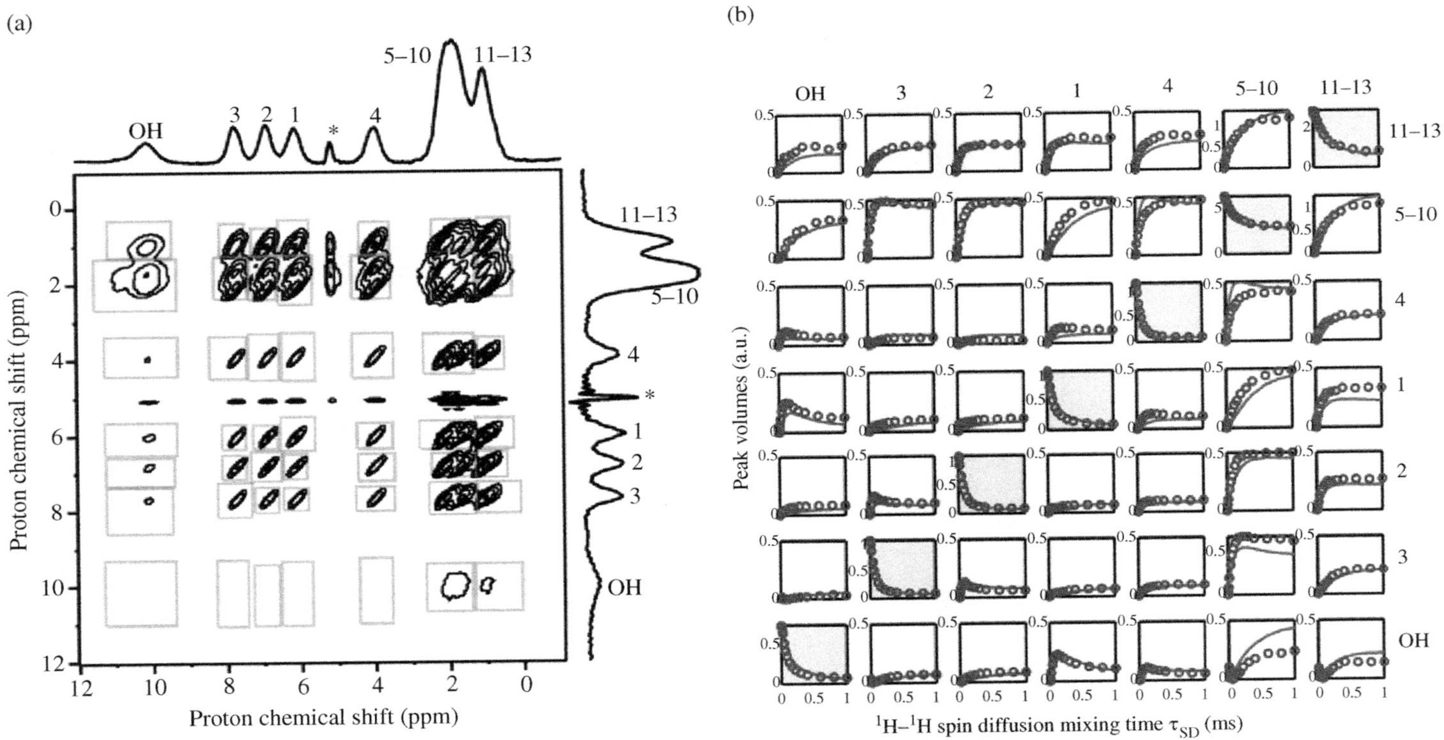

FIGURE 12.16 (a) 2D ^{1}H–^{1}H spin-diffusion spectrum of thymol. Boxes indicate the regions used for integration of each cross-peak. Asterisks indicate the carrier frequency artifacts (b) ^{1}H–^{1}H spin-diffusion build-up curves. Experimental data points are represented by circles; the best fit from the rate matrix analysis using the X-ray structure is shown using solid lines. Adapted with permission from Ref. [71]. Copyright 2009, RSC Publishing.

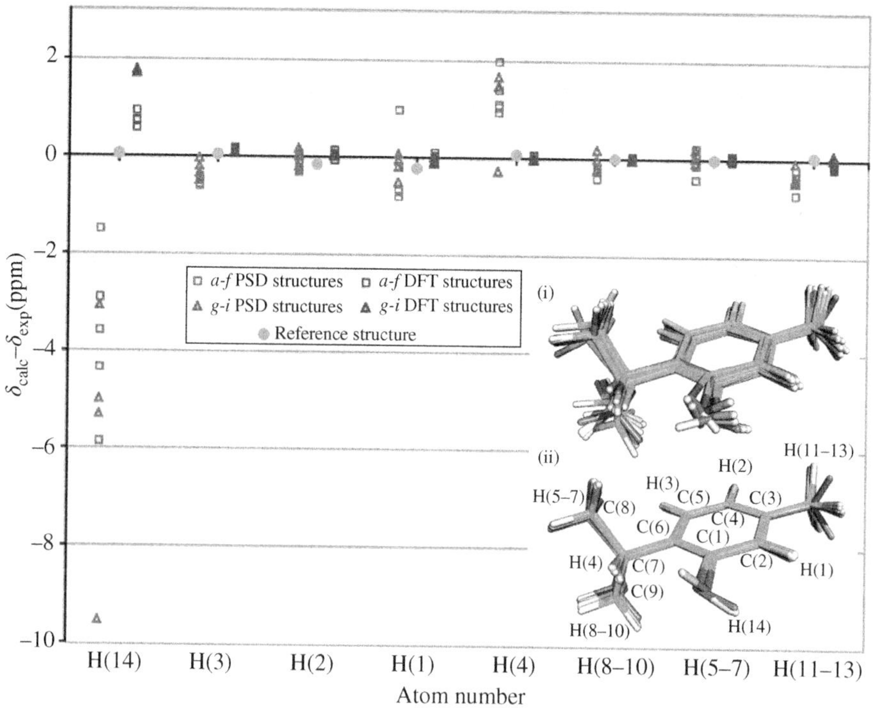

FIGURE 12.17 Comparison between calculated and experimental ^{1}H CSs for each resonance of thymol. Red (left) corresponds to the structures before the DFT-optimization (the nine structures are displayed in i) and blue (right) after optimization (same nine structures after periodic DFT optimization represented in ii, where the orange structure is the reference structure). Squares and triangles indicate structures where the orientation of the hydroxyl proton H(14) is respectively similar to (structures a–f) or different from (structures g–i) the reference structure (represented by green circles), respectively. Adapted with permission from Ref. [71]. Copyright 2009, RSC Publishing. (*see insert for color representation of the figure.*)

shown in Figure 12.16b represents in fact a sum of different curves arising from dipolar interaction with many different equivalent molecules [73]. Each experimental and calculated cross-peak volume obtained from the PSD spectra is used to calculate χ^2_{PSD}. Globally, the reported molecular modeling approach discussed earlier encompasses three main steps, starting from an initial population of 3000 trial structures and ending in the 42 lowest energy PSD-optimized structures (E_{PSD} < 2925 kJ/mol^{-1}) [71]. To further optimize the agreement with the XRD structure from the CSD database, a subset of nine structures (labeled *a–i*) were chosen as representative of the ensemble and geometry optimized using planewave DFT. The comparison of the nine PSD optimized structures before and after DFT geometry optimization is shown in Figure 12.17(i) and 12.17(ii), respectively; geometry optimization considerably improves the agreement toward the XRD reference structure. One notable feature resides in the OH(14) bond length, which can be oriented in two ways: (i) one where the O–H bond is parallel to the aromatic ring (structures *a–f*), similar to the XRD structure, and (ii) another where the O–H bond is approximately perpendicular to the ring plane (structures *g–i*) (see Fig. 12.17(ii)). Indeed, structures in the situation of (i) showed a lower RMSD. The difference between calculated and experimental CSs of the nine PSD-optimized structures for each proton (Fig. 12.17) before and after DFT geometry optimization is illustrated. The green circles represent the CS difference for the reference structure for comparison.

12.5.4.2 **Ab Initio *Crystal Structure Determination*** The method described earlier was successfully applied to a small and rigid molecule (thymol). However, application of this protocol to larger drugs with several degrees of freedom may prove difficult, due to limitations in proton resolution and computing power. To overcome this problem, a protocol for *ab initio* crystal structure determination of natural abundance powdered solids has been introduced [72] combining CSP computational approaches and involving measured and calculated ^{1}H ICSs. This method was also applied to thymol, and the obtained structure was also identical to the reference structure with an improved RMSD compared to the previous PSD-optimization protocol. The same protocol has been also applied more recently to four small drug molecules with higher flexibility: cocaine, flutamide, flufenamic acid, and theophylline [74]. The correct structure of these four compounds was successfully determined based on the RMSD values resulting from the comparison of experimental and DFT calculated ^{1}H ICSs.

One of the main conclusions from this study is related to the ^{1}H ICSs being much more sensitive to structural changes than ^{13}C ICSs. This may be explained by the higher errors associated to the calculation of ^{13}C ICSs. For cocaine, flutamide, and flufenamic acid, the ^{1}H ICSs lead to a clear structure determination by comparison with the ensemble of structure predictions, whereas in all these cases the ^{13}C CSs were not sufficiently sensitive to lead to structure determination. Figure 12.18 presents the ^{1}H RMSD values between experimental and calculated ICSs for the 30 lowest energy predicted crystal structures for cocaine. Structure 1 (Fig. 12.18a) was considered to be the correct crystal structure based on the agreement between calculated and experimental ^{1}H CSs as well as on the comparison between calculated ICSs for the previously known single-crystal X-ray structure and the experimental ICSs (Fig. 12.18a, gray stem).

The all-atom RMSD between the molecular geometry of the structure determined in this work and the previously known structure of cocaine was found to be 0.069 Å, and Figure 12.18b shows the two structures superimposed. The unit cell dimensions all agree to within 2.3% and the volume difference between the two structures is 0.8% (3.29 Å per molecule).

12.5.4.3 ***Chemical Shift–Driven Crystal Structure Determination*** Our group has recently proposed a new strategy for NMR crystallography of multiple-component molecular crystals in which ^{1}H NMR CSs enter directly in the structure generation step, governed by a genetic algorithm [75]. The methodology developed in this work avoids the use of time-consuming high-level DFT calculations and uses CSs in the

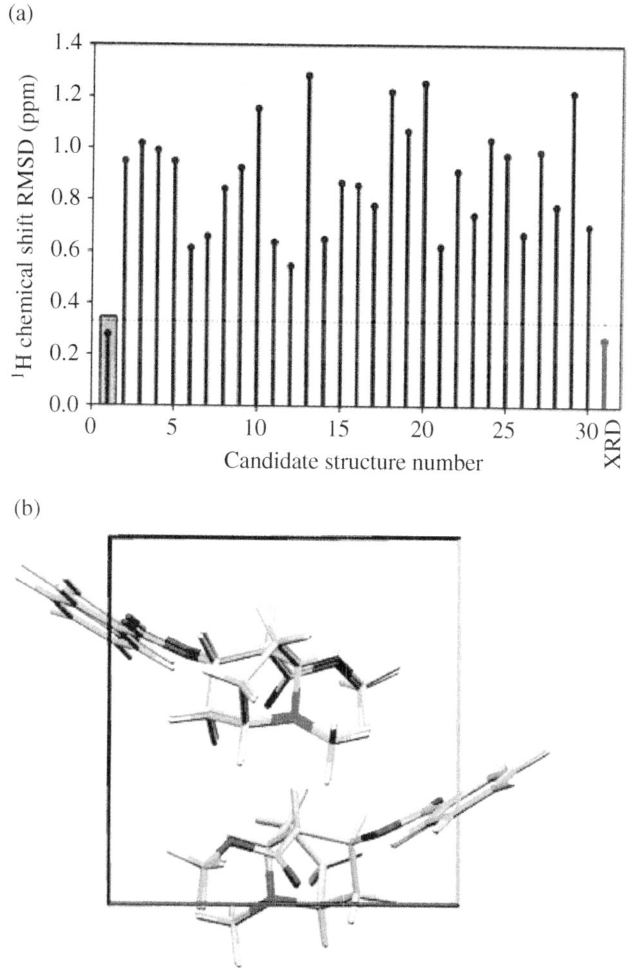

FIGURE 12.18 (a) Comparison between experimental and calculated ^{1}H CSs for cocaine. The comparison is made using assigned experimental CSs. Predicted structures are ordered by increasing calculated lattice energies (decreasing predicted stability). The comparison with the crystal structure determined from single crystal XRD is shown on the far right. The dotted horizontal black line shows the mean RMSD error as described in the main text and the horizontal grey shaded zone indicates the expected limits of the RMSD in CS. (b) Comparison between the structure of cocaine determined by powder ^{1}H NMR and computational modeling and the single crystal XRD structure. Adapted with permission from Ref. [74]. Copyright 2013, RSC Publishing.

structure-refinement step as pseudoforces acting on the models, leading to the lowest energy structure. The protocol was applied to powdered amoxicillin trihydrate, a widely used β-lactamic antibiotic. It should be noted that until quite recently, NMR crystallography essentially followed a two-step strategy: (i) structure proposal and

(ii) structure validation: once the set of possible structures is available, they are ranked and validated against experimental SSNMR data. In this way, the available NMR data plays a passive role in the sense that is solely used as a static validation criteria, rather than being actively used as driving force in the construction of the structure model. The proposed protocol takes the unit cell parameters, space group information (both taken from PXRD experiments) and solid-state ^{1}H ICSs as input information and processes the data in three steps: (i) generation of an ensemble of structures, (ii) refinement of the generated structures, (iii) high-level energy minimization of the refined structures. Step (i) employs a genetic algorithm where the fitness (E_{total}) of each structure is defined as the sum of the lattice energy ($E_{lattice}$) of the system with a term accounting for the deviation between the experimental (δ_{exp}) and the on-the-fly calculated (δ_{calc}) ^{1}H CSs ($E\delta$):

$$E_{total} = E_{lattice} + E_{\delta}$$

$$E_{\delta} = \sum_{h} \begin{cases} k\left(\delta^{h}_{calc} - \delta^{h}_{\exp} + \varepsilon_{\delta}\right)^2, & \delta^{h}_{calc} \le \delta^{h}_{\exp} - \varepsilon_{\delta} \\ 0 & \left|\delta^{h}_{calc} - \delta^{h}_{\exp}\right| < \varepsilon_{\delta} \\ k\left(\delta^{h}_{calc} - \delta^{h}_{\exp} - \varepsilon_{\delta}\right)^2, & \delta^{h}_{calc} \ge \delta^{h}_{\exp} - \varepsilon_{\delta} \end{cases}$$

$E_{lattice}$ is described by the classical biomolecular force field (GAFF), which comprises intramolecular (bonds, angles, and dihedrals) and intermolecular (Coulomb and van der Waals) components, the latter having contributions from all unit cells. In E_{δ}, the sum runs over all the protons h for which experimental CSs (δ_{exp}) are available. The parameter ε_{δ} controls the width of the flat-well potential and defines a threshold for the accepted error in the δ_{calc} values. The on-the-fly calculation of δ_{calc} requires a fast methodology, amenable to integration within a classical MM/dynamics code. The calculation of δ_{calc} is inspired by previous protein solution NMR work [76–78] and is parametrically calculated according to

$$\delta_{calc} = \delta_{local} + \delta_{ef} + \delta_{m} + \delta_{rc}$$

where δ_{local} is a constant particular to each proton (assumed proportional to its atomic partial charge), and δ_{ef}, δ_{m}, and δ_{rc} are contributions due to electric field effects [79], group magnetic anisotropy [80, 81] and aromatic ring currents [78, 82]. The genetic algorithm yields, in step (i), a set of generated structures, which moderately obey the input data. In step (ii), these are refined by means of molecular dynamics simulated annealing. Finally, in step (iii) the set of refined structures is energy minimized using tight-binding DFT. In the whole process, $E_{lattice}$ is responsible for the physical and chemical validity of the generated structures, whereas E_{δ} is responsible for imposing the experimental constraints derived from SSNMR experiments.

Figure 12.19 shows the evolution of the fitness (E_{total}) of the fittest individual during generations over five distinct runs, with and without inclusion in the fitness function of the E_{δ} term, selected such that the final structures are comparable to the reference X-ray structure and the final fitnesses (E_{total}) are similar for all runs.

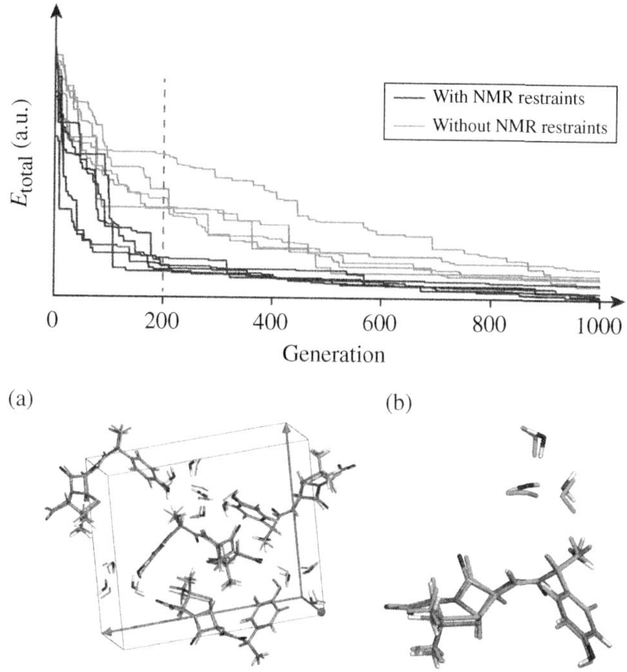

FIGURE 12.19 Evolution of E_{total} of the fittest individual during the course of the genetic algorithm over five distinct runs, with and without inclusion of CS restraints (grey and black curves, respectively). Apart from the inclusion of the CS restraints, all running parameters are the same in the two sets; (top) and predicted best-crystal packing of amoxicillin trihydrate (bottom): (a) entire unit cell and (b) asymmetric unit. The reference X-ray structure is shown as solid gray. RMSD between both structures is 0.608 Å (or 0.576 Å if only the asymmetric unit is considered). Reprinted with permission from Ref. [75]. Copyright 2013, American Chemical Society.

The inclusion of experimental data (E_δ) accelerates genetic algorithm convergence toward the minimum energy, so that much fewer (hundreds) generations are required to obtain a reasonable set of molecular packings. In the shown data set, the runs with NMR restraints converge to the global energy minimum basin in approximately 200 generations, whereas runs without NMR restraints require significantly more generations and yield a higher E_{total}. The procedure yielded a structure that closely matches the single-crystal X-ray structure (RMSD ~ 0.608 Å).

12.6 SUMMARY

Solid-state NMR spectroscopy is a well-established method for studying interactions based on the magnetic and electric properties of the molecules, and is currently an important tool in the pharmaceutical industry, largely owing to its unique capabilities in distinguishing between polymorphic, anhydrate, hydrate, and other solvate forms of APIs. Solid-state NMR is also highly sensitive to strong and weak crystal packing interactions and is a powerful partner of X-ray diffraction in crystal structure determination. In this chapter, we have highlighted the main applications of the technique combined with theoretical calculations of NMR parameters, emphasizing the computation of chemical shifts, providing structural insight into a broad family of small pharmaceutical molecules.

The case studies presented (published in the past decade) were carefully selected taking into consideration the most interesting applications at an academic and industrial perspective.

REFERENCES

[1] MacKenzie, K.J.D.; Smith, M.E. *Multinuclear Solid-State NMR of Inorganic Materials*; Pergamon Materials Series Vol. 6, Cahn, R.W., Ed.; Elsevier: Oxford, 2002.

[2] Schmidt-Rohr, K.; Spiess, H. W. *Multidimensional Solid-State NMR and Polymers*; Academic Press, London, 1994.

[3] Koller, H.; Weiss, M. *Top. Curr. Chem.* 2012, *306*, 189–228.

[4] van Wüllen, C. Chemical shifts with Hartree-Fock and Density Functional Methods. In *Calculation of NMR and EPR Parameters: Theory and Applications*; Kaupp, M.; Bühl, M.; Malkin, V. G., Eds. Wiley-VCH Verlag GmbH & Co.: Weinheim, 2004.

[5] Sebastiani, D.; Parrinello, M. A. *J. Phys. Chem. A* 2001, *105*, 1951–1958.

[6] Bloch, F. *Z. Physik* 1929, *52*, 555–600.

[7] Kittel, C. *Introduction to Solid State Physics*; Wiley: New York, 2003.

[8] Mauri, F.; Pfrommer, B.; Louie, S. *Phys. Rev. Lett.* 1996, *77*, 5300–5303.

[9] Schindler, M. *J. Chem. Phys.* 1982, *76*, 1919–1933.

[10] van Wüllen, C.; Kutzelnigg, W. *J. Chem. Phys.* 1996, *104*, 2330–2340.

[11] London, F. *J. Phys. Radium* 1937, *8*, 397–409.

[12] Ditchfield, R. *J. Chem. Phys.* 1972, *56*, 5688–5691.

[13] Keith, T. A.; Bader, R. *Chem. Phys. Lett.* 1993, *210*, 223–231.

[14] Keith, T. A.; Bader, R. F. W. *Chem. Phys. Lett.* 1992, *194*, 1–8.

[15] Cheeseman, J. R.; Trucks, G. W.; Keith, T. A.; Frisch, M. J. *J. Chem. Phys.* 1996, *104*, 5497–5509.

[16] Pickard, C.; Mauri, F. *Phys. Rev. B* 2001, *63*, 245101(1-13).

[17] Blöchl, P. *Phys. Rev., B: Condens. Matter* 1994, *50*, 17953–17979.

[18] Thonhauser, T.; Ceresoli, D.; Mostofi, A. A.; Marzari, N.; Resta, R.; Vanderbilt, D. *J. Chem. Phys.* 2009, *131*, 101101(1-4).

[19] Dumez, J.-N.; Pickard, C. J. *J. Chem. Phys.* 2009, *130*, 104701(1-8).

[20] Mafra, L.; Santos, S. M.; Siegel, R.; Alves, I.; Paz, F. A. A.; Dudenko, D.; Spiess, H. W. *J. Am. Chem. Soc.* 2012, *134*, 71–74.

[21] Webber, A.; Emsley, L.; Claramunt, R. M.; Brown, S. P. *J. Phys. Chem. A* 2010, *114*, 10435–10442.

[22] Filip, X.; Borodi, G.; Filip, C. *Phys. Chem. Chem. Phys.* 2011, *13*, 17978–17986.

[23] Harris, R.; Hodgkinson, P.; Larsson, T.; Muruganantham, A.; Ymen, I.; Yufit, D.; Zorin, V. *Cryst. Growth Des.* 2008, *8*, 80–90.

[24] Ragué Schleyer, von, P.; Maerker, C.; Dransfeld, A.; Jaio, H.; van Eikema Hommes, N. *J. Am. Chem. Soc.* 1996, *118*, 6317–6318.

[25] Johnson, C. E.; Bovey, F. A. *J. Chem. Phys.* 1958, *29*, 1012–1014.

[26] Chen, Z.; Wannere, C. S.; Corminboeuf, C.; Puchta, R.; Schleyer, P. V. R. *Chem. Rev.* 2005, *105*, 3842–3888.

[27] Dudenko, D. V.; Yates, J. R.; Harris, K. D. M.; Brown, S. P. *Cryst. Eng. Comm.* 2013, *15*, 8797–8807.

[28] Geppi, D.; Mollica, G.; Borsacchi, S.; Veracini, C. *Appl. Spectrosc. Rev.* 2008, *43*, 202–302.

[29] Berendt, R. T.; Sperger, D. M.; Isbester, P. K.; Munson, E. J. *TrAC-Trends Anal. Chem.* 2006, *25*, 977–984.

[30] Harris, R. K. *J. Pharm. Pharmacol.* 2007, *59*, 225–239.

[31] Vogt, F. G. *Future Med. Chem.* 2010, *2*, 915–921.

[32] Yates, J. R.; Dobbins, S. E.; Pickard, C. J.; Mauri, F.; Ghi, P. Y.; Harris, R. K. *Phys. Chem. Chem. Phys.* 2005, *7*, 1402–1407.

[33] Webber, A. L.; Elena, B.; Griffin, J. M.; Yates, J. R.; Pham, T. N.; Mauri, F.; Pickard, C. J.; Gil, A. M.; Stein, R.; Lesage, A.; Emsley, L.; Brown, S. P. *Phys. Chem. Chem. Phys.* 2010, *12*, 6970–6983.

[34] Yates, J. R. *Magn. Reson. Chem.* 2010, *48*, S23–S31.

[35] Mifsud, N.; Elena, B. N. D.; Pickard, C. J.; Lesage, A.; Emsley, L. *Phys. Chem. Chem. Phys.* 2006, *8*, 3418–3422.

[36] Sardo, M.; Siegel, R.; Santos, S. M.; Rocha, J.; Gomes, J.; Mafra, L. *J. Phys. Chem. A* 2012, *116*, 6711–6719.

[37] Bryce, D. L.; Sward, G. D.; Adiga, S. *J. Am. Chem. Soc.* 2006, *128*, 2121–2134.

[38] Bryce, D. L.; Bultz, E. B. *Chem. Eur. J.* 2007, *13*, 4786–4796.

[39] Hamaed, H.; Pawlowski, J.; Cooper, B.; Fu, R.; Eichhorn, S.; Schurko, R. *J. Am. Chem. Soc.* 2008, *130*, 11056–11065.

[40] Harris, R. K.; Becker, E. D.; Cabral De Menezes, S. M.; Granger, P.; Hoffman, R. E.; Zilm, K. W. *Pure Appl. Chem.* 2008, *80*, 59–84.

[41] Mason, J. *Solid State Nucl. Magn. Reson.* 1993, *2*, 285–288.

[42] O'Dell, L. A.; Schurko, R. W. *J. Am. Chem. Soc.* 2009, *131*, 6658–6659.

[43] Cavadini, S.; Vitzthum, V.; Ulzega, S.; Abraham, A.; Bodenhausen, G. *J. Magn. Reson.* 2010, *202*, 57–63.

[44] Tatton, A. S.; Pham, T. N.; Vogt, F. G.; Iuga, D.; Edwards, A. J.; Brown, S. P. *Mol. Pharmaceut.* 2013, *10*, 999–1007.

[45] Tatton, A.; Pham, T. N.; Vogt, F. G.; Iuga, D.; Edwards, A.; Brown, S. P. *Cryst. Eng. Comm.* 2012, *14*, 2654–2659.

[46] Kong, X.; Shan, M.; Terskikh, V.; Hung, I.; Gan, Z.; Wu, G. *J. Phys. Chem. B* 2013, *117*, 9643–9654.

[47] Bonhomme, C.; Gervais, C.; Folliet, N.; Pourpoint, F.; Coelho Diogo, C.; Lao, J.; Jallot, E.; Lacroix, J.; Nedelec, J.-M.; Iuga, D.; Hanna, J. V.; Smith, M. E.; Xiang, Y.; Du, J.; Laurencin, D. *J. Am. Chem. Soc.* 2012, *134*, 12611–12628.

[48] Facey, G.; Ratcliffe, C.; Ripmeester, J. *J. Phys. Chem.* 1995, *99*, 12249–12256.

[49] Smith, J. R.; Xu, W.; Raftery, D. *J.Phys. Chem. B* 2006, *110*, 7766–7776.

[50] Heider, E.; Harper, J.; Grant, D. M. *Phys. Chem. Chem. Phys.* 2007, *9*, 6083–6097.

[51] Alderman, D. W.; McGeorge, G.; Hu, J. Z.; Pugmire, R. J.; Grant, D. M. *Mol. Phys.* 1998, *95*, 1113–1126.

[52] Hamilton, W. C. *Statistics in Physical Science*; The Ronald Press Company: New York, 1964.

[53] Wolniak, M.; Oszmiański, J.; Wawer, I. *Magn. Reson. Chem.* 2008, *46*, 215–225.

[54] Vogt, F. G.; Clawson, J. S.; Strohmeier, M.; Edwards, A.; Pham, T. N.; Watson, S. *Cryst. Growth Des.* 2009, *9*, 921–937.

[55] Wang, W. D.; Gao, X.; Strohmeier, M.; Wang, W.; Bai, S.; Dybowski, C. *J. Phys. Chem. B* 2012, *116*, 3641–3649.

[56] Liu, S.-F.; Mao, J.-D.; Schmidt-Rohr, K. *J. Magn. Reson.* 2002, *155*, 15–28.

[57] Babinský, M.; Bouzková, K.; Pipíška, M.; Novosadová, L.; Marek, R. *J. Phys. Chem. A* 2013, *117*, 497–503.

[58] Bouzková, K.; Babinský, M.; Novosadová, L.; Marek, R. *J. Chem. Theory Comput.* 2013, *9*, 2629–2638.

[59] Uldry, A.; Griffin, J. M.; Yates, J. R.; Perez-Torralba, M.; Maria, M. D. S.; Webber, A. L.; Beaumont, M. L. L.; Samoson, A.; Claramunt, R. M.; Pickard, C. J.; Brown, S. P. *J. Am. Chem. Soc.* 2008, *130*, 945–954.

[60] *Guidance for Industry: Regulatory Classification of Pharmaceutical Co-Crystals*; Food and Drug Administration: Silver Spring, MD, December 2011.

[61] Li, Z. J.; Abramov, Y.; Bordner, J.; Leonard, J.; Medek, A.; Trask, A. V *J. Am. Chem. Soc.* 2006, *128*, 8199–8210.

[62] Khan, M.; Enkelmann, V.; Brunklaus, G. *J. Am. Chem. Soc.* 2010, *132*, 5254–5263.

[63] Khan, M.; Enkelmann, V.; Brunklaus, G. *Cryst. Eng. Comm.* 2011, *13*, 3213–3223.

[64] Yates, J. R.; Pham, T. N.; Pickard, C. J.; Mauri, F.; Amado, A. M.; Gil, A. M.; Brown, S. P. *J. Am. Chem. Soc.* 2005, *127*, 10216–10220.

[65] Dudenko, D. V.; Williams, P. A.; Hughes, C. E.; Antzutkin, O. N.; Velaga, S. P.; Brown, S. P.; Harris, K. D. M. *J. Phys. Chem. C* 2013, *117*, 12258–12265.

[66] Smith, E. D. L.; Hammond, R. B.; Jones, M. J.; Roberts, K. J.; Mitchell, J. B. O.; Price, S. L.; Harris, R. K.; Apperley, D. C.; Cherryman, J. C.; Docherty, R. *J. Phys. Chem. B* 2001, *105*, 5818–5826.

[67] Vogt, F. G.; Katrincic, L. M.; Long, S. T.; Mueller, R. L.; Carlton, R. A.; Sun, Y. T.; Johnson, M. N.; Copley, R. C. B.; Light, M. E. *J. Pharm. Sci.* 2008, *97*, 4756–4782.

[68] Delley, B. *J. Chem. Phys.* 1990, *92*, 508–517.

[69] Delley, B. *J. Chem. Phys.* 2000, *113*, 7756–7764.

[70] Filip, X.; Grosu, I.-G.; Miclăuş, M.; Filip, C. *Cryst. Eng. Comm.* 2013, *15*, 4131.

[71] Salager, E.; Stein, R.; Pickard, C. J.; Elena, B.; Emsley, L. *Phys. Chem. Chem. Phys.* 2009, *11*, 2610–2621.

[72] Salager, E.; Day, G.; Stein, R.; Pickard, C. J.; Elena, B.; Emsley, L. *J. Am. Chem. Soc.* 2010, *132*, 2564–2566.

[73] Elena, B.; Emsley, L. *J. Am. Chem. Soc.* 2005, *127*, 9140–9146.

[74] Baias, M.; Widdifield, C. M.; Dumez, J.-N.; Thompson, H. P. G.; Cooper, T. G.; Salager, E.; Bassil, S.; Stein, R. S.; Lesage, A.; Day, G. M.; Emsley, L. *Phys. Chem. Chem. Phys.* 2013, *15*, 8069–8080.

[75] Santos, S. M.; Rocha, J.; Mafra, L. *Cryst. Growth Des.* 2013, *13*, 2390–2395.

[76] Hunter, C. A.; Packer, M. J.; Zonta, C. *Prog. Nucl. Mag. Res. Sp.* 2005, *47*, 27–39.

[77] Cross, K. J.; Wright, P. E. *J. Magn. Reson.* 1985, *64*, 220–231.

[78] Osapay, K.; Case, D. A. *J. Am. Chem. Soc.* 1991, *113*, 9436–9444.

[79] Wolff, R.; Radeglia, R. In *eMagRes*; Harris, R. K.; Wasylishen, R. L., Eds. John Wiley & Sons, Ltd: Chichester, 2007.

[80] Buckingham, A. D. *Can. J. Chem.* 1960, *38*, 300–307.

[81] Grayson, M.; Raynes, W. T. *Magn. Reson. Chem.* 1995, *33*, 138–143.

[82] McConnell, H. M. *J. Chem. Phys.* 1957, *27*, 226–229.

13

MOLECULAR DYNAMICS SIMULATIONS OF AMORPHOUS SYSTEMS

Bradley D. Anderson and Tian-Xiang Xiang

Department of Pharmaceutical Sciences, College of Pharmacy, University of Kentucky, Lexington, KY, USA

13.1 INTRODUCTION

Molecular dynamics (MD) simulations may have utility in predicting various properties of amorphous excipients, amorphous drugs, and amorphous drug formulations. Both thermodynamic and kinetic properties can be explored, and in many instances the results compare favorably to experimental observations. Among the thermodynamic properties of particular interest that lend themselves to exploration by MD simulations are the underlying structural features of the excipients themselves, drug–excipient interactions, moisture uptake under various storage conditions, crystalline drug solubility and amorphous drug phase miscibility with amorphous excipients, and solubility or dissolution rate enhancement potentially attainable through the use of amorphous systems and its impact on oral bioavailability. Dynamic properties of greatest interest based on the literature are the T_g and its relevance to pharmaceutical product performance, molecular mobility as characterized by various relaxation parameters (e.g., α- and β-relaxations), the nature of molecular motions underlying relaxation processes that may also be probed by various experiments, and the relationship between molecular mobility and the

Computational Pharmaceutical Solid State Chemistry, First Edition. Edited by Yuriy A. Abramov.

physical and chemical stability of amorphous drug–excipient mixtures. In addition to their predictive potential, MD simulations can provide molecular insights relating to such questions as heterogeneity of drug–excipient distributions, water distribution, molecular interactions that lead to favorable or unfavorable mixing, and molecular events coupled to various relaxation processes.

While there are many possible types of computer simulations, this chapter emphasizes fully atomistic MD simulations of pharmaceutically relevant, amorphous glasses consisting either of small drug molecules, small molecule or polymeric excipients, or drug–excipient mixtures with or without small percentages of water. These are among the systems that have been generally explored by the authors, which also include simulations of lipid bilayers in order to better understand drug binding to membranes and membrane permeability [1] and simulations of lipid vehicles including triglycerides and triglyceride/monoglyceride mixtures found in lipid-based formulations of poorly water-soluble drugs [2]. With the exception of one example illustrating water clustering and diffusivity in lipids, these studies are not part of this chapter.

Full-atom simulations have been of particular interest because of their relevance to real pharmaceutical systems in that the molecular structures are constructed in full atomic detail. In this chapter, simulations of amorphous systems at temperatures below and in some cases above the glass transition temperature (T_g) will be explored. While amorphous pharmaceutical systems are generally at temperatures below T_g, validation of simulations may be more readily done by comparisons at temperatures near or above the T_g where molecular mobility may be sufficiently high that both the simulated and experimental systems may be closer to equilibrium. Above the T_g certain dynamic processes are sufficiently rapid that they can be studied both experimentally and in the time frames accessible by current MD simulations (typically <1 μs).

13.2 MD SIMULATION METHODOLOGY

Since the first simulation of a liquid argon in 1964 [3], MD simulations have grown in popularity and complexity, as described in several comprehensive reviews over the past several decades [4]. The simulations involve numerically integrating Newton's equations of motion for each of the N particles or atoms in an assembly to obtain information on the position, velocity, and acceleration of each atom and their variation over time. The system to be simulated requires initial construction of the molecules of interest. Structures may be constructed by hand using commercial software (e.g., xLeap within AMBER [5]) or found by searching databases crystal or NMR structures. The assembled molecules are then packed into a periodic cell. Interactions between the atoms are defined by the "force field"—an equation for potential energy, U, that includes terms for intermolecular interactions involving covalently bonded atoms (i.e., bond length, bond angle, and torsional contributions) and non-bonded interactions (e.g., van der Waals, electrostatic interactions) shown as follows:

$$U = \sum_{\text{bonds}} \frac{k_{ij}^{\text{b}}}{2}\left(r_{ij} - r_{ij}^{\text{eq}}\right)^2 + \sum_{\text{angles}} \frac{k_{ijk}^{\theta}}{2}\left(\theta_{ijk} - \theta_{ijk}^{\text{eq}}\right)^2 + \sum_{\text{dihedrals}} k^{\phi}\left[1+\cos\left(n\left(\phi-\phi^{\text{eq}}\right)\right)\right]$$
$$+\sum_{i<j}\left[\frac{A_{ij}}{r_{ij}^{12}} - \frac{B_{ij}}{r_{ij}^{6}} + \frac{q_i q_j}{4\pi\varepsilon_{\text{o}} r_{ij}}\right]$$

The force constants for the covalent bonds, bond angles, and dihedrals, respectively, in the given equation are k^{b}, k^{θ}, and k^{ϕ} where r_{ij} represents the distance between two atoms i and j; θ_{ijk} is the i, j, k bond angle; and ϕ represents a dihedral angle. A_{ij} and B_{ij} are the Lennard–Jones parameters for non-covalent interactions and q is a partial charge. A variety of different force fields are available for MD simulation including AMBER and GAFF [5, 6], COMPASS [7], CHARMM [8], DREIDING [9], CVFF [1a, b], GROMOS [10], and OPLS-AA [11]. Generally, the parameterization of these force fields is based on quantum mechanical calculations and fits to experimental data (e.g., spectroscopic data) with adjustments for specific applications based on comparisons of experimental data to simulation predictions.

Once the molecular assembly is constructed, energy minimization is necessary to reduce steric contacts and the system is heated to a desired temperature (corresponding to velocities in the Maxwell–Boltzmann distribution) to generate initial positions and velocities of the atoms. During dynamic runs, changes in position and velocity in response to the force field according to Newton's laws of motion are determined with the calculations repeated every 1 or 2 fs.

An important limitation of current fully atomistic MD simulations is the simulation time needed to explore various relaxation processes. Some of the dynamic processes of greatest interest in real amorphous pharmaceutical systems include changes in composition and properties due to moisture uptake at varying relative humidity or during dissolution of amorphous dispersions after ingestion, physical instability during manufacturing and storage (e.g., lyophile collapse, phase separation of amorphous drug–excipient mixtures, nucleation, and crystal growth), and chemical instability (e.g., inter- and intramolecular chemical reactions often involving water or oxygen and disproportionation of pharmaceutical salts in systems containing acidic or basic excipients). Because many of the aforementioned stability-related phenomena are accelerated by moisture, the influence of water on these kinetic processes has been a focal point of many investigations.

For the most part, the timescales for the aforementioned kinetic processes are well beyond the accessible timescale for fully atomistic MD simulations. Local dynamics such as rotation of a methyl group or a polymer side chain can certainly be explored. For example, in a polymer melt at a temperature of 100K above the T_g, the timescale for methyl-group rotations is about 1 ps and approximately 1–10 ns for segmental α-relaxation in a polymer [4b]. Diffusion for even a small molecule such as water in a polymer below T_g may exhibit anomalous non-Einsteinian diffusion behavior up to and beyond 100 ns [12], while random long-range diffusion of relatively low-molecular-weight drug molecules may not be accessible by MD simulation without coarse-graining [13]. For the purpose of validating dynamic processes explored in MD simulations, experimental methods that probe local dynamics are therefore most applicable.

If the amorphous structures are reliable, then thermodynamic information generated from MD simulations may have utility in terms of predicting such properties as the tendency for amorphous phase separation in mixtures, drug solubility in amorphous solids and solubility enhancement upon dissolution, equilibrium moisture content as a function of relative humidity, and the like. Generally, amorphous assemblies are prepared from their melts at temperatures several hundred degrees above T_g to ensure equilibrium. (One should bear in mind, however, that the required timescales to achieve equilibrium structures for well-entangled dense polymers may be beyond current computational capabilities [14].) Rapid cooling of these assemblies to a temperature below T_g approximates experimental preparation methods such as melt quenching, but the rates of cooling in MD simulations are much faster. Experimental verification of structural details observed by simulation is therefore of considerable interest and is the topic of the section that follows.

13.3 POLYMER PROPERTIES—MD SIMULATION VERSUS EXPERIMENT

13.3.1 Glass Transition Temperature (T_g)

Validation of the structural and dynamic features of amorphous excipients and drug–excipient mixtures from MD simulations may be complicated by the fact that amorphous solids are not in their equilibrium state. Upon cooling a molten glass-forming material from a high temperature, the translational motion of the excipient molecules slows and the system undergoes thermal contraction accompanied by decreases in enthalpy and increases in density. Above the glass transition temperature (T_g), molecular motion is sufficient to maintain equilibrium as temperature is decreased, but near T_g the translational and orientational motions that enable the system to continue further compaction are slowed to the point that the system falls out of equilibrium and below T_g solidifies to form a glass [15]. Experimentally, T_g is usually determined calorimetrically or by dilatometry. In MD simulations, the usual approach is to determine T_g dilatometrically by the intersection of linear fits of plots of volume versus temperature in the melt and glass regions of the curve.

Because of the perceived importance of the T_g in relation to physical or chemical stability of amorphous pharmaceutical formulations [16], experimental data are available for a large number of excipients and drug–excipient combinations including most systems chosen for MD simulation. Therefore, published MD simulations of glasses often include comparisons of T_g values from MD simulation and experiment. As the T_g does not reflect a first-order phase transition such as the melting point for a crystalline compound, the value of the reported T_g depends on the rate of cooling (i.e., the timescale of the experiment or simulation). (It should be noted, however, that the apparent melting point can also vary with the rate of heating.)

The values of T_g typically reported from MD simulations are in reasonable agreement with experimental values, though often higher due to the more rapid cooling rates in simulations. Examples of linear correlations between simulated and experimental T_g values have been published for representative polymers [17] and also glasses of low-molecular-weight organic compounds [18]. Barrat et al. [19] noted that cooling rates are typically 10^{12} K/min in simulations, whereas experiments to determine T_g such as DSC typically employ heating rates of approximately 10 K/min, a difference of 12 orders of magnitude. They suggested a change in T_g of 3 K for a 10-fold change in cooling rate to estimate the shift in T_g resulting from more rapid cooling in MD simulations, or approximately 35–40 K for the aforementioned rate differences.

Systematic studies of the effect of cooling rate on T_g for a number of low-molecular-weight and polymer glasses using either MD simulation [14, 17, 20] or experiments have established that T_g varies nonlinearly with the logarithm of the cooling rate [21]. The standard procedure is to fit T_g data as a function of cooling rate to the following equation:

$$T_g(\gamma) = \frac{T_0 - B}{\log(A\gamma)} \tag{13.1}$$

where γ is the cooling rate and T_0, A, and B are fitting parameters. The equation is based on the assumption that the dependence of the structural relaxation time of a system follows the Vogel–Fulcher–Tammann relationship [22]. The value of T_0 has been interpreted as the value of T_g at the limit of infinitely slow cooling [16, 21, 23].

Given the shifts in T_g with cooling rate, MD simulated glasses cooled to a given temperature below T_g are expected to be at a higher energy than experimental glasses of the same material and consequently at a somewhat lower density. This is generally consistent with the results for densities of polymer and small-molecule glasses simulated by the authors, which have been observed to gradually increase with simulation time over the time frame of simulations (typically 100 ns) as energies gradually decreased [12b, 24], indicating that aging (volume and enthalpy relaxation) continues well below T_g in simulations as it does in real systems. In amorphous polymers, α-relaxation is generally attributed to cooperative segmental relaxations that dramatically increase with an increase in temperature near T_g, while β-relaxations are prominent well below T_g and reflect local mobility due to polymer side group rotations, side group conformational changes, or localized motions in the polymer chain. The increases in density observed in MD simulations below T_g likely reflect local relaxation (β-relaxation) as timescales for structural relaxation near T_g are on the order of 10^2 s or greater. Current explicit all-atom simulation methods cannot adequately probe cooperative structural (α-) relaxations at temperatures at or below T_g. Experimental time frames allow much larger changes to be monitored [16, 25].

At temperatures below the T_g, physical or chemical degradation may be coupled to the molecular mobility associated with structural relaxation to varying degrees depending on the nature of the reaction and its mobility requirements [26]. Therefore, the difference between the storage temperature of a formulation and T_g is often considered in estimating stability, with a general rule of thumb being that the temperature of a pharmaceutical product should be 50 K below T_g in order to slow molecular motions to the point that their impact on shelf life is negligible [16]. Consequently, the extent to which certain formulation components such as moisture or the amorphous drug itself lower the T_g may be very useful to predict. Similarly, small-molecule or polymeric excipients that can increase T_g may be sought to improve stability of amorphous formulations containing a drug having a low T_g when present in amorphous form. MD simulations have potential utility in screening drug–excipient mixtures with T_g's that are likely to provide more stable formulations [27].

The effects of plasticizers such as water on T_g have been examined in numerous experimental studies [26a, 28] and a few MD simulations. Such comparisons provide a means of assessing the reliability of MD simulations for predicting possible effects of moisture on stability as influenced by molecular mobility associated with structural relaxation.

Typically, the effect of water sorption on the T_g for an amorphous solid is estimated using the Gordon–Taylor equation [28a]:

$$T_g = \frac{f_1 \cdot T_{g1} + K \cdot f_2 \cdot T_{g2}}{f_1 + K \cdot f_2} \tag{13.2}$$

where f_1 and f_2 are the weight fractions of each component and T_{g2} (for water) = 135 K. The value of K is determined by the ratios of the component densities (ρ) and their thermal expansivities ($\Delta\alpha$):

$$K = \left(\frac{\rho_1}{\rho_2}\right) \cdot \left(\frac{\Delta\alpha_2}{\Delta\alpha_1}\right) \tag{13.3}$$

For water uptake, the Simha–Boyer rule (i.e., $\Delta\alpha \cdot T_g \sim$ constant) is often applied to estimate the value of K because $\Delta\alpha$ is difficult to measure at the low T_g of water. Then K becomes

$$K = \left(\frac{\rho_1}{\rho_2}\right) \cdot \left(\frac{T_{g1}}{T_{g2}}\right) \tag{13.4}$$

Alternatively, the value of K can be estimated from the ratio of change in heat capacities of the two components at the T_g as derived by Couchman and Karasz [29].

Several simulation studies have now examined the effects of water content on T_g. Xiang andAnderson simulated HPMCAS polymers at three different water contents between 0.7 and 13.2% w/w water [12c] over which range they observed a reduction in T_g of ~80°—similar to experimental results for HPMCAS-LG and HPMCAS–MG [30]. A change in water content in polyvinylpyrrolidone (PVP) from 0.5 to 10%

reduced T_g by approximately 90 K in simulations reported by Xiang and Anderson [24a], corresponding closely to the reduction in T_g reported experimentally by Hancock and Zografi over the same range of water contents, a range within which the changes in T_g were well described by the Gordon–Taylor equation (13.2) [28a]. The effects of plasticization by water in simulations resembled those observed experimentally even though the T_g values themselves were generally higher in the MD simulations due to the rapid cooling. Gupta et al. recently compared T_g values obtained in MD simulations for sucrose at water contents of 0–5% to those obtained experimentally as well as to estimates of T_g using either the Gordon–Taylor or the Fox equation in which the densities of the two components are assumed to be unity [27]. A reasonable agreement was reported between the simulated T_g values and those obtained from the other approaches, though again the absolute T_g values were systematically greater than experimental T_g values.

The antiplasticization effect afforded by the addition of high-T_g excipients to formulations of amorphous drugs having a lower T_g has been illustrated in several experimental studies, but there are thus far few MD simulation studies that have systematically probed this effect. Binary mixtures of amorphous indomethacin (IMC) and PVP serve as one set of results examined in both experimental and MD simulation studies, as compared in the following text. Matsumoto and Zografi [31] explored the effects of PVP90, PVP30, PVP17, and PVP12 on the T_g of binary mixtures with indomethacin. They reported an onset T_g for indomethacin of 315 K and T_g values for PVP90, PVP30, PVP17, and PVP12 of 445, 429, 409, and 372 K. Thus, all of the PVP polymers evaluated would be expected to function as antiplasticizers for indomethacin. The results for T_g observed experimentally for PVP90 and PVP12 are shown in Figure 13.1 versus predictions based on the Couchman–Karasz approach. Also shown are the MD simulated T_g values for indomethacin–PVP reported by Xiang and Anderson [32] versus the Couchman–Karasz prediction based on heat capacity values from Crowley and Zografi [33]. Though limited, the MD simulations seem to provide reasonable predictions of the antiplasticizing effect of PVP on amorphous indomethacin.

13.3.2 Amorphous Structure and Dynamics

MD simulations can be exploited to generate detailed molecular information relating to local structure including polymer configurations, main-chain conformations and side group orientations, the nature and probabilities of various intermolecular interactions, nanosegregation of side groups or added solutes (e.g., water), as well as information on the dynamics of certain relaxation and diffusional processes. The availability of experimental results from X-ray diffraction (XRD), quasielastic neutron scattering, solid-state NMR, or various spectroscopic measurements that can be compared to those from simulations can serve to validate the simulations, while the simulations can provide molecular-level insights to aid in interpreting the experiments. Dynamic processes of interest, particularly structural relaxation events that contribute to local and global mobility and small-molecule diffusion may also be probed.

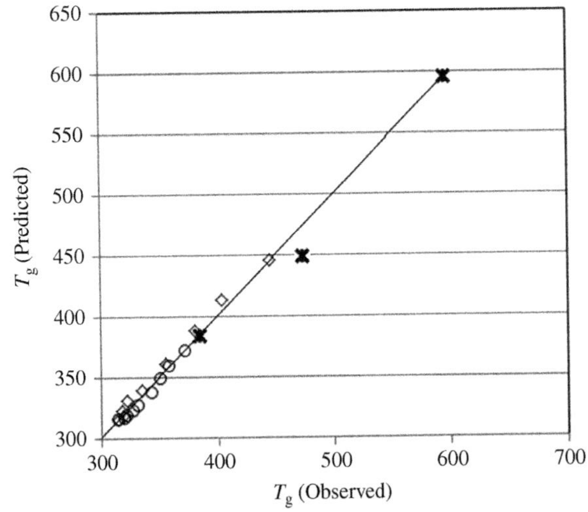

FIGURE 13.1 Experimental T_g values resulting from the antiplasticization effect of PVP90 (◇), PVP12 (○), or MD simulated values for PVP (∗) in amorphous mixtures with indomethacin. T_g predicted is based on the Gordon–Taylor equation with the value of K estimated using the Couchman and Karasz relationship.

In the polymer field, XRD and neutron scattering are the two most widely used techniques for validating results from MD simulations. Whereas XRD experiments can provide structural features over the relevant length scales, neutron scattering can reveal both structural and dynamic properties. Studies that combine MD simulations with one or both of these experimental tools have been relatively few in number, but interest in this approach is growing because of obvious synergies.

13.3.2.1 X-Ray Diffraction Conversion of a crystalline material to an amorphous solid is accompanied by a change from the distinct X-ray powder diffraction (XRPD) pattern characteristic of long-range order in crystalline solids to broad "halo" patterns indicating an X-ray amorphous material [34]. X-ray scattering intensity is generally measured either as a function of the scattering angle 2θ or the magnitude of the scattering vector Q in units of $Å^{-1}$ ($Q = 4\pi\sin(\theta)/\lambda$). A recent study of amorphous PVP by Busselez et al. illustrates the approach and its utility in validating results from MD simulations [35].

PVP (Fig. 13.2) is widely used in pharmaceutical formulations, including amorphous dispersions, and often selected as a model system for studies of amorphous excipients and drug–excipient mixtures [36]. The amide functional group in the pyrrolidone side group lacks a donatable hydrogen but serves as a good hydrogen acceptor to facilitate hydrogen bond (HB) formation with drugs having hydrogen-donating functional groups, thus promoting miscibility. PVP and PVP–drug combinations will be referred to repeatedly in this chapter.

Shown in Figure 13.2 is the wide-angle XRD pattern for PVP measured at 550 K. The PVP in this example had an average molecular weight of 160,000 Da and a T_g of 445 K by DSC. The short-range order can be characterized by the experimentally

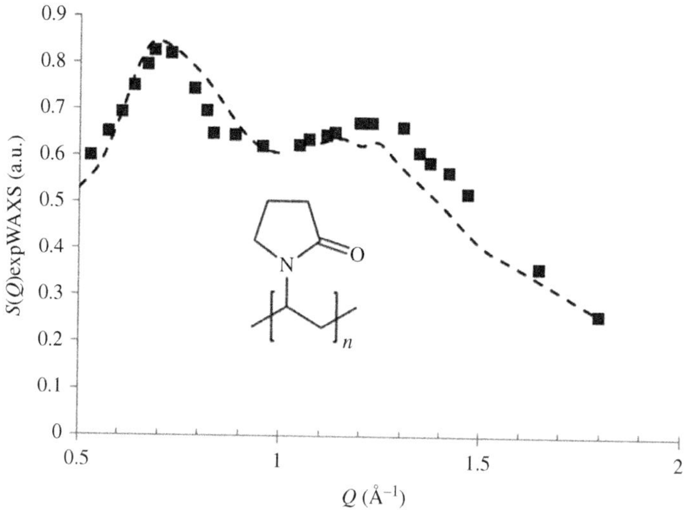

FIGURE 13.2 Comparison of measured PVP X-ray diffraction pattern at 550 K(■) with that obtained from MD simulation (- -). Source: Adapted from Busselez et al. [35]. Reproduced with permission of AIP Publishing LLC.

observable total static structure factor $S(Q)^{exp}$ which is derived from the coherent elastically scattered intensity, $I^{coh}(Q)$, where Q is the scattering vector and c_i and f_i are the atom concentration and scattering factor for atoms of type i [37]:

$$S(Q)-1=\frac{I^{coh}(Q)-\Sigma c_i\left|f_i(Q)\right|^2}{\Sigma c_i\left|f_i(Q)\right|^2} \tag{13.5}$$

The pair distribution function ($G(r)$), which is proportional to the probability of finding an atom at a position r relative to a reference atom chosen to be at the origin, can be determined from the Fourier transform of the experimental structure factor, $S(Q)^{exp}$, over a sufficiently wide range of reciprocal space ($Q_{min} \rightarrow Q_{max}$) within which the X-ray measurements are obtained:

$$G(r)-1=\frac{1}{2\pi^2 r\rho_0}\int_{Q_{min}}^{Q_{max}} Q\left[S(Q)-1\right]\sin(Qr)\,dQ \tag{13.6}$$

where r is the radial distance and ρ_0 is the average atomic number density in Å^3. The pair distribution function (PDF) or radial distribution function ($g(r)=4\pi r^2 G(r)$) can also be generated by MD simulation and converted to the structure function $S(Q)$, therefore establishing a link between MD simulations and XRD measurements to provide a rigorous test of structures obtained by simulation.

Two peaks were observed in the XRD pattern in Figure 13.2 at Q values of approximately 0.7 and 1.2 Å^{-1}. As shown in the figure, structure factors generated from MD simulations agreed reasonably well with the X-ray pattern and exhibited similar temperature dependence, leading the authors to conclude that their simulations adequately reproduced the local structure of PVP. To obtain a molecular interpretation of the two peaks in the X-ray pattern, Busselez et al. [35] generated partial structure factors from the MD simulations by grouping the atoms in PVP as either main chain or side group atoms. This enabled them to conclude that the "prepeak" reflects spacing between polymer backbones, which was estimated to be 10.5Å in PVP due to the bulky pyrrolidone side groups, while the peak at $Q = 1.2\,Å^{-1}$ was attributed to side group correlations consistent with a characteristic spacing of side group centers of mass of approximately 5Å. Overall, the structural analyses (which also included neutron diffraction results) indicated nanosegregation of the backbone and ring domains within PVP glasses above and below T_g.

13.3.2.2 Neutron Scattering Increasingly, neutron scattering techniques are being exploited to validate the results of MD simulations as recently reviewed [4b]. Neutron scattering detects the positions and motions of atoms, and therefore has advantages over XRD where energies involved are too high to explore relaxation processes. Moreover, neutron diffraction probes length and timescales that overlap with those accessible in MD simulations.

The scattering cross-sections obtained as neutrons pass through a sample and interact with atomic nuclei contain both coherent (dependent on the direction of the momentum transfer or scattering vector (Q)) and incoherent (uniform in all directions) contributions. The coherent elastic scattering contribution can provide information regarding the static structure factor $S(Q)$ which can be Fourier transformed to produce radial distribution functions for the relative positions of different nuclei. Tracing the coherent scattering function over time relates the scattering to changes in all atom configurations.

Thus, quantities such as pair distribution functions and mean squared displacement versus time profiles frequently reported in MD simulations may be compared to results from neutron scattering experiments. For polymer and small-molecule amorphous glass formers of greatest interest for pharmaceutical applications consisting largely of C, H, N, and O atoms, incoherent scattering by hydrogen atoms tends to dominate the neutron diffraction pattern [38]. In fully deuterated samples, the scattering is mainly coherent and all-pair correlations are nearly equally weighted. Coherent dynamic structure factors generated from quasielastic neutron scattering techniques such as neutron spin echo experiments are particularly useful in analyzing structural relaxation.

Narros et al. recently compared results of a fully atomistic MD simulation with neutron scattering data for real samples of deuterated 1,4-polybudadiene [39]. They compared results from neutron spin echo experiments and MD simulations at temperatures near and above the T_g (~180K) in order to explore the molecular motions taking place in the temperature region where the β- and α-relaxation processes merge. Though 1,4-polybutadiene is not a pharmaceutical excipient, the structural features and dynamic processes observed are relevant. The normalized dynamic structure factor profiles determined by MD simulation and neutron scattering versus time were compared as a function of temperature as illustrated in Figure 13.3.

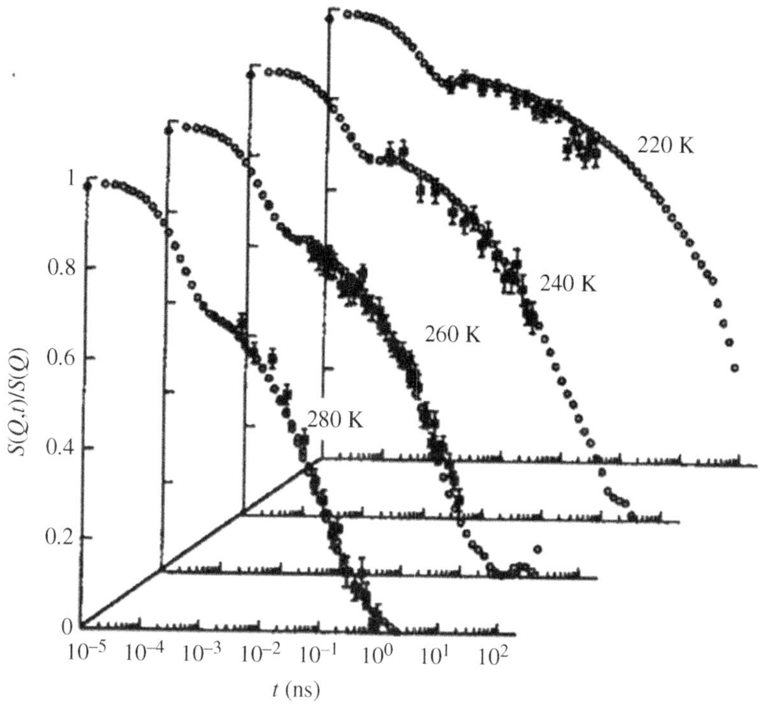

FIGURE 13.3 Comparison of results (data with error bars) for normalized dynamic structure factors near Q_{max} obtained from neutron spin echo experiments on deuterated samples with MD simulation data (open circles) at the same Q value [39]. Timescales for the experimental data were shifted −15 K in T_0 (see text). Source: Narros et al. [39]. Reprinted with permission of AIP Publishing LLC.

At the lowest temperature (220 K, but above T_g) where the α- and β-relaxation processes merge, extremely long simulation times (1 μs) were necessary to include even a portion of the time needed for structural (α) relaxation. Both the simulated and experiment results versus time (t) after 2 ps could be fitted to the well-known Kohlrausch–Williams–Watts (KWW) stretched exponential equation:

$$\frac{S(Q_{max,t})}{S(Q_{max})} = \text{DWF}\exp\left[-\left(\frac{t}{\tau_s}\right)^{\beta}\right] \quad (13.7)$$

where the Debye–Waller factor (DWF) accounts for the fraction of elastic scattering, τ_s is the structural relaxation time, and β is the stretching exponent. The simulation data revealed slightly faster decay in structure factors obtained by MD simulation than those obtained by neutron spin echo experiments. Assuming

Vogel–Fulcher–Tammann behavior they found that the temperature dependence of the relaxation for both could be described by the following relationship [40]:

$$\tau_s = \frac{A}{T}\exp\left[\frac{B}{(T-T_0)}\right] \tag{13.8}$$

when A and B were fixed and a difference of 15 K in the T_0 values was assumed. The data indicated that the α-relaxation dynamics from the MD simulations and spin echo experiments were superimposable when the timescales for the experiments were shifted to faster values corresponding to a difference of 15 K in T_0.

A second relaxation process that at the lower temperatures (particularly at 200 and 240 K) was nearly complete was attributed to β-relaxation reflecting local segmental motions within the polymer often described as local jumps within a confined space. The local motions within this region were characterized as local jumps between two positions separated by approximately 2.5 Å. Narros et al. [39] exploited a powerful capability of simulations that would only be possible experimentally by selective deuteration—the ability to distinguish between the types of hydrogen within 1,4-polybutadiene. Specifically, they distinguished *cis* and *trans* methyne hydrogens, *cis* and *trans* methylene hydrogens as well as vinyl hydrogens. The rapid hopping between positions separated by approximately 2.5 Å could be attributed to the *trans*-hydrogens which relocated by reorientation of double bonds via cooperative counter-rotation of their two flanking —CH—CH_2 bonds. At 200 K, the relaxation times for *trans*-hydrogen hopping and structural relaxation were well separated (2 ns vs. 300 ns). The relaxation time for *cis*-hydrogens (70 ns) was much longer and appeared to be coupled to structural relaxation.

13.4 HYDROGEN BONDING PATTERNS, WATER UPTAKE, AND DISTRIBUTION IN AMORPHOUS SOLIDS

Moisture uptake and the effect of water on amorphous system properties have now been explored both experimentally and by MD simulations for several pharmaceutically useful amorphous polymer and small-molecule excipients as well as drug–excipient mixtures. Thus, it is becoming possible to evaluate the reliability of the simulations in some cases and to begin exploration of the molecular basis for many of the important water-related property changes in amorphous solids. In contrast to the nature of water uptake in crystalline solids, which is often characterized in terms of surface adsorption, it is possible for water to dissolve in disordered amorphous solids [41], leading to profound changes such as alterations in the dynamics that affect properties such as the glass transition temperature and molecular mobility and changes in thermodynamic properties such as the Flory–Huggins interaction parameter (χ) [42].

Water present in amorphous solid formulations can influence both physical and chemical stability in a variety of ways [43]. The most direct influence of sorbed water is when it acts as a nucleophile in hydrolysis reactions occurring in amorphous solids [44], but it may also promote degradation by solubilizing reactants, altering the local pH of the microenvironment, or stabilizing transition states through hydrogen bonding. Water also functions as a plasticizer resulting in increases in local or global molecular mobility in amorphous solids. Solid-state drug degradation may be coupled to water induced increases in molecular mobility to varying degrees depending on the type of reaction. In particular cases, bimolecular reactions may be increasingly favored with increases in translational mobility of reactants, resulting in changes in degradation product profiles with increasing moisture content [26c, 44c, e, 45]. Physical changes induced by water may include lyophile collapse due to a reduction in T_g, phase separation resulting from mobility increases or alteration in drug–excipient miscibility in the presence of water, drug crystallization, and disproportionation of pharmaceutical salts due to water's activation of counterion exchange processes. As a result of the above myriad possibilities, the extent of water uptake and its influence on solid-state product stability are recurring themes in the pharmaceutical literature.

MD simulations appear to be potentially useful in providing molecular-level insights to aid in understanding the properties of amorphous excipients and amorphous drug formulations such as the following: (i) What molecular factors govern the equilibrium water content in amorphous formulations as a function of relative humidity? (ii) How is water distributed in the amorphous matrix, and how does its distribution affect other properties? (iii) What properties of the amorphous solid determine water's diffusion, and how does water affect mobility of the drug and excipients? (iv) To what extent does water uptake influence amorphous or crystalline drug solubility in the solid? (v) How does matrix relaxation (e.g., annealing) influence the aforementioned properties?

Shown in Figure 13.4 are moisture uptake profiles as a function of relative humidity for a number of amorphous systems including polymers, an amorphous drug, and drug–polymer mixtures that have also been examined in MD simulations [12b, 42d, 46]. The changes in water content vary considerably with relative humidity depending on the polymer and with the drug–excipient ratio within mixtures. In the examples shown, PVP exhibits the highest propensity for moisture uptake, while poly(D,L)lactide (PLA) has the lowest.

13.4.1 Poly(D,L)lactide

Xiang and Anderson [12b] recently developed an explicit all-atom computational model for amorphous PLA (see monomer unit structure in Fig. 13.4) and conducted MD simulations to predict certain properties, including the water sorption isotherm for PLA at 25°C. Water sorption was determined using the particle insertion method of Widom [47], which has been successfully employed to predict distributions of water and other solutes in lipid bilayers [1f, 48] and solubility of gasses and water in other polymers [48, 49]. The method involves calculating the ensemble average for the change in potential energy resulting from the random insertion of water molecules from the vapor phase into the simulated PLA assembly. The quantity of interest

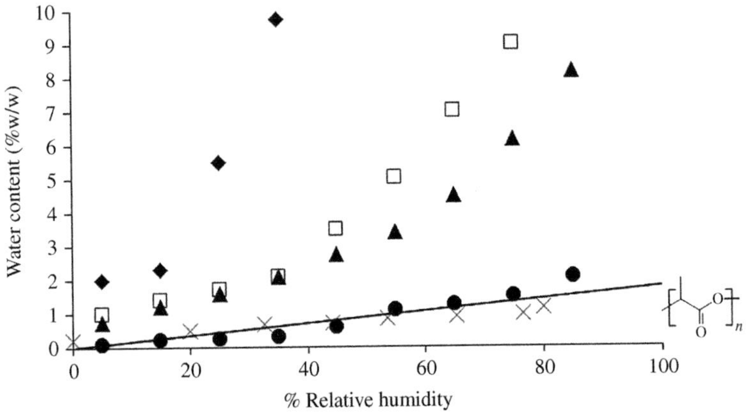

FIGURE 13.4 Moisture uptake profiles as a function of relative humidity for a number of amorphous systems including polymers (PVP, ◆; HPMCAS, ▲; PLA, X), amorphous indomethacin (IMC, ●) and a 50% IMC–PVP amorphous mixture (□) [42d, 46]. MD simulations for similar systems at selected water compositions are discussed in this chapter. The solid line represents the predicted moisture uptake profile for PLA from a recent MD simulation [12b]. The structure shown adjacent to the solid line is the monomer unit within PLA.

is the excess chemical potential for water in PLA, $\mu^{exc}(w|PLA)$, which is the Gibbs free energy change due to interaction between inserted water molecules and the surrounding PLA polymer molecules. At a sufficiently low moisture content, the sorbed water content (ϕ_w) is a linear function of relative humidity ($RH = P/P_s$):

$$\phi_w = K\left(\frac{P}{P_s}\right) \tag{13.9}$$

where K is the Henry's constant. The Henry's constant can be approximated from the difference in the excess chemical potentials of water in PLA and in pure water ($\mu^{exc}(w|l) = -5.74\,\text{kcal/mol}$) [50], where d_l ($=1.0\,\text{g/cm}^3$) is the density of pure liquid water and d_{PLA} ($=1.23\,\text{g/cm}^3$) was determined from the MD simulation:

$$K = e^{-\left(\mu^{exc}(w|PLA) - \mu^{exc}(w|l)\right)/RT}\left(\frac{d_l}{d_{PLA}}\right) \tag{13.10}$$

The water content (% w/w) in PLA versus relative humidity based on the value of K determined in MD simulations is presented in Figure 13.4 along with experimental results reported previously [46] for comparison. The agreement was within the error limits of the computational results. Previous studies have shown that for PLA, the L/D ratio and degree of crystallinity had little influence on the water sorption [51], which is generally less than 2% w/w within the 0–100% relative humidity range. This can be attributed to significantly weaker interactions between water and PLA compared to the water–water interactions in pure liquid water.

Snapshots of water molecules taken either in newly formed PLA simulated glasses containing 0.6% w/w water content at 298 K or after 100 ns aging indicated that most water molecules were not associated with each other. Radial distribution functions, $g(r)$, between the hydrogen atoms in water molecules and the oxygen atoms in the ester groups on PLA revealed a strong peak at 1.85 Å indicating hydrogen bonding between water molecules and polar ester groups in the PLA polymers, while the absence of a peak for $g(r)$ at a distance of less than 3 Å between the water hydrogen atoms and nonpolar methyl/methylene hydrogen atoms in PLA indicated that water molecules are not localized near the PLA backbone hydrogen atoms. The average interaction energy for water in the PLA glass (−12.3 kcal/mol) was much smaller than that in bulk water at 298 K (−22.3 kcal/mol). Both H-bonded (83%) and free (17%) water molecules were found in the simulated PLA glass. Most were bound to the carbonyl oxygen atoms in PLA forming either one (33%) or two (50%) H-bonds by forming a bridge between neighboring lactate units. The presence of the two types of H-bonding were also proposed in a recent study of water uptake into PLA films by time-resolved ATR-FTIR [52], although the authors also suggested the existence of water clusters.

13.4.2 Polyvinylpyrrolidone

The substantial water uptake observed in PVP contrasts markedly with the results in PLA, as shown in the water sorption profiles in Figure 13.4. The PVP sorption profile is also distinctly nonlinear, which may be the result of plasticization at high water contents or water self-association or clustering [53]. At water concentrations above approximately 25% the T_g of PVP falls below 25°C [28a], so an upturn at RH > 60% could certainly be a consequence of polymer swelling accompanying plasticization, but concave upward curvature is present at much lower concentrations, suggesting water self-association. Xiang and Anderson [12a, 24a] conducted MD simulations of a PVP polymer having a molecular weight close to that in Kollidon K12, a commercially available PVP polymer, to explore such properties as the distribution of water, water diffusion, and plasticization effects of water in assemblies containing either 0.5 or 10% w/w water.

Snapshots of the distribution of water molecules in the simulated PVP assembly containing 0.5 and 10% w/w water at 298 K are shown in Figure 13.5. Whereas most of the water molecules were monomeric and distributed uniformly throughout the simulation cell at 0.5% w/w water, clusters of self-associated water molecules were prevalent at 10% w/w water. In PVP containing 10% w/w water, an average of two water molecules were located within 3.4 Å of any given water molecule with a broad standard deviation (see combined figure 5 from Xiang and Anderson [24a]) while an average of 0.86 water molecules were within the same distance of a PVP carbonyl oxygen atom. Nearly 2/3 of the PVP carbonyl oxygens were found to be within a distance of 3.4 Å from a neighboring water molecule. These averages would be consistent with the formation of linear hydrogen-bonded chains of water molecules with each water molecule also hydrogen bonded to a PVP carbonyl oxygen. There was no evidence of the presence of any bulk water in this simulation. Radial distribution

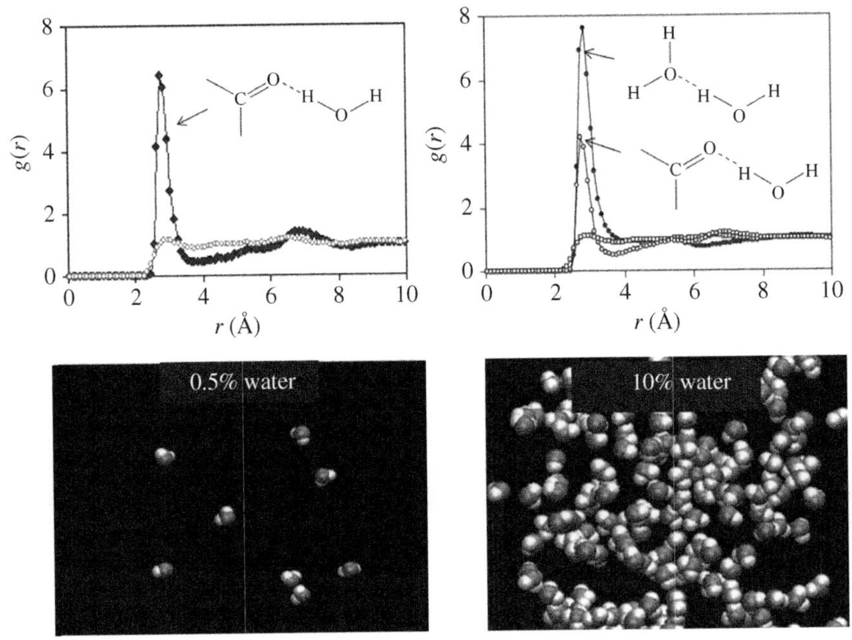

FIGURE 13.5 (Upper panel) Radial distribution functions, $g(r)$, for water molecules indicating hydrogen bonding to other water molecules or PVP carbonyl oxygen atoms (indicated by the arrows) at water contents of 0.5% (left) and 10% w/w (right). Lower curves (open squares) represent $g(r)$ between the oxygen atoms of water molecules and PVP methylene groups. (Lower panel) Snapshots of the distribution of water molecules in a simulated PVP assembly containing 0.5 and 10% w/w/ water at 298 K. Source: Xiang and Anderson [12a]. Reproduced with permission of John Wiley & Sons. Xiang and Anderson [24a]. Reproduced with permission of Springer Science+Business Media.

functions, $g(r)$, also shown in Figure 13.5 indicated a higher probability for water molecules to form HBs with the PVP carbonyl oxygen atom at a water content of 0.5% (relative to methylenes in PVP shown for comparison) and to other water molecules in addition to PVP carbonyl oxygen atoms at a water content of 10% w/w.

Several lines of experimental evidence lend support to the aforementioned model for water clustering in PVP. Lebedeva et al. [54] noted in the FTIR spectra of PVP samples containing sorbed water at low water contents that the band for unbound—OH groups was absent, leading them to conclude that the water molecules present were hydrogen bonded to oxygen atoms in PVP carbonyls. Taylor et al. examined peak shifts in the Raman spectra for PVP samples equilibrated at various relative humidities [28b]. At low water concentrations they suggested that the first one or two water molecules hydrogen bonded directly to the PVP side chain, but with increasing water content the additional water molecules increasingly hydrogen bonded to each other. Teng et al. [55] explored the effect of water content on local structure of PVP K90 and K12 as monitored by XRPD at ambient temperature. Recall that two broad halos were observed by Busselez et al. [35] at a temperature above T_g as shown in Figure 13.2, where the low angle halo was attributed to a spacing of 10.5Å between polymer backbones in PVP. Teng et al. monitored the characteristic scattering distance as a means of detecting changes in structure as a function of water content. They found that the low angle halo decreased with a decrease in temperature, indicating a characteristic scattering distance of approximately 7–8Å in both PVP K12 and K90 glasses at 0% water. This spacing remained relatively constant up to less than 12.5% w/w water, which Teng et al. interpreted as suggesting no apparent increase in free volume of the polymer within this range of water contents, thus consistent with the observation in MD simulations by Xiang and Anderson [24a] that changing from a water content of 0.5–10% w/w had only a minor effect on overall polymer free volume. Above T_g, this characteristic spacing increased more dramatically. The characteristic distance corresponding to the high angle halo of approximately 4.5Å at 0% water content (which Busselez et al. attributed to pyrrolidone side group spacing [35]) was found to transition smoothly to the spacing found for the primary halo in bulk water at high water contents well above T_g.

More recently, Busselez et al. [56] generated neutron scattering cross-sections for hydrated PVP containing either 44% w/w H_2O, 22% w/w D_2O, or 44% w/w/ D_2O. From the results in 44% w/w H_2O, they inferred that water swells the polymer chains and induces disorder in the nanosegregated ring domains. Deuteration enhanced the visibility of water–water correlations from which they found evidence for water strands or clusters in support of the MD simulations by Xiang and Anderson [24a]. One cross correlation peak at approximately 1.1 Å^{-1} was attributed to the correlation between atoms of one PVP side group and a water molecule hydrogen bonded to the carbonyl of a neighboring ring. Thus, the experimental evidence available from several types of studies appears to be consistent with the results of MD simulations.

13.4.3 Hydroxypropylmethylcellulose Acetate Succinate (HPMCAS)

Hydroxypropylmethylcellulose acetate succinate (HPMCAS) has recently become quite popular in the pharmaceutical industry due to its frequently demonstrated ability to produce stable amorphous drug dispersions [30, 57]. Its high T_g in its unionized state [30] tends to reduce molecular mobility and improve physical stability of amorphous drug dispersions. The hydrophobic regions within HPMCAS aid in solubilization of poorly water-soluble, lipophilic drugs, while its amphiphilic nature promotes dissolution and inhibits drug crystallization. In some comparison studies, drug/HPMCAS solid dispersions have been shown to be more effective than other polymers investigated [57, 58] at maintaining drug supersaturation in simulated intestinal fluid and to provide greater enhancement in bioavailability *in vivo* [30, 59].

As illustrated in Figure 13.4, HPMCAS exhibits a tendency to absorb a significant quantity of water (e.g., 0–12% w/w water between 0 and 95% RH) in comparison to the relatively simple PLA polymer, although it is much less hygroscopic than PVP. Because water plays such a critical role in determining the stability and drug release from drug/HPMCAS solid dispersions, Xiang and Anderson recently conducted a

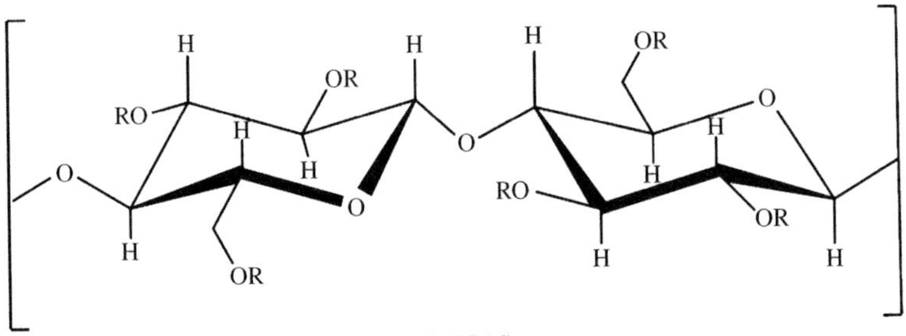

HPMCAS

R = $-CH_3$ (methyl)
$-CH_2CH(CH_3)OH$ (HP)
$-CH_2CH(CH_3)OCH_3$ (HP Me)
$-COCH_3$ (acetate)
$-COCH_2CH_2COOH$ (succinate)
$-CH_2CH(CH_3)OCOCH_3$ (HP acetate)
$-H_2CH(CH_3)OCOCH_2CH_2COOH$ (HP succinate)

FIGURE 13.6 Representative HPMCAS substructure and the variety of R-substituents that may be found at each of the indicated oxygens.

full atomistic simulation of an HPMCAS polymer at different water contents to explore both the distribution of water molecules and the influence of water on molecular mobility [12c].

Construction of a molecular model for HPMCAS is complicated by the complex and varied substitution patterns possible as evident in the representative substructure shown along with the variety of R-substituents that may be found at each of the indicated oxygens (Fig. 13.6). As recently observed by Porter III et al. [60], a limited variety of HPMCAS products are available, and they cover a relatively small subspace of the entire allowed compendial space. In particular, they observed that the ratio of acetyl to succinyl substitution may have a dramatic impact on the ability of HPMCAS to form supersaturated solutions as measured by areas under the solution concentration versus time profiles (AUCs). Supersaturation profiles are highly dependent on the HPMCAS composition and also very drug specific. MD simulations may ultimately contribute to understanding of the molecular basis for the relationship between HPMCAS molecular structure and dispersion performance. Clearly, HPMCAS polymer assembly in terms of composition and substitution pattern requires careful attention.

Prior to the simulation by Xiang and Anderson, no previous MD simulations of HPMCAS had been reported in the literature. Therefore, the molecular model developed in terms of substituent distribution was one that resembled the typical values in a commercial product (HPMCAS-MF, Shin Etsu Chemical Co.)—23% methoxy, 7% hydroxypropoxy, 9% acetyl, and 11% succinyl [57]. The percentages in the simulated HPMCAS polymer were 25% methoxy, 4% hydroxypropoxy, 11% acetyl, and 13% succinyl by weight. In each simulation cell, there were six HPMCAS polymer chains within which a total of 50 residue patterns were present. The polymer molecular weight was 5213 Da. The densities obtained for the simulated

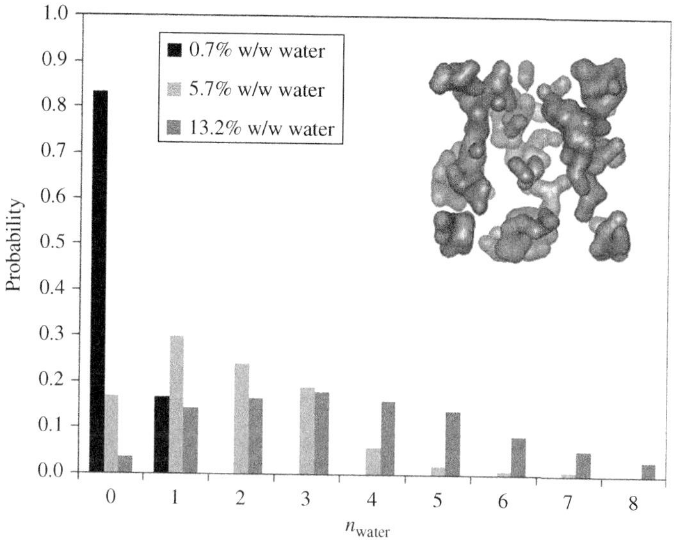

FIGURE 13.7 Probabilities for a certain number of water molecules (n_{water}) within a distance of 3.4 Å of a given water molecule in simulated HPMCAS glasses at 298 K and different water contents. Inset shows a snapshot of water clustering at a water content of 13.2% w/w. Source: Xiang and Anderson [12c]. Reprinted with permission of American Chemical Society.

amorphous HPMCAS at 298 K were 1.295, 1.287, and 1.276 g/cc at 0.7, 5.7, and 13.2% w/w water, which compared favorably with the reported density of 1.285 g/cc for the *AQOAT® AS-MF* commercial HPMCAS. The reduced densities with increased water content were consistent with polymer swelling upon increasing water content.

The synthetic route for producing HPMCAS from HPMC results in blocking of a significant fraction of the hydrogen-bond-donating sites in HPMC by acetyl groups. Thus, in the HPMCAS polymer assembly constructed for MD simulations, only 0.47 of a hydrogen donor group was present in each glucose unit, on average. Of these H-donors, at a water content of 0.7% w/w, approximately 2/3 were involved in an HB. Of these HBs, nearly 90% were formed between individual glucose units either within a chain or between chains. The remaining HPMCAS hydrogen donors (5) formed HBs with water molecules. There were 12 water molecules present in the simulation at 0.7% water content having 24 hydrogen donor sites. Of these, 14 formed HBs with acceptor atoms in HPMCAS and none of them were involved in HBs with other water molecules, although there is a small probability for dimer formation even at 0.7% w/w water content as evident in the probability distributions shown in Figure 13.7.

At a higher water content of 5.7%, most of the water molecules were involved in dimers or water clusters with less than 20% isolated. This appears to contrast with the water distribution in the more hydrophilic polymer poly(vinyl)alcohol where at 5.2% w/w water the water distribution was homogeneous although the temperature in that study (502 K) was much higher [61]. At 13.2% w/w water in the present study, less than 5% of water molecules were isolated. A representative snapshot in Figure 13.7 shows the extensive water clustering at 13.2% water. The degree of self-association of water molecules can be assessed by the probability distribution reflecting the number of water molecules (n_{water}) within the solvation shell (3.4 Å in the present study) of a given water molecule as shown in Figure 13.7. The averages for the number of water neighbors at a distance of 3.4 Å from a given water molecule in HPMCAS at water contents of 0.7, 5.7, and 13.2% w/w were 0.17, 1.8, and 3.5, respectively. These are significantly below the average in bulk water of 8.9 determined in the same manner, indicating that there is no evidence for complete phase separation up to 13.2% w/w water.

13.4.4 Amorphous Indomethacin

Indomethacin (1-(*p*-chlorobenzoyl)-5-methoxy-2-methylindole-3-acetic acid, IMC), can exist in several crystalline forms (e.g., α-, γ-, and δ-polymorphs and others) [34, 62]. Due to the poor water solubility of the most stable γ-form [63], IMC has been a popular choice as a model compound to explore the benefits of amorphous solid dispersions for enhancing solubility [64], thereby improving dissolution rates and oral bioavailability [65].

The structure of IMC, shown in Figure 13.8, suggests possible reasons for the existence of multiple polymorphic forms based on the variety of potential packing motifs. IMC contains hydrophobic indole and chlorobenzyl groups as well as multiple polar substituents that provide HB donating and/or accepting capacity. In γ-crystals, hydrogen-bonded dimers form between carboxylic acid groups of adjacent IMC molecules while in metastable α-crystals two IMC molecules form a carboxylic acid dimer, while the —COOH from a third molecule forms an HB to the amide carbonyl of one member of the dimer [66]. This polymorphism may stem from the conformational diversity [66, 67], arising from hindered rotation of the partial double bond between the indole N1 nitrogen and C9 carbonyl carbon of the amide linkage (ϕ_1) in IMC (Fig. 13.8), which leads to the prevalence of different conformations in different crystalline forms. IMC exists exclusively as the Z-isomer at $\phi_1 \sim -32°$ in γ-crystals, while in the metastable α-crystals three different conformations coexist [66]. The distribution of conformations depends on the environment as the Z-isomer is favored in solution while the E-isomer is the form preferred in inclusion complexes with β-cyclodextrin [68].

MD simulation may be ideally suited for unraveling molecular structures in amorphous solids that may lead to instability or favor crystallization of one polymorphic form over another. Xiang and Anderson [24b] recently conducted an MD simulation study of amorphous IMC assemblies containing 0.6% w/w water content to explore both structural properties (e.g., IMC hydrogen bonding patterns, distribution of various IMC conformations, and the nature of water–IMC interactions) as well as dynamic processes (e.g., relaxation processes within the IMC molecule, and water diffusion). To explore the entire ϕ_1/ϕ_2 conformational space by MD simulation, it was necessary to reduce the torsional barrier for ϕ_2 from 29.0 to 8.0 kcal/mol in order

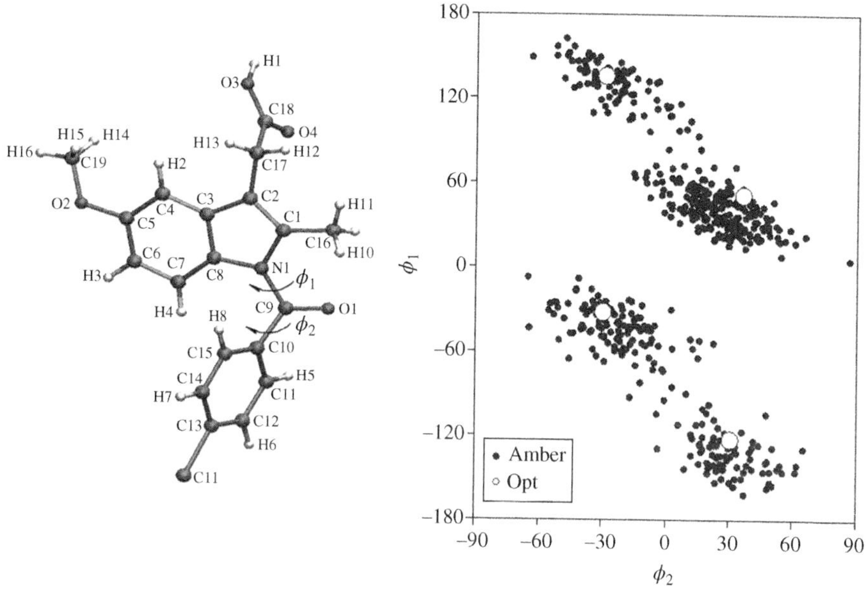

FIGURE 13.8 Molecular structure of indomethacin (IMC, left) and a plot of the dihedral angle pairs (ϕ_1, ϕ_2) represented in simulated amorphous IMC (●) and the preferred conformations (○) obtained by *ab initio* quantum mechanical calculations [67a]. Source: Xiang and Anderson [24b]. Reprinted with permission of American Chemical Society.

to achieve sufficiently rapid interconversions for equilibration between the different ϕ_1 states within a reasonable time frame. As shown in Figure 13.8 (right panel), the sampling points in an amorphous solid simulation included all diastereomeric structures and were concentrated around the optimized structures obtained from *ab initio* calculations [67a] (indicated by large open circles).

Curiously, the properties of amorphous IMC dispersions often depend on the method of preparation (e.g., melt-cooling, milling, spray drying, and vapor deposition) [62a, 69]. Of particular interest are the underlying structural and dynamic properties that lead to IMC crystallization. Crystallization near or below T_g may lead to either the γ-form or the α-form depending on such factors as the rate of cooling [70], water content [71], or the method of preparation (e.g., ball milling, cryo-milling, melt quenching, or spray drying) [69e].

Shown in Figure 13.9 are snapshots taken from the simulated amorphous IMC assemblies that illustrate the diversity of hydrogen bonded structures present. Two structures of particular interest, shown in the upper panels, are the hydrogen-bonded carboxylic acid dimer in which a third molecule forms an HB to the amide of one of the IMC molecules involved in the dimer as found in the asymmetric unit of the α-crystalline form (I) and the hydrogen-bonded carboxylic acid dimer as found in the γ-polymorph (II) [66]. More prevalent were various larger hydrogen-bonded cyclic aggregates (III) or chains (IV) linked through their carboxyl residues as illustrated in the lower panels of Figure 13.9. Such nanoaggregates may lead to

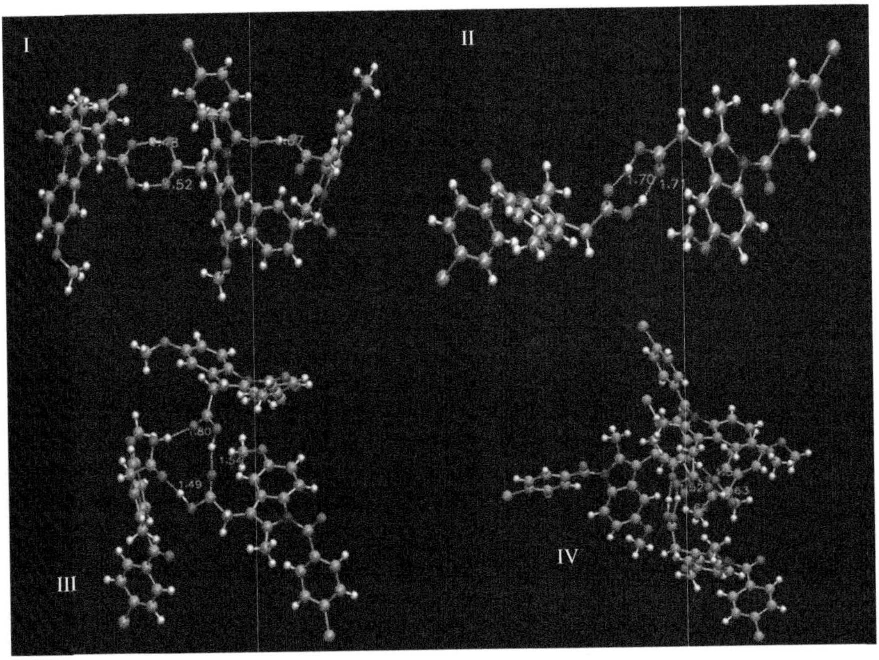

FIGURE 13.9 Representative H-bonding patterns in a simulated IMC glass: (I), H-bonded trimer similar to that in the unit cell of the α-crystalline polymorph; (II) H-bonded dimer similar to that in the unit cell of the γ-polymorph; (III) H-bonded cyclic trimer; and (IV) H-bonded tetramer. Source: Xiang and Anderson [24b]. Reprinted with permission of American Chemical Society. (*see insert for color representation of the figure.*)

spatial heterogeneity in terms of the distribution of polar and nonpolar residues that could influence the distribution of other molecules such as water.

Probability distributions were generated for the number of HBs in which each IMC molecule participated. Most (79 ± 3%) were found to participate in at least one HB, while approximately 21 ± 3% of the molecules were not involved in an HB in the simulated IMC glass. Considering the IMC–COOH as a hydrogen donor, approximately 73 ± 6% of the HBs that were formed were with the carboxylic acid carbonyl group on another IMC molecule while 17 ± 6% were formed with a benzamide carbonyl and 9 ± 2% with water molecules. These quantitative results can be compared to experimental data obtained by various methods, including XRPD, infrared and Raman spectroscopy, and solid-state NMR. Generally, the experimental results do not provide the necessary detail to enable quantitative comparisons. For example, Yuan et al. could only conclude from x-ray amorphous powder diffraction that the most favorable packing pattern was one that preserved the cyclic dimer [34]. Taylor and Zografi suggested based on infrared and Raman spectra that amorphous IMC consists largely of cyclic dimers with a "small proportion" of molecules hydrogen bonded to form a chain [64d]. The latter conclusion was based on the presence of a shoulder that was attributed to non-hydrogen-bonded carbonyls at the end of chains of varying length. Thus far, vibrational spectroscopic methods have not distinguished between cyclic dimers and larger cyclic aggregates such as the cyclic trimer (III) depicted in the snapshot in Figure 13.9. Solid-state NMR offers the potential to provide more quantitative estimates of the various species present. Munson et al. [72] examined the distribution of species in melt quenched amorphous indomethacin by solid-state NMR using indomethacin labeled with ^{13}C at the —COOH group. This enabled them to subtract partially overlapping peaks from the amide carbonyl. They found that 92.5% of the —COOH residues were hydrogen bonded in their sample of amorphous IMC and 7.5% were free (water content was not reported), in reasonable agreement with the aforementioned results from MD simulation. Of those carboxyl groups participating in HBs, 74% were formed with another carboxyl group on another IMC molecule, while 18.9% were formed with benzamide carbonyls.

Snapshots of the distribution of the 12 water molecules present in the simulated IMC assembly, representing a concentration of 0.6% w/w, indicated that most, but not all, of the water molecules were monomeric with occasional dimer formation also detected. More than 90% of the water molecules were hydrogen bonded. Of these, 36 ± 13% were involved in a single HB, while 28 ± 7% and 29 ± 10% participated in two or three HBs, respectively. When water was the hydrogen donor, 39 ± 8% of its HBs were formed with the benzamide carbonyl, 37 ± 12% were with an IMC carboxylic acid carbonyl, 12 ± 6% were with the —OCH_3 group, and 12 ± 9% were with other water molecules.

Glasses produced by MD simulation as well as most of those generated experimentally are metastable systems that continue to undergo structural relaxation with aging as they gradually drift toward equilibrium [25a, b]. Over the 100 ns dynamics simulation of the IMC glass described herein, density increased by 0.073% and the overall potential energy decreased by 0.092%, indicating that further relaxation could be observed even over this short time frame. The strong tendency for IMC carboxyl groups to form cyclic dimers or other self-associated aggregates combined with the presence of approximately 20% non-hydrogen-bonded —COOH residues in the simulated IMC assembly (but only 7.5% in the sample probed in solid-state NMR studies [72]) leads one to speculate that additional hydrogen bonding or strengthening of existing HBs may occur over time after glass formation. Indeed, the greater extent of hydrogen bonding found by solid-state NMR in the earlier example comparing a simulation and an experimentally prepared (melt quenched) sample suggests that dramatic differences in rates of cooling may produce significant structural differences. Further supporting this contention is evidence that highly stable amorphous indomethacin glasses prepared by vapor phase deposition appear to differ significantly in certain properties [69c]. Most striking, perhaps, is the substantially reduced moisture uptake of the more stable indomethacin glass. Dawson et al. [69c] suggested and the earlier simulations seem to confirm that water sorption in IMC occurs because of hydrogen bonding with IMC. They speculated that their more stable glass formed more HBs between IMC molecules resulting in fewer available sites for water molecules. They also observed higher water diffusivity in the more stable glass which they attributed to less stable water binding. Similarly, aging of trehalose for 120 h at a temperature of 373 K reduced its tendency to absorb moisture from 1.4 to 0.8% at 10% relative humidity, indicating that water uptake in amorphous solids may not depend solely on the functional groups present but also their availability for interaction and competition between various HB donors and acceptors [73]. Amorphous dispersions at a higher energy appear to have increased water uptake relative to the same dispersions after annealing [74]. These are findings that highlight the complexity and dynamic nature of amorphous systems that will hopefully continue to stimulate further interest in MD simulations and experiments that can provide detailed molecular interpretations for these phenomena.

13.5 AMORPHOUS DRUG–POLYMER BLENDS

13.5.1 Molecular Interactions Probed by MD Simulation

Amorphous drug formulations have attracted increasing interest for improving oral bioavailability of poorly soluble compounds because the amorphous form is the highest energy solid form of a compound and, therefore, should have the greatest advantage in terms of solubility and dissolution rate [64a, 65]. Using indomethacin as an example of a poorly soluble, hydrophobic API, Hancock and Parks estimated the solubility enhancement of amorphous indomethacin relative to its most stable γ-crystalline form to range from 25- to 104-fold depending on the value selected for the heat capacity difference (ΔC_p) of the two forms [65]. Murdande et al. [75] combined consideration of the effects of water sorption into the amorphous solid and drug ionization with a more accurate value for ΔC_p, to obtain an improved estimate of the solubility enhancement of amorphous IMC of sevenfold. This value was much closer to the apparent solubility enhancement of 4.9-fold obtained from dissolution experiments.

Thermodynamically, amorphous systems are supersaturated relative to the crystalline state so the tendency over time is for crystallization to occur, particularly at elevated storage temperatures or at high relative humidity [76]. This is perceived to be a major drawback of such systems. Moreover, while higher energy amorphous forms often do provide improved bioavailability, this is not always the case [77]. An additional significant challenge is inhibiting the crystallization that may occur during dissolution, which negates the solubility and potential bioavailability advantage of amorphization [36a, 78].

Physical stability (i.e., inhibition of crystal growth) during storage and maintenance of supersaturation during dissolution *in vivo* may be greatly improved by mixing amorphous APIs with other excipients such as certain polymers, including PVP or PVP/PVA (polyvinylacetate) copolymers, hydroxypropylmethyl cellulose (HPMC) or acetate/succinate esterified HPMC (HPMCAS), cyclodextrins, etc. Intimate mixing of a low T_g amorphous drug such as indomethacin ($T_g = 320$ K [16]) with polymer excipients that have a high T_g can greatly improve physical stability of the amorphous dispersion due to the antiplasticizing effect of the polymer. The stability enhancement in such a case may be at least partially attributable to the limited molecular mobility of the system, but thermodynamic factors that govern the driving force for drug crystallization are also important.

Typically such amorphous dispersions are prepared by melt extrusion (i.e., cooling from a melt) [79], spray drying [80], or lyophilization [81]. Although the reliable determination of drug solubility in amorphous excipients can be challenging [82], it is generally the case stemming from practical reasons (e.g., dose) that the concentration of API in amorphous pharmaceutical dispersions exceeds the saturation solubility of the most stable crystalline form. Thus, crystallization is likely from a thermodynamic viewpoint. Several recent examples have supported the notion that the tendency for crystallization cannot be reliably predicted solely from mobility considerations [31, 42e, 82e, 83]. Specific polymer–drug interactions such as hydrogen bonding that favor miscibility can contribute to physical stability of amorphous dispersions by preventing or reducing the tendency for phase separation, thereby reducing the driving force for crystal nucleation and growth, as crystallization from an amorphous solid dispersion requires a critical concentration of drug molecules with the proper orientation to form crystallites [42e]. On the other hand, as amorphous drug–polymer interactions become more favorable, reductions in the aqueous solubility enhancement that may be achievable must also be considered. To further complicate the situation, the degree of miscibility for a given drug–polymer has been shown to depend on temperature, composition, and water content [42e, 82e, 83]. Particularly intriguing in this regard are observations of moisture-induced phase separation that ultimately lead to drug crystallization.

The solubility of crystalline API, miscibility of the amorphous form, and the aqueous solubility enhancement that may be achieved for a given amorphous dispersion and the detailed atomic-level structural information that can enable a molecular understanding are often difficult to determine experimentally but potentially accessible by MD simulations. Most recent examples have focused primarily on solubility/miscibility properties [84]. The example highlighted in this section explores the underlying molecular interactions contributing to drug miscibility/solubility and the aqueous solubility enhancement predicted as a function of composition by combining MD simulations with Flory–Huggins theory. The model API chosen for this study was indomethacin and PVP was the model polymer excipient. Both components have been discussed previously in this chapter and their combinations have been the subject of numerous experimental investigations probing the molecular interactions, miscibility, solubility enhancement, molecular mobility, and many other properties [16, 31, 36a, 42d, 64d, 82c, d, 83b, c, 85].

As shown previously in Figure 13.8, IMC possesses several polar groups (i.e., —COOH, —CONH—, and CH_3O—) that can act either as HB donors and/or acceptors. As a result and as discussed previously, numerous hydrogen-bonded species including H-bonded chains and cyclic aggregates coexist in amorphous IMC. PVP is relatively hydrophilic due to the presence of the pyrrolidone side group amide, but it lacks an HB donor group, and therefore can only participate in hydrogen bonding as an HB acceptor. The increasing HB acceptor concentration accompanying increases in the concentration of PVP in amorphous IMC–PVP dispersions would be expected to result in some exchange of IMC-IMC HBs for IMC–PVP hydrogen bonding. The extent of this migration and its influence on the enthalpy of mixing are critical factors in determining the solubility of crystalline IMC in PVP, the miscibility of amorphous IMC in PVP, and the potential reduction in aqueous solubility enhancement in IMC–PVP amorphous dispersions relative to pure amorphous IMC.

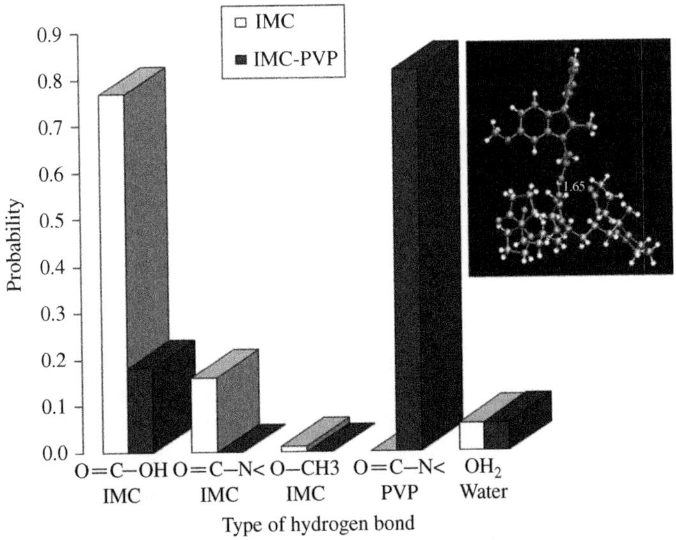

FIGURE 13.10 Probabilities for various types of hydrogen bonds (HBs) between an IMC–COOH donor and various functional groups in simulated IMC (white) and IMC–PVP (58%:41%, w/w) (black) glasses at 298 K. A simulated structure of the IMC–PVP HB is also shown. Source: Xiang and Anderson [32]. Reproduced with permission of John Wiley & Sons.

In addition to aforementioned MD simulations in amorphous IMC and PVP, Xiang and Anderson recently reported additional simulations in amorphous IMC–PVP dispersions at PVP percentages of 13, 41, and 74% w/w [32]. All of the simulated assemblies contained a small percentage of water ranging from 0.6 to 0.8%. After energy minimization, the assemblies were equilibrated at 600 K then cooled to 200 K over 12 ns. Microstructures obtained after cooling were then used as starting configurations for 100 ns aging dynamic runs at 298 K in order to explore the hydrogen bonding patterns as a function of percentage PVP and to determine relative solubility/miscibility of IMC and the aqueous solubility enhancement likely as a function of IMC dilution in PVP.

Shown in Figure 13.10 is a comparison of the probability distributions for the carboxylic acid (—COOH) HB donor in IMC to form HBs at different HB acceptor sites in amorphous IMC (white bars) and in an amorphous IMC-PVP (58 : 41) dispersion (black bars) obtained in the MD simulations at 298 K [32]. A representative structure of an IMC-PVP HB is also shown (only a portion of a PVP chain is included). Previously it has been suggested that the inhibition of cyclic dimer formation that is necessary for nucleation of the most stable γ-crystalline form in the presence of PVP is a major factor in the inhibition of IMC crystallization in IMC–PVP amorphous dispersions [31]. The results of MD simulations appear to support this argument. The probability that the HB from a given IMC carboxylic acid is to another IMC molecule was found to shift from greater than 90% in amorphous IMC to less than 20% in the presence of 41% w/w PVP. Nearly 80% of the —COOH HBs in 58:41 IMC–PVP were with PVP. HBs to water accounted for approximately 9% of the HBs in both formulations and approximately 1/3 of the carboxylic acid groups appeared to be non-hydrogen bonded in both amorphous IMC and the 58:41 IMC-PVP simulations. The latter result was somewhat unexpected because with increasing PVP content (and more hydrogen acceptor groups) an increase in the fraction of IMC participating in HBs would also be expected. However, this may have been a result of the rapid cooling employed to form amorphous glasses in the MD simulations. In cooling the IMC melt from 600 to 200 K over a longer time frame of 12 ns the percentage of non-hydrogen-bonded IMC gradually declined (unpublished results) suggesting that with further aging the fraction of non-hydrogen-bonded IMC would decrease. Establishing complete equilibrium with respect to hydrogen-bonded structures may require longer simulation times in the presence of PVP because of its longer chain length and higher T_g.

Qualitatively, the hydrogen bonding patterns observed in amorphous IMC–PVP dispersions mimic those reported experimentally. Taylor and Zografi [64d] reported that the infrared peak at 1710 cm^{-1} assigned to the cyclic dimer was replaced by a new peak at 1726 cm^{-1} assigned to the non-hydrogen-bonded carbonyl stretch of the —COOH group of IMC that is hydrogen bonded to a PVP molecule. A new peak at 1636 cm^{-1} emerged, which was assigned to the hydrogen-bonded PVP carbonyl. Newman et al. [85b] generated the pair distribution function from X-ray scattering data for a 50 wt% IMC-PVP dispersion sample and compared it to the best fit for a linear combination of the PDFs of the individual amorphous IMC and PVP components. Clear differences were seen in a plot of residuals, which the authors interpreted as evidence for a miscible dispersion, as was also indicated by the finding of a single T_g for the amorphous mixture. Pham et al. [86] examined a 30% w/w dispersion of IMC in PVP by proton solid-state NMR (^{1}H SSNMR), contending that this technique offers a more direct approach for demonstrating the absence of the cyclic dimer. 2D rotor-synchronized ^{1}H double-quantum broadband back-to-back (DQ-BABA) magic-angle spinning (MAS) experiments were performed to generate 2D spectra for the 30% IMC-PVP dispersion and the γ-crystalline form of IMC. The absence of an autocorrelation peak indicative of the cyclic dimer in the amorphous dispersion that could be clearly observed in the γ-polymorph suggested the absence of dimeric carboxylate hydrogen bonding at 30% w/w IMC/PVP. Munson et al. [72] employed ^{13}C SSNMR using IMC that was ^{13}C enriched at the carboxylic acid carbon in order to eliminate interfering peaks from PVP. Difference spectra obtained for a 50% w/w IMC–PVP melt-quenched dispersion demonstrated the predominance of a peak corresponding to the IMC carboxylic acid hydrogen bonded to PVP.

13.5.2 Solubility and Miscibility Prediction

Ascertaining whether or not any given drug would be miscible with a given excipient at a particular concentration using computational or theoretical approaches to predict miscibility as a function of amorphous dispersion composition would be quite useful. One popular (though not universally reliable) method for doing so is the solubility parameter method based on Hildebrand's regular solution theory [87]. The solubility parameter for a given material is the square root of the cohesive energy density (CED = $\Delta E_v/V$) where ΔE_v and V are the energy of vaporization at a given temperature and V is the volume of the material (or assembly of interest in a simulation). Accordingly,

$$\delta = \sqrt{\text{CED}} = \sqrt{\frac{\Delta E_v}{V}} \tag{13.11}$$

The total solubility parameter is often expressed in terms of separate contributions including nonpolar dispersive interactions (δ_d) and electrostatic components (δ_e) including both polar and hydrogen bonding interactions as developed by Hansen [88]:

$$\delta^2 = \delta_d^2 + \delta_e^2 = \frac{\Delta E_d}{V} + \frac{\Delta E_e}{V} \tag{13.12}$$

Regular solution theory assumes that specific interactions such as hydrogen bonding are absent, and therefore Hildebrand solubility parameters are generally applicable only to systems containing relatively nonpolar constituents. It is important to remember that neither of these approaches are theoretically justified for mixtures in which specific solvation interactions are important.

Gupta et al. [84a] generated solubility parameters for indomethacin (IMC), polyethylene oxide (PEO), sucrose (SUC), and glucose (GLU) using MD simulations with the goal of predicting whether or not blends of indomethacin with each of these excipients would be miscible or immiscible. The solubility parameter values obtained (in $\text{MPa}^{0.5}$) were 23.9 ± 0.3, 22.2 ± 0.2, 29.9 ± 0.5, and 34.8 ± 0.2 for IMC, PEO, SUC, and GLU, respectively. The solubilities were then predicted from the differences in solubility parameters:

$$\ln X_s = \ln X_0 - \frac{\left(\delta_{\text{IMC}} - \delta_{\text{excipient}}\right)^2 V_{\text{IMC}} \Phi^2_{\text{excipient}}}{RT} \tag{13.13}$$

where X_s and X_0 are the mole fraction solubility of IMC in a selected excipient and its ideal mole fraction solubility, respectively, $\Phi_{\text{excipient}}$ is the volume fraction of excipient, and V is the molar volume of indomethacin.

Forster et al. [89] found that drug–excipient pairs (including IMC-PVP mixtures) having a $\Delta\delta < 2.0\ \text{MPa}^{1/2}$ were miscible, while combinations with a $\Delta\delta > 10.0\ \text{MPa}^{1/2}$ were generally immiscible. Following the same guideline, Gupta et al. predicted that IMC/PEO mixtures should be miscible, while IMC/GLU dispersions would be immiscible. Since $\Delta\delta$ for IMC/SUC was $6.0\ \text{MPa}^{1/2}$, they predicted borderline miscibility for this combination. Differential scanning calorimetry indicated that the melting point of polyethylene oxide was progressively depressed and broadened with increasing concentrations of indomethacin, confirming miscibility, while blends of IMC with either glucose or sucrose were incompletely miscible.

Xiang and Anderson [32] generated solubility parameters for amorphous indomethacin and PVP (each containing <1% water), obtaining values of 25.5 and 19.0 for IMC and PVP, respectively, suggesting a small positive free energy of mixing

($\Delta\delta = 6.5\,\text{MPa}^{1/2}$) and borderline miscibility. More importantly, however, they found that predictions using solubility parameters were opposite to those from Flory–Huggins lattice theory. This and other recent reports indicate that more reliable predictions are likely to result from the use of Flory–Huggins theory as now described.

The Flory–Huggins theory takes into account the intermolecular interactions between the components in a mixture in evaluating the free energy of mixing, ΔG_{mix} [90]. For a binary mixture such as IMC–PVP, the relationship between ΔG_{mix} and the interaction parameter, $\chi_{IMC-PVP}$, can be expressed by

$$\frac{\Delta G_{mix}}{RT} = \frac{\Delta H_{mix} - T\Delta S_{mix}}{RT} = n_{IMC} \ln \Phi_{IMC} + n_{PVP} \ln \Phi_{PVP} + n_{IMC}\Phi_{PVP}\chi_{IMC-PVP} \tag{13.14}$$

where Φ indicates a volume fraction and n refers to the number of moles of the indicated component. The first two terms in equation 13.4 constitute the entropy contribution, ΔS_{mix}, and are always favorable (i.e., $-T\Delta S_{mix}$ is negative), while the third term is related to the enthalpy of mixing, ΔH_{mix}. In MD simulations, the interaction parameter $\chi_{IMC-PVP}$ was determined from the following equation:

$$\chi_{IMC-PVP} = \frac{V_{site}\Delta H_{mix}}{RT} \tag{13.15}$$

where V_{site} was assumed to equal the size of a PVP monomer, R is the gas constant, and T is temperature (K). Because the component of ΔH_{mix} attributable to volume changes was less than 1%, ΔH_{mix} could be determined from cohesive energy densities where again the Φ values are volume fractions.

$$\Delta H_{mix} = \left(\frac{\Delta E_v}{V}\right)_{IMC-PVP} - \Phi_{IMC}\left(\frac{\Delta E_v}{V}\right)_{IMC} - \Phi_{PVP}\left(\frac{\Delta E_v}{V}\right)_{PVP} \tag{13.16}$$

The value of ΔH_{mix} averaged over the entire 100 ns simulation was -2.3 ± 0.9 cal/mol indicating favorable interaction between IMC and PVP. From this value, an interaction parameter of $\chi_{IMC-PVP} = -0.61 \pm 0.25$ was obtained at 298 K. The predicted free energies of mixing obtained from this $\chi_{IMC-PVP}$ are plotted in Figure 13.11 as a function of the volume fraction of PVP and compared to the results predicted from solubility parameters. Figure 13.11 illustrates various thermodynamic transfer quantities of interest. The free energy of transfer of IMC from amorphous IMC to an IMC–PVP mixture is indicated to be favorable by the relative levels of IMC and IMC–PVP in the figure. The use of χ interaction parameters is likely to be more reliable than the solubility parameter approach as it accounts for the favorable hydrogen bonding interactions between IMC and PVP. The prediction of complete miscibility over the entire composition range is consistent with experimental findings of a single T_g for IMC–PVP dispersions over the entire range of PVP concentrations [85d, 91]. The value of $\chi_{IMC-PVP}$ is expected to vary with composition and temperature, therefore the predictions in Figure 13.11 that are based on a value of $\chi_{IMC-PVP}$ determined at a single composition may need further refinement.

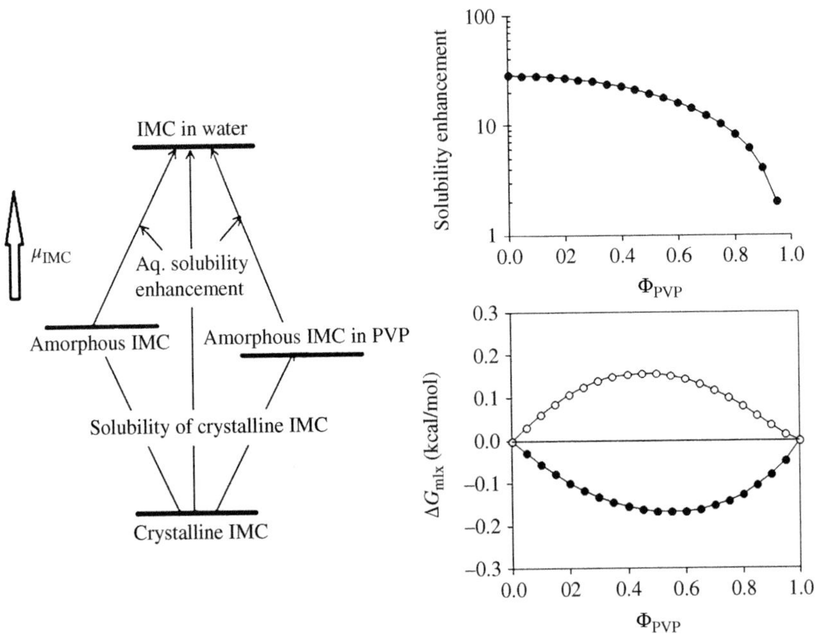

FIGURE 13.11 Left: Transfer energies (μ_{IMC}) of interest for amorphous IMC–PVP formulations. Right (Upper): Theoretical IMC aqueous solubility enhancement in simulated IMC–PVP relative to γ-crystalline IMC at 298 K. (Lower): Free energy of mixing of PVP with amorphous IMC (per mole of lattice sites) according to Flory–Huggins theory using $\chi_{IMC-PVP}$ obtained directly from MD simulations of the interaction energies (•) or from solubility parameters (°). Source: Xiang and Anderson [32]. Reproduced with permission of John Wiley & Sons.

Returning to Figure 13.11 illustrating the thermodynamic quantities of interest in the design of an amorphous dispersion, it was possible to estimate the enhancement in aqueous solubility of IMC likely as a function of composition as well as the solubility of crystalline IMC in PVP [32]. These calculations incorporated a previously reported experimental estimate for the difference in free energy between amorphous IMC and the pure crystalline γ-form, $\Delta G^{\circ}_{crystal \rightarrow amorphousIMC} = 1.99$ kcal/mol [75]. The free energy difference between the γ-crystalline form of IMC and amorphously dispersed IMC at a given PVP composition is as follows:

$$\Delta G^{\circ}_{crystal \rightarrow amorphousIMC/PVP} = \Delta G^{\circ}_{crystal \rightarrow amorphousIMC} + \frac{\partial \Delta G_{mix}}{\partial n_{IMC}} \tag{13.17}$$

$\partial \Delta G_{mix} / \partial n_{IMC}$ is the change in chemical potential for mixing pure amorphous IMC and PVP obtained by taking the derivative of ΔG_{mix} from Flory–Huggins theory with respect to n_{IMC} where V_{IMC} and V_{PVP} are the IMC and PVP molar volumes, respectively:

$$\frac{\partial \Delta G_{mix}}{\partial n_{IMC}} = RT\left[\ln\left(1-\Phi_{PVP}\right)+\left(1-\frac{V_{IMC}}{V_{PVP}}\right)\Phi_{PVP} + \Phi^{2}_{PVP}\chi_{IMC-PVP}\right] \tag{13.18}$$

The overall IMC solubility enhancement from an amorphous PVP dispersion compared to that for the γ-crystalline form of IMC is then:

$$S = \exp\left(\frac{\Delta G^{\circ}_{\text{crystallineIMC}\rightarrow\text{amorphousIMC/PVP}}}{RT}\right)$$
$$= \exp\left(\frac{1.99}{RT} + \ln\left(1-\Phi_{\text{PVP}}\right) + \left(1-\frac{V_{\text{IMC}}}{V_{\text{PVP}}}\right)\Phi_{\text{PVP}} + \Phi^2_{\text{PVP}}\chi_{\text{IMC-PVP}}\right) \quad (13.19)$$

The calculated solubility enhancement values are plotted as a function of composition in Figure 13.11. The solubility enhancement of pure amorphous IMC in comparison to the γ-crystalline form of 29-fold does not take into account the sorption of water and partial ionization of IMC that would also need to be included to match experimental conditions. Both of these factors would reduce the solubility enhancement, although the latter quantity would not be necessary to consider at the low pH found in the stomach. The plot illustrates that the solubility enhancement estimated from MD simulations remains relatively high at volume fractions of PVP less than about 0.5, declining only approximately 30% at equal volume fractions ($\Phi_{\text{IMC}} = \Phi_{\text{PVP}}$) from the value obtained for pure amorphous IMC but then dropping more precipitously at higher PVP volume fractions.

The solubility of crystalline IMC in PVP can also be estimated once a value of the χ interaction parameter is found. The key assumption is that at equilibrium the standard chemical potential for crystalline IMC equals that of IMC in an amorphous IMC–PVP dispersion ($\mu^{\circ\text{crystal}}_{\text{IMC}} = \mu^{\text{PVP}}_{\text{IMC}}$). The chemical potential of IMC in the amorphous dispersion can also be expressed as follows:

$$\mu^{\text{PVP}}_{\text{IMC}} = \mu^{\circ\text{crystal}}_{\text{IMC}} + \Delta G^{\circ}_{\text{crystal}\rightarrow\text{amorphousIMC}} + \frac{\partial \Delta G_{\text{mix}}}{\partial n_{\text{IMC}}} \quad (13.20)$$

This leads to the following:

$$\Delta G^{\circ}_{\text{crystal}\rightarrow\text{amorphousIMC}} + \frac{\partial \Delta G_{\text{mix}}}{\partial n_{\text{IMC}}} = 0 \quad (13.21)$$

Solving this equation using the value for $\Delta G^{\circ}_{\text{crystal}\rightarrow\text{amorphousIMC}} = 1.99$ from Murdande et al. [75] gave a solubility from crystalline γ-form of IMC of 27 mg/g. This illustrates that even for a dispersion in which amorphous IMC and PVP appear to be completely miscible, the system would be supersaturated with respect to the most stable crystalline form except at very low IMC concentrations.

13.5.3 Molecular Mobility and Small-Molecule Diffusion in Amorphous Dispersions

The translational and orientational mobility of small molecule APIs as well as other solutes such as water, oxygen, hydrogen peroxide, or low-molecular-weight acids or bases may influence both physical and chemical stability in pharmaceutical glasses.

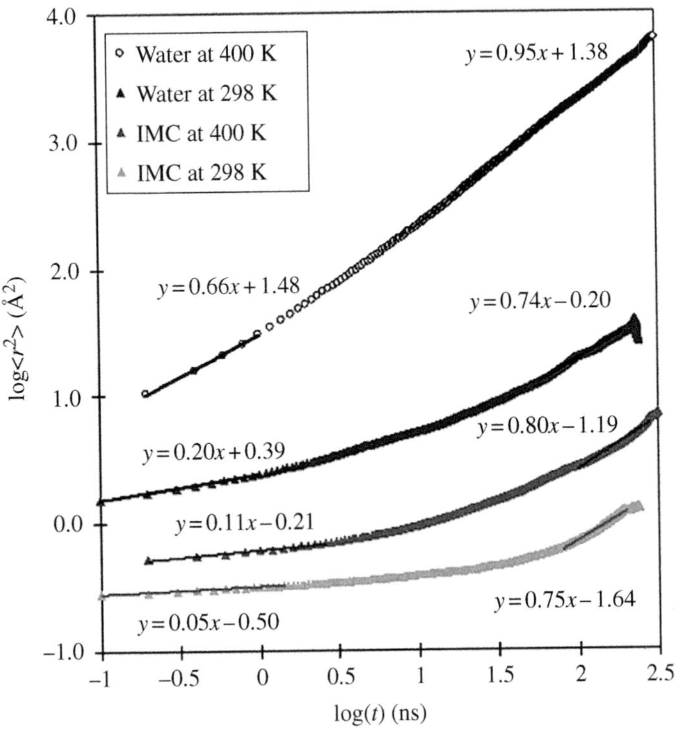

FIGURE 13.12 Log–log plots of mean squared displacements, $\langle r^2 \rangle$, versus time for water and IMC in simulated amorphous IMC at 298 and 400 K. Source: Xiang and Anderson [24b]. Reprinted with permission of American Chemical Society.

For example, water diffusion may facilitate chemical degradation by acting as a nucleophile in hydrolytic processes [44a–c], oxygen and hydrogen peroxide diffusion may influence the rates of oxidative processes as well as the products formed [45b, 92], and the diffusion of small-molecule acids and bases may be involved in physical instability such as salt disproportionation [93]. Through MD simulations, it is possible to monitor displacement trajectories of individual molecules over time ($|r(t) - r(0)|$) from which mean squared displacements $\left(\text{MSD} = \left\langle |r(t) - r(0)|^2 \right\rangle \text{or} \left\langle r^2 \right\rangle\right)$ can be generated for the entire ensemble.

The main limitation of fully atomistic simulations currently is the timescale that is feasible to explore in MD simulations, usually in the 10–100 ns range, in relation to the timescales for most relaxation and diffusion processes that are of interest in pharmaceutical systems. A representative example of the timescale problem is provided in Figure 13.12 where log–log plots of the mean squared displacements versus time for water and indomethacin diffusion in amorphous IMC containing 0.6% w/w water are shown at temperatures above T_g (i.e., at 400 K compared to the simulated T_g of 384 K) and well below T_g (i.e., 298 K).

According to the Einstein diffusion equation, the MSD for a randomly diffusing molecule in a given material should be proportional to time over a sufficiently long period of time, allowing an estimate of the molecule's diffusivity:

$$D = \lim_{t \to \infty} D_t = \lim_{t \to \infty} \frac{\left\langle \left| r(t_0 - t) - r(t_0) \right|^2 \right\rangle}{6t} \tag{13.22}$$

Log–log plots of MSD versus time should approach linearity with a slope of 1 for truly Fickian diffusion. Clearly, none of the plots in Figure 13.12 are linear over the entire time frame of these simulations although the diffusion of water in amorphous IMC at 400 K is nearly linear at later time points (simulations at 400 K were carried out for 350 ns) and the terminal slope is quite close to one. In the amorphous IMC glass at 298 K, neither MSD plot approaches a slope of one although both exhibit an increase in slope at later times. The mean displacement of water over the 230 ns simulation at 298 K was only approximately 5 Å, while all of the IMC molecules were virtually confined in a very small volume over the entire time. Even at 400 K, the MSD for IMC (=7.8 Å^2) corresponds to negligible displacement. Thus, extracting reliable diffusivities for drug molecules in amorphous glasses or even their melts may be beyond the capabilities of current fully atomistic simulations whereas such information does appear to be obtainable for small molecules such as water, at least under certain conditions.

To obtain estimates of diffusivity from the simulations, Xiang and Anderson [24b] adapted the Kohlrausch–Williams–Watts (KWW) stretched exponential function [94] to fit the D_t decay profiles using the following equation:

$$D_t = D + C\exp\left[-\left(\frac{t}{\tau}\right)^{\beta}\right] \tag{13.23}$$

The value of the diffusion coefficient for water in amorphous IMC at 298 K as determined from a fit of the simulated D_t versus time data to the above equation was $2.7 \times 10^{-9}\,\mathrm{cm^2/s}$, close to published experimental values of $0.9-2.0 \times 10^{-9}\,\mathrm{cm^2/s}$ obtained in melt-quenched IMC at the same temperature.

A similar approach has been employed to estimate diffusion coefficients of water at 298 K in simulated amorphous PLA solids [12b]. The value obtained in PLA from fitting simulation data to equation 13.23 adapted from the KWW stretched exponential function was $1.3 \times 10^{-8}\,\mathrm{cm^2/s}$, closer to reported experimental values than previous estimates from simulations [95]. In the PLA simulation study, a KWW stretched exponential function also provided a good fit of the average correlation functions for the rotational relaxation of PLA carbonyl groups. Remarkably, the KWW parameters generated from fits of the PLA carbonyl rotations ($\tau = 0.21$ ns and $\beta = 0.23$) could also describe the diffusivity profile for water over the 100 ns time frame of the simulation, suggesting that the non-Einstenian diffusion behavior of water was coupled to the local β-relaxation of the polymer.

A more comprehensive exploration of the molecular factors influencing the mobility of water was carried out by MD simulations conducted in PVP containing 0.5% water at 298 K by Xiang and Anderson [12a]. In this study, MD simulations of water's self-diffusion in bulk water were also conducted at the same temperature. The self-diffusion of water in bulk water at 298 K was Einsteinian over the 20 ns simulation resulting in a self-diffusion coefficient of 3.4×10^{-5} cm²/s which compared favorably to the value of 2.6×10^{-5} cm²/s reported in another simulation at 300 K [96] and reported experimental values (2.2×10^{-5} and 2.3×10^{-5} cm²/s) [97]. In PVP at 0.5% water, the diffusion of water was non-Einsteinian over the entire 100 ns time frame with a slope in the terminal portion of the log (MSD) versus log(t) plot of 0.89. While the diffusion of single water molecules in bulk water appeared to reflect continuous random Brownian motion, water's displacement in PVP was characterized by small back-and-forth "rattling" within confined spaces over relatively long periods with occasional large dislocations or "hopping" events to new locations. These hopping events are likely to reflect the jumping of water molecules from one free volume pocket to another (and sometimes back) within microdomains surrounded by the rigid polymer matrix. Such hopping events have been shown in some cases to be highly coordinated with changes in polymer configuration [98]. The diffusion coefficient for water in PVP at 0.5% water was estimated in MD simulations to be approximately 1×10^{-7} cm²/s, significantly higher than values reported experimentally [53, 99]. Oksanen et al., for example, reported a value of 6.5×10^{-9} cm²/s in PVP K90 at 25°C at 1.5% w/w water [99b]. This disparity may be attributable to the lower density (i.e., higher free volume) of the polymer glass in simulations due to the more rapid cooling. It may also reflect the fact that the simulation time (100 ns) was insufficient for water's diffusion to reach Einsteinian behavior and thus the apparent diffusion coefficient obtained was amplified by the localized "hopping" of water molecules between interconnected microdomains.

Given the earlier discussion of the importance of hydrogen bonding interactions between water and PVP carbonyl oxygen atoms, an interesting "experiment" performed in an MD simulation involved setting the partial charges on the hydrogen and oxygen atoms in water molecules to zero to explore the possible influence of hydrogen bonding of water to PVP carbonyl groups on diffusion [12a]. The apparent diffusivity of water molecules without partial charges increased by approximately sixfold in comparison to TIP3P water molecules with partial charges intact. Most of the jumps observed for either case were coupled with a jump back to the original position, suggesting that such dislocations involved rattling within a cage consisting of interconnected microdomains within the surrounding frozen polymer glass. While the frequency of jumps did not change, the length of jumps increased from 3.8 Å for water molecules with partial charges to 6.9 Å for water molecules with partial charges removed. A likely explanation for these results is that water molecules with partial charges are hydrogen bonded to carbonyl oxygen atoms of PVP side groups and the hopping events consist of jumps from one PVP carbonyl oxygen atom to an adjacent one. The hopping distance appears to be consistent with the observation by Busselez [35] from XRD patterns that the pyrrolidone side group centers of mass in PVP are approximately 5 Å apart. With partial charges removed, water molecules appear to have greater freedom of movement, although most jumps still reflect confinement within a relatively small "cage."

The critical importance of molecular size was clearly revealed in the PVP simulations by comparing the mobility of water (MW = 18) to that of a single tripeptide molecule (Phe-Asn-Gly, MW = 336) in the PVP glass [12a]. While water molecules exhibited a mean displacement of nearly 20 Å over the 100 ns time course of the simulation the larger tripeptide remained virtually frozen in place. Thus, while typical drug molecules are likely to be nearly immobilized within pharmaceutical glasses, smaller molecules such as water appear to be largely confined within small free volume cavities only over very short time frames but able to move more freely within glasses over longer timescales. Consistent with this generalization, Oksanen and Zografi noted based on absorption/desorption measurements that only 0.76 s would be required for a water molecule to diffuse 1 mm in PVP at 298 K [99b].

13.5.4 Plasticization by Water Clusters

Numerous studies have shown that water acts as a plasticizer when it is absorbed into amorphous solids, resulting in reductions in T_g [42a], decreases in NMR relaxation times [99b, 100], and increases in translational diffusion of water [99b] and other solutes [100b]. Plasticization of amorphous solids by water has also been a subject of several MD simulations where increasing water content has been associated with increases in polymer mobility and increases in diffusivities of water or other low-molecular-weight solutes [56, 61, 96, 101].

In PVP containing 10% w/w water, as shown previously in MD simulations (Fig. 13.5), water molecules tend to self-associate to form clusters or chains [24a]. This tendency for water to self-associate as its concentration increases has been reported in several other polymer glasses, including HPMCAS, polyimide [102], polyethylene [103], as well as in sugars above their T_g [96] and in monoglyceride/triglyceride lipid vehicles [2b].

In MD simulations, the apparent diffusion coefficient of water in PVP increased approximately fivefold at a water content of 10% w/w [24a], comparable to increases observed experimentally [99b]. A plot of water diffusivity in PVP at a water content of 10% w/w versus n_w, the average number of neighboring water molecules within 3.4 Å of a given water molecule, was found to be linear with a slope indicating an increase of 2.6×10^{-7} cm^2/s per neighboring water molecule. Mechanistically, these results suggest that the enhancement in water's diffusivity with increasing water content may be closely related to the formation of water clusters that expand the cavity size and mutually facilitate hopping events through hydrogen bonding and coordinated movements. Shown in Figure 13.13 are snapshots taken over an approximately 100 ps time span of a water hopping event in an MD simulation in PVP containing 10% w/w. During this period a water molecule (tagged in yellow) underwent a rapid displacement from one free volume pocket (on the right) to another located approximately 20 Å to the left. (The displacement of the tagged water molecule ($|r(t) - r(0)|$ Å) versus time is also shown in the accompanying panel where the starting time for the snapshots is marked with an arrow.) Within these two

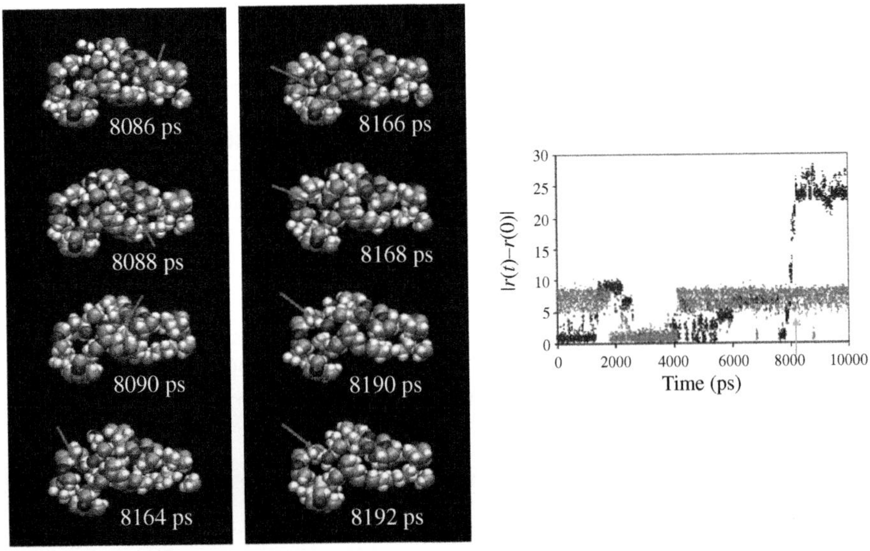

FIGURE 13.13 (Left): Snapshots of a water hopping event in simulated PVP containing 10% water over a approximately 100 ps time span. The yellow-tagged water molecule underwent a rapid displacement from one free volume pocket to another approximately 20 Å to the left. The red arrow tracks a second water molecule that co-migrated with the first water molecule. Right: The displacement of the yellow-tagged water molecule versus time is marked with an arrow. (*see insert for color representation of the figure.*)

interconnected microdomains were two water molecules, the yellow tagged water molecule, and a second water molecule shown in red and highlighted with a red arrow in each panel. Note that these two molecules tended to co-localize and appeared to remain hydrogen-bonded to each other or in close proximity much of the time. While this snapshot represents a single event, it may shed light on the observed increases in diffusivity in water as the number of water neighbors increases. The likelihood that breaking an HB between a water molecule and PVP carbonyl oxygen is necessary for a hopping event implies a significant energetic barrier that could be minimized by a coordinated translocation of two or more water molecules. This tandem migration could preserve hydrogen bonding in the transition state as passage from one carbonyl oxygen atom to the next occurs and may account for the increases in water diffusivity observed as a function of the number of neighboring water molecules.

Busselez et al. [56] used neutron scattering with H/D substitution labeling to examine the structure and dynamics of glassy PVP and water in hydrated PVP systems, from which they identified two types of water motion. Consistent with the earlier simulations, they obtained structural evidence for the existence of water clusters, nanosegregation of PVP side-groups, and swelling and disorder within ring nanodomains in the presence of water. The Q-dependence observed for the relaxation time of water molecules indicated that water relaxation was a diffusive-like process in water-rich domains while between water clusters subdiffusive motions prevailed.

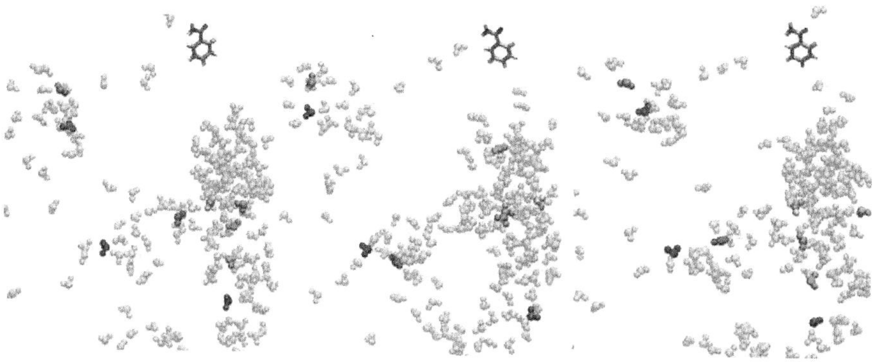

FIGURE 13.14 Three snapshots of water clusters at different times in a simulation of a water saturated 60% tricaprylin/40% monocaprylin lipid mixture. A few water molecules are color coded to allow tracking of their facile movement within water clusters. A single molecule of benzamide remains nearly stationary throughout this time frame. (*see insert for color representation of the figure.*)

A visual depiction of the organization and translation of water molecules within water clusters is shown in Figure 13.14 for an MD simulation of a 60% tricaprylin/40% monocaprylin lipid mixture containing an amount of water corresponding to that present at saturation (~6.5% wt/wt) [2b]. The water at this concentration was mostly found in clusters that were heterogeneously distributed and partially interconnected through water channels. A single drug-like solute molecule (benzamide) was also included in the simulation, as shown in each panel. A sequence of snapshots taken during an MD –simulation at 298 K illustrates the heterogeneous nature of water diffusion and its high relative mobility in comparison to the benzamide molecule which is virtually stationary. Water molecules tended to move relatively freely within water clusters as indicated by comparing the relative positions of those water molecules that are color-coded in the three panels while jumps of water molecules from one cluster to another were infrequent.

13.6 SUMMARY

With continuing advances in technology, the variety of amorphous system properties and processes being probed by MD simulations also continues to expand. In this chapter, we have hopefully demonstrated that MD simulations can provide useful insights relating to pure amorphous excipients, amorphous drugs, and drug–excipient mixtures that may complement experimental findings. While both thermodynamic and kinetic properties are of interest, the timescale accessible in full-atom simulations limits the types of mobility and diffusive-related processes that can be characterized. On the other hand, structural features and intermolecular interactions that underlie various thermodynamic properties of interest (e.g., drug solubility, amorphous miscibility, and water sorption isotherms) as revealed in simulations may have greater applicability, although one should always be cognizant of the fact that the simulated amorphous systems formed by cooling from a melt are always likely to be at higher energy than their experimental counterparts due to the rapid cooling rates in simulations. A great advantage that MD simulations of amorphous systems appear to have over experiments is that while experiments on such systems are often complicated by their metastable nature and limited to measurements of averages reflecting macroscopic or microscopic domains, simulations enable one to monitor single molecules, characterize distributions of intermolecular interaction types, accumulate information on conformational ensembles, etc., and relate this information to overall system energies. We are optimistic that advances in both computational and analytical techniques will further increase the relevance of MD simulations to real systems.

REFERENCES

[1] (a) Tejwani, R. W.; Davis, M. E.; Anderson, B. D.; Stouch, T. R., *J. Pharm. Sci.* 2011, *100*, 2136–46; (b) Tejwani, R. W.; Davis, M. E.; Anderson, B. D.; Stouch, T. R., *Mol. Pharm.* 2011, *8*, 2204–15; (c) Xiang, T.-X., *Biophys. J.* 1993, *65*, 1108–20; (d) Xiang, T.-X., *J. Phys. Chem. B* 1999, *103*, 385–94; (e) Xiang, T.-X.; Anderson, B. D., *Biophys. J.* 1994, *66*, 561–73; (f) Xiang, T.-X.; Anderson, B. D., *J. Chem. Phys.* 1999, *110*, 1807–18; (g) Xiang, T.-X.; Anderson, B. D., *Biophys. J.* 2002, *82*, 2052–66.

[2] (a) Rane, S. S.; Anderson, B. D., *Mol. Pharm.* 2008, *5*, 1023–36; (b) Rane, S. S.; Anderson, B. D., *Adv. Drug Deliv. Rev.* 2008, *60*, 638–56.

[3] Rahman, A., *Phys. Rev.* 1964, *136*, 405–11.

[4] (a) Allen, M. P., Introduction to Molecular Dynamics Simulation. In *Computational Soft Matter: From Synthetic Polymers to Proteins, Lecture Notes*; Attig, N.; Binder, K.; Grubmuller, H.; Kremer, K., Eds.; John von Neumann Institute for Computing: Juelich, NIC Series, Vol. *23*, 2004; pp 1–28; (b) Arbe, A.; Alvarezab, F.; Colmenero, J., *Soft Matter* 2012, *8*, 8257–70; (c) Kotelyanski, M.; Theodorou, D., *Simulation Methods for Polymers*. Marcel Dekker: New York, 2004; (d) Xiang, T.-X.; Anderson, B. D., *Adv. Drug Deliv. Rev.* 2006, *58*, 1357–78.

[5] Case, D. A.; Darden, T. A.; Cheatham, T. E., III; Simmerling, C. L.; Wang, J.; Duke, R. E.; Luo, R.; Walker, R. C.; Zhang, W.; Merz, K. M.; Roberts, B.; Hayik, S.; Roitberg, A.; Seabra, G.; Swails, J.; Goetz, A. W.; Kolossváry, I.; Wong, K. F.; Paesani, F.; Vanicek, J.; Wolf, R. M.; Liu, J.; Wu, X.; Brozell, S. R.; Steinbrecher, T.; Gohlke, H.; Cai, Q.; Ye, X.; Wang, J.; Hsieh, M.-J.; Cui, G.; Roe, D. R.; Mathews, D. H.; Seetin, M. G.; Salomon-Ferrer, R.; Sagui, C.; Babin, V.; Luchko, T.; Gusarov, S.; Kovalenko, A.; Kollman, P.A., *AMBER 12*. University of California: San Francisco, 2012.

[6] (a) Cornell, W. D.; Cieplak, P.; Bayly, C. I.; Gould, I. R.; Merz, K. M.; Ferguson, D. M.; Spellmeyer, D. C.; Fox, T.; Caldwell, J. W.; Kollman, P. A., *J. Am. Chem. Soc.* 1995, *117*, 5179–97; (b) Wang, J.; Wolf, R. M.; Caldwell, J. W.; Kollman, P. A.; Case, D. A., *J. Comput. Chem.* 2004, *25*, 1157–74.

[7] Sun, H., *J. Phys. Chem. B* 1998, *102*, 7338–64.

[8] MacKerell, A. J.; Bashford, D.; Bellott, M.; Dunbrack, R. J.; Evanseck, J. D.; Field, M. J.; Fischer, S.; Gao, J.; Guo, H.; Ha, S.; Joseph-McCarthy, D.; Kuchnir, L.; Kuczera, K.; Lau, F. T. K.; Mattos, C.; Michnick, S.; Ngo, T.; Nguyen, D. T.; Prodhom, B.; Reiher, W. E.; Roux, B.; Schlenkrich, M.; Smith, J. C.; Stote, J.; Straub, R.; Watanabe, M.; Wiorkiewicz-Kuczera, J.; Yin, D.; Karplus, M., *J. Phys. Chem. B* 1998, *102*, 3586–616.

[9] Mayo, S. L.; Olafson, B. D.; Goddard, W. A. *J. Phys. Chem.* 1990, *94*, 8897–909.

[10] Scott, W. R. P.; Huenenberger, P. H.; Tironi, I. G.; Mark, A. E.; Billeter, S. R.; Fennen, J. A.; Torda, E.; Huber, T.; Krueger, P.; van Gunsteren, W. F., *J. Phys. Chem. A* 1999, *103*, 3596–607.

[11] Jorgensen, W. L.; Maxwell, D. S.; Tirado-Rives, J., *J. Am. Chem. Soc.* 1996, *118*, 11225–36.

[12] (a) Xiang, T.-X.; Anderson, B. D., *J. Pharm. Sci.* 2004, *93*, 855–76; (b) Xiang, T.-X.; Anderson, B. D., *J. Pharm. Sci.* 2014, *103,* 2759–71; (c) Xiang, T.-X.; Anderson, B. D., *Mol. Pharm.* 2014, *11*, 2400–11.

[13] Baschnagel, J.; Varnik, F., *J. Phys. Condens. Matter* 2005, *17*, R851–953.

[14] Lyulin, S. V.; Gurtovenko, A. A.; Larin, S. V.; Nazarychev, V. M.; Lyulin, A. V., *Macromolecules* 2013, *46*, 6357–63.

[15] Donth, E., *The Glass Transition*. Springer: Berlin-Heidelberg, 2001.

[16] Hancock, B. C.; Shamblin, S. L.; Zografi, G., *Pharm. Res.* 1995, *12*, 799–806.

[17] Han, J.; Gee, R. H.; Boyd, R. H., *Macromolecules* 1994, *27*, 7781–4.

[18] Plante, A.; Palato, S.; Lebel, O.; Soldera, A., *J. Mater. Chem. C* 2013, *1*, 1037–42.

[19] Barrat, J.-L.; Baschnagel, J.; Lyulin, A., *Soft Matter* 2010, *6*, 3430–46.

[20] (a) Vollmayr, K.; Kob, W.; Binder, K., *Phys. Revi. B Condens. Matter* 1996, *54*, 15808–27; (b) Buchholz, J.; Paul, W.; Varnik, F.; Binder, K., *J. Chem. Phys.* 2002, *117*, 7364–72.

[21] Bruning, R.; Samwer, K., *Phys. Revi. B Condens. Matter* 1992, *46*, 11318–22.

[22] (a) Fulcher, G. S., *J. Am. Ceram. Soc.* 1925, *8*, 339–55; (b) Tamman, G.; Hesse, W., *Z. Anorg. Allg. Chem.* 1926, *156*, 245–57; (c) Vogel, H., *Phys. Z.* 1921, *22*, 645–6.

[23] Lyulin, A. V.; Balabaev, N. K.; Michels, M. A. J., *Macromolecules* 2003, *36*, 8574–5.

[24] (a) Xiang, T.-X.; Anderson, B. D., *Pharm. Res.* 2005, *22*, 1205–14; (b) Xiang, T.-X.; Anderson, B. D., *Mol. Pharm.* 2013, *10*, 102–14.

[25] (a) Hutchinson, J. M.; Montserrat, S., *Thermochim. Acta* 1995, *286*, 263–96; (b) Tribone, J. J.; O'Reilly, J. M.; Greener, J., *Macromolecules* 1986, *19*, 1732–9; (c) Liu, J.; Rigsbee, D. R.; Stotz, C.; Pikal, M. J., *J. Pharm. Sci.* 2002, *91* (8), 1853–62; (d) Rault, J., *J. Phys. Condens. Matter* 2003, *15*, S1193–213; (e) Slobodian, P.; Lengalova, A.; Saha, P., *Polym. J.* 2004, *36*, 176–81.

[26] (a) Bell, L. N.; Touma, D. E.; White, K. L.; Chen, Y.-H., *J. Food Sci.* 1998, *63*, 625–8; (b) Bell, L. N.; White, K. L., *J. Food Sci.* 2000, *65*, 498–501; (c) Shamblin, S. L.; Hancock, B. C.; Pikal, M. J., *Pharm. Res.* 2006, *23*, 2254–68.

[27] Gupta, J.; Nunes, C.; Jonnalagadda, S., *Mol. Pharm.* 2013, *10*, 4136–45.

[28] (a) Hancock, B. C.; Zografi, G., *Pharm. Res.* 1994, *11*, 471–7; (b) Taylor, L. S.; Langkilde, F. W.; Zografi, G., *J. Pharm. Sci.* 2001, *90*, 888–901; (c) Turner, D. T.; Schwartz, A., *Polymer* 1985, *26*, 757–62; (d) Zhang, J.; Zografi, G., *J. Pharm. Sci.* 2001, *90*, 1375–85.

[29] Couchman, P. R.; Karasz, F. E., *Macromolecules* 1978, *11*, 117–9.

[30] Friesen, D. T.; Shanker, R.; Crew, M.; Smithey, D. T.; Curatolo, W. J.; Nightingale, J. A., *Mol. Pharm.* 2008, *5*, 1003–19.

[31] Matsumoto, T.; Zografi, G., *Pharm. Res.* 1999, *16*, 1722–8.

[32] Xiang, T.-X.; Anderson, B. D., *J. Pharm. Sci.* 2013, *102*, 876–91.

[33] Crowley, K. J.; Zografi, G., *J. Pharm. Sci.* 2002, *90*, 2150–65.

[34] Bates, S.; Zografi, G.; Engers, D.; Morris, K.; Crowley, K.; Newman, A., *Pharm. Res.* 2006, *23*, 2333–49.

[35] Busselez, R.; Arbe, A.; Alvarez, F.; Colmenero, J.; Frick, B., *J. Chem. Phys.* 2011, *134*, 054904.

[36] (a) Alonzo, D. E.; Gao, Y.; Zhou, D.; Mo, H.; Zhang, G. G.; Taylor, L. S., *J. Pharm. Sci.* 2011, *100*, 3316–31; (b) Bühler, V., Soluble Polyvinylpyrrolidone (Povidone). In *Polyvinylpyrrolidone Excipients for Pharmaceuticals*, Springer: Berlin, 2005; pp 5–124; (c) Bühler, V., Registration and Toxicological Data. In *Polyvinylpyrrolidone Excipients for Pharmaceuticals*, Springer: Berlin, 2005; pp 221–5.

[37] Petkov, V.; Billinge, S. J. L.; Shastri, S. D.; Himmel, B., *J. Non-Cryst. Solids* 2001, *293–295*, 726–30.

[38] Arrese-Igor, S.; Arbe, A.; Alegria, A.; Colmenero, J.; Frick, B., *J. Chem. Phys.* 2004, *120*, 423–36.

[39] Narros, A.; Arbe, A.; Alvarez, F.; Colmenero, J.; Richter, D., *J. Chem. Phys.* 2008, *128*, 224905.

[40] Richter, D.; Frick, B.; Farago, B., *Phys. Rev. Lett.* 1988, *61*, 2465.

[41] Ahlneck, C.; Zografi, G., *Int. J. Pharm.* 1990, *62*, 87–95.

[42] (a) Oksanen, C. A.; Zografi, G., *Pharm. Res.* 1990, *7*, 654–7; (b) Hancock, B. C.; Zografi, G., *Pharm. Res.* 1993, *10*, 1262–7; (c) Zhang, J.; Zografi, G., *J. Pharm. Sci.* 2000, *89*, 1063–72; (d) Rumondor, A. C. F.; Konno, H.; Marsac, P. J.; Taylor, L. S., *J. Appl. Polym. Sci.* 2010, *117*, 1055–603; (e) Marsac, P. J.; Rumondor, A. C.; Nivens, D. E.; Kestur, U. S.; Stanciu, L.; Taylor, L. S., *J. Pharm. Sci.* 2010, *99*, 169–85.

[43] (a) Hageman, M. J., *Drug Dev. Ind. Pharm.* 1988, *14*, 2047–70; (b) Hageman, M. J., Water Sorption and Solid State Stability of Proteins. In *Stability of Protein Pharmaceuticals: Chemical and Physical Pathways of Protein Degradation*, Ahern, T. J.; Manning, M. C., Eds.; Plenum: New York, 1992, 273–309.

[44] (a) DeHart, M. P.; Anderson, B. D., *J. Pharm. Sci.* 2012, *101*, 3142–56; (b) Dehart, M. P.; Anderson, B. D., *J. Pharm. Sci.* 2012, *101*, 3096–109; (c) Dehart, M. P.; Anderson, B. D., *Pharm. Res.* 2012, *29*, 2722–37; (d) Strickley, R. G.; Anderson, B. D., *Pharm. Res.* 1996, *13*, 1142–53; (e) Strickley, R. G.; Anderson, B. D., *J. Pharm. Sci.* 1997, *86*, 645–53.

[45] (a) Qiu, Z.; Stowell, J. G.; Morris, K. R.; Byrn, S. R.; Pinal, R., *Int. J. Pharm.* 2005, *303*, 20–30; (b) Luo, D.; Anderson, B. D., *Pharm. Res.* 2006, *23*, 2239–53; (c) Luo, D.; Smith, S. W.; Anderson, B. D., *J. Pharm. Sci.* 2005, *94*, 304–16.

[46] Mucha, M.; Ludwiczak, S., Water Sorption by Biodegradable Chitosan/Polylactide Composites., Polish Chitin Society, 2007, Monograph XII, pp 41–8.

[47] Widom, B., *J. Chem. Phys.* 1963, *39*, 2808–12.

[48] (a) Jedlovszky, P.; Mezei, M., *J. Am. Chem. Soc.* 2000, *122*, 5125–31; (b) Marrink, S. J.; Berendsen, H. J. C., *J. Phys. Chem.* 1994, *98*, 4155–68.

[49] (a) Canales, M.; Aradilla, D.; Alemán, C., *Polym. Phys.* 2011, *49*, 1322–31; (b) Ostwal, M. M.; Sahimi, M.; Tsotsis, T. T., *Phys. Rev. E: Stat. Nonlin. Soft Matter Phys.* 2009, *79* (6 Pt 1), 061801; (c) Boshoff, J. H. D.; Lobo, R. F.; Wagner, N. J., *Macromolecules* 2001, *34*, 6107–16.

[50] Hermans, J.; Patbiaseril, A.; Anderson, A., *J. Am. Chem. Soc.* 1988, *110*, 5982–6.

[51] (a) Siparsky, G. L.; Voorhees, K. J.; Dorgan, J. R.; Schilling, K., *J. Environ. Polym. Degrad.* 1997, *5*, 125–36; (b) Yoon, J. S.; Jung, H. W.; Kim, M.-N.; Park, E.-S., *J. Appl. Polym. Sci.* 2000, *77*, 1716–22.

[52] Jin, Y.; Wang, W.; Su, Z., *Polym. Chem.* 2012, *3*, 2430–5.

[53] Rodriguez, O.; Fornasiero, F.; Arce, A.; Radke, C. J.; Prausnitz, J. M., *Polymer* 2003, *44*, 6323–33.

[54] Lebedeva, T. L.; Kuptsov, S. A.; Feldstein, M. M.; Plate, N. A., Molecular Arrangement of Water Associated with Poly (N-vinyl pyrrolidone) in the First Hydrate Shell. In *Water Transport in Synthetic Polymers*, Iordanskii, A. L.; Startsev, O. V.; Zaikov, G. E., Eds.; Topchiev Institute of Petrochemical Synthesis, Russian Academy of Sciences: Moscow, 2003; pp 69–93.

[55] Teng, J.; Bates, S.; Engers, D. A.; Leach, K.; Schields, P.; Yang, Y., *J. Pharm. Sci.* 2010, *99*, 3815–25.

[56] Busselez, R.; Arbe, A.; Cerveny, S.; Capponi, S.; Colmenero, J.; Frick, B., *J. Chem. Phys.* 2012, *137*, 084902.

[57] Curatolo, W.; Nightingale, J. A.; Herbig, S. M., *Pharm. Res.* 2009, *26*, 1419–31.

[58] (a) Konno, H.; Handa, T.; Alonzo, D. E.; Taylor, L. S., *Eur. J. Pharm. Biopharm.* 2008, *70*, 493–9; (b) Murdande, S. B.; Pikal, M. J.; Shanker, R. M.; Bogner, R. H., *J. Pharm. Sci.* 2011, *100*, 4349–56.

[59] Shah, N.; Iyer, R. M.; Mair, H. J.; Choi, D. S.; Tian, H.; Diodone, R.; Fahnrich, K.; Pabst-Ravot, A.; Tang, K.; Scheubel, E.; Grippo, J. F.; Moreira, S. A.; Go, Z.; Mouskountakis, J.; Louie, T.; Ibrahim, P. N.; Sandhu, H.; Rubia, L.; Chokshi, H.; Singhal, D.; Malick, W., *J. Pharm. Sci.* 2013, *102*, 967–81.

[60] Porter III, W. W.; O'Donnell, K.; Schmitt, B.; Petermann, O.; Guillaudeu, S.; Brackhagen, M.; Sprehe, M. Impact of HPMCAS Polymer Properties on the in vitro Performance of Spray Dried Dispersions., *AAPS Annual Meeting*, 2013, San Antonio, Texas.

[61] Karlsson, G. E.; Gedde, U. W.; Hedenqvist, M. S., *Polymer* 2004, *45* (11), 3893–900.

[62] (a) Crowley, K. J.; Zografi, G., *J. Pharm. Sci.* 2002, *91*, 492–507; (b) Surwase, S. A.; Boetker, J. P.; Saville, D.; Boyd, B. J.; Gordon, K. C.; Peltonen, L.; Strachan, C. J., *Mol. Pharm.* 2013, *10*, 4472–80.

[63] Patel, D. D.; Joguparthi, V.; Wang, Z.; Anderson, B. D., *J. Pharm. Sci.* 2011, *100*, 2623–41.

[64] (a) Hancock, B. C.; Zografi, G., *J. Pharm. Sci.* 1997, *86*, 1–12; (b) Leuner, C.; Dressman, J., *Eur. J. Pharm. Biopharm.* 2000, *50* (1), 47–60; (c) Serajuddin, A. T., *J. Pharm. Sci.* 1999, *88* (10), 1058–66; (d) Taylor, L. S.; Zografi, G., *Pharm. Res.* 1997, *14*, 1691–8.

[65] Hancock, B. C.; Parks, M., *Pharm. Res.* 2000, *17*, 397–404.

[66] Chen, X.; Morris, K. R.; Griesser, U. J.; Byrn, S. R.; Stowell, J. G., *J. Am. Chem. Soc.* 2002, *124*, 15012–9.

[67] (a) Aubrey-Medendorp, C.; Swadley, M. J.; Li, T., *Pharm. Res.* 2008, *25*, 953–9; (b) Masuda, K.; Tabata, S.; Sakata, Y.; Hayase, T.; Yonemochi, E.; Terada, K., *Pharm. Res.* 2005, *22*, 797–805.

[68] Fronza, G.; Mele, A.; Redenti, E.; Ventura, P., *J. Org. Chem.* 1996, *61*, 909–14.

[69] (a) Savolainen, M.; Heinz, A.; Strachan, C.; Gordon, K. C.; Yliruusi, J.; Rades, T.; Sandler, N., *Eur. J. Pharm. Sci.* 2007, *30*, 113–23; (b) Swallen, S. F.; Kearns, K. L.; Mapes, M. K.; Kim, Y. S.; McMahon, R. J.; Ediger, M. D.; Wu, T.; Yu, L.; Satija, S., *Science* 2007, *315*, 353–6; (c) Dawson, K. J.; Kearns, K. L.; Ediger, M. D.; Sacchetti, M. J.; Zografi, G. D., *J. Phys. Chem. B* 2009, *113*, 2422–7; (d) Kaarsholm, N. C.; Ludvigsen, S., *Receptor* 1995, *5*, 1–8; (e) Karmwar, P.; Graeser, K.; Gordon, K. C.; Strachan, C. J.; Rades, T., *Int. J. Pharm.* 2011, *417* (1–2), 94–100; (f) Kearns, K. L.; Swallen, S. F.; Ediger, M. D.; Wu, T.; Yu, L., *J. Chem. Phys.* 2007, *127* (15), 154702.

[70] (a) Carpentier, L.; Decressain, R.; Desprez, S.; Descamps, M., *J. Phys. Chem. B* 2006, *110*, 457–64; (b) Yoshioka, M.; Hancock, B. C.; Zografi, G., *J. Pharm. Sci.* 1994, *83*, 1700–5.

[71] X. Yuan, T.-X.; Xiang, B.D.; Anderson, and E.J. Munson, *Mol. Pharmaceutics.* 2015, (DOI 10.1021/acs.molpharmaceut.5b00705)

[72] Yuan, X.; Munson, E. J. AAPS Annual Meeting and Exposition, 2013, Poster No. W4203.

[73] Surana, R.; Pyne, A.; Suryanarayanan, R., *Pharm. Res.* 2004, *21*, 1167–76.

[74] Bhardwaj, S. P.; Suryanarayanan, R., *Pharm. Res.* 2013, *30*, 714–20.

[75] Murdande, S. B.; Pikal, M. J.; Shanker, R. M.; Bogner, R. H., *J. Pharm. Sci.* 2010, *99*, 1254–64.

[76] (a) Bhugra, C.; Pikal, M. J., *J. Pharm. Sci.* 2008, *97*, 1329–49; (b) Bhugra, C.; Rambhatla, S.; Bakri, A.; Duddu, S. P.; Miller, D. P.; Pikal, M. J.; Lechuga-Ballesteros, D., *J. Pharm. Sci.* 2007, *96*, 1258–69; (c) Bhugra, C.; Shmeis, R.; Krill, S. L.; Pikal, M. J., *Pharm. Res.* 2006, *23*, 2277–90.

[77] Newman, A.; Knipp, G.; Zografi, G., *J. Pharm. Sci.* 2012, *101*, 1355–77.

[78] Alonzo, D. E.; Zhang, G. G.; Zhou, D.; Gao, Y.; Taylor, L. S., *Pharm. Res.* 2010, *27*, 608–18.

[79] Breitenbach, J., *Eur. J. Pharm. Biopharm.* 2002, *54*, 107–17.

[80] Dobry, D. E.; Settell, D. M.; Baumann, J. M.; Ray, R. J.; Graham, L. J.; Beyerinck, R. A., *J. Pharm. Innov.* 2009, *4*, 133–42.

[81] Shamblin, S. L.; Taylor, L. S.; Zografi, G., *J. Pharm. Sci.* 1998, *87*, 694–701.

[82] (a) Marsac, P. J.; Li, T.; Taylor, L. S., *Pharm. Res.* 2009, *26*, 139–51; (b) Marsac, P. J.; Shamblin, S. L.; Taylor, L. S., *Pharm. Res.* 2006, *23*, 2417–26; (c) Sun, Y.; Tao, J.; Zhang, G. G.; Yu, L., *J. Pharm. Sci.* 2010, *99*, 4023–31; (d) Tao, J.; Sun, Y.; Zhang, G. G.; Yu, L., *Pharm. Res.* 2009, *26*, 855–64; (e) Paudel, A.; Van Humbeeck, J.; Van den Mooter, G., *Mol. Pharm.* 2010, *7*, 1133–48; (f) Tian, Y.; Booth, J.; Meehan, E.; Jones, D. S.; Li, S.; Andrews, G. P., *Mol. Pharm.* 2013, *10*, 236–48.

[83] (a) Rumondor, A. C.; Ivanisevic, I.; Bates, S.; Alonzo, D. E.; Taylor, L. S., *Pharm. Res.* 2009, *26*, 2523–34; (b) Rumondor, A. C.; Marsac, P. J.; Stanford, L. A.; Taylor, L. S., *Mol. Pharm.* 2009, *6*, 1492–505; (c) Rumondor, A. C.; Taylor, L. S., *Mol. Pharm.* 2010, *7*, 477–90.

[84] (a) Gupta, J.; Nunes, C.; Vyas, S.; Jonnalagadda, S., *J. Phys. Chem. B* 2011, *115*, 2014–23; (b) Huynh, L.; Grant, J.; Leroux, J. C.; Delmas, P.; Allen, C., *Pharm. Res.* 2008, *25*, 147–57; (c) Yang, H.; Ze-Sheng, L.; Qian, H.-J.; Yang, Y.-B.; Zhang, X.-B.; Sun, C.-C., *Polymer* 2004, *45*, 453–7.

[85] (a) Lu, Q.; Zografi, G., *Pharm. Res.* 1998, *15*, 1202–6; (b) Newman, A.; Engers, D.; Bates, S.; Ivanisevic, I.; Kelly, R. C.; Zografi, G., *J. Pharm. Sci.* 2008, *97*, 4840–56; (c) Reddy, R.; Chang, L. L.; Luthra, S.; Collins, G.; Lopez, C.; Shamblin, S. L.; Pikal, M. J.; Gatlin, L. A.; Shalaev, E. Y., *J. Pharm. Sci.* 2009, *98*, 81–93; (d) Yoshioka, M.; Hancock, B. C.; Zografi, G., *J. Pharm. Sci.* 1995, *84*, 983–6.

[86] Pham, T. N.; Watson, S. A.; Edwards, A. J.; Chavda, M.; Clawson, J. S.; Strohmeier, M.; Vogt, F. G., *Mol. Pharm.* 2010, *7*, 1667–91.

[87] Hildebrand, J. H., *J. Am. Chem. Soc.* 1916, *38*, 1452–73.

[88] Hansen, C. M., *Ind. Eng. Chem. Prod. Res. Dev.* 1969, *8*, 2–11.

[89] Forster, A.; Hempenstall, J.; Tucker, I.; Rades, T., *Int. J. Pharm.* 2001, *226*, 147–61.

[90] (a) Flory, P. J., *J. Chem. Phys.* 1941, *9*, 660–1; (b) Huggins, M. L., *J. Chem. Phys.* 1941, *9*, 440.

[91] Chokshi, R. J.; Sandhu, H. K.; Iyer, R. M.; Shah, N. H.; Malick, A. W.; Zia, H., *J. Pharm. Sci.* 2005, *94*, 2463–74.

[92] Luo, D.; Anderson, B. D., *J. Pharm. Sci.* 2008, *97*, 3907–26.

[93] Rohrs, B. R.; Thamann, T. J.; Gao, P.; Stelzer, D. J.; Bergren, M. S.; Chao, R. S., *Pharm. Res.* 1999, *16*, 1850–6.

[94] (a) Kohlrausch, R., *Pogg. Ann. Phys. Chem.* 1854, *91*, 179–214; (b) Williams, G.; Watts, D. C., *Trans. Faraday Soc.* 1970, *66*, 80–5; (c) Williams, G.; Watts, D. C.; Dev, S. B.; North, A. M., *Trans. Faraday Soc.* 1971, *67*, 1323–35.

[95] (a) Entrialgo-Castaño, M.; Salvucci, A. E.; Lendlein, A.; Hofmann, D., *Macromol. Symp.* 2008, *269*, 47–64; (b) Gautieri, A.; Mezzanzanica, A.; Motta, A.; Redealli, A.; Vesentini, A., *J. Mol. Model.* 2012, *18*, 1495–502.

[96] Roberts, C. J.; Debenedetti, P. G., *J. Phys. Chem. B* 1999, *103*, 7308–18.

[97] Hayduk, W.; Laudie, H., *AIChE J.* 1974, *20*, 611–5.

[98] Xi, L.; Shah, M.; Trout, B. L., *J. Phys. Chem. B* 2013, *117*, 3634–47.

[99] (a) Chang, M. J.; Myerson, A. S.; Kwei, T. K., *J. Appl. Polym. Sci.* 1997, *66*, 279–91; (b) Oksanen, C. A.; Zografi, G., *Pharm. Res.* 1993, *10*, 791–9.

[100] (a) Luthra, S. A.; Shalaev, E. Y.; Medek, A.; Hong, J.; Pikal, M. J., *J. Pharm. Sci.* 2012, *101*, 3110–23; (b) Aso, Y.; Yoshioka, S.; Zhang, J.; Zografi, G., *Chem. Pharm. Bull.* 2002, *50*, 822–6.

[101] Bermejo, J. S.; Ugarte, C. M., *J. Chem. Phys.* 2008, *129*, 154907.

[102] Marque, G.; Neyertz, S.; Verdu, J.; Prunier, V.; Brown, D., *Macromolecules* 2008, *41*, 3349–62.

[103] Fukuda, M., *J. Chem. Phys.* 1998, *109*, 6476–85.

14

NUMERICAL SIMULATIONS OF UNIT OPERATIONS IN PHARMACEUTICAL SOLID DOSE MANUFACTURING

EKNEET KAUR SAHNI[2], SHIVANGI NAIK[4], AND BODHISATTWA CHAUDHURI[1,3]

[1] *Department of Pharmaceutical Sciences, University of Connecticut, Storrs, CT, USA*
[2] *Pfizer Manufacturing Science and Technology, McPherson, KS, USA*
[3] *Institute of Material Sciences, University of Connecticut, Storrs, CT, USA*
[4] *Research and Development, Kamat Pharmatech, NJ, USA*

14.1 INTRODUCTION

Solid dose manufacturing is a critical step in procuring safe medications as it comprises sequence of complicated operations before the drug product is marketed. However, each unit operation is expected to modify the material characteristics resulting in the desired properties of the final product. Moreover, minimizing the production cost without compromising the product quality is another challenge in manufacturing. An improved understanding of these unit operations with the aid of first-principle models would help in circumventing some of the process development concerns (discarded batches and delayed development). Process modeling approaches are particularly beneficial for such intricate processes since predictions can readily be made from simulations that would be difficult to obtain experimentally. In addition to parametric studies, the effect of equipment design on a given process can also be examined. The focus of this chapter is limited to three unit operations: contact drying [1, 2], film coating [3, 4], and milling [5–8]. Hence, the objective of this study was to elucidate and address the effect of process variables and materials properties using

Computational Pharmaceutical Solid State Chemistry, First Edition. Edited by Yuriy A. Abramov.

experimentally validated discrete element model (DEM) for all the unit operations. As DEM explicitly considers interparticle and particle–boundary interactions along with kinematics of primary and secondary particles it provides an effective tool to model pharmaceutical unit operations such as drying, coating, and fragmentation processes.

14.2 NUMERICAL METHOD

The DEM, a popular tool to simulate particle dynamics, was originally developed by Cundall and Strack [9]. The granular material is considered as a collection of frictional inelastic spherical particles. The contact model used in this work is based on the work of Walton and Braun [10, 11] where each particle may interact with its neighbors or with the boundary only at the contact points through normal and tangential forces. Details of DEM algorithm can be found in the earlier publication of the authors [12, 13]. The forces and torques acting on each of the particles are calculated in the following way:

$$\Sigma F_i = m_i g + F_{ij}^{\mathrm{N}} + F_{ij}^{\mathrm{T}} \tag{14.1}$$

$$\Sigma T_i = R_i \, X \, F_{ij}^{\mathrm{T}} \tag{14.2}$$

The normal forces (F_{ij}^{N}) and the tangential forces $\left(F_{ij}^{\mathrm{T}}\right)$ in inter-particle or particle–wall collision are calculated using the "latching spring model" and "incrementally slipping model," respectively, developed by Walton and Braun [10, 11]. The force on each particle is given by the sum of gravitational and interparticle (normal and tangential), as indicated in Equation 14.1. The corresponding torque on each particle is the sum of the moment of the tangential forces arising from interparticle contacts (Eq. 14.2).

14.2.1 Contact Drying in an Agitated Filter Dryer

14.2.1.1 Model Assumptions Numerical model developed is based on the following assumptions:

1. Interstitial gas is neglected.
2. Solvent is initially equally distributed within the particles of the granular bed.
3. Boundary wall and impeller temperature remains constant.
4. Physical properties such as heat capacity, thermal conductivity and Young's modulus are constant.
5. During each simulation time step the temperature is uniform in each particle (Biot's number well below unity).

14.2.1.2 Incorporation of Liquid Bridge Force Model This force model for simulating the forces due to the formation of liquid bridge within particles in a wet granular bed was taken from previous publication of the authors as shown in Figure 14.1 [12].

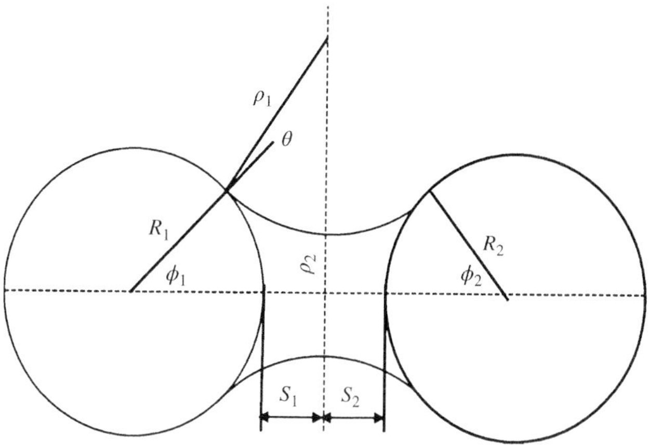

FIGURE 14.1 Toroidal approximation of a liquid bridge formed between two spherical particles. Adapted from Ref. [12] with permission. Copyright 2013, Elsevier.

The forces due to the liquid bridge comprise attractive capillary force and repulsive viscous force. The algorithms for the calculation of these forces were implemented in the DEM model along with the collisional forces that arise during interparticle contact.

14.2.1.3 Modeling Heat Transfer Thermal particle dynamics (TPD) primarily introduced by Vargas and McCarthy [14] incorporated both the contact mechanics and contact conductance theories to model the flow dynamics and heat conduction through dry granular materials. The details of the model can be found in Sahni et al. [12]. Heat transport is simulated accounting for the initial material temperature, wall temperature, heat capacity, heat transfer coefficient, and flow properties using a linear model. The flux of heat transported across the mutual boundary between two particles i and j in contact is described as follows:

$$Q_{ij} = H_c\left(T_j - T_i\right) \tag{14.3}$$

Here, T_i and T_j are the temperatures of two particles and the inter-particle conductance H_c [14] is given by

$$H_c = 2k_s \sqrt[3]{\left[\frac{3F_N r^*}{4E^*}\right]} = 2a \tag{14.4}$$

where k_s is the thermal conductivity of the solid material, E^* is effective young's modulus for the two particles, and r^* is the geometric mean of the particle radii. The evolution of temperature of particle i from its neighbor j is

$$\frac{dT_i}{T_j - T_i} = \frac{Q_i}{\rho_i C_i V_i} \tag{14.5}$$

Here Q_i is the sum of all heat fluxes involving particle i and $\rho_i C_i V_i$ is the thermal capacity of particle i. The change in mass of the solvent is computed from the coupled heat and mass transfer equation as

$$Q\,dt = m\Delta H_v + m_s c\Delta T \tag{14.6}$$

where Q is the total heat, m is the mass of the solvent, ΔH_v is the heat of vaporization, m_s is the mass of the sample, c is the specific heat capacity, and ΔT is the change in temperature. At the boiling point, ΔT goes to zero making the second term zero and the remaining equation will then be used to calculate the change in the mass of the solvent. Hence, there is no contribution from the second term at the boiling point in the model.

14.2.2 Coating in a Conventional Pan Coater

14.2.2.1 Model Assumptions The numerical model for coating was developed based on the following assumptions:

1. Presence of coating fluid was not considered.
2. Physical properties such as Young's modulus and density were kept constant during the simulation.

As very small amount of coating solution was introduced into the bed that dried immediately, the DEM model developed simply deals with a dry granular bed. Instead, the particles were color-coded based on their residence time under the spray zone for visual purposes.

DEM simulations provide the temporal variation of velocity, orientation, and position of each particle within the coater, allowing true measurements of the time that each particle spends in the spray zone. The particles were assigned random initial nonoverlapping positions and then allowed to settle inside the coater under the force of gravity. This initial configuration was equilibrated for a period of 5 s and then the coater boundary begins to rotate at a fixed revolution per minute (rpm) value. In DEM as the 3D position of every particle was known, it was easy to perform sampling of particles from specified locations and then counting the number of each species. A circular spray zone on the surface of the granular bed was assumed. The three-dimensional coordinate of the center of the spray and the radius of the spray zone on the granular bed ascertained from the experiments was used to determine the residence time distribution of all the particles in the bed from the particle trajectory produced by the simulation. The coating variability was therefore dependent on the residence time spent by the nonpareils under the spray zone. Any variation in residence time was reflected as variation in the coating distribution of a coated batch of particles. Simulation of coating process was achieved by postprocessing the particle dynamics data.

DEM simulations were performed using 40,000 particles (corresponding to 21% fill fraction) where the spray duration was 1 s and the time difference between the initial and the next spray was 3 s. The residence time for each particle was subsequently estimated. The particles were color coded based on the residence time in the spray zone for the interpretation of coating in the simulation. Quantities estimated in each run of the parametric study include the total number of particles passed through the spray zone and frequency distribution of the residence time of the coated particles. The following parameters were investigated:

1. *Vessel tilt*: 0°, 16°, and 32°
2. *Vessel speed*: 10 rpm, 20 rpm, and 30 rpm

14.2.3 Modeling of milling in a Wiley Mill

A more fundamental approach toward modeling comminution was by the application of fracture mechanics to particle size reduction. In essence, an energy balance is applied to the process of crack extension within a particle by equating the loss of energy from the strain field within the particle to the increase in surface energy when the crack propagates. From the concepts of fundamentals of crack propagation, Grady derived a model of dynamic fragmentation [15–17]. Based on Grady's model, the particle size of fragment following fracture is given by Equation 14.7:

$$D = \frac{4.472 \times K_{\mathrm{lc}}}{\rho V_{\mathrm{p}} S_{\mathrm{r}}} \tag{14.7}$$

D is the diameter of the resultant progeny particle, K_{lc} is fracture toughness of material, (Pa·m$^{0.5}$), V_{p} is propagation velocity (m/s) of longitudinal elastic waves in the material, S_{r} is the induced strain rate. The strain rate in the sample is defined as the difference in particle velocity at the point of impact and at the opposite end of the sample.

The strain rate is calculated from the rotational velocity of the impeller and the radial distance between the center of the rotation and the impact point [18].

$$S_{\mathrm{r}} = \frac{V_{\mathrm{pp1}} - V_{\mathrm{pp2}}}{L} \tag{14.8}$$

where V_{pp1} and V_{pp2} are particle peak velocities at the impact point and at the free end of the particle, respectively, and L is the radius of the particle sample (m). The peak velocity V_{pp1} at the interface of the impeller and particle is equal to the impeller velocity V_{i}, and that V_{pp2} at the free end of the particle sample is of the opposite sign of V_{pp1}, thus the strain rate can be represented as follows:

$$S_{\mathrm{r}} = \frac{2V_i}{L}$$

Using DEM, the breakage algorithm was modified such that the particle breaks only when the contact forces between granule/hammer and granule/wall exceed the breakage force obtained from DMA studies [18].

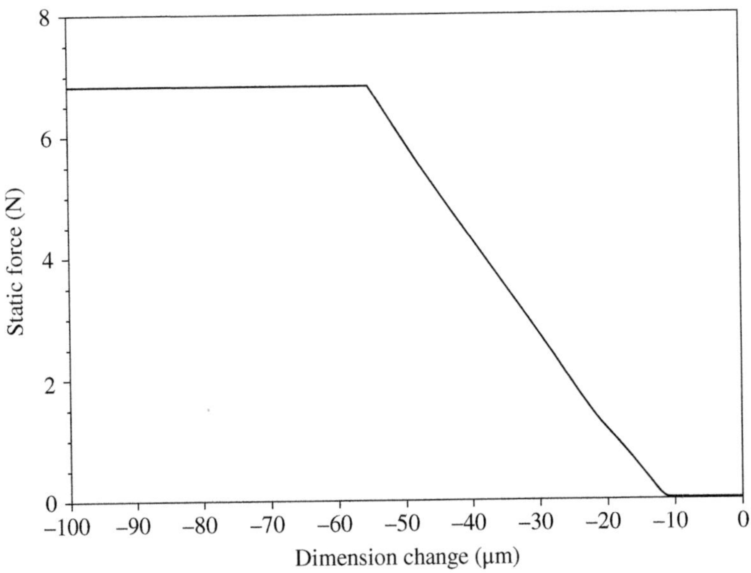

FIGURE 14.2 Breakage of lactose granules under static compression. Adapted from Ref. [18] with permission. Copyright 2013, Elsevier.

14.3 EXPERIMENTAL METHOD FOR MILLING

Single-particle breakage studies were performed in Dynamic Mechanical Analyzer (DMA) from TA instruments to determine the breakage forces to be used in DEM simulation of the milling process. From the DMA studies the breakage force (Fig. 14.2) was found to be 6.5 ± 0.5 N. Additional studies for creep and high strain effects revealed no damage accumulation for the material under investigation.

To facilitate model verification, experiments have been performed under similar conditions. The milling equipment (Thomas Wiley Mill, Thomas Scientific, Swedesboro, NJ) used in this study is a variable speed, digitally controlled, direct drive mill that provides continuous variation of cutting speeds from 600 to 1140 rpm with constant torque maintained throughout the speed range. Parametric studies were conducted with lactose granules to study the effect of impeller wall tolerance and feed rate on particle size reduction.

14.4 RESULTS AND DISCUSSION

14.4.1 Simulation of Contact Drying

To elucidate the effect of various parameters, TPD simulations were performed under similar operative conditions as that of the experiments [12] with variation of one single parameter while keeping the others unchanged. In order to maintain the constant condition of 0.2 bar pressure, the boiling point of ethanol was set to 45°C. Color-coding as specified before was done in order to visually track the evolution of particle temperature in simulations. Particles with temperature equal to 293 K were colored dark blue; between, 293 and 295 K dyed sky blue; 295 and 297 K colored cyan; 297 and 299 K as bottle green, 299–301 K as bright green, 301–303 K as light green, 303–305 K as yellow, and 305–307 K mustard orange. Particles with and above 307 K were colored red.

14.4.1.1 Influence of Wall Temperature To study the effect of different wall temperature, the simulations were implemented at 5 rpm impeller speed for 45% fill load. Simulations were operated under similar conditions as that of experiments in order to establish a correlation by checking the trends. Simulating the flow and heat transfer of 67,200 particles accommodated the percentage fill for lactose–ethanol system similar to the experimental study [12]. The thermo-physical properties of lactose monohydrate and ethanol used in the model are stated in Tables 14.1 and 14.2, respectively. Figure 14.3a–c depicts the matrix of axial snapshots to compare the temperature distribution for 318 K, 328 K, and 338 K (45°C, 55°C, and 65°C), respectively at T=0, 10, and 20 s from left to right. The temperature of the bed increases with time, and this increase is much faster for a higher jacket temperature. The cold blue (T=0 s) color changes to warmer versions of blue. Further, for 318 K a green layer can be seen propagating which exhibits a higher temperature than the blue zones. It can be noticed that this green zone increases for higher

TABLE 14.1 Thermo-Physical Properties of the Model Compounds

Property	Glassbeads	Lactose
Number of particles	9,000–25,000	40,000–100,000
Density	2,500 kg/m^3	1985.7 kg/m^3
Poisson's ratio	0.12	0.2
Coefficient of restitution	PP: 0.7	PP: 0.5
	PW: 0.5	PW: 0.4
Coefficient of friction	PP: 0.4	PP: 0.7
	PW: 0.2	PW: 0.5
Spring's constant	6,000 N/m	6,000 N/m
Young's modulus	2.6 * 10^6 Pa	1.7 * 10^7 Pa
Thermal conductivity	1.1 W/mK	0.9 W/mK
Thermal diffusivity	5.1 * 10^{-7} m^2/s	1.25 * 10^{-7} m^2/s
Time step	1.25 * 10^{-6}	1.25 * 10^{-6}

TABLE 14.2 Input Properties for Ethanol

Property	Glassbeads
Heat of vaporization	855 kJ/kg
Viscosity	0.0012 Pa·s
Surface tension	0.022 N/m
Thermal conductivity	0.2 W/mK

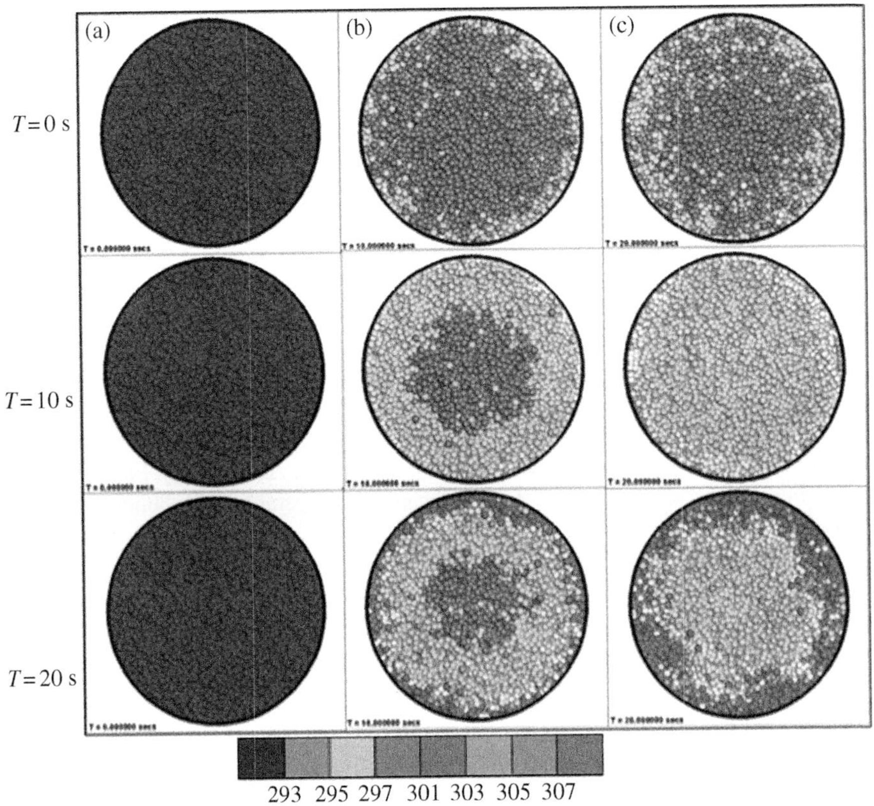

FIGURE 14.3 Shows the snapshots comparing the temperature profile for the effect of wall temperature (a) 318 K, (b) 328 K, and (c) 338 K, for $T=0$, 10, and 20 s; $\omega=5$ rpm and $f=45\%$. Adapted from Ref. [12] with permission. Copyright 2013, Elsevier. (*see insert for color representation of the figure.*)

wall temperatures of 328 and 338 K. For 338 K, as the bed temperature will be much higher at the same time point due to higher driving force (T_w-T_b), red zones also start to appear at $T=10$ s with temperature greater than or equal to 307 K. Moreover as the green, yellow, and red colored particles start to appear, the blue zones seem to decrease with time.

Figure 14.4a and 14.4b depicts the average bed temperature and normalized solvent content plots, respectively, for the same conditions till time, $T=30$ s. As the temperature of the bed increases with wall temperature (Fig. 14.4a), solvent content decreases correspondingly to a higher drying rate (Fig. 14.4b).

14.4.1.2 Influence of Dryer Load The effect of dryer load was studied for three different fills: 25%, 45%, and 65% which were simulated under fixed operating conditions with 42,600; 67,200, and 91,800 particles, respectively. Input factors include 5 rpm impeller speed with the jacket heated at 338 K. Figure 14.5a and b shows the temporal variation of average bed temperature and normalized solvent concentration respectively for the lactose–ethanol system for the abovementioned simulation conditions. The bed

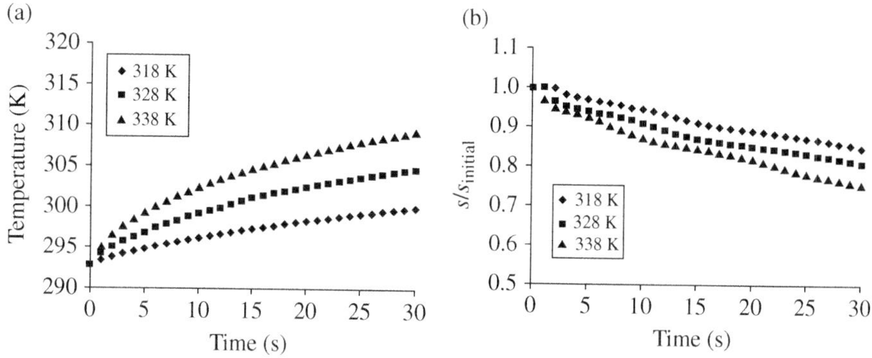

FIGURE 14.4 Shows the effect of temperature with (a) bed temperature profile and (b) loss of solvent content with time; ω=5rpm and f=45%. Adapted from Ref. [12] with permission. Copyright 2013, Elsevier.

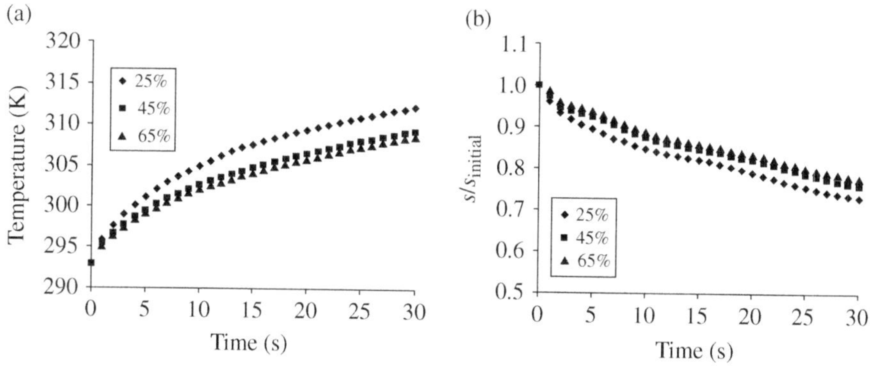

FIGURE 14.5 Illustrates the effect of bed depth by showing the (a) temperature profile of the bed and (b) loss of solvent content; ω = 5 rpm and T_w = 338 K. Adapted from Ref. [12] with permission. Copyright 2013, Elsevier.

temperature rises quicker for lower fill. Correspondingly the decrease in the solvent is much faster as for lower fill volume as temperature distribution becomes uniform with time.

14.4.1.3 Influence of Impeller Speed Agitation is known to improve the heat and mass transfer rates by enhancing mixing of particles and the particle renewal rates at the heat transfer surface. Moreover, most solids are thermolabile, which further limits the drying temperature at which they can be operated. Hence, the role of agitation becomes even more important in such situations. This effect of agitation speed was studied at 353 K wall temperature and 45% fill. Figure 14.6a and b shows the average bed temperature and normalized solvent content plots, respectively, for the same conditions till T=30 s. Figure 14.6a illustrates that the temperature of the bed increases with increase in speed and subsequently the solvent content dries out faster for a higher impeller speed of 25 rpm. As expected, the solvent drying is faster for 25 rpm. Increasing the speed resulted in an increase in the drying rate and therefore reduction in the overall drying time.

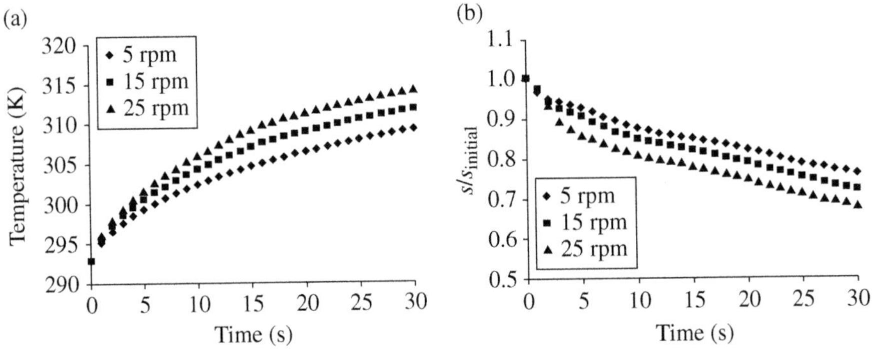

FIGURE 14.6 Illustrates the impact of speed using the temporal profiles of (a) bed temperature and (b) solvent content for different agitation speeds (5 rpm, 15 rpm, and 25 rpm); $T_w = 353$ K, and $f = 45\%$. Adapted from Ref. [12] with permission. Copyright 2013, Elsevier.

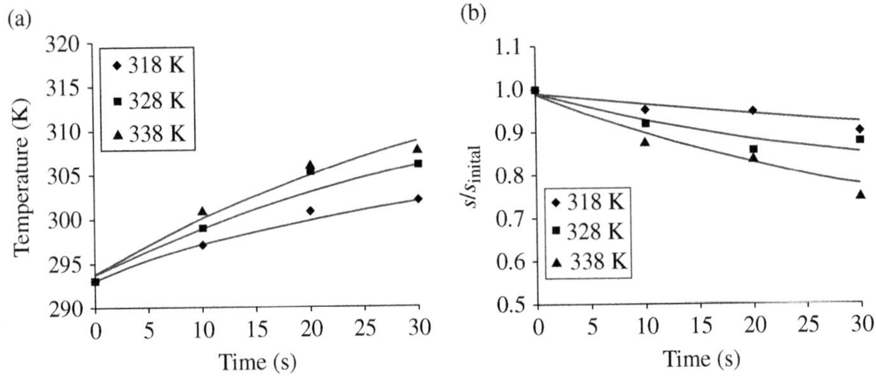

FIGURE 14.7 Experimental plots for the effect of wall temperature (318, 328, and 338 K) showing temporal profiles of (a) bed temperature and (b) solvent content; $\omega = 5$ rpm and $f = 45\%$. Adapted from Ref. [12] with permission. Copyright 2013, Elsevier.

14.4.1.4 Experimental Verification of Simulation Results Figures 14.7 and 14.8 compare the experimental results with simulation respectively for the effect of wall temperature (318 K, 338 K, and 353 K). The simulation data (Figure 14.8a and b) for an earlier time period displays a trend similar to the experimental data (Figure 14.7a and b) for increase in the bed temperature (Figures 14.7a and 14.8a) and decrease in the solvent content (Figures 14.7b and 14.8b), although simulations somewhat over-predicted the effect of wall temperature. These deviations were mainly attributed to the nature of soft particle algorithm and particle size distribution in the bed.

14.4.2 Simulation of Tablet Coating

14.4.2.1 Effect of Pan Tilt on Coating Variability From the previous studies [13] on mixing, we have learned that the pan tilt enhances axial mixing of the bed resulting in better coating. The DEM simulations of the pan coating process were performed using 40,000 frictional spherical particles (of properties as listed in Table 14.3) with

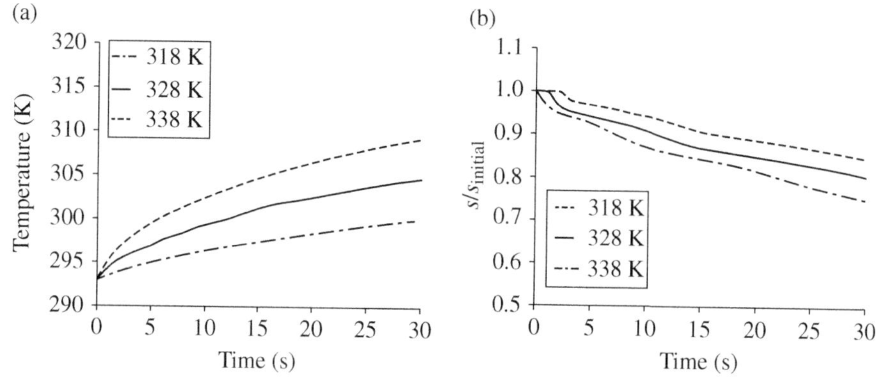

FIGURE 14.8 Simulation results for the effect of wall temperature (318, 328, and 338 K) showing temporal profiles of (a) bed temperature and (b) solvent content; $\omega = 5$ rpm and $f = 45\%$. Adapted from Ref. [12] with permission. Copyright 2013, Elsevier.

TABLE 14.3 List of Parameters Used in DEM Simulations

Parameters	Values
Total number of particles	40,000–90,000 (drying), 10,000 (milling)
Radius of the particles	1.7 mm (drying), 0.25–1 mm (milling)
Density of the particles	1.6 g/cm^3
Frictional coefficients	
Particle/particle	0.7
Particle/wall	0.4
Coefficient of restitution	
Particle/particle	0.5
Particle/wall	0.4
Normal stiffness coefficient	
Particle/particle	6000 N/m
Particle/wall	6000 N/m
Fracture toughness	0.5–0.8 MPa·m$^{0.5}$
Time step (Δt)	2.0 * 10^{-6} s

the spray duration and interval of 1 and 3 s, respectively. Figure 14.9a shows the simulation results of the temporal variation of percentage of coated particles for different orientations of the coater vessel. The percentage of coated particles is higher for the case of 32° tilt as compared to the other tilts, and Figure 14.9b shows the corresponding frequency distribution for the same. As the tilt increases, the distribution becomes more compact approaching toward the Gaussian distribution suggesting better uniformity. Therefore, the frequency distribution plot cons the decrease in variability with increase in tilt up to 32° as compared to 0° and 16° tilt, where the distribution is broad due to wide variation in the particle diameter.

14.4.2.2 Effect of Rotational Speed on Coating Variability

Rotational speed has an impact on the movement of the particles affecting the time spent under the spray zone.

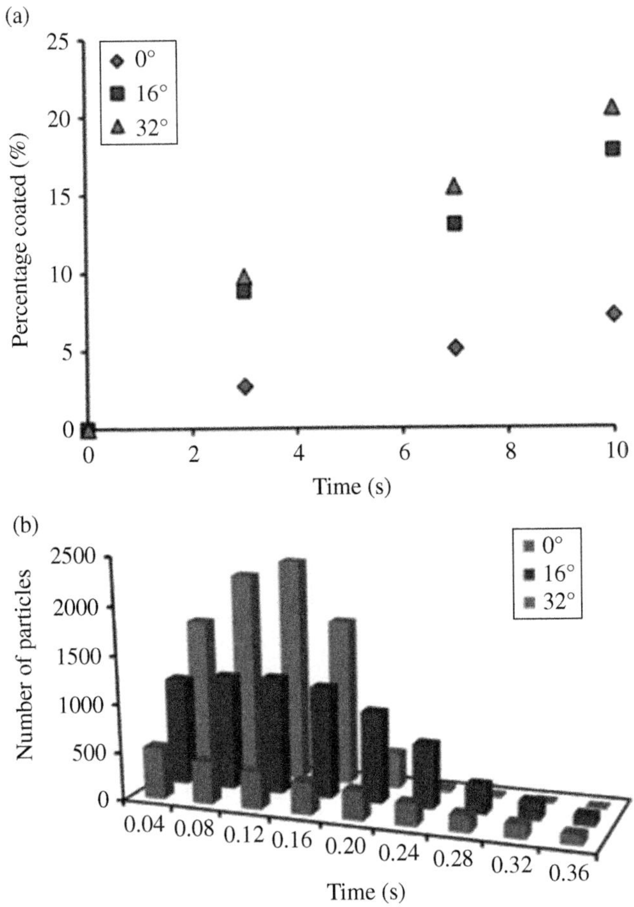

FIGURE 14.9 (a) Illustrates the simulation results for the effect tilt on percentage coated at pan rotational speed of 30 rpm at spray rate of 2.316 ml/min. (b) Shows the frequency distribution of the residence time of the coated particles for the effect of different tilt at 30 rpm. Adapted from Ref. [13] with permission. Copyright 2011, Elsevier.

Too low speeds may cause localized over wetting of the particle mass leading to agglomeration, whereas higher vessel speed may result breakage of the particles and also lack of enough drying time of the particles before their re-introduction in the spray area.

From previous results on mixing, high optimum vessel speed resulted in good mixing of the granular bed thereby improving the coating performance [13]. Figure 14.10a shows the relation of percentage of coated particles with time for different vessel speeds. The percentage coated was observed to be greater at higher speeds. Figure 14.10b shows the corresponding frequency distribution plot. As the vessel speed increased, the histogram approached the bell-shaped Gaussian distribution signifying better uniformity of coating as the standard deviation decreased with time. Frequency distribution plots in Figure 14.10b further corroborated the decrease in variability with increase in speed. A wider distribution was observed for 10 rpm;

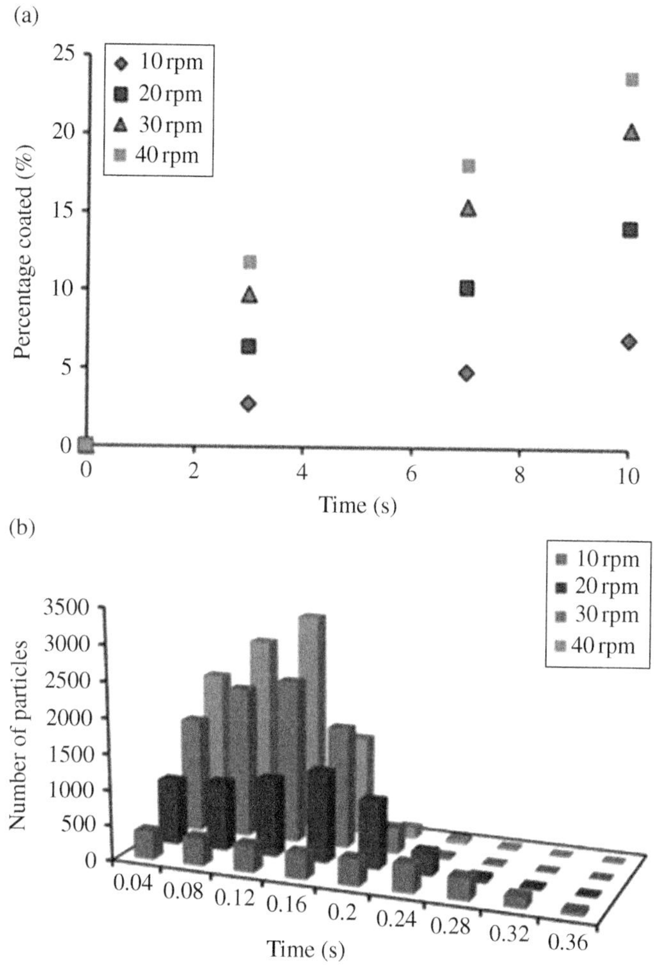

FIGURE 14.10 (a) The effect of rotational speed on coating variability (*CV*) at spray rate of 2.316 ml/min and (b) the frequency distribution of the residence time of the coated particles for the effect of speed at 32° tilt. Adapted from Ref. [13] with permission. Copyright 2011, Elsevier.

however, the diameter distribution became tighter as the speed increased. Therefore, high pan speeds resulted in better coating uniformity.

14.4.3 Simulation of Size Fragmentation (Milling)

14.4.3.1 Effect of Feed Rate Initially 10,000 spherical, inelastic, frictional particles with material properties described in Table 14.3 were deposited in the hopper with a closed outlet. After deposition, particles were discharged from the hopper into the mill. The blades/impellers were then set into motion at prescribed rotational velocity of 600 or 1140 rpm.

In the snapshots illustrated in Figure 14.11, color-coding is based on the particle size to visually track its evolution trajectory in simulations. Similar to experimental

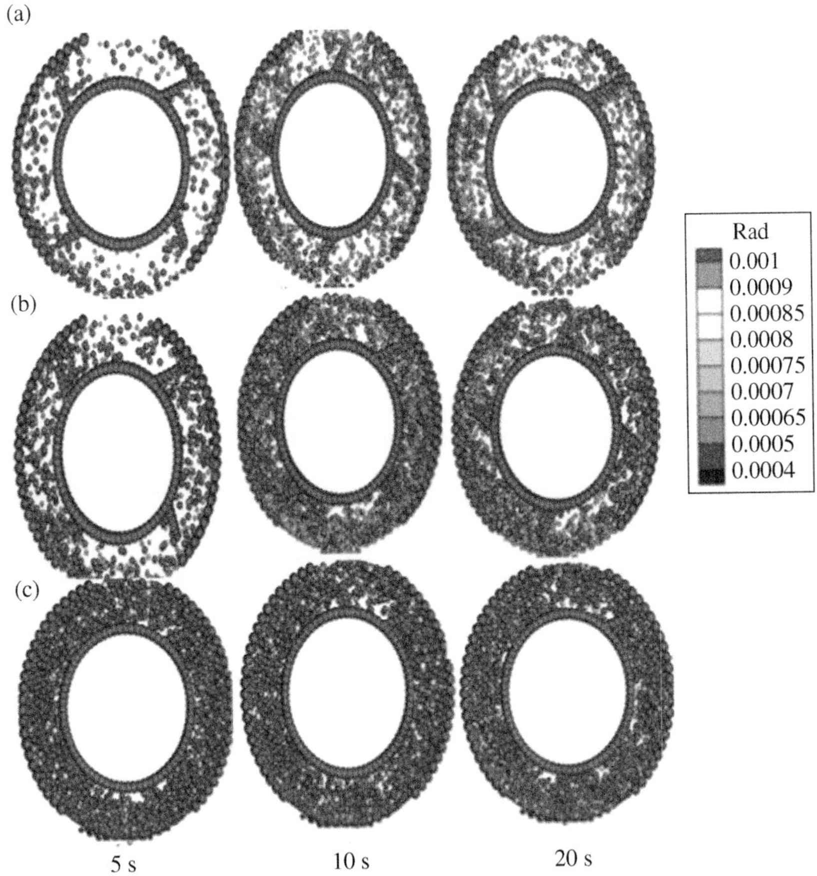

FIGURE 14.11 Temporal distribution of particle size under the impeller speed of 600 rpm at feed rates of (a) 60 g/min, (b) 120 g/min, and (c) 200 g/min. Adapted from Ref. [18] with permission. Copyright 2013, Elsevier. (*see insert for color representation of the figure.*)

results, the size distribution was found to become coarse as the feed rate was increased. The size distribution was found to increase from (580–620 μm) at 60 g/min to around (750–900 μm) at 200 g/min. There was also build of some smaller particles (<600 μm) observed at feed rate of 200 g/min.

This perhaps could be as a result of flood feed conditions due to which the movement of powder bed gets impeded. As a result, the same particles might come in contact with the hammer that could result in attrition type conditions. From the simulation data the particle size obtained at 5, 10, 15, and 20 s was compared to experimental results for the three different feed rates at 600 rpm. As the parent particle size in simulation (2 mm) differs from experimental (1.1 mm), a dimensionless size was computed and plotted against time. The simulation illustrated a similar trend (Fig. 14.12) to the experimental observations.

To further explore the effect of feed rate, the velocity of the particles was compared at different time points (Fig. 14.13). In these snapshots, the particles have

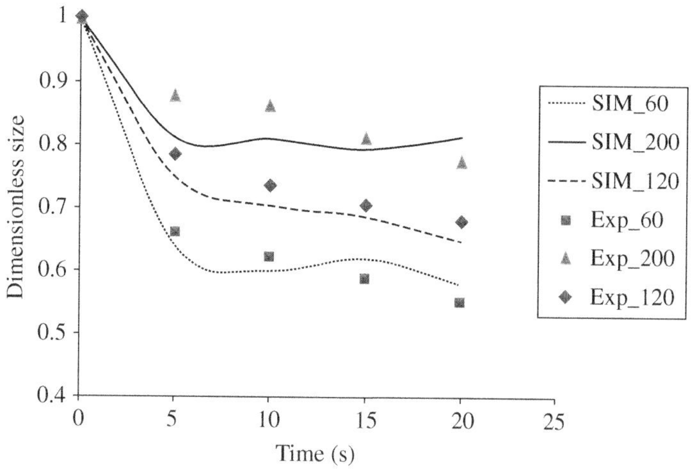

FIGURE 14.12 Experimental and DEM model comparison at an impeller speed of 600 rpm for three different feed rates—60, 120, and 200 g/min. Adapted from Ref. [18] with permission. Copyright 2013, Elsevier.

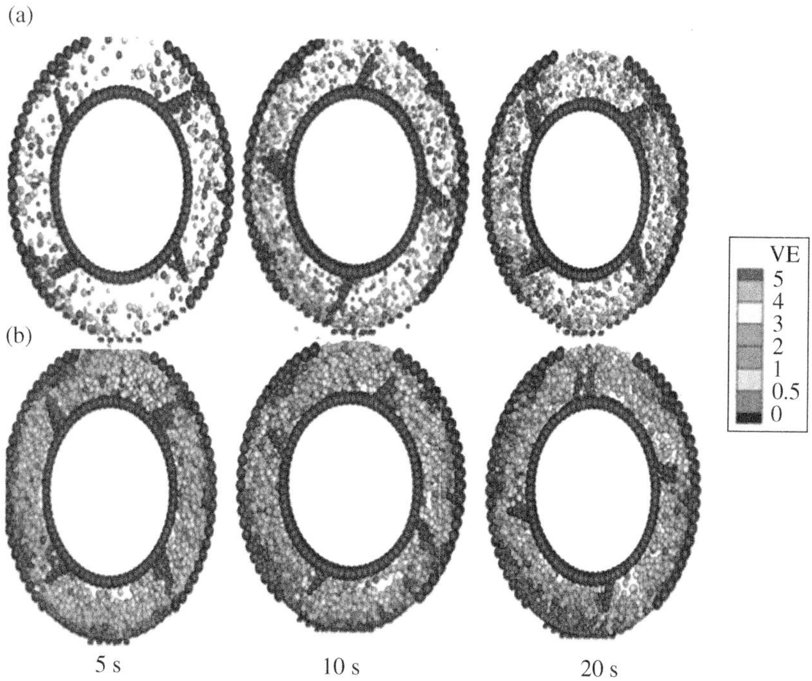

FIGURE 14.13 Temporal distribution of velocity under impeller speed of 600 rpm at feed rates (a) 60 and (b) 200 g/min. Adapted from Ref. [18] with permission. Copyright 2013, Elsevier. (*see insert for color representation of the figure.*)

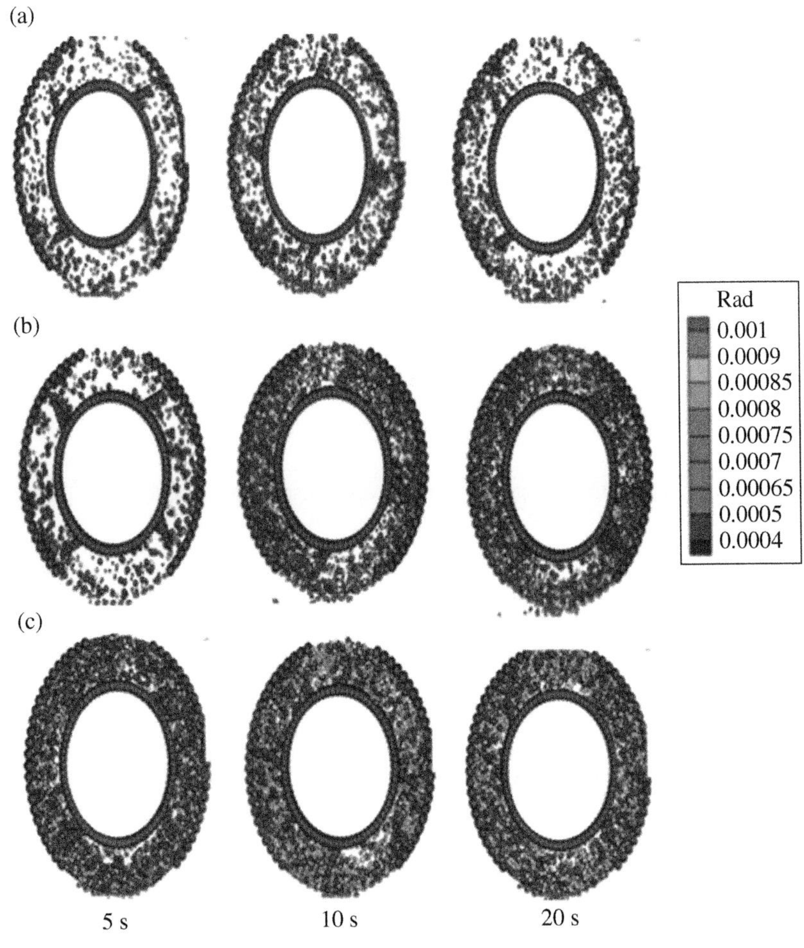

FIGURE 14.14 Temporal distribution of particle size under impeller speed of 1140 rpm at feed rates of (a) 60, (b) 120, and (c) 200 g/min. Adapted from Ref. [18] with permission. Copyright 2013, Elsevier. (*see insert for color representation of the figure.*)

been color coded for velocity where VE in contour chart represents velocity of particles. From the snapshots, one can observe that as the feed rate was increased the velocity of the particle decreased gradually with time. As a result there was a decline in strain rate. Moreover, higher feed rates at 600 rpm perhaps also decrease the mean free path length and thus reduce the overall comminution rate.

At 1140 rpm lower feed rates generated a much narrower size distribution (400–580 μm). As the feed rate was increased there was some decrease in breakage rate; however, no flood feed conditions were observed (as shown in Fig. 14.14). This suggests that due to greater centrifugal forces movement of powder bed is not obstructed, and hence there is no much accumulation of particles.

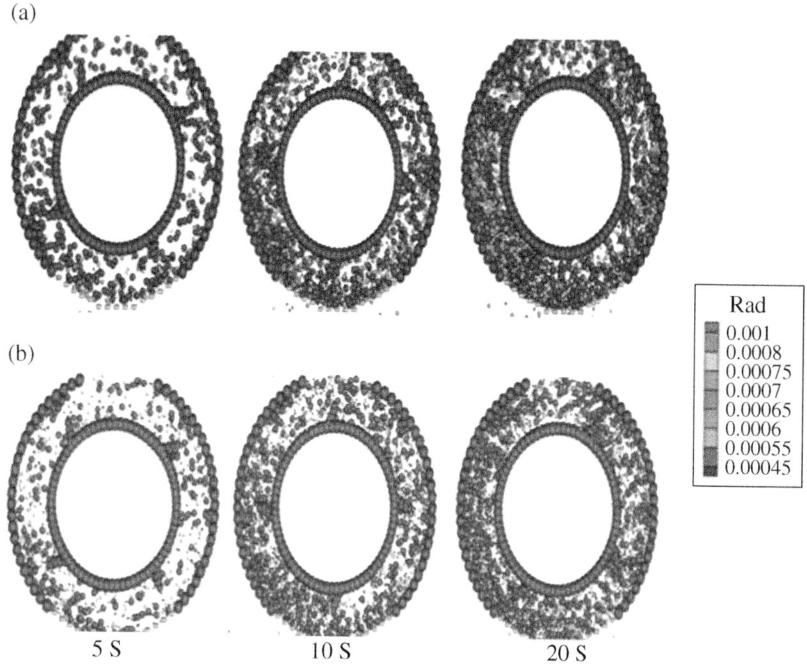

FIGURE 14.15 Temporal distribution of particle size for impeller wall tolerance 3.9 mm at different impeller speeds of (a) 600 rpm and (b) 1140 rpm. Adapted from Ref. [18] with permission. Copyright 2013, Elsevier. (*see insert for color representation of the figure.*)

14.4.3.2 Effect of Impeller Wall Clearance The effect of impeller wall tolerance was approximated by creating long hammers (1800 glued particles) and short hammers (3200 glued particles), thereby providing a tolerance of 2 and 3 mm, respectively. The operational speed during the simulation is 600 and 1140 rpm. As the impeller tolerance was increased, the comminution rate slowed down. The snapshots illustrate (Fig. 14.15) that there is accumulation of large particles near the base of mill. This is because as length of blade is reduced the particles do not strike the surrounding chamber/blade easily especially at lower speeds as the centrifugal forces go down. Thus the powder bed tends to collapse easily. This effect was more significant at 600 than 1140 rpm.

14.5 SUMMARY AND CONCLUSIONS

Numerical simulations of granular flow and heat transfer in an agitated filter dryer are presented using TPD. The effects of operating parameters (wall temperature and impeller speed) and batch size on drying behavior have been systematically studied and compared with experimental data. Detailed analysis of temporal profiles of temperature and solvent content in the bed was performed. Under constant fill, both the wall temperature and the impeller speed have a remarkable influence on the drying rates in the examined range. The variation of the operating parameters revealed that increasing the drying temperature enhances the drying rate and hence the drying time is reduced. Particle temperature and solvent content profiles obtained from the simulations are in good agreement with those obtained experimentally.

A mechanistic first principle-based model to study the coating variability in a pan coater is being developed. The maximum coating efficiency with minimal variability was accomplished at the optimal coating condition of 32° vessel tilt and rotational speed of 30 rpm, in both experiments and simulations. The coating variability decreased with time under all conditions.

Finally, the experimental and computational methods were developed to study comminution of granular bed in a Wiley mill. Granular flow and mechanical properties of lactose are taken into account in order to develop a fundamental understanding of their effect on milling performance. DEM simulations were used to examine the breakage rate in the granular materials processed in a lab scale Wiley mill. It was found that the speed and mass flow rate play an important role in comminution of powder bed. Very high mass flow rates resulted in flood feed conditions at lower speeds that caused the mode of breakage to change from fragmentation to attrition. Increasing the impeller wall clearance reduced particle size reduction at both speeds. However, the effect was more significant at lower speeds. The low velocity of particles attained at lower speeds caused them to settle at the base of mill than impact the wall thereby reducing fragmentation. In these simulations, the breakage algorithm does not depict the effect of attrition on particle size distribution as it does not account for mechanism of surface removal. Nevertheless, the simulations could still capture the difference in breakage behavior upon changing the experimental conditions. This suggests that simulations can be used to direct future experimental design hence providing a better understanding of the process.

REFERENCES

[1] Lekhal, A.; Girard, K. P.; Brown, M. A.; Kiang, S.; Khinast, J.G.; Glasser, B.J. *Int. J. Pharm.* 2004, *270*, 263–277.

[2] Mujumdar, A.; Lim, L., *Chem. Ind. Digest* 2006, *19*, 54–62.

[3] Kalbag, A.; Wassgren, C., *Chem. Eng. Sci.* 2009, *64*, 2705–2717.

[4] Pandey, P.; Katakdaunde, M.; Turton, R. *AAPS PharmSci.* 2006, 7, E1–E10.

[5] Snow, R.H.; Allen, T.; Ennis, B.G.; Litster, J.D. Size reduction and size enlargement. In: *Perry's Chemical Engineering Handbook*. Perry, R.H. and Green, D.W., Eds., 7th edition. McGraw-Hill: New York, 1997; pp. 10–21.

[6] Parrott E.L. J. *Pharm. Sci.* 1974, *63*, 813–829

[7] Heywood, H. J. *Imp. Coll. Chem. Eng. Soc.* 1950, *6*, 26–28.

[8] Piret, E. L. *Chem. Eng. Prog.* 1953, *49*, 56–63.

[9] Strack, O.D.,; Cundall, P.A. *Geotechnique* 1979, *29*, 47–65.

[10] Walton, O., Braun, R. J. *Rheol.* 1986, *30*, 949–980.

[11] Walton, O.R. *Mech. Mater.* 1993, *16*, 239–247.

[12] Sahni, E.; Chaudhuri, B. *Chem. Eng. Sci.* 2013, *97*, 34–49.

[13] Sahni, E.; Yau, R.; Chaudhuri, B. *Powder Technol.* 2011, *205*, 231–241.

[14] Vargas, W.L.; McCarthy, J.J. *AIChE J.* 2001, *47*, 1052–1059.

[15] Vogel, L; Peukert, W. *Powder Technol.* 2003, *129*, 101–110.

[16] Naik, S.; Malla, R.; Shaw, M; Chaudhuri, B. *Powder Technol.* 2013, *237*, 338–354.

[17] Capriz, G. and Mariano, P.M., Eds. *Advances in Multifield Theories for Continua with Substructure*, Springer Science: New York, 2004.

[18] Djordjevic, N.; Shi, F.N.; Morrison, R.D. *Minerals Eng.* 2003, *16*, 983–991.

INDEX

Computational Pharmaceutical Solid State Chemistry, First Edition. Edited by Yuriy A. Abramov.

Printed and bound by CPI Group (UK) Ltd, Croydon, CR0 4YY
16/09/2026
14963425-0001